T0143198

The Transmutations of Chymistry

Synthesis

A series in the history of chemistry, broadly construed, edited by Carin Berkowitz, Angela N. H. Creager, John E. Lesch, Lawrence M. Principe, Alan Rocke, and E. C. Spary, in partnership with the Science History Institute

The Transmutations of Chymistry

Wilhelm Homberg and the Académie Royale des Sciences

LAWRENCE M. PRINCIPE

The University of Chicago Press
Chicago and London

The University of Chicago Press, Chicago 60637
The University of Chicago Press, Ltd., London
© 2020 by The University of Chicago
Published 2020
Printed in the United States of America

29 28 27 26 25 24 23 22 21 20   1 2 3 4 5

ISBN-13: 978-0-226-70078-6 (cloth)
ISBN-13: 978-0-226-70081-6 (e-book)
DOI: https://doi.org/10.7208/chicago/9780226700816.001.0001

Library of Congress Cataloging-in-Publication Data

Names: Principe, Lawrence, author.
Title: The transmutations of chymistry : Wilhelm Homberg and the Académie royale
    des sciences / Lawrence M. Principe.
Description: Chicago ; London : The University of Chicago Press, 2020. | Series:
    Synthesis | Includes bibliographical references and index.
Identifiers: LCCN 2019054881 | ISBN 9780226700786 (cloth) | ISBN 9780226700816
    (ebook)
Subjects: LCSH: Homberg, Wilhelm, 1652–1715. | Académie royale des sciences
    (France) | Chemists—France—Biography. | Scientists—France—Biography. |
    Chemistry—History—17th century. | Chemistry—History—18th century. |
    Alchemy—History—17th century. | Alchemy—History—18th century. | LCGFT:
    Biographies.
Classification: LCC QD22.H74 P75 2020 | DDC 540.92 [B]—dc23
LC record available at https://lccn.loc.gov/2019054881

♾ This paper meets the requirements of ANSI/NISO Z39.48–1992 (Permanence
of Paper).

Contents

Figures and Tables

Abbreviations

A D S : Archives de l'Académie des Sciences, Paris

A N : Archives Nationales, Paris

B M H N : Bibliothèque du Muséum National d'Histoire Naturelle, Paris

B N F : Bibliothèque Nationale de France, Paris

B O U R D E L I N D I A R Y : Diary of Claude Bourdelin *fils*; BNF, MS NAF 5148

B O U R D E L I N E X P E N S E A C C O U N T : "Memoire de la despence faitte par Monsieur Bourdelin pour le Laboratoire de l'Academie Royalle des sciences" (1667–99); BNF, MS NAF 5147

H A R S : *Histoire de l'Académie Royale des Sciences*; see Note on Sources

H E L L O T A R S E N A L N O T E B O O K S : BNF, Bibliothèque de l'Arsenal, MSS 3006–8; see Note on Sources

H E L L O T C A E N N O T E B O O K S : Caen, Bibliothèque Municipale, MS in-4to 171, 9 vols.; see Note on Sources

H M A R S 1 6 6 6 – 9 9 : *Histoire et mémoires de l'Académie Royale des Sciences*, 11 vols. (Paris, 1729–33); see note on sources

J D S : *Journal des Sçavans*

L B R : Leibniz Briefe, Leibniz Archiv, NLB

M A R S : *Mémoires de l'Académie Royale des Sciences*; see Note on Sources

M S : Manuscript

N A F : Nouvelles acquisitions françaises

N L B : Niedersächsische Landesbibliothek, Hannover

P V : Registres des procès-verbaux des séances, Archives de l'Académie des Sciences, Paris

V M A : Voenno-Meditsinskoi Akademii (Library of the Military-Medical Academy), Boerhaave Archive, St. Petersburg, Russia

Acknowledgments

This book took far longer to complete than ever I imagined when I began the project. What I initially proposed as a five- to six-year project ended up taking nearly twenty years to complete. Throughout the time this project has been ongoing, I have been helped in various ways by many colleagues, archivists, librarians, and funding agencies.

The project was begun with a five-year CAREER grant from the National Science Foundation, #9984106, "The Formation and Support of Chemists and Chemistry." Without that research grant, this project never would have gotten started, and the research it initiated has produced much more than this book. Besides the shorter publications and numerous talks that also emerged from this project, the grant also allowed me expand my historical research skills—I had previously worked primarily with materials in British archives—by learning how to work in French (and other) archives and gaining familiarity with the historical landscape of early modern France. Thus, this NSF grant did indeed promote the development of a major dimension of my subsequent career. Thereafter, part of the Francis Bacon Prize I was awarded in 2005 funded a semester at Caltech and allowed me to convene an international workshop there. The edited volume *New Narratives in Eighteenth-Century Chemistry* (2007) came out of that workshop, and my own article in it contained several of the basic ideas pursued further in the present book. In 2015–16 I was supported by a generous John Simon Guggenheim Memorial Fellowship, and in 2016 I received a Rausing Fellowship for a sabbatical at Uppsala University. These last two fellowships provided the undisturbed thinking space needed to finish writing the book manuscript. I am extremely grateful to all of these patrons.

Many colleagues helped, supported, critiqued, and encouraged in various ways over the years. Three sets of special thanks are in order. First, when this project was in its infancy, Alice Stroup provided invaluable guidance to help get it started in the right direction. Alice and her husband, Tim, welcomed me at their apartment in Paris, and the three of us shared many convivial lunches in the early days of my research. Alice introduced me to the staff at the Archives of the Académie des Sciences, and offered freely of her enormous experience of working in French archives. She shared photocopies and notes from her own pioneering research on the Académie and gave pointers to resources I should track down, which included a document by Duclos in Strasbourg and a reference to the Hellot material in Caen, which has proven so significant for this project. The next special mention is for my friend and colleague Bernard Joly and the circle of scholars he collected at his long-running seminar on early modern chymistry at the Université de Lille III. Bernard and I shared many visits and conversations about our overlapping interests. I presented the first products of this project at his seminar, and these were also among my first attempts to present papers in French; I could not have found a more patient and helpful audience. Many heartfelt thanks to Bernard and his circle of *chymistes* for their kindness, intellectual contributions, and companionship.

The third set of special thanks involves a long story. In 1996, as the first paper I ever wrote about Homberg was taking shape, I attended a workshop at the Dibner Institute at MIT. There I met John Powers, at that time a graduate student writing his dissertation. As generally happens when academics meet, we asked about each other's current research. When I mentioned Homberg, John recalled that he had seen Homberg's name in a manuscript catalogue during his work in St. Petersburg, and he promised to check his notes. A few days later he sent me the shelfmark for a manuscript called "Elemens de chimie" stored with the Boerhaave Papers at the Military-Medical Academy. The most probable and obvious interpretation of the catalogue entry was that this manuscript was simply a transcription of some or all of Homberg's "Essais de chimie," published serially in the Académie's *Mémoires* in the first decade of the eighteenth century. But I knew from a remark made by Bernard de Fontenelle in 1715 that Homberg had gone further with the text, leaving behind at the time of his death something that was "ready for the press" but that was never published. Although it would have been an incredible long shot for the St. Petersburg manuscript to be related to that lost document, I thought it would be worth checking the manuscript if it was not too much trouble to do so. I asked John if he could contact the archivists and ask for photocopies of the first few pages. They replied that the archive had just burned.

For the next two or three years we tried to get information about *how much* of the archive had burned. Colleagues visiting St. Petersburg sent back descriptions of an abandoned building with boarded-up windows surrounded by fencing. At last, we were told that the fire had been confined to the mailroom, and that it had been started by one of the archivists themselves who, in the disarray of the post-Soviet years, had not been paid for many months. The apparent reasoning was that if a fire closed down the building, the archivists could keep their positions but not have to work until the building was repaired, and meanwhile they could get jobs that actually paid. After a couple more years the archive reopened, but obtaining photocopies was then impossible because the photocopier had been, of course, in the mailroom and had been crushed by a ceiling that collapsed during the fire. Yet worse, by this time—the early 2000s—the doors opened by the dissolution of the Soviet Union were slamming shut again, and getting cooperation became steadily more difficult. At this time, a colleague trying to see the Boerhaave papers was turned away at the door of the archive every day for two weeks and never allowed in. Several emissaries were sent on my behalf to the archive, and more than one plan to get access to the manuscript was put into action, but all came to naught. While technically now "open," the archive was functionally as closed as it had been when boards covered the windows. These mounting obstacles, rather than dissuading me, made me all the more determined to see that manuscript—pyrotechnic archivists and geopolitical nonsense be damned.

In 2004, I went to the first meeting of the European Society for the History of Science, held in Maastricht, determined to find someone who could help me get the manuscript. Among the various people I spoke with about it was Ernst Homburg, who, delighted to hear about the almost-eponymous chymist I was researching, promised to find someone to help. A few days later he put me in touch with Elena Zaitseva, who in turn introduced me electronically to Igor Dmitriev, director of the Mendeleev Museum and professor at the University of St. Petersburg. Igor was more than I could have hoped for. Without his unparalleled help and enormous generosity, this book would be a very different and much poorer one. Igor invited me to St. Petersburg and promised to get access to the archive for me. During two delightful weeks in St. Petersburg in June 2005 I enjoyed warm hospitality and wonderful conversations with him and his wife, Natalya. True to his word, Igor cleared away one obstacle after another. After three days of the chief archivist instructing by telephone to "call back tomorrow," he personally took me to the archive. There we found a door conveniently fitted with both a disconnected doorbell and a sign reading "No Admittance" in six languages. With the help of a set of keys borrowed from a nearby office, we got the locked door open, greatly surprising a circle of archivists having cake and tea, and even more greatly annoying the head archivist by our entrance. The head archivist, who resembled in more ways than one the caricature of a Soviet functionary from a Cold War–era American farce, told us brusquely that there was no such manuscript in the collection. But upon our showing her a catalogue of the Boerhaave collection made by Dutch researchers in the 1950s, she grudgingly went to fetch it. The manuscript turned out to be far more important than I could have imagined, making the nine-year odyssey to acquire it completely worthwhile; chapters 3 and 4 of the present book draw upon it. Over the course of a week in the archive, I managed to photograph the entire document (nearly 300 folios) while pondering the various things that could go wrong with an extremely unwelcome American photographing documents in a Russian military archive. (Exactly how I managed to do so, along with several other features of this adventure, are things to be told only over drinks and not in print.) I will deposit a copy of these photographs at the Archives of the Académie des Sciences in Paris, where the document really should have ended up all along.

There are many other colleagues to whom further thanks are due. Close to home at Johns Hopkins, my friend and colleague Dan Todes gave useful advice about Russian archives and for years has been my regular companion for lunches over which we have discussed the problems and joys we both encountered in writing biographies, tackling research questions, and much else. Thanks also to Jean-Olivier Richard and Justin Rivest, two wonderful colleagues and scholars, both former graduate students in our program at Johns Hopkins. Their research frequently overlapped with mine, and the three of us held a seminar on early modern French science and medicine that was a true exchange of ideas, expertise, and primary sources. Additionally, Jean-Olivier repeatedly deployed his linguistic skills to make my awkward French prose elegant, and Justin provided me on several occasions with references to and photographs of valuable documents in the Archives Nationales. My thanks also to my further Johns Hopkins colleagues Michael Kwass and Gianna Pomata.

Many other colleagues helped in various ways. Christine Lehman shared freely of her knowledge and findings about burning lenses, the Académie, and early modern French chymistry. My friend and colleague Didier Kahn and I met frequently when I visited Paris, and he welcomed me to his home on several occasions; we often exchanged documents, ideas, and information and he commented on several parts of this manuscript. Bill Newman read the entire manuscript in draft and offered useful advice on revisions. My thanks go also to Catherine Abou-Nemeh, Peter Anstey, Bernadette Bensaude-Vincent,

Marco Beretta, Paola Bertucci, Victor Boantza, Patrice Bret, Jed Buchwald, Michael Bycroft, Clarinda Calma, Kevin Chang, Dennis Des Chene, Maria Pia Donato, Moti Feingold, Hjalmar Fors, Rémi Franckowiak, Margaret Garber, Corinna Guerra, Philippe Hamou, Michael Hunter, Joel Klein, Sy Mauskopf, Maggie Osler, François Pépin, Miguel Lopez Perez, Luc Peterschmitt, Albert Philipse, Evan Raglund, Jennifer Rampling, Alan Rocke, Otto Sibum, David Sturdy, Brigitte Van Tiggelen, and Alexander Wragge-Morley, all of whom contributed in various ways through their conversation, collegiality, and sharing of ideas and source materials. It is extremely heartening to know that the early modern *république des lettres* continues to thrive even in our days.

In terms of archives, my thanks go first to the staff at the Archives of the Académie des Sciences, where I spent so many enjoyable hours over the course of two decades. The former head of the archives, Florence Greffe, was always welcoming, helped guide my first steps there, and never hesitated to go out of her way to be helpful in locating and suggesting documents. Mme. Mine and the former *documentaliste* M. Leroi were always kind and helpful, as have been the rest of the staff, both past and present. For special help in accessing manuscripts, books, and other documents in various libraries, archives, and other respositories, there are several people to whom I owe thanks: for documents in Caen, my thanks to Isabelle Laboulais and Marie-Noël Vivier; at the Università di Padova, Emilia Veronese; at the Université de Bordeaux 3-Lettres, Paul-Henri Allioux; in Magdeburg at the Landeshauptarchiv Sachsen-Anhalt, Ralf Lusiardi; at the Niedersächische Landesbibliothek Hannover, Anke Hoeltzer; at Universitätsbibliothek Leipzig, Cornelia Bathke; at the Chemical Heritage Foundation in Philadelphia, James Voelkel; at the Historical Collection of the Institute for the History of Medicine at Johns Hopkins, Christine Ruggere; in Special Collections at Claremont College, Carrie Marsh; in Special Collections at Cornell University, David Corson; and at the Wellcome Library, Christopher Hilton, Holly Peel, and Rada Vlatkovic.

Finally, this book might well not have been completed without the constant support and encouragement from my partner of now nearly thirty years. Kip read the entire manuscript not once but three times and created figures 2.6 and 4.5. He identified weak arguments, offered suggestions about how to resolve problems and clarify tortured prose, and helped figure out what the book was actually about in the end. He also pointed out several places where I innocently assumed too much familiarity on the reader's part with the intricacies of seventeenth-century chymical theory and practice. For all this and so much more, my love and gratitude.

Introduction

In the not-so-distant past, the years between the death of Robert Boyle (1627–91) and the work of Antoine-Laurent Lavoisier (1743–94) were often thought a rather dull and largely sterile period for the history of chemistry. General histories of chemistry tended to skip lightly over the period, usually with a mention only of Stephen Hales's pneumatics and Georg Ernst Stahl's phlogiston—both of them included mainly as "setups" for the advent of Lavoisier. Some historians saw the period as a time of "stagnation" for chymical theory and claimed that "no remarkable genius" had appeared on the scene between these two reckoning points.[1] Interestingly, the relative neglect, even denigration, of this period appears already in some rhetorical posturings of the later eighteenth century itself, and more recent historians have often repeated those expressions without deeper investigation. Dismissal of this period's significance is one factor that allowed the notion of a "postponed" revolution in chemistry to persist—if not explicitly then implicitly—in part by permitting the phlogiston-oxygen controversy to virtually monopolize the period.[2] It is also true, of course, that prominent figures, such as Boyle and Lavoisier, naturally tend to overshadow surrounding characters who have not been as celebrated or dramatized. Broad treatments, dating from a time when the notion of "revolutions in science" was popular, tended to look for (and to

---

1. Robert Siegfried, *From Elements to Atoms: A History of Chemical Composition* (Philadelphia: American Philosophical Society, 2002), 56–73; and Herbert Butterfield, *The Origins of Modern Science* (New York: Macmillan, 1959), 202. Seymour Mauskopf points out that the 1949 edition of Butterfield's book lacks the paragraph containing this assessment; see his "Reflections: 'A Likely Story,'" in *New Narratives in Eighteenth-Century Chemistry*, ed. Lawrence M. Principe (Dordrecht: Springer, 2007), 177–93.

2. See Mauskopf, "Reflections."

some extent create) great moments of change, or connected up already canonized figures and events rather than investigating more closely the details of the complex and often surprising routes of scientific development.

More recent studies have, however, begun to shed new light on the period, exploring some of its unexpected dimensions and revealing more of its character and its denizens. Such work has started to change former impressions by taking the practices and ideas of the period more seriously, and by reading the texts from a broader range of practitioners more carefully and contextually.[3] As I endeavored to sketch out more than a decade ago, this period actually witnessed several significant and lasting transformations of the discipline of chemistry—intellectually, socially, practically, and institutionally—even though such changes had been little noticed, or had been obscured by assumptions and preexisting narratives borrowed from other fields.[4] We are now in a position to reevaluate this period more thoroughly and to reassess its place in the broader sweep of the history of science generally and the history of chemistry specifically. This book contributes to that ongoing reevaluation by focusing on a character who by any standard would rank as a "remarkable genius" and on the institution where he spent his mature career. This chymist devised a comprehensive and experimentally based theory of chymistry that was widely celebrated, adopted, and adapted, and his institution played a central role in the development of the discipline throughout the rest of the eighteenth century and beyond.[5]

Wilhelm Homberg (1653–1715) is a truly remarkable character, even if he currently remains not very well known even among historians of chemistry. When I first encountered him more than twenty years ago, I was fascinated by his extraordinary life and intrigued by his chymical investigations, his original theorizations, and his vision for chymistry as a natural philosophical discipline. At the same time, his continuing pursuit of metallic transmutation,

3. An early example of this reevaluation was Frederic L. Holmes, *Eighteenth-Century Chemistry as an Investigative Enterprise* (Berkeley, CA: Office of the History of Science and Technology, 1989). Since that time we have benefited from, among others, John Powers on Boerhaave, Kevin (Ku-Ming) Chang on Stahl, Ursula Klein on practitioners in Germany, and Christine Lehman on mid-century French chemists.

4. Lawrence M. Principe, "A Revolution Nobody Noticed? Changes in Early Eighteenth Century Chymistry," in *New Narratives*, ed. Principe, 1–22.

5. For those unfamiliar with the usage, the odd spelling of *chymist* and *chymistry* is intended to express the totality of the discipline before what we today understand as *chemistry* and *alchemy* (that is, transmutational endeavors) were clearly differentiated. See William R. Newman and Lawrence M. Principe, "Alchemy vs. Chemistry: The Etymological Origins of a Historiographic Mistake," *Early Science and Medicine* 3 (1998): 32–65, at 59–61.

lightly veiled in his official publications, indicated his position as a bridge between traditions and as an important participant in the transformations of chymistry in the period just after Boyle.[6] I soon realized that Homberg deserved a book-length study. This book took far longer to complete than I ever imagined, even though the generous Monsieur Homberg and his colleagues provided ample material for a series of talks and publications while this book slowly took shape in the background. One reason for this book's long gestation is that my endeavor to find new primary sources that could provide fresh documentary evidence about the historical actors seemed at times to rival the search for the philosophers' stone. As will soon become apparent, this study relies heavily on a wealth of manuscript and other archival material, some of it discovered or correctly identified only during the research for this book, the reward for optimistic and persistent visits to archives and libraries in ten countries.

The success of such archival spelunking—even if realized at a snail's pace—allowed this book to become considerably more than I had originally planned. Rather than the more or less straightforward biography of Homberg and study of his chymistry that I initially envisioned, it has become also a biography of chymistry itself during the years from about 1670 until about 1730, particularly in France. Homberg and his work provide new insights into the significant changes that chymistry underwent during these years. Because Homberg was also a member of the Parisian Académie Royale des Sciences, the evolution of that institution forms a third "biographical" theme for this study. Not only was the Académie the premier scientific institution of Homberg's day, it was also arguably the most important institutional locus for the development of chymistry during the eighteenth century. Immediately upon its founding in 1666, its first research questions involved chymistry, and then it undertook a project that involved an innovative chymical component and that would extend over the next thirty years. In 1699, its new regulations established eight positions specifically for chymists, thus creating a permanent institutional home for chymists and chymistry like nowhere else and never before. Throughout the eighteenth century, many of the most notable contributors to the history of chemistry—including, of course, Lavoisier—worked within the context of the institution. This book, therefore, has much to say about the Académie Royale, its role in shaping the contours of chy-

6. Lawrence M. Principe, "Wilhelm Homberg: Chymical Corpuscularianism and Chryso-poeia in the Early Eighteenth Century," in *Late Medieval and Early Modern Corpuscular Matter Theories*, ed. Christoph Lüthy, John E. Murdoch, and William R. Newman (Leiden: Brill, 2001), 535–56.

mistry, and the shifting array of chymical workers affiliated with it. Many of these early academicians have not been subjected to close study until recently, and the new documentary sources used throughout this book provide a wealth of new or corrected biographical information about many of them.[7] Homberg was thus situated at both the right time and the right place for the purposes of this study. These three biographies—human, disciplinary, and institutional—are necessarily interwoven in the following chapters.

Chapter 1 follows Homberg from his birth and childhood in a distant colonial outpost, through his training and brief career as a lawyer in Germany, followed by many years and thousands of miles of travel throughout Europe, until his admission to the Académie Royale des Sciences in 1691 allowed him to settle down for the first time, at the age of thirty-eight. This chapter explores Homberg's intellectual formation as a chymist and presents a radically new portrait of his early life. This new image of Homberg departs significantly from the one given in his official éloge, and which almost all subsequent biographical notices have copied. That éloge, written in 1715 by the Académie's perpetual secretary, Bernard de Fontenelle (1657–1757), portrays Homberg as a heroic figure trained by serial apprenticeships with the "great men" of the day and destined to become the ideal (or rather *Fontenelle's* ideal) of the "enlightened" academician of the eighteenth century. But the real story of Homberg's life is far more interesting. It reveals much about the character of chymistry and chymists during the late seventeenth century. Homberg's eight-year flirtation with the Académie in the hopes of gaining admission highlights the changing fortunes of the institution during the seventeenth century and how these ups and downs were determined by its various governmental administrators. It also reveals what I call the institution's "penumbra," a more diffuse circle of hopefuls, collaborators, artisans, visitors, laborers, and others arrayed around the core of its officially enrolled and better recognized members. Finally, it presents new information about otherwise invisible factors at play in whether or not a prospective academician was admitted.

Chapter 2 examines the nature and status of chymistry at the seventeenth-century Académie, that is, before Homberg's admission and during his first decade as a member. Central to this analysis is the institution's communal project to compile a "History of Plants" that included the chymical analysis

---

7. I have also provided an overview of the primary sources relating to the Académie Royale des Sciences in the note on sources at the end of this book, which I hope will prove a useful summary and orientation for those interested in carrying out further work on the institution.

of plant materials as its most innovative feature. Although this grand under-taking has been the subject of excellent previous studies, particularly by Alice Stroup, in the present context I use it to illustrate the tug of war between two competing visions for chymistry at the Académie.[8] One, narrow and rather pedestrian, directed chymistry toward producing practical and im-mediately applicable results for medicine and pharmacy. The other, grander and more ambitious, envisioned a central role for chymistry in producing natural philosophical knowledge about the world, its material composition, and its transformations. Initially conceived and directed by the brilliant (but often misunderstood and misrepresented) Samuel Cottereau Duclos (1598– 1685), the project received more than one redirection over the course of its thirty or more years, one of which took place at Homberg's hands in the 1690s. Throughout this time, a divide persisted between those academicians, predominantly apothecaries and practicing physicians, who held a narrow view of chymistry's domain and those, notably Duclos and Homberg, who saw chymistry expansively as the best means for gaining solid knowledge about the workings of the natural world. This chapter also includes a survey of all the academicians during the 1690s who were active in chymical matters (broadly defined), thus providing a more detailed portrait of what chymistry meant "on the ground" within the institution at the end of the seventeenth century.

Chief among Homberg's long-term goals was the formulation of a com-prehensive and experimentally based theory of chymistry. I am particularly interested in questions of how scientific theories develop, how practitioners craft their individual and multifarious observations into broader explana-tory theories and extended research programs, and what further sources they draw upon in doing so. Such questions are particularly interesting in the case of chymistry, thanks to the widely varied and complicated results that chymi-cal experiments ordinarily provide, full of qualitative data and, as Homberg would insist, direct *sensory* information of sight, smell, touch, taste, and oc-casionally sound. Consequently, an important part of this study focuses on the changing content and dynamic character of scientific ideas and explana-tions in response to experimental results. Accordingly, chapter 2 begins the study of Homberg's endeavor to establish sound theoretical foundations for chymistry during the 1690s, and chapters 3 and 4 follow this project through the rest of his life. A close analysis of his constantly changing ideas is made

8. Alice Stroup, *A Company of Scientists: Botany, Patronage, and Community at the Seventeenth-Century Parisian Royal Academy of Sciences* (Berkeley: University of California Press, 1990).

possible by the survival of four sequential versions of his "textbook" of chymistry, spanning a period of twenty-five years, virtually his entire career at the Académie. Two of these sources have long been known but not previously subjected to close comparison. The other two, both lost for nearly three hundred years, I recovered while carrying out the research for this book. Significantly, some of the most profound changes to Homberg's chymical system can be clearly and convincingly tied to the results of specific experiments. Chapter 3 provides a close examination of Homberg's first attempt to write a textbook, in the 1690s. In writing it, Homberg diverged significantly from the established French didactic tradition of the seventeenth century, which highlights the distinctiveness of Homberg's own vision for chymistry. But Homberg eventually gave up on this text before it was finished, and started over in the opening years of the eighteenth century with a new plan and new ideas formulated in response to his experimental results.

Chapter 4 recounts how Homberg's social and financial position changed dramatically at the start of the eighteenth century, and how this change transformed both his ability to pursue chymistry and the foundations of his chymical ideas. At this time he developed a close relationship with Philippe II, Duc d'Orléans, nephew of Louis XIV and future Regent of France. Philippe became not only his patron but also his scientific collaborator. Letters and diaries from observers in the French court and the Orléans household, where Homberg quickly became an esteemed figure, provide a compelling portrait of Homberg's character and engaging personality, and his relationship with the Duc d'Orléans. With Homberg's guidance, Philippe transformed part of his residence at the Palais Royal into what observers called the finest chymical laboratory the world had ever seen. Philippe worked side by side with Homberg in this laboratory on an array of chymical endeavors for a decade, and several other academicians worked there as well under Homberg's supervision. When building the laboratory, Philippe equipped it not only with the usual array of furnaces and chymical vessels, but also with the most expensive and extraordinary scientific instrument of the day: a massive convex lens that could direct concentrated sunlight onto chymical substances to reveal astonishing new phenomena. Homberg's first experiments with this device caused him to abandon his new textbook just a few months after he had begun writing it, and to immerse himself in a fervent program of tightly focused research, at the end of which he completely revised his chymical theory with a stunning claim about the very foundations of chymical change and activity.

One of the major transformations of chymistry during the early eighteenth century was the apparent eclipse of the search for metallic transmutation, a topic that had been central to chymistry since late classical antiquity.

Yet how and why chrysopoeia (gold-making) went into steep decline, and why at this particular time, has remained either unclear or obscured by trivial triumphalist responses. Chapter 5, therefore, examines the complicated status of transmutational chymistry at the Académie. It reveals a striking divide between administrative and governmental prohibitions against the pursuit of chrysopoeia at the Académie, and the continuing transmutational endeavors of several of its most prominent members, notably Duclos, Homberg (with the collaboration and support of Philippe), and Homberg's protégé Étienne-François Geoffroy, another of the Académie's most distinguished chymists of the period. Traditional chrysopoetic—that is, "alchemical"—processes stand at the core of Homberg's mature theory of chymistry. His system both draws evidentiary support from his reported transmutation of mercury into gold and provides new theoretical and explanatory foundations for such transmutations. Homberg and Geoffroy appear to have divided the work of exploring the internal composition of the metals during their extensive collaboration; indeed, the collaboration between the two was much more substantial than has previously been recognized. Nevertheless, they sometimes disagreed on the interpretation of specific experimental results, and thus for a time held divergent theoretical explanations of such results. Yet Homberg's chymical system, which he constantly adjusted in the aftermath of new results and observations, eventually incorporated both his own and Geoffroy's interpretations.

Chapter 6 completes Homberg's biography and moves outward from Homberg's life and work at the Palais Royal laboratory to broader horizons. It examines the networks of correspondence, collaboration, and support that connected Homberg and Philippe to wider worlds, intellectual, geographical, and social. These connections portray the broad range of early modern chymical practitioners, the dynamics of the exchange of scientific knowledge, and the peculiar difficulties encountered in transmitting practical chymical know-how successfully. This chapter engages also with chymistry's unique "image problem," that is, its ambivalent reputation due both to its laborious artisanal character and to its connections to unsavory practitioners and potentially illegal practices, including poisoning, counterfeiting, and fraud. I explore the contours of these practices and practioners, and the continuing fate of chrysopoeia, through the surprising and hitherto barely recognized attempts by multiple arms of the French state to seek out potential transmuters in the early eighteenth century. Government ministers walked a narrow line, simultaneously fearful that such practitioners posed a threat to social and political order *and* hopeful that they could solve the dire financial condition into which the kingdom had fallen. Many such figures ended up under surveillance or in the Bastille, and the extensive archival documents recording these events have much to say about the schizophrenic way in which chymistry was viewed at the time. Homberg himself narrowly escaped imprisonment when he was implicated in palace gossip and intrigue over the royal succession. Although he was spared imprisonment, the affair—directly traceable to the poor public reputation of chymists and chymistry—tragically ended his ability to continue his chymical research.

The final chapter explores Homberg's impact and legacy for both chymistry and the Académie. His work was closely followed, and occasionally repeated, by a host of characters around Europe. Some adopted his system, others mined his results, and some took issue with his claims and endeavored to refute them. The roles of Homberg's two closest collaborators—Philippe, declared Regent just a week before Homberg's death, and Geoffroy, who inherited Homberg's position at the Académie—prove especially significant. The former took direct control of the Académie, and the latter published two famous and influential papers, one on affinity theory, which I can here connect more closely to Homberg's own earlier ideas, and the other, in 1722, on the frauds related to transmutation. The second of these papers has often been cited as a death knell for transmutational alchemy, but chrysopoeia actually continued to be pursued by a new cadre of academicians for at least the next forty years—a surprising fact revealed only by the discovery of new archival sources. Finally, this last chapter brings this book's three interwoven stories of Homberg, chymistry, and the Académie back full circle when a team of academicians, some sixty years after Homberg's death, pulled Homberg's and Philippe's great scientific instrument out of storage in order to repeat his experiments on the composition of metals. One of those experimenters, a twenty-nine-year-old Antoine-Laurent Lavoisier, inspired by lingering questions about the results they obtained, would then use Homberg's instrument and ideas to elaborate and advance his own research program, which led directly in turn to his own new and "revolutionary" theory for chemistry.

1

A Merchant of the Marvelous

On Wednesday, 22 April 1705, the virtuosi and curiosi of Paris gathered at the
Palais du Louvre for one of the public assemblies of the Académie Royale des
Sciences. Since its reorganization in 1699, the Académie had offered the Parisian public twice-annual glimpses of the workings of Europe's premier scientific institution. Ordinarily, the Académie met privately on Wednesdays and
Saturdays in rooms at the Louvre, but the first meetings after Easter and after
the feast of St. Martin (11 November) were open to the public. These public
meetings received regular coverage in both popular Parisian periodicals, such
as the monthly *Mercure galant*, and in more erudite serials like the *Nouvelles
de la république des lettres* and the *Journal de Trévoux*. By all accounts they
were well-attended events.[1]

The program on that spring day in 1705 promised four papers by emi-
nent academicians. The astronomer Jacques Cassini, son of Gian Domenico
Cassini, would present a new approach to the perennial problem of determining longitude. Philippe de la Hire would outline his experiments with
magnets and the compass. Joseph Pitton de Tournefort would describe his
recent Académie-sponsored research voyage to the Greek archipelago. But
the first academician to present a paper on this day was a short, thin, fair-

1. For accounts of this particular meeting, see *Mercure galant* (May 1705): 168–72; *Nouvelles de la république des lettres* 7 (June 1705): 699–703; and *Journal de Trévoux* (August 1705):
1415–26. In terms of attendance, the *Nouvelles*, describing a public assembly of the previous
year, remarked that "il y eut un grand concours de personnes de tout sexe & de toute qualité"
(6 [July 1704]:105). For the Académie's meeting rooms at the Louvre, see Camille Frémontier-
Murphy, "La construction monarchique d'un lieu neutre: l'Académie royale des sciences au
palais du Louvre," in *Règlement, usages et science dans la France de l'absolutisme*, ed. Christiane
Demeulenaere-Douyère and Éric Brian (Paris: Lavoisier Tec & Doc, 2002), 169–203.

complexioned man whose subject was chymistry.[2] He was well known to the assembled crowd, some of whom, if they had attended his earlier public presentations, possibly held their breath in anticipation. For this speaker was a good showman—he always seemed to have remarkable observations to reveal and experiences to share, and some of his presentations included demonstrations of the sort chymists (then as now) particularly enjoy, full of fire, smoke, and explosions.[3] This man did not have to travel far to the meeting, for he lived in spacious apartments within the Palais Royal—just across the rue Saint-Honoré from the Louvre—as first physician to Philippe II, Duc d'Orléans, nephew to King Louis XIV.

Today this academician was scheduled to offer the next installment in his series of essays on the principles of chymistry. He had presented the first part three years earlier, in 1702, but since that time nothing more had been heard of this much anticipated project. In the first installment he had argued that compound substances were composed of five chymical principles, or classes of substances: mercury, sulphur, salt, earth, and water. There was nothing surprising or novel about this claim. It followed the basic formulations of French chymical theory back at least to Étienne de Clave some sixty years earlier. Today would prove much more surprising.

The academician rose and began to speak slowly in his soft voice. Today, he began in German-accented French, he would reveal the true nature of the sulphur principle. He then reminded his audience that a whole class of oily substances could be separated from compound bodies by chymical analysis, and that chymists routinely called these oily materials "sulphurs." But these various sulphurs were themselves compound substances and not fundamental principles. During the past several years, he recounted, he had struggled to discover experimentally what simpler substance all these sulphurs contained in common, a more basic substance that gave them their particular properties and distinguished them from the other products of analysis. In short, he had

2. This description of physical attributes comes from Elisabeth Charlotte von der Pfalz to Raugräfin Louise, 7 March 1720, in *Briefe der Herzogin Elisabeth von der Pfalz aus dem Jahre 1720,* ed. Wilhelm Ludwig Holland, *Bibliothek des Litterarischen Vereins Stuttgart* 144 (1879), 66: "ein klein, blundt, mager mäntgen."

3. For the April 1701 public assembly, he mixed concentrated nitric acid with cinnamon oil (both pure and combined with gunpowder) to show how the mixture spontaneously bursts into flames after a few minutes; see Bourdelin Diary, fol. 33v. Apparently, audiences liked this sort of thing—the December 1705 public assembly was criticized as being "too speculative" because "la plûpart de ceux qui y vont pour satisfaire leur curosité voudroient non seulement entendre, mais voir des choses, qui surprennent agréablement"; *Nouvelles de la république des lettres* 8 (December 1705):700.

been laboring to isolate and identify what he called the "true sulphur princi-ple," an elemental substance upon which more accurate theories of chymical composition could at last be based. His endeavor to isolate the true sulphur principle had proven extremely difficult, for he had found that "the more the artist struggles to separate it, the less he finds it." His many laborious experi-ments finally brought him to a startling but inescapable conclusion: "our sul-phur principle, and the sole active principle of all compound substances . . . is the matter of light itself."[4]

Who was this small man with the German accent and the surprising an-nouncement? His name was Wilhelm Homberg, and in 1705 he was at the height of his powers, position, and notoriety. Born in 1653 in a distant outpost surrounded by tropical jungle, trained as a lawyer, and seized with an insa-tiable desire for travel, study, and experimentation, Homberg had risen to be-come a respected authority in scientific circles across Europe, the chief chy-mist of the Académie Royale des Sciences, and a favored person in the royal household of France. His is undoubtedly one of the most remarkable biog-raphies in the history of chemistry. But how did he rise to the high positions he held in 1705? By what paths did he get there? How, where, by whom was he trained? How did one—intellectually and socially—become recognized as a serious natural philosopher, in particular one devoted to chymistry, toward the end of the seventeenth century? What did being a "chymist" mean at that time, and what did Homberg consider to be the domain and goals of chymis-try? The expectations for a natural philosopher and for chymistry itself were changing significantly in the years around 1700, and Homberg's life and work situated him squarely in the middle of these changes. For his part, Homberg had very clear ideas of what a chymist should be and do, and his ideas on this score were often at odds with those of his colleagues and of the larger public.

Homberg's Biography: Facts and Fictions

Ever since Homberg's death in 1715, virtually the sole source for his biography has been the "Éloge de M. Homberg" written by Bernard le Bouyer de Fon-tenelle (1657–1757), perpetual secretary of the Académie Royale des Sciences.[5] Nearly all of the many outlines of Homberg's life that have appeared since

---

4. Wilhelm Homberg, "Suite des essays de chimie: Article troisième: Du souphre principe," *MARS* (1705): 88–96, at 89; oral presentation, PV 24 (22 April 1705): 125r–130v.

5. Bernard de Fontenelle, "Éloge de M. Homberg," *HARS* (1715): 82–93. It was reprinted frequently in collected editions of the éloges Fontenelle wrote for academicians during his long tenure as perpetual secretary.

have been little more than extracts and paraphrases (often very sloppy) of Fontenelle's essay.[6] But the accuracy and purposes of Fontenelle's account need to be scrutinized carefully. On the one hand, Fontenelle was acquainted with Homberg for at least eighteen years. Fontenelle presumably heard Homberg make autobiographical remarks during that time, perhaps even held an interview with him to gather material for the eventual éloge. Yet the closeness of the biographer and his subject should not blind us to the errors, both unintentional and intentional, of both omission and commission, that occur not only here, but throughout virtually *all* the éloges that Fontenelle wrote. Acting as the major crafter of the Académie's public image, Fontenelle did not write merely as a chronicler, nor were the éloges intended primarily to provide historically accurate biographical accounts. Fontenelle's éloges were publicly presented rhetorical performances intended to convey carefully crafted depictions of the Académie's members and their work, and they often tell us less about their subjects than about what Fontenelle considered to be the proper characteristics of an early eighteenth-century natural philosopher. Thus, the éloges omit details that (at least to Fontenelle) might reflect badly on the Académie, and offer biographies and character sketches that exemplify Fontenelle's idealized portrait of an academician. But Fontenelle's view of what constitutes "proper" science, "proper" pedigree, and "proper" pursuits and method was frequently at odds with the reality of an academician's work and life. This proves especially the case with Homberg. There is a persistent and misleading disconnect between the idealized, even fictionalized, Homberg presented by Fontenelle—widely repeated and accepted ever since 1715—and the real Homberg that this study reveals.

Homberg emerges from Fontenelle's pen as a figure endowed with classical "heroic virtues" and as the archetype of what the perpetual secretary believed a "modern" natural philosopher should be. A similar archetypal image stands in for the real Nicolas Lémery, another Académie *chimiste* who died in the same year as Homberg. Despite the widely different work, life, interests, and character of the two men, they are depicted largely uniformly: both broke from the murky and disreputable past of chymistry, both over-

6. For example, Jean-Pierre Niceron, *Mémoires des hommes illustres dans la république des lettres*, 42 vols. (Paris, 1727), 14:151–67; Paul-Antoine Cap, "Guillaume Homberg, naturaliste, 1652–1715," *Journal de pharmacie et de chimie* 3d ser., 44 (1863): 406–18; Cap, *Études biographiques pour servir à l'histoire des sciences*, 2 vols. (Paris: 1857–64), 2:214–32; and most of Marie Boas Hall's entry for Homberg in the *Dictionary of Scientific Biography*. The biographical material on Homberg in J. R. Partington, *A History of Chemistry*, 4 vols. (London: Macmillan, 1961), 3:42–47, is full of errors, many of them stemming from reliance on Cap's sloppy and largely fictionalized essays, and whatever does not fall into that category is taken from Fontenelle.

came obstacles and misfortunes to achieve greatness, and so on. In terms of enhancing "status," Fontenelle claims, for example, that Lémery earned the rank of "Maître apothiquaire" when he was actually only an "apothicaire privilégié," a purchased office, and he asserts that Lémery discovered valuable pharmaceutical preparations that were in fact well known long before Lémery ever used them.[7] Such rhetorical realignments have been pointed out in other éloges, and even Homberg's has been used to showcase the "scientific qualities" that Fontenelle imposed upon his subjects.[8] In terms of specific factual errors, Alice Stroup was the first to observe, in 1979, that some of Fontenelle's statements about Homberg are certainly false. She noted, for example, that Fontenelle's claim that Homberg studied in the Netherlands with the Dutch anatomist Reinier de Graaf around the year 1678 cannot possibly be true since de Graaf had died six years earlier.[9]

I have therefore sought for independent verification of Fontenelle's claims wherever possible, with the result that many features of Homberg's long-standing biographical details must be corrected. Most significantly, these new archival discoveries reveal a very different Wilhelm Homberg, one much more solidly rooted in the social context and intellectual practices of the late seventeenth century, and, indeed, a character far more interesting to historians than Fontenelle's "cleaned-up" version. These corrections likewise throw Fontenelle's particular preoccupations into relief, highlighting in exactly what sorts of cases—and thus for what reasons—Fontenelle "tweaked" the historical data. Identifying these features provides a better picture of Fontenelle's influential vision for the sciences, and what that vision meant for the identity that was being crafted for chymistry in the early eighteenth century. It would be rash, however, to discard Fontenelle's "Éloge de M. Homberg" entirely; it remains an important source, although one that must be used circumspectly and provisionally. It covers, for example, some parts of

7. Michel Bougard, *La chimie de Nicolas Lémery* (Turnhout: Brepols, 1999), 21–67, especially 32 and 40.

8. Volker Kapp, "Les qualités du scientifique et le prestige des sciences dans les éloges académiques de Fontenelle," in *Fontenelle: Actes du colloque tenu à Rouen*, ed. Alain Niderst (Paris: Presses Universitaire de France, 1989), 441–53, especially 445–47. See also Peter France, "From Eulogy to Biography: The French Academic Éloge," in *Mapping Lives: The Uses of Biography*, ed. Peter France and William St. Clair (Oxford: Oxford University Press, 2002), 83–102; Charles B. Paul, *Science and Immortality: The Éloges of the Paris Academy of Sciences (1699–1791)* (Berkeley: University of California Press, 1980); and Susan Delorme, "Des éloges de Fontenelle et de la psychologie des savants," in *Mélanges G. Jamati* (Paris: CNRS, 1956), 91–100.

9. Alice Stroup, "Wilhelm Homberg and the Search for the Constituents of Plants at the 17th-Century Académie Royale des Sciences," *Ambix* 26 (1979):184–202, at 185–86.

Homberg's life for which no independent sources have yet come to light. Fontenelle's éloges must simply be read critically against the backdrop of his role as public spokesman for the Académie, and checked wherever possible against other sources.

Homberg's Birth, Family, and Early Life, 1653–77

Wilhelm Homberg was unusual from the start, for he was born at Batavia (now Jakarta) on the island of Java on 8 January 1653. Fontenelle, and consequently all subsequent authors, place Homberg's birthdate a year earlier, in 1652, but the baptismal registers of the Dutch Church at Batavia clearly record that Wilhelm was baptized on 16 January 1653, and it is implausible that his parents would have waited a year before his baptism. Since the records were entered into the register in chronological order, the year 1653 cannot be a simple slip of the pen. The error might be Fontenelle's, or perhaps Homberg himself had a false impression of the year of his birth.[10]

Wilhelm's father, Johann (Hans) Homberg, was a native of Quedlinburg, an important Saxon town near the Harz mountains in what is now Sachsen-Anhalt. According to Fontenelle, Johann was displaced by Swedish incursions during the Thirty Years War. Indeed, Quedlinburg was first plundered by imperial troops in 1622–23, and then more seriously by the Swedes (presumably the event referred to by Fontenelle) in October 1631. Almost continuously from 1632 until 1650, troops of one army or another were quartered in Quedlinburg, and frequent "contributions" were raised from the citizenry.[11] Amid this military turmoil, the town was also struck by two severe outbreaks of plague, first in 1626 and then again in 1636. Given so doleful a state of affairs, it is hardly surprising that Johann left Germany to seek his fortunes abroad.

Johann Homberg went to the Netherlands and took up a commission with the Dutch East India Company, eventually becoming commander of the Company's substantial arsenal at its entrepôt at Batavia on the island of Java. There he met Barbara Beer, the widow of Antoni Beer, another officer in the Company's garrison, and married her on 21 September 1645. Hom-

10. Den Haag, Nationaal Archief, genealogical section, baptismal register for the Dutch Church at Batavia [*Hollandische Doopboek*], 16 January 1653. The discrepancy might possibly stem from calendrical issues, since some early modern systems considered the new year to begin in March; under this system, January 1653 would have been reckoned as January 1652.

11. The records of these *Kontributionen* for some years are preserved in Magdeburg; for the records that survive for the years 1632, 1633, and 1638, no one with the family name of Homberg is recorded. Magdeburg, Landeshauptarchiv Sachsen-Anhalt, A20, XXIV, #20.

berg's mother has always been assumed (following Fontenelle) to have been Dutch; the marriage register gives her hometown as "Hattenburg," probably the Dutch town of Hattem, and her maiden name as Geelis.[12] The Batavian church records indicate that the couple baptized six children: Andreas, Anna, Catharina, Wilhelm, Maria, and Elisabeth.[13] The eldest, Andreas, eventually distinguished himself as a physician in Germany, but all that is recorded of Wilhelm's other siblings is that, according to Fontenelle, one of the sisters is reputed to have been married in Java at the age of eight and to have given birth at age nine. This implausible claim is possibly a borrowing from (or conflation with) reports of the customary child marriages among the Javanese natives that circulated in Europe.[14] In any event, Fontenelle deploys this surprising account to demonstrate how the body matures more rapidly in tropical climes than in temperate zones, although this enhanced physical development occurs at the expense of the mind, since study is impossible in such oppressive heat. This unsettling story about Wilhelm's sister is not Fontenelle's only claim of Johann Homberg's accelerated expectations for his children—in 1657 he supposedly purchased the office of corporal for the then four-year-old Wilhelm.[15]

12. Den Haag, Nationaal Archief, genealogical section, marriage register for the Dutch Church at Batavia [*Hollandstrouwboek 1616–1652*], 21 September 1645: "Hans Homberg van quedelenburg Jongman Burger met Barbar geelis van Hattenburg weeduw van Antonii Beer." Barbara's maiden name is given by Fontenelle as van Hedemar ("Éloge de Homberg," 82), the origin of which is a mystery. Hattenburg might instead refer to the Alsatian town of Hatten, although less plausibly given the Dutch family name Geelis. Oddly, however, in recording the baptism of the couple's last child born in Java, the *Doopboek* (19 December 1655) gives the mother's maiden name as Jaques; this entry surely refers to the same family as it is extremely unlikely that two Hans Hombergs, both married to women named Barbara, lived at Batavia in the 1650s, although the discrepancy in maiden name is as puzzling as the origin of Fontenelle's "van Hedemar" (see also note 16, below).

13. Den Haag, Nationaal Archief, genealogical section, baptismal register of the Dutch Church at Batavia: Andreas (2 August 1648), Anna (17 October 1649), Catharina (23 April 1651), Wilhelmus (16 January 1653), Maria (10 May 1654), and Elisabet (19 December 1655). Fontenelle claims the family had only four children, of whom Wilhelm was the second ("Éloge de Homberg," 83); this conflicting evidence suggests that Wilhelm's two elder sisters died in infancy, leaving him as the apparent second child (see note 16, below). The burial registers for Batavia dating before 1727 do not survive.

14. For example, W. Glanius, *A New Voyage to the East Indies* (London, 1682), 61.

15. Fontenelle, "Éloge de Homberg," 83. There is no marriage of any Homberg daughter recorded by the Dutch Church in Batavia, and no records survive regarding the purchase of an office for young Wilhelm. Only Anna could have reached the age of nine while the family was still in Java, and given the reuse of her name for the couple's last child, it is likely she died in infancy. Additionally, Fontenelle claims that Wilhelm was the second child of the family when he

The family embarked from Java for the approximately year-long voyage back to Europe sometime between 1657 and 1659. They had certainly settled in Amsterdam by early 1660, since in March of that year Hans and Barbara presented their seventh child to the city's Lutheran Church for baptism.[16] This dating means that Wilhelm Homberg would have been six years old at most when he left Java. Nevertheless, he later recalled observations about the East Indies, the inhabitants, and their language. Much of this information must have been heard from elder family members after their return to Europe. For example, in 1707 Homberg recounted to the Académie that European women in Java produce milk that is so salty their children will not nurse, and so the infants must be fed by native women, as he himself was. (Homberg explained that this phenomenon was due to the heat of the climate, which dilates the lactating vessels in European women, who are physically unaccommodated to the tropics, thus preventing the vessels from straining out the salts as they would normally do in temperate Europe.)[17] He reported that the lead used to cover roofs at Batavia lasts only a short time.[18] He also described how Javanese natives use burns to cure themselves of various illnesses, including jaundice and colic, and explained how he had himself used boiling water as they do to cure himself of a whitlow, thus "following in some things the customs of his homeland."[19] Additionally, in her fond reminiscences of Homberg, the Princesse Palatine, Elisabeth Charlotte von der Pfalz, mother of Homberg's eventual employer and collaborator, Philippe II, Duc d'Orléans, and sister-in-law of Louis XIV, recorded that Homberg knew some of the language of the Javanese natives.[20]

---

was in fact the fourth, a discrepancy that can be explained readily if Wilhelm's two elder sisters died in infancy.

16. Stadsarchief Amsterdam, Archief van de Burgerlijke Stand, Doopboek van de Evange-lische Luthersche Gemeente 1657–1660, 5001/1/144, p. 337 (21 March 1660). The child was bap-tized Anna Marija, which is a further indication that her elder sisters Anna and Maria had died. The parents are given as Hans Homberg and Barbara Jacques, using the maternal maiden name found for the couple's previous child born in Java in 1655.

17. *HARS* (1707): 10–11.

18. Bourdelin Diary, fol. 18v (3 March 1700): "Le Plomb qui couvroit les maisons dans les indes a Batavia n'a duré que 2 ou 3 ans a ce que raporta M. Homberg."

19. Wilhelm Homberg, "Observations sur quelques effets d'une légère brulure," PV 27, fols. 85r–88v (17 March 1708); partly summarized by Fontenelle in *HARS* (1708): 46–48.

20. Elisabeth Charlotte to Raugräfin Louise, 7 March 1720 in *Briefe aus dem Jahre 1720*, ed. Holland, 66: "Er konte gar viel sprachen, Teutsch, die sprache von Batavia, Latein, Frantzösch, Holländisch, Spanisch; ich glaube, daß er auch Grichisch wuste." Spanish is probably a mistake for Italian. Fontenelle, "Éloge de Homberg," 92, adds Hebrew to Homberg's polyglot abilities, and he certainly could read English as well, since he cited Henry Bond's work on longitude (see chapter 3, note 28).

The two male Homberg children eventually embarked on professional careers. Andreas studied medicine, first at the University of Jena, where he matriculated on 7 May 1669 and defended a dissertation on *tentigo* (a sexual disorder in women) in January 1671. He chose this little discussed disorder because it was "common enough in certain places, especially in the native soil of the East Indies that gave rise to me"; the famous chymist and critic of transmutation Werner Rolfinck contributed a congratulatory poem to the event.[21] Later that year, Andreas moved to the University of Helmstedt, where he defended his thesis in 1672.[22] Thereafter Andreas transferred to the renowned medical school at Wittenberg in July 1673, gained his medical license in October, and gave his inaugural disputation on cranial fractures. On that occasion, his twenty-year-old brother Wilhelm contributed a congratulatory poem in Latin. The verses refer to their East Indian origins and suggest that Andreas's future medical education lay behind the family's journey back to Europe: "Hygeia . . . ordered you over the sea through devious waters to go to see with me the rivers of Germany." The poem marks Wilhelm's first appearance in print, and the only Latin text we have from him.[23] Andreas received his MD degree from Wittenberg in September 1676, and then began to practice medicine—perhaps in his ancestral hometown of Quedlinburg, to which the Homberg family had relocated.[24]

Jena was Wilhelm's first educational destination as well, but for the study of law; he matriculated there in summer 1672.[25] After three years he moved

21. *Die Matrikel der Universität Jena*, ed. Reinhold Jauernig and Marga Steiger, 5 vols. (Weimar: Hermann Böhlaus Nachfolger, 1977), 2:422: "Homberg, Andr. Indus Orientalis Batavia Javanus, 7. Mai 1669"; Andreas Homberg, *De tentigine disputationem* (Jena, 1671).

22. *Die Matrikel der Universität Helmstedt*, ed. Werner Hillebrand, 3 vols. (Hildesheim: Verlag August Lax, 1981), 2:191; Andreas Homberg, *Exercitatio medica de cephalgia* (Helmstedt, 1672).

23. Andreas Homberg, *Disputatio inauguralis de fracturis cranii* (Wittenberg, 1673), p. ult.: "Hygeia . . . jussit, per mare TE, per vada devia / Mecum Teutoniae visere flumina."

24. *Album Academiae Vitebergensis, Jüngere Reihe*, ed. Fritz Juntke, 4 vols. (Halle: Selbstverlag der Universitäts- und Landesbibliothek, 1952), *Part 2:1660–1710*, 180; and Hans Bekker, "Bemerkungen zu den Dr.-Dissertationen des Wilhelm Homberg," *Wissenschaftliche Zeitschrift der Technischen Hochschule "Otto von Guericke" Magdeburg* 31 (1987): 63–64. Quedlinburg had a thriving medical community—the town published its own pharmacopeia with fixed prices for the various materia medica, *Quedlimburgica officina pharmaceutica* (Quedlinburg, 1665); it was reprinted regularly into the eighteenth century. I have found few indications of the family's life in Quedlinburg, save that Johann Homberg is mentioned as a town commissioner in a document dated 24 November 1676 (Magdeburg, Landeshauptarchiv Sachsen-Anhalt A20, XXIV, #24, unfoliated section), and again as a "Revision und Peraequatio Verordend Commissarius" in 1689 and 1690 (fols. 79r, 80r, 105r).

25. *Matrikel der Universität Jena*, 2:422.

to the University of Leipzig, registering and paying his fee of twenty-two gro-
schen in the summer semester of 1675, and less than a year later, on 9 March
1676, he defended his dissertation (on diffidation, the repudiation of oaths
of allegiance).[26] Wilhelm signed himself on his law dissertation as "the Javan
East Indian," and Andreas had likewise referred to himself as "the Javan" in
his 1671 disputation. While it was standard for disputants to cite their hometown in this way, Wilhelm continued to wear his exotic East Indian origins
with pride. His self-identification as a "native" of Java extended for years to
come, and he was called the "Indian Gentleman" in several contexts. In response to an inquiry about Homberg's nationality in 1696, Gottfried Wilhelm
Leibniz wrote that "I've often wanted to laugh that he does not wish to be a
German but instead has himself labeled an Indian gentleman."[27] Leibniz was
once again obliged, after Homberg's death in 1715, to explain that Homberg
"was a German, with origins in Halberstadt, but since he was born in the East
Indies, he sometimes called himself an Indian."[28] As this chapter will show,
Homberg delighted in collecting unusual and wondrous natural philosophical materials, and displayed them with the flair of a showman. Styling himself
an East Indian virtuoso—even though he had left Java when he was no more
than six years old—made him more than just a possessor of curious exotica,
it made him an exotic curiosity himself.

Homberg began to practice law in Magdeburg, the major administrative center near Quedlinburg. No records of Homberg's time in Magdeburg
have been located, but Fontenelle claims that Homberg was "received as a
lawyer" there in 1674. But given that he matriculated at Leipzig only in 1675
and did not receive his law degree until 1676, this claim is clearly erroneous.
Homberg could not have begun work as a lawyer in Magdeburg until 1676.
There Homberg's life permanently took a new direction. Fontenelle recounts
elegantly that there "he felt that there was something else to learn about in
this world besides the arbitrary laws of men; the spectacle of nature—always
present before the eyes of everyone and yet almost never noticed—began to
draw his attention and to interest his curiosity." According to this account,

26. *Die Jüngerer Matrikel der Universität Leipzig, 1559–1809*, ed. Georg Erler, 3 vols. (Leipzig:
Giesecke & Devrient, 1909), 2:192; Wilhelm Homberg, *Disputatio juridica de diffidationibus,
vulgo vom Befehden*, (Leipzig, 1676); a second edition appeared in Leipzig in 1716.

27. Leibniz to Rudolf Christian von Bodenhausen, 3/13 January 1696, in Wilhelm Gottfried
Leibniz, *Sämtliche Schriften und Briefe*, 3d ser., vol. 6 (Berlin: Akademie Verlag, 2004), 607.

28. Gottfried Wilhelm Leibniz to Nicolas-François Rémond, 6 December 1715; NLB, LBr
768, fols. 49r–v, on fol. 49r; "il estoit Allemand originaire de Halberstat, mais comme il estoit
né aux indes orientales, il sappelloit quelquefois Indien." Halberstadt is just a few miles from
Quedlinburg.

Homberg turned first to botany. Astronomy also attracted him, and here he is reputed to have shown the earliest signs of the mechanical ingenuity that would reassert itself throughout his later career. According to Fontenelle, he constructed a miniature "planetarium" using a hollow sphere pierced with pinholes representing the brightest stars. When a candle was placed inside, its light projected images of the constellations onto the walls of a darkened room, and by turning the sphere on its axis, the spots of light would move across the walls, mimicking the diurnal motions of the heavens.[29]

In Magdeburg, Homberg came into contact (perhaps through his legal functions) with one of the city's mayors, Otto von Guericke (1602–86), famous for his mechanical contrivances and especially for his air pump and his celebrated demonstrations of the power of air pressure using the "Magdeburg spheres."[30] Multiple contemporaneous sources testify to the significance of the connection that developed between the young Homberg and the elderly von Guericke. Homberg's study of the vacuum and his construction of air pumps over the following twenty years is surely connected with von Guericke's early influence, as are other lines of research he would later pursue. Equally significantly, von Guericke entrusted Homberg with the secrets of several of his ingenious mechanical contrivances and, it seems, instilled in the young Homberg not only a love of natural philosophy but also his own attitude toward the acquisition and management of secret knowledge. The significance of their relationship may explain Fontenelle's impossibly early dating of Homberg's move to Magdeburg—it would have kept Homberg within the sphere of von Guericke's influence for several years.[31] Nevertheless, Homberg could not have been in Magdeburg for more than about eighteen months. In order to accommodate Homberg's subsequent activities before the next well-documented milestone in his life—a trip to Berlin in

29. Fontenelle, "Éloge de Homberg," 83–84.

30. Von Guericke held the post of Bürgermeister (one of the four) in Magdeburg from 1646 to 1676 and sat on the city council from 1662 to 1678. On von Guericke, see *Die Welt im leeren Raum: Otto von Guericke, 1602–1686* (Munich: Deutscher Kunstverlag, 2002); Friedrich Wilhelm Hoffmann, *Otto von Guericke, Bürgermeister der Stadt Magdeburg* (Magdeburg, 1874); *Otto von Guerickes Neue (sogenanntes) Magdeburger Versuche über den leeren Raum*, ed. Hans Schimank with Hans Gossen, Gregor Maurach, and Fritz Krafft (Düsseldorf: VDI Verlag, 1968); Fritz Krafft, *Otto von Guericke* (Darmstadt: Wissenschaftliche Buchgesellschaft, 1978); and Ditmar Schneider, *Otto von Guericke: Ein Leben für die Alte Stadt Magdeburg* (Stuttgart: Teubner, 1995).

31. Another possibility is that Fontenelle merely extrapolated from the (erroneous) birth year of 1652 he had for Homberg to make him twenty-two years old when he began to practice law, which would have been a reasonable estimate. Both Homberg brothers, however, began university education rather late by seventeenth-century standards—perhaps a result of delayed primary education due to their childhood at the East India Company's outpost in Java.

mid-1679—he must have left Magdeburg in 1677 or early 1678. Homberg's departure from Magdeburg may well correlate with von Guericke's resignation from the town council and his withdrawal to a more private life, which occurred at just this time.[32] The short time required for Homberg's conversion from law to natural philosophy suggests that his heart was never really in the legal profession, and that it was little more than a parental choice.

What certainly happened next is that Homberg left Magdeburg and went to Italy. The causes and purpose of this voyage, however, remain open to question. Fontenelle claims that when Homberg's new interest in natural philosophy began to interfere with his legal career, his friends endeavored to get him married in order to force him to settle down in the legal profession—so Homberg fled to Italy.[33] Such a response would seem rather an overreaction on Homberg's part. The éloge, however, consistently represents Homberg's family and friends as obstacles to his "destiny" as an academician. In each case, Homberg heroically—and in one instance virtually miraculously—overcomes these challenges. In the context of the éloge, a flight from matchmaking friends interfering with his study of natural philosophy does more to contribute to Fontenelle's triumphant and heroic character sketch than to explain Homberg's departure for Italy.

Fontenelle does offer some remediation to this tale by asserting that Homberg then spent a year studying medicine at Padua. Homberg's name is, however, absent from the university's records. It is possible that he attended some classes there irregularly, but he certainly never matriculated formally, meaning that if he was ever at Padua, he was not seeking a degree.[34] If medical training had really been his goal, he could much more easily have enrolled in a nearby German medical school, as did his elder brother, rather than trekking across the Alps. Fontenelle's preoccupations may again explain the matter; he frames much of his éloge around giving Homberg formal education and sequential apprenticeships with famous figures, thereby creating an impressive formational pedigree—one he felt appropriate for an academician. In terms of medical training, Fontenelle further claims that after Padua, Homberg received an MD degree from Wittenberg; yet once again, there is no trace of him in the university rolls. Fontenelle's assertion that Homberg

---

32. Hoffmann, *Guericke*, 180; Krafft, *Guericke*, 7.

33. Fontenelle, "Éloge de Homberg," 84.

34. *Matricula nationis Germanicae artistarum in gymnasio Patavino (1553–1721)*, ed. Lucia Rossetti (Padua: Editrice Antenore, 1986); *Acta nationis germanicae artistarum (1663–1694)*, ed. Lucia Rossetti and Antonio Gauba (Padua: Editrice Antenore, 1999). My thanks to Emilia Veronese for her assistance in checking for Homberg's name among the student records in the archives of the University of Padua.

studied anatomy in the Netherlands with Reinier de Graaf (1641–73), who died at least six years before Homberg could possibly have met him, runs in the same vein of creating medical credentials. The perpetual secretary's repeated insistence upon giving Homberg formal medical training is probably an attempt to justify retroactively Homberg's later position as first physician to Philippe II, Duc d'Orléans. Between Homberg's 1704 appointment by Philippe and the time Fontenelle wrote the éloge in 1715, the qualifications of royal physicians had come under scrutiny, resulting in the 1707 Edict of Marly that required all physicians to the royal household to hold approved medical degrees. Without a formal degree, Homberg would have been considered (at least by the Parisian medical faculty) an "empiric," a status that by 1715 would not have been fitting (at least in Fontenelle's mind) for a leading light of the Académie Royale des Sciences.[35] Professional expectations and qualifications for Parisian medical practitioners had shifted over the span of Homberg's life, and Fontenelle may have been trying to compensate for this change by giving Homberg formal medical training that he never had. Homberg's formal education ended with his 1676 law degree from Leipzig.

Italy and the Bologna Stone, 1678–79

There is a plausible alternative explanation for Homberg's travel to Italy that accords well with what is reliably known about his subsequent years. The éloge remarks that "in Bologna, [Homberg] worked on the stone that carries the name of that city," a reference to the Bologna Stone, a chymical marvel of the seventeenth century. Homberg corroborates this statement in a paper read to the Académie in 1694 where he notes that "I applied myself with great care to [the Bologna Stone] when I was in the country where it is commonly found."[36]

The Bologna Stone was the first artificially prepared, persistently luminescent material, and in the seventeenth century it was a highly sought-after natural philosophical wonder. Its discovery and preparation were due to Vincenzo Casciarolo, a cobbler of Bologna, who around 1603 became intrigued by the potential of unusually dense stones he found on the nearby slopes of Monte Paderno. Hoping to obtain precious metal from the stones, either by

---

35. Alexandre Lunel, *La maison médicale du roi* (Seyssel: Champ Vallon, 2008), 226–34.

36. Fontenelle, "Éloge de Homberg," 84; Wilhelm Homberg, "Experiences sur la piere de bologne," AdS, pochette de séances 1694, fol. 1r; "je m'y suis appliqué avec soin lorsque j'ay esté dans le pays ou on la trouve communement." The memoir was read on 12 May (PV 14, fol. 14v), and an abbreviated version appeared eventually in *HMARS 1666–99*, 2:213–17.

smelting them or by producing a transmuting powder from them, he put the stones in a fire and heated them vigorously. Although heating the mineral provided no precious metal, it more than recompensed the cobbler's efforts, for when a cooled stone was exposed to light and carried into a dark chamber, it emitted a soft orange-red light like a glowing ember. Thus, while not lucriferous, Casciarolo's calcined stone turned out to be quite literally luciferous.

Casciarolo distributed samples of his "sponge of light" to several interested parties around Bologna. Multiple descriptions of the process needed to render the mineral (which is not itself luminescent) capable of glowing in the dark were published later, yet few practitioners could get the process to work—especially outside of Bologna—leading to the conclusion by mid-century that the true method had been lost. In 1666, for example, the *Philosophical Transactions* lamented "the loss of the Way to prepare the *Bononian Stone* for shining" even though "several Persons have pretended to know the Art of preparing and calcining it."[37] In 1668, Robert Boyle noted how it had "grown very rare, even in its own Countrey."[38] In 1669, Henri Justel reported from Paris that "since the priest who had the secret of the thing has died, no one knows how to calcine it so as to retain the light."[39] Thus, although multiple processes for making the stone luminescent were published on several occasions, efforts to reproduce the process generally failed, highlighting the significant gap between merely having a written chymical recipe and actually knowing how to make it work in practice.[40]

37. *Philosophical Transactions* 1 (1666): 375. It has been suggested that this note came from Robert Boyle; see E. Newton Harvey, *A History of Luminescence* (Philadelphia: American Philosophical Society, 1957), 314n15.

38. Robert Boyle, "New Experiments Concerning the Relation Between Light and Air," *Philosophical Transactions* 2 (1668): 581–600, 605–12; reprinted in *The Works of Robert Boyle*, ed. Michael Hunter and Edward B. Davis, 14 vols. (London: Pickering & Chatto, 1999–2000), 6:3–25, at 10. A stone was promised to Boyle by John Evelyn in 1659, and another brought to him from Italy by Sir Richard Southwell in 1660; Evelyn to Boyle, 1 December 1659, and Southwell to Boyle, 10 October 1660, in *The Correspondence of Robert Boyle*, ed. Michael Hunter, Antonio Clericuzio, and Lawrence M. Principe, 6 vols. (London: Pickering & Chatto, 2001), 1:394–95, 429–31. Interestingly, both Evelyn and Southwell obtained their samples from Ovidio Montalbani (1601–71), a professor at Bologna and keeper of Ulisse Aldrovandi's cabinet, who wrote about the Bologna Stone in *De illuminabili lapide Bononiensi epistolae* (Bologna, 1634). See Boyle, *Aerial Noctiluca* (1680), in *Works*, 9:265–304, at 269–70.

39. Henri Justel to Henry Oldenburg, 13 February 1669, in *The Correspondence of Henry Oldenburg*, ed. A. Rupert Hall and Marie Boas Hall, 13 vols. (vols 1–9: Madison: University of Wisconsin Press, 1965–73; vols. 10–11: London: Mansell Press, 1975–77; vols. 12–13: London: Taylor & Francis, 1986), 5:401–3.

40. For a more in-depth study of the Bologna Stone, the nature of its "secret," Homberg's work with it, and the successful modern reproduction of Homberg's method, see Lawrence M.

By the 1660s and 1670s the apparent loss of the method of making the Bologna Stone had become nearly as celebrated as the material itself. It was at this time—probably in 1678—that Homberg went to Italy and there learned how to prepare the Bologna Stone successfully. He developed what he learned into a method that produced stones more brightly and reliably luminous than had ever been seen before. The earliest references to him by name in the official annals of the Académie Royale des Sciences, in 1687 (four years before his admission as an academician), record that he showed the assembly his prepared Bologna Stones whose light was "incomparably brighter than the ordinary ones." Shortly thereafter he demonstrated his method in the Académie's laboratory.[41]

It seems unlikely that the young Homberg, with little or no chymical experience, would have decided on his own to give up legal practice and travel to Italy to attempt to rediscover a lost chymical secret. But it is plausible that he was encouraged, perhaps even sent to Italy, by Otto von Guericke. Von Guericke was himself deeply interested in light and its interaction with matter, specifically by what he called the *virtus lucens* in matter; an interest that, like the air pump, he transmitted to Homberg, as illustrated by the latter's surprising 1705 claim with which this chapter opened. Von Guericke's studies of light included a specific interest in the Bologna Stone, which he mentioned in the context of his invention of a device that elicited light from a ball of sulphur. This sphere, cast from molten sulphur and fixed on an iron axle, emitted a cool blue glow when turned rapidly and touched by the hand. For von Guericke, this device was far more than a curiosity; he considered it a manifestation of the *virtutes mundanae* that maintained the whole cosmological system. It drew attention from many natural philosophers, including Leibniz and members of the Royal Society.[42] Homberg knew this device well; he prepared one himself and in 1695 displayed its phenomena to the Académie.[43] It is worth mentioning in passing that the place given to von Guericke in standard "Scientific Revolution" narratives, particularly Anglophone treatments, tends to overemphasize his showy Magdeburg Spheres and work with the air pump—perhaps because these fit neatly as a prelude to Boyle's work.

Principe, "Chymical Exotica in the Seventeenth Century, or, How to Make the Bologna Stone," *Ambix* 63 (2016): 118–44.

41. PV 12, fols. 38r–39r (7 May 1687) and 60r–v.

42. Otto von Guericke, *Experimenta nova (ut vocatur) Magdeburgica de vacuo spatio* (Amsterdam, 1672), 141–42 (Bologna Stone) and 147–50 (sulphur ball); Krafft, *Guericke*, 34–37, 70–73; "Was die Welt zusammenhält," in *Die Welt im leeren Raum: Otto von Guericke 1602–1686* (Munich: Deutscher Kunstverlag, 2002), 90–107; Schneider, *Guericke*, 126.

43. PV 14, fol. 47r (26 January 1695); see also *HMARS* 2:233–34.

But this emphasis occludes von Guericke's other, probably more fundamental, interests in astronomy, physics, and the establishment of a new cosmological system.

Once in Bologna, Homberg learned how to prepare the Bologna Stone, probably by finding someone who knew how to do it, since it is improbable that he could have discovered it independently. Marcello Malpighi reported in 1673 that there was one resident of Bologna who still knew how to prepare it, and there might well have been others.[44] But chymical secrets were precious commodities in the seventeenth century; they were traded and collected, bought and sold, but rarely given away freely. Von Guericke himself asked in regard to one of his wondrous devices: "What benefit would it be to me if I should communicate this secret, the discovery of which I made at my own great expense, to anyone at all and without recompense?"[45] Fontenelle explicitly acknowledges the importance of such exchanges for Homberg, noting that after Homberg had acquired "a considerable store of scientific curiosities, he thought about trading them and acquiring new ones by means of exchange."[46] It is likely, therefore, that Homberg had to barter for the secret of the Bologna Stone, and if his later (and better documented) methods of acquiring knowledge are any guide, it is probable that he traded some other secret for knowledge of the Bologna Stone—perhaps a secret entrusted to him by von Guericke specifically for this purpose. Many well-documented examples of such exchanges of secrets occurred throughout Homberg's life, as will become abundantly clear in what follows.

Another possibility is that Homberg acquired the secret not in Bologna but in Rome. Fontenelle writes that Homberg (after obtaining the secret of the Bologna Stone) traveled to Rome, where he apprenticed himself to the astronomer, mathematician, and mechanist Marco Antonio Cellio.[47] Such a meeting seems very plausible, given the interests Homberg had picked up from von Guericke and the fact that when Homberg's prospects in Paris dimmed in 1685, he did not hesitate to relocate directly to Rome, implying that he knew he would have a good reception there, and at a time when he was unquestionably in close contact with Cellio (see below). Intriguingly, Cellio published a small book on the Bologna Stone in 1680, just a year or so after

44. Principe, "Chymical Exotica," 124.

45. Otto von Guericke, *Experimenta nova*, 196: "Quid mihi inde gratiae, si ego arcanum illud, cujus experimenta magno meo sumptu feci, cuivis gratis communicarem?" This comment was made in regard to his *Wettermännchen*; see below.

46. Fontenelle, "Éloge de Homberg," 85.

47. Biographical details on Cellio are scanty; see *Nouvelle dictionnaire biographique*, 9:358.

FIGURE 1.1. The special furnace for calcining the Bologna Stone. On the left, as depicted in Marco Antonio Cellio, *Il Fosforo o' vero la pietra Bolognese* (Rome, 1680), 103; on the right, as depicted according to Wilhelm Homberg and published in Nicolas Lémery, *Cours de chymie*, 7th ed. (Paris, 1690), 661. Author's collection.

Homberg would have met him in Rome.[48] The method Cellio published is nearly identical to the one Homberg used; even the design specifics for the special furnace needed to heat the stones correctly are essentially identical between Cellio and Homberg (fig. 1.1). It is impossible that Cellio and Homberg arrived at such similar methods independently, yet neither makes reference to the other. Two explanations are possible. First, Homberg simply obtained the details from Cellio in Rome, possibly in exchange for some other secret. Second, Cellio obtained the details from Homberg who had previously obtained them elsewhere, presumably in Bologna. In this second case, Homberg might have exchanged the secret of making the Bologna Stone for training in optics and mechanics from Cellio, knowledge for which Cellio was renowned and that Homberg was later known to possess to a high degree. Whatever the exact source, Homberg improved the Bologna Stone process

48. Marco Antonio Cellio, *Il fosforo, o'vero la pietra bolognese* (Rome, 1680). The book was reviewed in Hooke's *Philosophical Collections* 3 (10 December 1681): 77–79.

on his own, and made significant new discoveries about the phenomenon of persistent luminescence.[49]

Fontenelle mentions that, while in Rome, Homberg developed connoisseurship in the fine arts. Homberg did have a developed taste for fine arts, as witnessed by the significant art collection he later amassed. This collection included originals of a now-lost painting by Rembrandt; two paintings by Rubens; a pair of portraits by Van Dyck; an "Alexander and Darius" by LeBrun; various Netherlandish works; copies of multiple works by Carracci, Le Sueur, Poussin, and others; pastels; drawings; and further items.[50] Homberg also commissioned a portrait of himself in 1697 from the famous portraitist of the French court Hyacinthe Rigaud (1659–1743), although this painting does not seem to have been in his apartment at the time of his death (it is not listed in the inventory of his goods or those of his widow); it had perhaps been given away as a gift. Homberg commissioned another portrait of himself, together with one of his wife, probably in 1708, by the court portraitist Pierre Gobert (1662–1744).[51] Both the Rigaud and the Gobert portraits are now lost or at best unidentified, hence we currently have no depiction of Homberg.[52]

49. For a further discussion of these points, see Principe, "Chymical Exotica," 127–33; and below in this chapter.

50. The clearest description of Homberg's art holdings comes from an inventory provided in the marriage contract of his widow: AN, Minutier Central, LX-206, Mariage 30 April 1716, fols. 4v–5r. Some of these paintings, plus others, are identifiable from the briefer descriptions given in Y//11647 (Homberg's Scellé après-décès, 24 September 1715) and LX-205 (Homberg's Inventaire-après-décès, 26 September 1715).

51. AN, Minutier Central, LX-206, Mariage 30 April 1716, fol. 4v: "Les feu s. Homberg et madlle dodart mariee en espagne originaux par gobert." The description "en espagne," presumably means à l'espagnole (since the two were never literally "in Spain")—that is, dressed in fashions of the sixteenth century, a style just then becoming popular at the French court, especially for masquerades. On fashions "à l'espagnol," see Emma Barker, "Mme Geoffrin, Painting and Galanterie: Carle Van Loo's Conversation espagnole and Lecture espagnole," Eighteenth-Century Studies 40 (2007): 587–614, especially 594–95.

52. The portrait Rigaud painted for Homberg (at the price of 140 livres) is recorded in the artist's account books, see Le livre de raison du peintre Hyacinthe Rigaud, ed. J. Roman (Paris: Henri Laurens, 1919), 58. The Gobert is listed in AN, Minutier Central, LX-206, Mariage 30 April 1716, fol. 4v. David Sturdy (Science and Social Status: The Members of the Académie des Sciences, 1666–1750 [Woodbridge, UK: Boydell Press, 1995], 232) cites Thomas Gobert as the artist, which must be a slip for Pierre; he also claims that Homberg owned a double portrait of his parents by Jean Jouvenet, but this is a misreading of LX-206, fol. 4v, which states that the Jouvenet painting was of "les pere et mere de lad[ite] d[am]e Homberg." The present author would be enormously grateful to any readers willing to notify him of portraits by Rigaud or Gobert that might conceivably be the missing Homberg.

Germany, England, and White Phosphorus, 1679–80

Homberg left Italy probably in 1679 and returned to Germany. There he further pursued his interests in chymical exotica, adding to his knowledge of the Bologna Stone that of other newly discovered luminous substances.[53] One of these was the "hermetic phosphorous" discovered by Christian Adolf Balduin during his attempts to capture the *spiritus mundi* for chrysopoetic purposes using hygroscopic salts as "magnets" to attract the *spiritus* out of the air.[54] Balduin described the substance in 1675 and wrote two letters to Henry Oldenburg about it in 1676, sending also a sample in a gilded silver box to be presented to Charles II. An account appeared in the *Philosophical Transactions*, and Balduin was made a Fellow of the Royal Society of London.[55] Although Oldenburg urged Balduin to reveal the secret of its preparation to the Royal Society "to be Recorded in their Register books, as a perpetual Monument of his ingeniosity and frankness," Balduin apparently did not find the invitation sufficiently appealing and kept the recipe secret. But Homberg was successful in acquiring the secret of its preparation by "purchasing it with another experiment," although it is unclear if he obtained it directly from Balduin or from someone else who had already extracted it from Balduin. This "hermetic phosphorus" was eventually revealed as having been produced from a chalky earth and aqua fortis, and so can be identified as a persistently luminescent form of fused calcium nitrate.[56]

53. The itinerary presented here conflicts with Fontenelle's claim that Homberg went from Rome to France, then to England (where he worked with Boyle), then to the Netherlands (where he would have apprenticed himself to the corpse of De Graaf), then picked up his supposed MD degree at Wittenberg before seeking the secrets of phosphorus; Fontenelle, "Éloge de Homberg," 85. There would not have been adequate time for these peregrinations, and there is either no evidence for them or clear evidence against them. Only the trip to England is demonstrably true, but new evidence (see below) shows that Fontenelle places it out of sequence. Other writers, building on Fontenelle's unsubstantiated claim that Homberg visited France at this time, assert that he attached himself to Nicolas Lémery in Paris, perhaps even attending his *cours de chimie*. This claim seems to have started with Cap's dreadful fictionalized "biography" ("Homberg," 410) and is repeated uncritically by Partington and others; there is no shred of evidence for it.

54. For a study of the background to Balduin's "phosphorus," see Vera Keller, "Hermetic Atomism: Christian Adolf Balduin (1632–1682), *Aurum aurae*, and the 1674 Phosphor," *Ambix* 61 (2014): 366–84.

55. Christian Adolf Balduin, *Aurum superius et inferius aurae superioris et inferioris hermeticum* (Amsterdam, 1675); *Oldenburg Correspondence*, 12:362–64 and 13:65–66, 126–27, 175–77, 219–21; "A Factitious Stony Matter or Paste, Shining in the Dark like a Glowing Coal," *Philosophical Transactions* 11 (1676–77): 788–89.

56. Fontenelle, "Éloge de Homberg," 86. Homberg may have obtained the secret from either Johann Kunckel or Leibniz, both of whom knew the preparation at this time, and both of whom

Of much greater importance was the chymical secret Homberg acquired in 1679 in Berlin from the more famous chymist Johann Kunckel (1630–1703).[57] Kunckel possessed the secret of making white phosphorus—perhaps the greatest "sensation" in late seventeenth-century chymistry. He had obtained it from its inventor, the rather obscure Hennig Brand, and in 1678, Kunckel had just published a slim volume on the subject.[58] Homberg was once again successful in obtaining the chymical secret he desired, and so became only the fourth person in the world (after Brand, Kunckel, and Leibniz) to possess the secret of making white phosphorus. Significantly, Homberg did not just obtain a written recipe or oral instructions from Kunckel. Instead, Homberg learned by *seeing and doing*: he and Kunckel prepared white phosphorus (from human urine) together in Kunckel's laboratory.[59] This hands-on teaching proved crucial. Although many other practitioners sooner or later obtained the same recipe for making phosphorus, nearly all of them, including several highly skilled chymical experimentalists, failed in their attempts to carry out the difficult procedure—as was also the case with preparing the Bologna Stone. Homberg could now prepare both of these coveted materials reliably and successfully. This situation underscores the host of problems that accompany the practical and manual operations of chymistry, and the difficulty of transmitting such knowledge by text alone. This hands-on experience with Kunckel suggests that perhaps this was the way that Homberg customarily gained practical knowledge during his travels: through hands-on training sessions negotiated as part of an exchange of secrets.

The precious knowledge of how to make white phosphorus came, as

---

Homberg visited in mid-1679. Kunckel tells the story of acquiring both Balduin's and Brand's secrets in his *Collegium physico-chymicum experimentale* (Hamburg and Leipzig, 1716), 656–65. Leibniz sent the recipe for Balduin's phosphor to the Académie, where Samuel Duclos tried to prepare it in May 1679. See Hellot Caen Notebooks, 3: fol. 272r, where the recipe is annotated as "communiqué à Mr. Duclos par Mr de Leybnitz"; and Bourdelin Expense Account, fol. 92v (19 May 1679): "donné a Mr. Duclos une livre desprit de nitre plus 1 livre de craye ces 2 pour essayer de faire un phosphore." Duclos's attempt was unsuccessful; see Christiaan Huygens to Leibniz, 22 November 1679, in Leibniz, *Sämtliche Schriften und Briefe*, 3d ser., 2:886–90, at 888.

57. Homberg provides the date 1679 in his "Maniere de faire le phosphore brûlant de Kunkel" (30 April 1692), *HMARS 1666–99*, 10:84–90, at 85.

58. Johann Kunckel, *Öffentliche Zuschrifft von dem Phosphoro mirabili* (Leipzig, 1678). A description of Kunckel's relations with Brand and his four-year effort to make the phosphorus from urine is related in Homberg, "Phosphore brûlant," 85–86, although the details of Homberg's historical account were disputed by Leibniz; see below.

59. Wilhelm Homberg to Elisabeth Charlotte, 13 January 1711, NLB, LBr 420, 2r–v, at 2r: "la maniere que je l'ay fait autres fois à Berlin avec Mr. Kunckel."

usual, at a price, which Homberg paid by dipping into his steadily accumulating treasury of secrets. In this case the "price" Homberg paid for the phosphorus process is known: it was the secret of the *Wettermännchen,* or "little
weatherman," a celebrated device designed by Otto von Guericke.[60] Homberg
had undoubtedly learned of this device while in Magdeburg when he was first
filling his intellectual purse with secrets from the famous Bürgermeister. This
"little weatherman" was a kind of barometer, although there has been much
confusion over exactly what sort of device it was. It is often conflated with
Guericke's 30-foot-long, water-filled barometer; this error is partly due to the
two devices being displayed in the same plate of the *Experimenta nova.*[61] According to eyewitness descriptions,—for example, by the traveler Balthasar
de Monconys, who saw it in 1663—the Wettermännchen was a small con-
traption able to fit in an interior room.[62] It displayed a small carved figure of
a man with one outstretched arm pointing to a scale. Changes of barometric
pressure caused the figure to move up and down in a glass cylinder, and thus
to point at different marks on the scale, each predicting a particular sort of
weather. It was sometimes also reputed to predict the appearance of comets.
Only the figure and the piston on which it stood were visible to the viewer;
the internal mechanism was hidden by an elaborate case, and von Guericke
kept the Wettermännchen's construction a closely guarded secret, thus provoking considerable speculation over how it actually operated.[63] Homberg's
knowledge of this renowned secret is witness either to von Guericke's high
regard for the young Batavian or to Homberg's mechanical ingenuity in divining the secret of its operation for himself.

Homberg's contact with Johann Kunckel proved long-lasting. In 1699,
when the Académie Royale was reorganized, each academician was asked
to name his correspondents outside Paris; the first name on Homberg's list

60. Fontenelle, "Éloge de Homberg," 86.

61. Von Guericke, *Experimenta nova,* 99; the confusion is recounted by Erich Moewes,
"Irrtümer und Unklarheiten: Bemerckungen zu Beschreibung der Guerickeschen Experimente," *Monumenta Guerickiana* 6 (1999): 41–50, who also suggests a possible mechanism for
its operation. A lengthy contemporary account of the mystery of the device (with illustrations)
is Claude Comiers, "L'Homme artificiel anemoscope, ou Prophete physique des changements
du temps," *Mercure galant* (March 1683):164–214, an account of this publication with a partial
Latin translation appeared in *Acta eruditorum* 3 (1684): 26–28.

62. Balthasar de Monconys, *Journal des voyages,* 3 vols. (Lyons, 1665–66), 2:232–33.

63. A figure floated on a mercury-filled barometer would not have displayed the changes of
height described, and the use of a water-filled one would have required a much larger apparatus.
Von Guericke's son declared in 1668 that his father had revealed the secret only to the Elector of
Brandenburg, to whom he had also presented one of the devices; see the letter printed in Stanis-
laus de Lubieniecki, *Theatrum cometicum,* 3 vols. (Amsterdam, 1668), 1:249–51.

was "Mr. Kunckel at Hamburg."[64] Unfortunately, no letters between the two men appear to have survived. Nonetheless, other scientific connections between them, besides phosphorus, are discernible. A key source for Kunckel's fame was glassworking, particularly his discovery—also in 1679, the year of Homberg's visit—of how to produce ruby glass by the addition of properly prepared gold, and the manufacture of artificial gems. In an unpublished manuscript written in the early 1690s, Homberg himself recorded a detailed process of how to prepare a purple gold powder (by detonating fulminating gold in a paper container) that "is very useful for making red crystal."[65] In the late 1690s, Homberg also worked extensively on the preparation of factitious gems, even announcing that such work, along with glassmaking in general, would be one of his major research projects during 1699. None of this work was ever published, although a substantial manuscript by Homberg on making artificial gems once existed.[66]

While in Berlin in 1679, Homberg also took time to visit the gardens that were then being expanded by the Elector of Brandenburg, and which developed into the Berlin botanical gardens. He later recalled having seen a curious fruit there that resembled both an apple and a pear. This visit would support Fontenelle's statement that one of Homberg's earliest scientific interests was botany, a claim for which there is otherwise very little evidence, since Homberg seems not to have continued this early interest later in his life, when records of his work are more abundant.[67]

From Berlin, Homberg now traveled, in summer 1679, to Hannover, where he met the next notable savant of his itinerary (and one rather surprisingly left unmentioned by Fontenelle), Gottfried Wilhelm Leibniz (1646–1716). Homberg's interaction with Leibniz also proved long-lasting; they exchanged letters and chymical secrets at least as late as 1712. At their first meeting, Homberg requested a letter of introduction to his next destination,

---

64. PV 18, fol. 149v (4 March 1699).

65. VMA MS 130, fol. 150v.

66. PV 15, fols. 74r–78v, 80r–82v, 142r; PV 18, fol. 142v (21 February 1699). There is also a lengthy account in Jean-Baptiste Duhamel, *Regiae scientiarum Academiae historia*, 2d ed. (Paris, 1701), 410–12. Excerpts of otherwise-lost Homberg recipes for artificial gems are transcribed in the Hellot Caen Notebooks, 8:18v–19v. Homberg displayed considerable experience in glassmaking, including a red glass made using gold, in "Observations sur quelques effets que l'or produit seul dans le grand feu," PV 14, fols. 274r–277v (22 February 1696).

67. PV 30, fol. 431r–v (16 December 1711). The gardens were expanded beginning in 1679 by the Great Elector, Friedrich Wilhelm (r. 1640–88), who was the father of the Elector reigning when Homberg shared his observation with the Académie, not the grandfather as the PV states. Fontenelle, "Éloge de Homberg," 83.

which is presumably something he did at each stop of his itinerary. Accordingly, Leibniz wrote Homberg a letter of introduction to Nehemiah Grew, Henry Oldenburg's successor as secretary of the Royal Society of London. This letter provides the earliest description of Homberg's abilities and corroborates important biographical details.

> Two young German men most worthy of your notice and who will sufficiently commend themselves are carrying this letter to you. . . . Mr. Homberg is not only very skilled in matters of mechanics and natural philosophy but is also extremely intelligent. For he was for a while with the most noble von Guericke, the Councillor of Magdeburg and first inventor of the air pump, and he told me how he knows of von Guericke's little man who indicates the weather. He also has the glowing phosphorus which, as you know, Heinrich Brand was the first to discover and from whom both Mr. Krafft (who was with you two years ago) and Mr. Kunckel (who improved upon it thereafter) do not deny to have benefitted.[68]

Armed with a letter of introduction from Leibniz, Homberg traveled from Hannover to England, presumably to London, in the latter part of 1679 or early 1680.[69] How long Homberg stayed in England is as unclear as what he did while there. Fontenelle reports that he "worked for a time" with Robert Boyle, "whose laboratory was one of the most learned schools of natural philosophy."[70] Despite the bulk of Boyle's publications and surviving papers, they preserve no clear record of Homberg's contact with Boyle. There is, however, strong indirect evidence that the two met and continued Homberg's customary practice of trading chymical secrets. In 1700, Homberg published an intricate and laborious process for purifying mercury using copper, an-

68. Gottfried Wilhelm Leibniz to Nehemiah Grew, in Leibniz, *Sämtliche Schriften und Briefe*, 3d ser., 2:802–3. The identity of Homberg's traveling companion is unknown, but one wonders for how long he accompanied Homberg on his many journeys. This letter was not actually sent; instead, it appears that Leibniz wrote a similar letter to Theodore Haak, who functioned as the Royal Society's unofficial liaison with German-speaking savants. See also Leibniz, *Sämtliche Schriften und Briefe*, 3d ser., 6:607, where Leibniz recalls Homberg's visit and request for a letter of introduction.

69. Dates from 1675 to 1678 have previously been proposed for Homberg's trip to England. Marie Boas Hall (*Robert Boyle and Seventeenth-Century Chemistry* [Cambridge: Cambridge University Press, 1958], 73–74 and 203) gives a date of "ca. 1675," presumably extrapolating from Fontenelle's erroneous dating and itinerary; this date is far too early. Stroup ("Homberg," 185–86) drew attention to this fact, and proposed 1677–78. I employed Stroup's reasonable dating in William R. Newman and Lawrence M. Principe, *Alchemy Tried in the Fire: Starkey, Boyle, and the Fate of Helmontian Chymistry* (Chicago: University of Chicago Press, 2004), 305–6; after discovery of the Leibniz letter, however, it is clear that the correct dating is late 1679 or early 1680.

70. Fontenelle, "Éloge de Homberg," 85.

timony, and reiterated distillations. I previously identified this recipe as the one given to Boyle by its inventor George Starkey (1628–65). Boyle prized this process as one of his greatest chymical arcana, the entrée to the philosophers' stone itself. The philosophical mercury it produced was identical with (or at least very closely related to) the "incalescent" mercury about which Boyle had published a paper in 1676 describing some of its properties while explicitly keeping its preparation hidden.[71] Since neither Starkey nor Boyle ever published the process in all its details, Homberg's most plausible source for it is Robert Boyle himself. A version of this recipe would achieve key importance in Homberg's mature chymistry; chapter 5 returns to a closer examination of it.

Homberg's acquisition of Boyle's prized arcanum could only have resulted from yet another exchange of secrets. Boyle's technique of maintaining a stock of chymical secrets to use in trade for other chymical secrets has been well documented, and his use of this technique was remarkably similar to Homberg's—which says something about the way special chymical knowledge was ordinarily obtained in the seventeenth century.[72] What Homberg gave Boyle in return cannot now be known with certainty, but given Boyle's interests and the secrets Homberg then had available for trade, it could have been information about the Bologna Stone, Balduin's phosphor, or, most likely, white phosphorus. Boyle's interest in white phosphorus had been piqued by Johann Daniel Krafft, a German chymist who visited England in September 1677 and displayed samples of the glowing material to Boyle and others (as Leibniz mentioned in his letter of introduction for Homberg).[73] Boyle was naturally eager to learn the method of preparing this new and remarkable luminescent substance, but Krafft was highly secretive, so the typical ritual of trading chymical secrets came into play. Boyle records that he offered Krafft "somewhat that I had discover'd about uncommon *Mercuries*,

71. Principe, *Aspiring Adept*, 153–79; and Lawrence M. Principe, "Wilhelm Homberg: Chymical Corpuscularianism and Chrysopoeia in the Early Eighteenth Century," in *Late Medieval and Early Modern Corpuscular Matter Theories*, ed. C. Luthy, J. E. Murdoch, and W. R. Newman (Leiden: Brill, 2001), 535–56, especially 546–49.

72. Michael Hunter, "Boyle and Secrecy," in Hunter, *Boyle Studies: Aspects of the Life and Thought of Robert Boyle* (Burlington, VT: Ashgate, 2015), 131–48; and Lawrence M. Principe, "Robert Boyle's Alchemical Secrecy: Codes, Ciphers, and Concealments," *Ambix* 39 (1992): 63–74.

73. *Lectures and Collections*, ed. Robert Hooke (London, 1678), 57–66; reprinted in Boyle, *Works*, 9:441–46. On Krafft and his visit to England, see Jan V. Golinski, "A Noble Spectacle: Phosphorus and the Public Culture of Science in the Early Royal Society," *Isis* 80 (1989): 11–39; and R. E. W. Maddison, "Studies in the Life of Robert Boyle, F.R.S.: Part I, Robert Boyle and His Foreign Visitors," *Notes and Records of the Royal Society* 9 (1951):1–35, at 29–32.

(which I had then communicated to but one Person in the World)," un-doubtedly a reference to the philosophical mercury. Krafft "in requital, con-fest to me at parting, that at least the principal Matter of his *Phosphorus's*, was somewhat that belong'd to the Body of Man."[74] Boyle correctly guessed this matter to be urine, and after many experiments, delays, and misadventures, he succeeded in preparing a minute quantity of phosphorus in mid-1680.[75] In October, he deposited at the Royal Society a document, closed with three wax seals, describing the method he used.[76]

Since Homberg knew how to make phosphorus before he visited Boyle, and given that Homberg's visit occurred in late 1679 or early 1680 just *before* Boyle's first successful preparation of the material, it is possible that Boyle re-ceived help from Homberg. Since Boyle had previously offered the recipe for philosophical mercury to Krafft in exchange for phosphorus, it seems plau-sible that Boyle made the same offer to Homberg, thus neatly explaining how Homberg got the Boyle/Starkey recipe. Furthermore, in his *Aerial Noctiluca*, Boyle notes that he received the key piece of information he needed during a visit by a "learned and ingenious Stranger," a German "who had newly made an Excursion into *England*, to see the Country." While this by itself seems a clinching piece of evidence that neatly describes Homberg, Boyle goes on to give the initials "A.G.M.D." to this German, whom scholars have identified as Ambrose Godfrey Hanckwitz, one of Boyle's operators who later made a successful business by preparing phosphorus. But while the first two initials do fit (Godfrey was not an MD), other descriptors do not. Godfrey, an émigré who at this time lived with his wife in London "in narrow means," was not a traveler coming to see the sights of England and who already knew how to prepare phosphorus. Nor did Godfrey ever leave London, as Boyle says his helpful visitor did shortly after relaying the crucial information. I would therefore suggest that the "ingenious Stranger" was really Homberg, and that

74. Boyle, *Aerial Noctiluca* (1680), in *Works*, 9:273. Krafft, later in 1677, acted on Boyle's or-ders to wrangle a secret out of another chymist whose name is given as Regius; Krafft to Boyle, in *Boyle Correspondence*, 4:482–84.

75. Boyle was still not fully successful in the preparation as of 8 July 1680, as expressed by his operator Frederick Slare in a letter to Leibniz, *Sämtliche Schriften*, 3d ser., 3:230–32, at 232.

76. Michael Hunter, *Boyle: Between God and Science* (New Haven, CT: Yale University Press, 2009), 188–89. A photo of the document is given in Hunter, *Boyle Studies*, pl. 3. Boyle distin-guishes between an "aerial" and an "icy" noctiluca; both have white phosphorus as the active ingredient, but the "aerial" or liquid form contains only minute particles of phosphorus dis-persed in water and muddy by-products of the distillation, while the "icy" is purer phosphorus in sufficient quantity that it appears as a solid mass resembling ice or wax. See Frederick Slare's report in *Philosophical Collections* (10 December 1681): 48–50.

Boyle inadvertently or purposely conflated two Germans, both related to the phosphorus.[77] It is further possible that the initials are an intentional misdirection that hides the true identity of Boyle's informant, thus protecting Homberg's reputation as one who could keep secrets.

Eastern Europe and Scandinavia, Mines and Metals: 1680–82

Homberg returned to the Continent from England probably in 1680. Having sought out Cellio, Kunckel, Leibniz, Boyle, and presumably unknown others in Italy, Germany, England, and elsewhere, and having greatly augmented his knowledge and his treasury of secrets, Homberg now pursued an interest in mining, metals, and metallurgy, and set out to visit important mining and refining centers. The study of metals was a key part of seventeenth-century chymistry, in terms of both practical commercial processes and chrysopoeia (metallic transmutation), and the mature Homberg would experiment and innovate extensively in both areas. His aims in 1680 turned from accumulating secrets and trading them with renowned individuals to seeing important manufacturing sites and learning from the skilled (but neither famous nor erudite) artisans who worked them. Homberg drew upon these experiences frequently at the Académie. In the succeeding years he amassed detailed knowledge and direct experience about the productive practices at a broad range of production sites. He often drew upon this knowledge later at the Académie. For example, he reported on a smith he watched make an iron bar glow red by hammering it, and compared the various types of brass made in Styria, Carinthia, Bavaria, the Low Countries, and elsewhere, and how gilders treat them differently.[78] He even wrote an entire treatise on refining and assaying, although it remained unpublished and is now lost.[79]

Homberg began these travels close to home, in Saxony, presumably in

77. On Godfrey, see R. E. W. Maddison, "Studies in the Life of Robert Boyle, F. R. S.: Part V, Boyle's Operator: Ambrose Godfrey Hanckwitz, F. R. S.," *Notes and Records of the Royal Society of London* 11 (1955): 159–88; Joseph Ince, "Ambrose Godfrey Hanckwitz," *Pharmaceutical Journal* 18, ser. 1 (1858): 126–30, 157–62, 215–22. A further datum is Lémery's claim (*Cours de chymie* [Paris, 1696], 685) that "Mr. Homberg, the German gentleman . . . saw [phosphorus] made by the author [of *Aerial Noctiluca*, that is, Boyle] himself." If this account is true, then Homberg would have been present for a very early—and possibly the very first—preparation of white phosphorus in Boyle's laboratory, implying that he was in England as late as summer 1680.

78. PV 14:5r (16 December 1693) and 110r–112v ("Maniere de faire du Latton sans Zink et sans Calamine," 11 May 1695).

79. Denis Dodart to Montesquieu, 28 December 1723, in *Oeuvres complètes de Montesquieu*, 19 vols. (Oxford: Voltaire Foundation, 1998), 18:73–75, at 74.

the Harz mountains (not far from Quedlinburg), an area renowned for its mineral riches. Thereafter he traveled farther afield, eastward and southward through Bohemia and into Hungary. He later recalled observations he had made about water mills along the Danube beyond Budapest.[80] After this time in Eastern Europe and the Balkans, Homberg turned north and journeyed to Scandinavia. He visited the famous mines and smelters of Sweden, quizzing the workmen about their practices, and was shown various surprising phenomena by them. He later used some of his observations about metal refining in Sweden to explain phenomena being discussed at the Académie.[81] Homberg's description of copper refining strongly suggest that he visited the immense mining operation in Falun—then the largest in Europe, with more than one hundred smelters in operation—and/or the refineries at Avesta, where the crude copper from Falun was purified, cast, and coined.[82] An unpublished manuscript by Homberg preserves an engaging first-person account of his time at one of these Swedish copper refineries. The workmen attempted to impress Homberg by throwing a piece of dry wood into the vat of molten copper and retrieving it with their bare hands, even dipping their fingers into the molten metal "up to the knuckles" without suffering the slightest burn. Suspecting a trick, Homberg recounts that "I thought that they had previously prepared [the wood and their hands] with some composition similar to that used by fire-eaters to astonish onlookers. For this reason I cut off a piece of my handkerchief, which another worker threw onto the molten copper and then retrieved and returned to me. I found it just as undamaged as if it had never been near the fire."[83] Homberg clearly loved the unusual, the exotic, and the surprising, but he also characteristically cast an inquisitive and skeptical eye on what he was shown and told.

It is generally asserted (following Fontenelle) that Homberg worked at this time with Urban Hjärne (1641–1724) in the royal laboratory newly founded in Stockholm. The Laboratorium Chymicum, however, was not founded until

80. Fontenelle, "Éloge de Homberg," 85; Wilhelm Homberg, "Observations et considerations touchant les causes des vents," PV 27, fols. 113r–121v (18 April 1708), at 117v.

81. For observations on molten copper made in Sweden, see Wilhelm Homberg, "Experience de l'evaporation de l'eau dans le vuide, avec des réflexions," HMARS 1666–99, 10:319–23, at 323; PV 14, fol. 5v (16 December 1693).

82. Daniels Sven Olsson, Falun Mine (Falun: Stiftelsen Stora Kopparberget, 2010); Sten Lindroth, Christopher Polhem och Stora Kopparberget (Uppsala: Almqvist & Wiksells, 1951); Sture Kristiansson, Strömningar till och från Store Kopparberget (Filipstad: Bronells, 1997), especially 61–82; Paul Adolf Kirchvogel, "Der Atlas von Falun," Kunst in Hessen und am Mittelrhein 22 (1982): 71–75.

83. VMA MS 130, fol. 212r–v.

1683, well after Homberg had left Sweden and become established in Paris.[84] While Hjärne did operate a private laboratory from 1680 to 1682 where Homberg might conceivably have worked, a 1692 letter to Hjärne from a Swedish visitor to the Académie clearly implies that Hjärne and Homberg had never met, although Homberg was well acquainted with the Uppsala professor Andreas Drossander (1648–96) and other Swedes.[85] Additionally, Hjärne's autobiography, which lists dozens of people he met, visited, worked with, or learned from, does not mention Homberg, even though it does mention many figures with whom Homberg would later associate.[86]

Fontenelle's citation of Hjärne and the Royal Chemical Laboratory in Stockholm represents another attempt to aggrandize, or at least "normalize," Homberg's intellectual pedigree by attaching him to important figures and places. It is accompanied by the claim that the "*Journal de Hambourg* of that time, printed in German, is full of memoirs that came from [Homberg]," with the implication that these memoirs were the result of his research with Hjärne.[87] But the *Journal de Hambourg*, founded by Leibniz, was published in French, not German, more than ten years after Homberg left Sweden (1694–96), and contains nothing by Homberg.[88] It seems odd that Fontenelle would have entirely invented so specific a claim. There is, however, one possible explanation, a solution that reveals Fontenelle's statement as an extravagant exaggeration at best. Only one journal with natural philosophical content was published in German at Hamburg during the period in question—a weekly entitled *Grösste Denckwürdigkeiten der Welt oder sogenannte Relationes curiosae*, a creation of the historical romance writer Eberhard Walter Happel

84. On Hjärne and the Royal Laboratory in Stockholm, see Hjalmar Fors, *The Limits of Matter: Chemistry, Mining, and Enlightenment* (Chicago: University of Chicago Press, 2015), 48–53.

85. Fors, *Limits*, 80, 182n17. The letter is Erich Odhelius to Urban Hjärne, 6 June 1692, in Carl Christoffer Gjörwell, *Det Swenska Biblioteket*, 2 vols. (Stockholm, 1757–62), 1:337–39; my thanks to Hjalmar Fors for bringing this letter to my attention. See also Sten Lindroth, "Urban Hjärne och Laboratorium Chymicum," *Lychnos* 57–58 (1946–47): 51–116, at 93; Lindroth already doubted Homberg's relation to Hjärne.

86. *Urban Hjärnes Själfbiografi*, ed. Henrik Schück, Äldre Svenska Biografier 5, Uppsala Universitet Årsskrift (Uppsala: Almqvist & Wiksells, 1916).

87. Fontenelle, "Éloge de Homberg," 86. Note that later editions alter the more meaningful "en Allemand" to the redundant "en Allemagne."

88. Stroup ("Wilhelm Homberg," 186) examined the first three of its four issues. On the journal, see Alfred Schröcker, "Gabriel d'Artis, Leibniz, und das *Journal de Hambourg*," *Niedersächsisches Jahrbuch für Landesgeschichte* 49 (1977): 109–29. On early German periodicals in general, see Joachim Kirchner, *Das deutsche Zeitschriftenwesen: Seine Geschichte und seine Probleme*, 2 vols. (Wiesbaden: Harrassowitz, 1958–62); Else Bogel and Elger Blühm, *Die deutsche Zeitungen des 17. Jahrhunderts*, 3 vols. (Bremen: Schünemann Universitätsverlag, 1971).

(1647–90).[89] This journal began publication in 1681, and was apparently popular in Sweden, since a Swedish translation appeared in 1682 (a year before the standard German collected edition) and the sole copy of the 1681 German edition I have located is in the Royal Library in Stockholm.[90] Homberg certainly could have been aware of it while living in Sweden. But it is impossible to identify what contribution he might have made, since *Grösste Denckwürdigkeiten* was a popular periodical, not a scientific journal, and none of its articles is signed. It does, however, describe natural and artificial wonders, most drawn from the books of Otto von Guericke and Athanasius Kircher, and thus just the kind of topics in which Homberg had a keen interest. But if Homberg was in any way connected with this journal, it would have been as a hack writer—perhaps a job that provided him with some income to continue his travels—not as a contributor of scientific memoirs.

Homberg's First French Residence: Colbert, Mariotte, and Bonnin, 1682–85

From Sweden, Homberg traveled to France, probably in early 1682 and probably stopping off somewhere in between (Fontenelle plausibly enough suggests the Netherlands).[91] In France he toured the French provinces, continuing to visit manufacturing sites: in Montpellier he observed the production

---

89. A brief notice of the journal appears in Kirchner, *Deutsches Zeitschriftwesen*, 1:47; on Happel, see Theo Schuwirth, *Eberhard Werner Happel (1647–1690): Ein Beitrag zur deutschen Literaturgeschichte des siebzehnten Jahrhunderts* (Marburg, 1908).

90. All bibliographical studies give a starting date of 1683 for *Grösste Denckwürdigkeiten*, but that is only the date of the most common collected edition. The first collected printing, previously unnoted, is *Die grösten Denkwürdigkeiten dieser Welt, oder so genante Relationes Curiosae* (Hamburg: Thomas von Wiering, 1681), containing 112 pages. This volume was followed by a 1682 edition of 304 pages (only one copy located, at Leopold-Sophien-Bibliothek in Überlingen), followed by *two* 1683 editions, *Gröste Denckwürdigkeiten der Welt, oder so genannt Relationes curiosae* (800 pages) and *Gröste Denckwürdigkeiten der Welt Oder so-genannte Relationes Curiosae* (744 pages), distinguished on the title page as "Zum andern Mahl auffgeleget." The foreword to the 1681 edition notes the journal's original weekly nature: "Wochentlich mit einer so genanten curieusen Relation zu bedienen / darin Er allemahl etwas von den raresten Materien / so bey den allerbesten Scribenten und bewehrtesten Leuten zu finden sind / antreffen wird." I have not been able to track down any surviving copies of the original weekly issues. The Swedish edition is *Denna werldennes största tänckwärdigheeter eller dhe så kallade Relationes curiosae* (Stockholm, 1682). See also Emil Key, *Försök till svenska tidningspressen historia* (Stockholm, 1883), 52–59; Bernhard Lundstedt, *Sveriges periodiska litteratur*, 2 vols. (Stockholm: Aktiebolag, 1895; reprint, Stockholm: Rediviva, 1969), 1:9.

91. Fontenelle, "Éloge de Homberg," 87.

of verdigris, in Guyenne he learned about the production of vinegar.[92] At the end of these journeys Homberg went to Paris. The three-year period in Paris that followed would prove crucial in shaping Homberg's future life and career. This critical period, however, was nearly ended before it began, according to Fontenelle's (surely somewhat dramatized) account. While Homberg was in Paris, in the fifth year of his travels throughout Europe, "his father finally lost patience, and made more serious and more insistent demands for his return home than ever before. Mr. Homberg obeyed. The day of his departure arrived, and he was ready to get into the carriage, when Mr. Colbert sought him out on behalf of the King. This minister, convinced that people of singular merit were resources to a state, made Mr. Homberg offers in order to detain him that were so advantageous that he begged a little time to come to a decision, and in the end decided to stay."[93]

What exactly these "advantageous offers" were remains unknown, and whether they were made so miraculously just as Homberg was stepping onto a carriage to leave Paris is open to question. It is clear from later events that Colbert did become Homberg's protector, and it is possible — and in keeping with Colbert's usual activities — that some sort of a pension, *gratification*, or other patronage was offered. Certainly, the experience that Homberg had acquired on his travels, particularly with respect to practical operations regarding mining, metal working, and refining, would have been of considerable interest to Colbert, who sought ways to improve the economic condition of France.[94]

In 1682, Homberg made another important decision—he converted to Catholicism. Fontenelle states that Homberg's growing affection for the Catholic faith was one of the reasons for Homberg's decision to remain in France,

92. VMA MS 130, fols. 221v and 223r.

93. Fontenelle, "Éloge de Homberg," 87.

94. There is a report (*Mémoires de Saint-Simon*, ed. A de Boislisle, 41 vols. [Paris: Hachette, 1879–1930], 22:385n2) that Colbert settled a pension of 1,600 livres upon Homberg at this time, but there is no record of Homberg's supposed stipend in "Gratifications faites par Louis XIV aux savants," published in *Lettres, instructions et mémoires de Colbert*, ed. Pierre Clement, 7 vols. (Paris, 1861–73), 5:466–98. This amount seems too large a sum for the young Homberg, as most Académie members were then receiving 1,500 livres per annum; on the amount of stipends see Stroup, *Company*, 230–42. On Colbert and the Académie, see Sturdy, *Science and Social Status*, especially 1–79; Stroup, *Company*, 13–61; David S. Lux, "Colbert's Plan for the *Grande Académie*: Royal Policy Towards Science, 1663–67," *Seventeenth-Century French Studies* 12 (1990): 177–88; and more broadly, Jacob Soll, *The Information Master: Jean-Baptiste Colbert's Secret State Intelligence System* (Ann Arbor: University of Michigan Press, 2009), 94–139. On Colbert's *gratifications* for savants, see Charles Perrault, *Mémoires de ma vie*, ed. Paul Bonnefon (Paris: Renouard, 1909), 48–49; on his endeavor to improve mining, 108.

and that his father disowned him as a consequence.[95] The latter event might well have happened; given Johann Homberg's experiences as a result of the Thirty Years War, he might indeed have found his son's embrace of Catholicism unacceptable. Alternatively, Fontenelle's claim might be yet another example of his rhetorical casting of Homberg's family as an impediment.

Some authors have viewed Homberg's conversion cynically as merely a political convenience. Certainly, conversions to Catholicism for the sake of political or economic expediency are hardly unknown in France in the years leading up to and especially after the 1685 Revocation of the Edict of Nantes. Homberg's future colleague at the Académie, Nicolas Lémery, converted (along with his entire family) in order to regain his pharmaceutical trade, which he had lost in 1683.[96] Nonetheless, there is good evidence that Homberg's conversion was a sincere religious act, even if it is undeniable that he gained various secular benefits from his choice, given the situation in late seventeenth-century France. David Sturdy has pointed to the significant number of Catholic theological and devotional texts present in Homberg's library; Saint-Simon refers to the depth and sincerity of Homberg's religiosity; and in fact, within a few years Homberg would seriously consider taking holy orders.[97]

Prospects in Paris must have appeared good from Homberg's perspective, for now he decided to end his travels and become a permanent resident of France; accordingly, in January 1683, at the age of thirty, he was naturalized as Guillaume Homberg.[98] It seems unlikely that the ever traveling Homberg would have naturalized without some clear prospect of a permanent position in France, suggesting that Colbert did indeed provide him with some sort of continuing support. It also indicates that events must have happened very quickly once Homberg got to Paris, since he naturalized only a few months after his arrival. How did he distinguish himself so quickly? Presumably he reached into his accumulated treasury of chymical and mechanical secrets, and showed them off to the right people, presumably members of the Académie Royale, who then gave a complimentary account of the young German savant to Colbert, who was always on the lookout for talent. Whatever Hom-

95. Fontenelle, "Éloge de Homberg," 87.

96. Bougard, *Lémery*, 41–49. Lémery's reacquisition of a pharmaceutical practice on 8 April 1686 is documented by AN, O/1/30, fol. 132r.

97. Sturdy, *Science and Social Change*, 230–31; Saint-Simon, *Mémoires*, ed. Yves Coirault, 8 vols. (Paris: Gallimard, 1983–88), 5:742; on his consideration of holy orders, see below.

98. AN, O/1/27, fol. 67r: "Naturalité en faveur de Guillaum Homberg natif de la Ville de Batavia en l'Isle de Java dans les indes Orientalles faisant profession de la R[eligion] C[atholique] A[postolique] et R[omaine] a Versailles au mois de Jan 1683."

berg did or said, it must have been quite impressive in order to attract Colbert so quickly. Unfortunately, I have not uncovered any reference to him in the months preceding his naturalization, so what he did so successfully and with whom must remain conjectural. Nevertheless, it is clear that from 1681 to 1683 events in Paris had prepared the perfect stage for Homberg's entrance: savants both inside and outside the Académie, as well as members of the public, had become fascinated by luminescent substances, especially white phosphorus. The learned excitement over this new substance and its public displays in London at this time have been well documented by Jan Golinski, but the analogous situation in Paris has not received the same attention.[99] Since this scene forms part of the context for Homberg's arrival in Paris and his quick rise to notoriety, it is worthwhile to take a moment to document it.

In January 1681, the secretary of the Académie Royale des Sciences, Jean-Baptiste Duhamel, received a letter from Boyle with recipes for white phosphorus and for Balduin's "hermetic phosphorus." In February, the Académie's chymist Samuel Cottereau Duclos (1598–1685) presented a paper on these topics. In April, his colleague Claude Bourdelin (1621–99) received a letter from one Lefevre, then in London and in contact with Boyle, that claimed (erroneously) that Boyle had still not succeeded in producing solid white phosphorus. Throughout May and June 1681, Duclos and Bourdelin labored to prepare phosphorus in the Académie's laboratory using the method sent by Boyle, but without success.[100] In late March 1682, the same Lefevre was in Paris, and sold to the Académie (through Bourdelin) "a little vial of phosphorus that the assembly desired" for nineteen livres; the sample was displayed on 1 April.[101] A few days later some academicians, when rubbing the material (with their bare hands!) on a hat to make it glow, succeeded instead in setting a bed and themselves on fire. Shortly thereafter, Jean Dominique Cassini accidentally set his handkerchief alight with the phosphorus

99. Golinski, "Noble Spectacle."

100. PV 10, fol. 58r (22 January 1681; Boyle's letter), fol. 61r (26 February 1681; Duclos's discourse); fol. 64r–v (16 April 1681; LeFevre's letter); fols. 70r–74r, at 73r–v (June 1681; summary report from Duhamel to Colbert); fol. 74r (25 June 1681; Duclos's report on the failure of Boyle's method); Bourdelin Expense Account, fol. 100v (7 and 17 June 1681; letter from England, purchase of sand to distill with urine to produce phosphorus according to Boyle's method). The Lefevre mentioned here is possibly the same Lefevre who lived across the street from Boyle on Pall Mall and who is mentioned as "apothecary to the king" in the context of Boyle's interactions with the French alchemist Georges Pierre des Clozets; see Principe, *Aspiring Adept*, 115–34; *Boyle Correspondence*, 5:13 and 28.

101. Bourdelin Expense Account, fol. 103v (30 March 1682): "payé a Mr. le febure medecin venu dangleterre pour une petite fiole de phosphore que l'assemblé a desiré"; PV 10, fol. 94v (1 April 1682).

and then burned a hole through his shoe trying to put out the fire.[102] In May, the Saxon nobleman, mathematician, and inventor Ehrenfried Walther von Tschirnhaus (1651–1708), who had been in Paris since late March, informed Leibniz that an unnamed Englishman had offered to sell the Académie the secret of making phosphorus, for fifty pistoles. Tschirnhaus begged Leibniz for the secret, explaining that he wished to present the process to Colbert, with the intention of being rewarded with a position in the Académie and a pension, and that he had begged the Abbé Jean Gallois (formerly Académie secretary) to stall negotiations with the unnamed Englishman in order to give Tschirnhaus time to obtain the recipe.[103] Leibniz sent Tschirnhaus the process (in exchange, of course, for two chymical secrets offered in return by the Académie's chymists), and Tschirnhaus presented it to the Académie in early July 1682. Tschirnhaus was promptly rewarded by being made a member of the Académie on 22 July, although vague promises of a stipend never materialized.[104] Duclos immediately set to work on this new process, bringing to the laboratory a few days later 400 pints of human urine collected on the rue du Seine.[105] But the attempt again failed.

In November 1682, still eager but unable to prepare the marvelous substance, Duclos presented the assembly with several methods he had received, and asked for advice on which one to pursue.[106] In December, Tschirnhaus sent yet another recipe for phosphorus to Abbé Edmé Mariotte; it was tried out by Joachim d'Alencé (1640?–1707) in the laboratory of the academician

102. Edmé Mariotte to Leibniz, 13 April 1682; Leibniz, *Sämtliche Schriften und Briefe*, 3d ser., 3:586–88. Cassini eventually extinguished the fire by smothering it with a copper ruler; the ruler itself then glowed for many days thereafter. See PV 9, fol. 159r (18 April 1682); Duhamel, *Historia*, 210; *HMARS 1666–99*, 1:343–44.

103. Tschirnhaus to Leibniz, 27 May and 31 May 1682; Leibniz, *Sämtliche Schriften und Briefe*, 3d ser., 3:624–25 and 639–41. This "Englishman" might again be Lefevre, who was settled in London.

104. PV 10, fol. 104r (1 July 1682; Tschirnhaus' presentation); PV 9, fols. 165r–166r (4 July 1682; recipe); PV 10, fol. 104v (8 July 1682; reading of Leibniz's recipe, translated from German by Blondel); Fontenelle, "Éloge de M. Tschirnhaus," *HARS* (1709): 114–24; *HMARS 1666–99*, 1:342–43, 346; Duhamel, *Historia*, 210; *Mercure galant* (July 1682): 308–9 (admission to Académie). Leibniz to Tschirnhaus, late June 1682; Leibniz, *Sämtliche Schriften und Briefe*, 3d ser., 3:651–63, at 660–63 (phosphorus recipe). In the same collection, Tschirnhaus to Leibniz, 27 July 1682, 3:685–91, at 686 ("obscure promesse" of a stipend and secrets); Mariotte to Leibniz, 22 June 1682, 3:648–50, at 649: "Nos Chymistes sont prests de troquer leurs plus beaux secrets contre le vostre du phosphore." Leibniz himself had tried sending samples of solid phosphorus to Huygens and Mariotte as early as 1680, but the quality did not respond to expectations; see Leibniz, *Sämtliche Schriften und Briefe*, 3d ser., 3:48–49.

105. Bourdelin Expense Account, fol. 106v (29 July 1682).

106. PV 9, fol. 180r (14 November 1682).

Jacques Borelly (d. 1689). Borelly displayed the result to the Académie on 16 December, although it is unclear from the records exactly what that product was; the brevity of the entry suggests it was not the avidly sought-after "phosphore sec."[107] Bourdelin and Duclos spent much of December on the same preparation, but failed to produce any phosphorus.[108] In January 1683, Étienne Hubin, the enameler to Charles II and a skilled instrument-maker, arrived from England with several kinds of "phosphorus"—presumably meaning various luminescent materials. He gave samples to Claude Comiers, the theologian, professor of mathematics, polymath, and co-editor of the popular *Mercure galant* from 1681 until his death in 1693. Comiers published a lengthy account of the various phosphori in the *Mercure galant*—an indication of the contemporaneous public interest in the subject.[109]

Thus, from 1681 to 1683, the academicians were keenly interested in white phosphorus but endlessly frustrated in their attempts to prepare it. Homberg, upon his arrival in Paris in 1682, would therefore have found an audience eager for his luminescent secrets and practical abilities. A candid statement from Leibniz reveals that this was *exactly* the situation Homberg had been looking for. In recalling Homberg's visit to him, Leibniz writes that Homberg intended "to make himself known by means of the phosphorus, and thus it has turned out to be the foundation of his good fortune since no one in the laboratory at Paris was able to make it even though they already had a description of [the process] for some years."[110] The continuing interest at

107. PV 10, fol. 117r–v (16 December 1682): "Mr. Borelli a apporté un Phosphore que Mr. Dalencé a fait chez Mr. Borelli suivant le memoire que Mr. Schirnouse avoit envoyé a Mr. Mariotte, et que la personne qui l'avoit apporté n'ayant pas trouvé Mr. Mariotte donna à Mr. Dalencé." On Borelly, see Pierre Chabbert, "Jacques Borelly (16. .–1689): Membre de l'Académie des Sciences," *Revue d'histoire des sciences* 23 (1970): 203–27; Joachim d'Alencé, *Traité des baromètres, thermomètres et notiomètres ou hygromètres* (Amsterdam, 1688).

108. PV 10, fol. 116v (9 December 1682; Bourdelin in the midst of the operation) and fol. 118v (30 December 1682; the exact procedures and results are recorded in Claude Bourdelin, "Analises chymiques," 6 vols., Cornell University Lavoisier MSS, QD B76++, 6:4925–34.

109. *Mercure galant* (January 1683): 164–69; (June 1683): 184–218 ("Traité des phosphores"); (July 1683): 187–98 ("Suite du traité des phosphores"). Hubin also sold two areometers to the Académie's laboratory during his visit; see Bourdelin Expense Account, fol. 110r (5 March 1683): "Mr. Hubin Emailleur pour deux areometres, 4 l[ivres]." Hubin was especially known for making artificial eyes; Martin Lister visited his shop in Paris in 1698: *Journey to Paris in the Year 1698*, 3d ed. (London, 1699; reprint, ed. Raymond Phineas Stearnes [Urbana: University of Illinois, 1967]), 145–46.

110. Leibniz to Rudolf Christian von Bodenhausen, 3/13 January 1696; Leibniz, *Sämtliche Schriften und Briefe*, 3d ser., 6:604–10, at 607: "Kurz als der phosphorus anfung bekandt zu werden, kame er nach Hannover und begehrte auch von mir adresse nach England und sonst, umb sich mit dem phosphoro bekand zu machen, wie er denn auch der grund seiner fortun ist,

Paris in pneumatics and air pumps also played in Homberg's favor, given the knowledge of the subject he had obtained from Otto von Guericke. Indeed, when accounts of Homberg's activities in Paris finally begin to appear in mid-1683, most of them relate to phosphorus, pneumatics, or both.

On 26 July 1683, the *Journal des sçavans* carried a detailed report about "new and curious" experiments Homberg "performed a few days ago in the presence of several gentlemen of the Académie Royale des Sciences."[111] Among other things, Homberg showed how the luminescence of phosphorus was dependent not only upon the quantity of air to which it had access, but also upon its temperature. Presumably Homberg used samples of phosphorus he had prepared himself, given the continuing failure of Académie chymists to make it.[112] Intriguingly, the *Journal* mentions Homberg without any introduction, as if he would be known to readers, suggesting that he had already made a splash in Paris. The article further notes that Homberg performed these experiments using an air pump of his own design, which was based on von Guericke's but was "much simpler and more exact." Making good use of what he had learned during his travels, Homberg riveted the attention of the academicians with his phosphorus, his air pump, and his chymical and mechanical prowess. The article ends with praise for Homberg's study of phosphorus, and notes (with a touch of secrecy probably reflecting Homberg's own limited disclosure of the matter) that he has discovered a new dissolvent for phosphorus, described only as a "highly fixed mineral liquor."[113] This mineral liquor "fort fixé" is almost certainly mercury, for Homberg later described to the Académie his production of a luminous combination of mercury and phosphorus; he was indeed the first to recognize that mercury would amalgamate with phosphorus.[114]

Unfortunately, the *Journal des sçavans* does not specify exactly which "gentlemen of the Académie" were present at Homberg's demonstration. The

denn keiner im laboratorio zu Paris ihn machen konnen, ob sie schohn die beschreibung davon vor etlichen jahren schohn gehabt."

111. *Journal des sçavans* (26 July 1683): 226–28; reprinted in *HMARS 1666–99*, 10:648–51.

112. Homberg certainly had a private laboratory in Paris at this time; he makes reference to it in a later paper (see below, note 134) and in an anecdote that dates from his early 1680s Parisian residence; see Wilhelm Homberg, "Experiences sur la pierre de bologne," AdS, pochette de séances 1694, fol. 3r–v.

113. *Journal des sçavans* (1683): 228. "M. Homberg a trouvé une liqueur Minerale fort fixé qui dissout le Phosphore, & qui le fait éclairer sans qu'il soit besoin de luy donner de l'air nouveau."

114. Homberg, "Phosphore brûlant," 89–90. See also the mention of a glowing material "en forme de mercure coulant," in Wilhelm Homberg, "Nouveau phosphore," *HMARS 1666–99*, 10:445–48, at 445.

academicians most interested in phosphorus at the time—Borelly, Bourde-lin, Duclos, and Cassini—might well have been in attendance. Further in-formation, however, appears in an account of the event hurriedly inserted in a new 1683 edition of Lémery's *Cours de chymie*. The experiments were carried out at the home of d'Alencé, whom we have already encountered in connection with phosphorus, and who also worked with pneumatic devices. Strikingly, this addition to Lémery's text must have been made at the very last minute, perhaps even after the manuscript had gone to the press, for the book cleared the press on 14 August, only three weeks after Homberg per-formed the experiments.[115]

This phosphorus experiment was not the only time Homberg displayed his skills before the academicians, although there is no record of his doing so at one of their formal meetings held at the Bibliothèque du Roi. It must, however, be recognized that the procès-verbaux provide very incomplete ac-counts of the weekly assemblies (see note on sources). In May or June 1683, Homberg conducted experiments to determine the ratio of the densities of air and water. Since this work used d'Alencé's air pump, it is probable that it too took place at d'Alencé's residence, and it may suggest some sort of relationship between him and Homberg at this time. Philippe de la Hire and Edmé Mari-otte were among the "several members of the Académie" present.[116] Mariotte reported the results (1 to 630) to the Académie formally on 9 June 1683.[117]

Mariotte and Homberg appear to have been collaborating on projects at this time. In 1692–93, after his admission to the Académie, Homberg pre-sented several papers that bear striking resemblances to subjects and re-search questions associated with Mariotte, and it seems likely that Homberg was either revisiting these experiments or publishing earlier results. One such experiment dealt with the anomalous expansion of water upon freez-ing. Mariotte had conjectured that the expansion was due to the release of dissolved air, and Homberg's experiment tests the idea. While Homberg's experiment—which required two years to complete—was published only in

115. Nicolas Lémery, *Cours de chymie* (Paris, 1683), 556–67; the date the printing was finished is given on sig. aiiii. Lémery's wording is dependent upon the *Journal des sçavans*, so he probably received an oral account supplementing the published report rather than having been present during the experiment.

116. Philippe de la Hire, "De la pesanteur de l'air," PV 14, fols. 243r–250r (26 January 1696), at 246r–v; "En 1683, M. Homberg fit une experience de la pesanteur de l'air en presence de plu-sieurs personnes de l'Académie, et où je me trouve."

117. The procès-verbaux cite only Mariotte for this work (PV 10, fol. 148r), while Duhamel (*Historia*, 222 and 401) and de la Hire's memoir cite only Homberg; Fontenelle's much later summary cites both (*HMARS 1666–99*, 1:361).

1693, it probably began in conjunction with Mariotte (who had died in May 1684). Indeed, a brief article published in the 31 January 1684 *Journal des sça-vans* is surely a contemporaneous notice of these experiments, even though neither Homberg nor Mariotte is mentioned there. Thus, Mariotte appears to have been Homberg's most frequent contact and link within the Académie, and their scientific collaboration was an important formative experience for the young Batavian.[118]

But Homberg's earliest known mention by name in Parisian print did not stem from his natural philosophical work or his collaboration with Mariotte. It deals with something far more pedestrian—almost amusingly so. The 3 May 1683 issue of the *Journal des sçavans* carried an advertisement that "Mr. Homberg, an Indian by birth" had invented a new and improved candleholder [*chandelier de cabinet*] which could be purchased at the shop of "Sieur Houdry on the rue de la Ferronnerie."[119] By means of a spring hidden under the candle, Homberg's invention kept the height of the flame above the table constant—the spring, as it expanded under the diminishing weight of the burning candle, pushed the candle up as it decreased in length. By holding the candle at an angle, the device also obviated the need to trim the wick because the wick gradually emerged from the *side* of the flame, thus burning up entirely in the external air. Homberg also fitted the lamp with a reflective shade to shield the eyes from glare and direct the light downward to the tabletop. The sale of this device not only recalls the inventiveness and mechanical ingenuity that characterizes much of Homberg's career, but also witnesses Homberg's interest in earning a few spare livres for his livelihood.

The Académie's "Penumbra" in Paris

Although Homberg was not an official member of the Académie at this period, he was in a sense a "shadow" member; that is to say, while there is no record of his having attended their assemblies, he did conduct experiments with academicians in other venues. Indeed, I suggest that the Académie of this period should not be envisioned as a sharply defined group, but instead as a heterogeneous assemblage with surprisingly diffuse boundaries. At the

118. "Experience curieuse et singuliere," *Journal des sçavans* (31 January 1684): 36. The later memoir covering similar materials is Homberg, "Expérience sur la glace dans le vuide," *HMARS 1666–99*, 10:255–62 (memoir dated 28 February 1693); see chapter 2 for more on these experiments. The closeness of Homberg and Mariotte at this time may have been assisted by the fact that Mariotte knew German (see PV 10, fol. 71v); Homberg's French at this time was probably still rather imperfect.

119. *Journal des sçavans* (3 May 1683): 120.

center stood the official academicians, who received a royal pension, but extending out from them existed a penumbra of others with varying degrees of affiliation to the formal institution and its members. How extensive this penumbra really was remains hard to discern. D'Alencé, for example, falls in quite close to the central group—while never an official academician himself, he associated regularly with the academicians, and worked from 1679 to 1684 to produce one of the institution's key publications, the *Connoissance des temps*. Alongside d'Alencé was Tschirnhaus who, while formally admitted in 1682, received no stipend and was not resident in Paris. Then there exist the numerous names that appear erratically in the procès-verbaux as present at a few meetings in order to showcase their knowledge—such as Abbé Antonio Felice Marsigli. Many others apparently attended one or more meetings or even came to deliver memoirs but are not listed in the records, such as Claude Comiers, who spoke on optics there in 1682; thus attendances and even contributions by Homberg might be similarly unrecorded.[120] Somewhat further out are instrument makers like Hubin, who was obviously on good terms with several academicians (like Homberg, he collaborated with Mariotte on several experiments) and supplied them with specialized equipment. On the periphery lie "hired hands" such as Jean Louette, Antoine le Duc, and Jean de la Rue, who sequentially stoked the fires and switched the distillation receivers in the chymical laboratory in exchange for a salary.[121]

Further complicating the picture are the meetings composed of both academicians and nonacademicians that took place in various locales, often domestic spaces, in addition to the Académie's official meetings at the Bibliothèque du Roi. I have already noted that some meetings took place at d'Alencé's house in mid-1683, where Homberg gave demonstrations; such meetings at d'Alencé house may have been regular occurrences. There were surely other venues as well. There are references to chymical laboratories maintained at the homes of at least Borelly and Bourdelin, as well as at Homberg's: did informal meetings and demonstrations sometimes occur in those spaces?[122] Such gatherings resemble the meetings of the earlier Montmor academy, held at Henri-Louis Habert de Montmor's house from

120. *Mercure galant*, Extraordinaire 1 (January 1683):118.

121. These "serviteurs du laboratoire" are listed in Bourdelin's Expense Account. Louette worked until late 1683, Le Duc in 1684–85, and De la Rue in 1686; fols. 112r, 114v, 116v, 118r.

122. Borelly's home laboratory is mentioned at PV 10, fol. 117r–v (16 December 1682) and was visited by Olaus Borrichius in 1664; *Itinerarium 1660–1665*, ed. H. D. Schepelern, 4 vols. (Copenhagen: Danish Society of Language and Literature, 1983), 3:366–67. Bourdelin's laboratory is mentioned in his Expense Account, fol. 120v, when he began doing his work for the Académie there (instead of at the Bibliothèque du Roi), in 1687.

1657 until about 1664, or the Cartesian conversations held first by Jacques Rohault and later by Pierre-Sylvain Régis, or the Monday-afternoon discussions sponsored by Pierre Michon, Abbé de Bourdelot, at his residence on the rue de Tournon from 1665 until his death in 1685.[123] Various members of the Académie are known to have attended each of these groups at one point or another. What is clear is that the founding of the Académie Royale des Sciences in 1666 did not cause these less formalized groups to stop meeting, nor did it sequester the new academicians from them. Indeed, these informal meetings may have provided key opportunities for newcomers to Paris— like Homberg—to meet the established figures in natural philosophy. It will be difficult to gauge the role and impact of these extra-Académie meetings, which have left only scanty and widely dispersed documentation, but their continuing occurrence sufficiently indicates that a significant portion of the Académie's work and discussions took place at scattered venues, and these are not recorded in the procès-verbaux.

While Homberg's connections with Mariotte in 1683 seem certain, it is impossible, given the lack of records, to ascertain the other academicians with whom he might have worked at this time, from whom he might have learned, and by whose aid he advanced intellectually and socially. His documented relations with Mariotte and d'Alencé suggest that Homberg probably met most of the other academicians during this period. His interest in chymistry and knowledge of phosphorus would certainly have invited a relationship with Duclos, Borelly, and Bourdelin, all of whom were at work on the phosphorus (and other chymical projects) at the time. Despite the excellent records of the Académie's laboratory, Homberg's name is nowhere to be found in them, nor in any of the other sources dating from the period that I have uncovered.

Given the associations he was developing, Homberg had reason to hope for his eventual admission to the Académie. His decision to end his travels and naturalize as a French subject implies that he anticipated staying in France. Yet Homberg did not spend out his treasury of secrets prodigally in trying to "buy" admission to the Académie, as Tschirnhaus did with Leibniz's (ineffective) recipe for phosphorus. Homberg displayed his knowledge and possession of phosphorus, but never taught the method of preparing it. Homberg revealed that secret only *after* he had been admitted to the august body in 1691—an act that can be seen as another kind of exchange. The

---

123. Harcourt Brown, *Scientific Organizations of Seventeenth-Century France* (Baltimore: Williams & Wilkins, 1934), 64–134 (Montmor Academy) and 231–53 (Bourdelot Academy); René Taton, *Les origines de l'Académie Royale des Sciences* (Paris: Palais de la Découverte, 1965), especially 13–29.

same is true of his knowledge of the Bologna Stone. Antonio Felice Marsigli displayed his own luminescent stones at the *assemblée* of 28 July 1683, and showed his method of calcining them to Bourdelin in the Académie's laboratory.[124] Yet even though this visit and demonstration happened at a time when Homberg was undeniably in close contact with the Académie—meaning that he presumably heard about it—Homberg apparently never divulged his own, superior method for calcining the stones.[125] His revelation of that secret would come only later, when it could do him the most good.

The rosy state of affairs Homberg enjoyed in the summer of 1683 faded quickly upon the death of Colbert on 6 September 1683. François Michel Le Tellier, Marquis de Louvois (1641–91) succeeded Colbert as the government minister in charge of the Académie, and since he considered Colbert a rival, he proved at best unwelcoming toward those whom Colbert had patronized. This negative attitude was surely extended to Homberg, who must have quickly realized that his prospects for becoming an academician were now significantly dimmed. The measures that Homberg then took further support the claim that the minister had been giving Homberg some kind of financial support. Homberg's response to the loss of Colbert was to seek a new patron, and he found one, apparently rather quickly. He now began his "grande liaison," as Fontenelle terms it, with Armand-Louis Bonnin, Abbé de Chalucet (1641–1712). Described in a contemporary reference as "un grand Naturaliste," Bonnin was very interested in chymistry and engaged Homberg to work on a chymical project in 1684.[126] That Bonnin hired Homberg for a chymical project indicates that Homberg had acquired a reputation for chymical projects and operations—more than just his possession of chymical

124. PV 11, fol. 4r; *HMARS 1666–99*, 1:362. Fewer than half of the stones were successfully rendered luminous by Marsigli's method. These accounts cite simply "Comte" Marsigli, which might be read as Count Luigi Ferdinando Marsigli, who later wrote on the Bologna Stone and became a foreign member of the Académie in 1715. In mid-1683, however, Luigi Ferdinando was enslaved by the Turks in Bosnia, and the reference is actually to his elder brother, the *Abbé* Marsigli, as he is correctly identified in Bourdelin's account of the process: "Analises chymiques de Monsieur Bourdelin," Cornell University Library, Lavoisier Bound MSS, QD B76++, 6 vols., "Calcination de la Pierre de Boulogne," 6:489–90. Bourdelin's Expense Account, fol. 110v, records that on 29 July 1683, Bourdelin spent 2 livres 3 sous for "a fat chicken, a larded pigeon, two pints of wine, three rolls, and six apricots for lunch and dinner for the valet de chambre of the Italian abbé de Marsigli who stayed eight hours in the laboratory to teach us how to calcine the Bologna Stone."

125. For part of this time Homberg was temporarily unsuccessful in carrying out the calcination process in Paris; see below.

126. Fontenelle, "Éloge de Homberg," 87; *Mercure galant* (June 1684): 243–44. Short biographical notices for Bonnin appear in the *Nouvelle biographie générale*, as well as in Jean-Pierre Papon, *Histoire générale de Provence*, 4 vols. (Paris, 1777–86), 1:380.

secrets—even if the published notices of him, cited above, refer more to his expertise in physics than to his work in chymistry.

A glimpse of the chymical expertise and investigative methodology Homberg had already developed by this time appears in a vivid narrative describing new discoveries he made about the Bologna Stone. In an unpublished memoir (fig. 1.2), Homberg recalls his frustration that his attempts to prepare

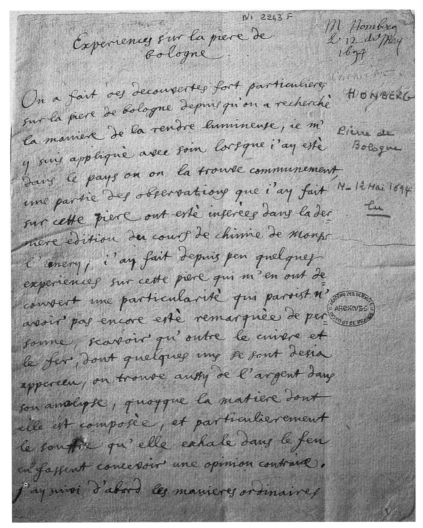

FIGURE 1.2. Homberg's unpublished autograph memoir on the Bologna Stone, describing his discovery of the role of trace iron as a luminescence poison and of trace copper as a dopant critical for luminescence. The endorsement in the upper righthand corner is in the hand of the Académie secretary, Abbé Jean-Baptiste Duhamel. Courtesy of the Archives of the Académie des Sciences, Paris.

luminous Bologna Stones always failed after his arrival in Paris, even though he had previously "rendered luminous a great quantity of these stones in more than a hundred different operations both in Italy and elsewhere."[127] One day, he ran into a friend to whom he had promised to teach the method. Unable to excuse himself, Homberg was led to the friend's house, where he found in readiness a furnace built according to his specifications and some uncalcined stones he had previously given his friend. "Being thus pressed, I began again the operation which had so often failed, and to speak the truth, I was trembling all the while, for I had not told him that I had always failed at it in Paris."[128] To Homberg's astonishment, the operation succeeded, and the stones glowed brilliantly. His curiosity aroused, the puzzled Homberg scrutinized every step of the process. A key part of Homberg's method required grinding some of the mineral into a powder, which he then used to coat other Bologna Stones before their calcination. The only, seemingly trivial difference he could identify in the process carried out at his friend's house was that the mortar he used there to grind the stones was made of bronze—as were the ones he had used elsewhere—while the one he had been using in his Paris laboratory was iron. So he tried again in his own laboratory using a bronze mortar and found that the process now worked perfectly. In response, Homberg designed and carried out an experimental program to explore this odd result. He gathered up mortars made of marble, porphyry, iron, bronze, and lead, as well as platters made of silver and tin, and a bowl made of copper. He used each vessel separately to prepare the powder. Only the powders prepared in the bronze mortar and the copper bowl produced luminescent stones. Homberg then reground each of the prepared powders in the bronze mortar—all of them then gave luminescent stones except for the one that had previously been ground in the iron mortar. Homberg thus concluded that a trace of copper was necessary for the luminescence, and that the least trace of iron completely prevented it. What Homberg's careful methodology discovered for the first time here in the years around 1684 is what modern chemists now know as iron's action as a "luminescence poison" as well as the need for copper as a "dopant" for persistent luminescence. These features were rediscovered independently only much later, partly in the 1890s and

---

127. The full account appears in Homberg's unpublished memoir, "Experiences sur la piere de bologne," AdS, pochette de séances 1694. Fol. 3r: "rendu lumineux une grande quantité de ces pieres en plus de cent differentes operations tant en italie qu'alieurs."

128. Homberg, "Experiences sur la piere de bologne," fol. 3v: "estant ainsy pressé je recommencay l'operation si souvent manquée, à la verité en tremblant, car je ne luy avois pas dit qu'à paris je l'avois tousjours manqué."

partly only in 2011.[129] For our present purposes, this account indicates the level of experimental prowess and methodological sophistication Homberg had achieved in chymical matters around the time he was engaged by Bonnin.

Fontenelle says little about Homberg's work for Bonnin. He merely contrasts Homberg with another chymist also in the employ of the abbé. Fontenelle then asserts—seemingly without provocation—that "Homberg was too capable to aspire to the philosophers' stone and too sincere to put such a vain idea into anyone's head." The modern reader would be right to think that the perpetual secretary protests too much in this regard. In fact, the quest for the stone and the transmutation of metals very much occupied Homberg's mind and hands throughout his life, as will be documented in chapter 5. Fontenelle writes that the other chymist in the abbé's employ was convinced of the reality of the philosophers' stone and its transmutatory powers, and in order to convince a supposedly dubious Homberg, gave him an ingot of gold—purportedly prepared by transmutation—worth 400 francs (more than the yearly salary of the Académie's laboratory operator). According to Fontenelle's apologetic reasonings, Homberg then not only left the company of Bonnin and his chymist but also fled from Paris, and even from France as well, in order to get as far away as possible from the baleful influence of this unnamed chrysopoeian.[130]

But what did Homberg actually do in 1684 while in the employ of the Abbé de Chalucet? Fontenelle gives no answer, leaving readers with only his suspiciously earnest insistence that it did *not* have to do with transmutation. Of course, that is exactly what it was. In 1711 and 1712 Homberg gave a full retrospective account of his work for the abbé. The story Homberg recounts is as remarkable as the famous material he finally produced. In 1711, Homberg recalled how "about thirty years" earlier, "a notable person" had repeatedly asked him to try to prepare an odorless and colorless oil from human feces.[131] The allure of this otherwise remarkably unsavory project was that the notable person in question, who is surely Bonnin given the rough dating Homberg provides, had been shown an experiment in which such an oil was used to transmute common mercury into silver. Homberg agreed to take on this project, which promised to "enrich us both." Since the exact

129. Homberg also discovered that while a trace of copper was necessary, more than just a trace decreased the luminescence; for more on this topic see Principe, "Chymical Exotica," 127–33.

130. Fontenelle, "Éloge de Homberg," 88.

131. Homberg, "Observations sur la matiere fecale," *MARS* (1711): 39–47.

method of preparing the oil was unknown, Homberg recognized that considerable experimentation would be required. So his first task was to secure a reliable supply of the starting material. Given its ubiquity in the streets of seventeenth-century Paris, one would not think this a problem, but Homberg was concerned about getting a standardized material, rather than one "collected by chance."[132] This precaution underscores the level of circumspection as a chymist that Homberg had already developed in 1684; many later examples further document Homberg's unusual and consistent degree of concern over the purity, consistency, and standardization (both qualitative and quantitative) of chymical substances. In this case, Homberg's solution was to "rent four robust and healthy young men" whom he locked up with himself for three months in a house specially rented for the purpose. Their job was to afford him a daily supply of the needed starting material. Further to ensure consistency of product, Homberg made them agree to eat only what he gave them, which for the three months of their productivity was only "bread of Gonesse" and the best wine from Champagne.[133]

Homberg trained one of his young men to collect the desired material each day, keeping each man's contribution separate (in case different men produced material with different properties), and taught him how to reduce it to a dry material by distilling off the aqueous part in an alembic. Homberg himself then distilled the dried feces at higher temperature in a retort, but obtained in this way only a red oil and a stinking black one. He then tried extracting fresh feces with hot water. After filtration and evaporation, he isolated an "essential salt of fecal matter," which resembled saltpeter. Upon destructive distillation, this salt gave an oil, but one that was stinking and red rather than odorless and colorless. Then he tried mixing various additives, such as vitriol, lime, and alum, with the dried feces before distilling them, but this too failed to provide the desired product. Finally, Homberg tried fermentation, "which is a gentle way" of separation that avoids the use of a strong fire. Knowing that this fermentation would occupy a long period of time, he gathered up all he would need from the rented young men, dried

132. Incidentally, when Claude Bourdelin decided to perform an analysis of human feces for the Académie in December 1678, he easily collected several pounds of the material "in the rue Vivienne, all along the walls of the Hôtel Mazarin, being in frozen lumps and produced just the day before"—right outside the Bibliothèque du Roi where the Académie met and had its laboratory. "Analises chymiques de Monsieur Bourdelin, " Cornell University Library, Lavoisier Bound MSS, 5:4377.

133. Homberg, "Matiere fecale," 40. The "pain de Gonesse" is a common white bread brought to Paris from the town of Gonesse, a few miles north of the city; it is described by Lister in his *Journey*, 148.

the material gently in an alembic and stored it as flasks of distilled liquid and jars of dried residue. He then released the men from their service, left the house he had rented for the project, and returned to his "usual laboratory" elsewhere in Paris.[134]

Once home, Homberg mixed dried feces with six times their weight of distillate, sealed the mixture in a flask, and exposed it to a gentle heat for six weeks. After this time the contents no longer smelled of their unsavory origin. He distilled off the liquid gently, and recommended it to "some persons" whose skin was rough and dry. They washed their face and arms with it once a day, and by means of its continued usage found their skin "considerably softer and whiter."[135] The distillation residue, to his surprise, smelled like perfume, in fact like ambergris, the costly ingredient of fine perfumes, and even the emptied distilling flask was so redolent of this fragrance that Homberg had to remove it from his lab. This remark recalls Paracelsus's *Zibetum humanum*, a highly fragrant material smelling like musk, supposedly prepared from human excrement; Homberg may also have had this substance in mind. Finally, he ground the now fragrant residue, put it into a retort, distilled it at high temperature, and at last obtained an odorless and colorless oil. He thereafter embarked on a long series of experiments trying to join this oil with mercury, but it steadfastly refused to be transmuted into silver.

While failing in the realm of argyropoeia, this research project unexpectedly brought Homberg some of his most lasting fame. For he noticed with considerable surprise that after he distilled a mixture of fresh excrement and alum, the residue spontaneously burst into flames when exposed to air.[136] Not one to let an unexpected result go unexplored, Homberg simplified the process to study this phenomenon. He took 4 ounces of feces "newly made," added an equal quantity of alum, and fried the mixture in an iron skillet, constantly stirring to keep it from sticking to the pan. He put the resulting dry powder into a long-necked flask and heated it gradually to red heat, stoppering the flask when the heating was complete. After the flask cooled, when a portion of its powdery contents was shaken out onto a sheet of paper, the material soon began to smoke and then burst into flames.[137] This substance, separated entirely from its transmutational origins (and given a less disagreeable starting material) became known in the chemical literature as "Hom-

134. Homberg, "Matiere fecale," 43.

135. Homberg, "Matiere fecale," 44.

136. Homberg, "Matiere fecale," 42, 46–47.

137. Homberg, "Phosphore nouveau, ou Suite des observations sur la matiere fecale," *MARS* (1711): 238–45.

berg's pyrophorus" and appeared as such in most chemical textbooks down
to the start of the twentieth century.[138]

Tallying up the time Homberg reports that he spent on these fertile operations indicates that he was in Bonnin's employ for a minimum of six months.[139]
Homberg remained on friendly terms with Bonnin long afterward—perhaps
the reason Fontenelle calls their relationship a "grande liaison," which would
seem to imply something more extended than several months in the mid-
1680s. After Bonnin had become bishop of Toulon and was having a new
parquet floor installed in his episcopal residence, he wrote to Homberg for
a method to keep it from being eaten by worms. Homberg suggested soak-
ing the wood in a weak solution of corrosive sublimate (mercuric chloride)
before installation, a suggestion that met with success.[140] In 1707, Homberg
presented a paper on spiders he had observed in a garden in Toulon, suggesting that he visited Bonnin at his residence.[141]

During his approximately three years in Paris, from about 1682 to 1685,
Homberg succeeded in attracting the attention of two patrons and in earning the respect of prominent natural philosophers both at the center and in
the penumbra of the Académie. But now, with his formal association with
his second patron ended—whether by the unsuccessful conclusion of their
project, by Bonnin's elevation as bishop of Toulon, or by a combination
of the two—Homberg was in need of a fresh way of making ends meet.[142]
His daily costs must have been fairly significant, since he had lodgings large
enough for maintaining a laboratory, and the cost of chemicals, instruments,
and fuel was always substantial. His choices at the time were limited. Increased association with the Académie was now unlikely, given the policies
of Colbert's successor Louvois; indeed, this period is generally seen as one of

138. For further material on Homberg's pyrophorus, see chapter 6; see also William B. Jensen, "Whatever Happened to Homberg's Pyrophorus?" *Bulletin of the History of Chemistry* 3
(1989): 21–24.

139. Three months in the rented house with rented men, six weeks for the fermentation of
the feces, and at least a month for the various subsequent distillations and attempts to combine
the oil with mercury.

140. *HARS* (1705): 38; note that here the bishop is described solely as "une personne de
qualité en Provence," but the procès-verbaux (PV 24, fol. 249v) clearly name "Mr. L'Evêque de
Toulon."

141. Homberg, "Observations sur les araignées," *MARS* (1707): 339–52. See below for further
probable connections with Bonnin.

142. Louis XIV named Bonnin as bishop of Toulon on 30 May 1684. See *Journal du Marquis
de Dangeau*, ed. Eudore Soulié et al., 19 vols. (Paris, 1854–60), 1:20–21; *Mercure galant* (June
1684): 243–44.

decline or at least stagnation for the Académie.[143] Homberg's closest scientific associates were now gone as well: Mariotte died in May 1684 and d'Alencé had left France for Holland earlier in the year.[144] With Paris no longer the promising venue it had been, Homberg packed up his phosphorus, his air pump, his Bologna Stones, and all his other secrets and set off back to Italy in 1685.

Homberg's Second Italian Period: Biological and Optical Work, 1685–87

The years that follow Homberg's departure from Paris remain the least well-documented period of his life. Fontenelle's account says almost nothing about the six years from 1685 to 1691, offering only that Homberg went to Rome in 1685 (to escape that pesky chrysopoeian) and supported himself there by practicing medicine. He does not specify if Homberg spent all six years in Rome, or occupied his time with anything other than medical practice.[145] New sources, however, indicate that while in Rome, Homberg associated with the Accademia Fisicomatematica Romana, a scientific society organized in 1677 by Giovanni Giustino Ciampini (1633–98) that met monthly at Ciampini's residence on the Piazza Navona.[146] Homberg likewise renewed his friendship with Marco Antonio Cellio, who was an active member of this academy. Indeed, the prospect of renewed association with savants he had met previously in Italy would explain why Homberg chose Rome as his destination when his opportunities in Paris dried up.

In a 1686 report on the Accademia's work, Ciampini describes two contributions made by "Signor Hombergh, the Indian Gentleman." The first was a new kind of tripod mount for the sliding tube-style microscopes made by

143. Elmo Stewart Saunders, *The Decline and Reform of the Académie des Sciences à Paris, 1676–1699* (PhD diss., Ohio State University, 1980); Stroup, *Company*, 51–56.

144. *Mercure galant*, Extraordinaire 2 (April 1684):127, notes that d'Alencé was "Secretaire du Roy, now in Visbourg." Duclos also died in 1685, but his relationship with Homberg remains unclear.

145. Fontenelle, "Éloge de Homberg," 88; Homberg's only explicit mention of this time in Rome refers to a physician he knew there who used realgar medicinally: *HARS* (1703), 52.

146. On this academy see W. E. Knowles Middleton, "Science in Rome, 1675–1700, and the Accademia Fisicomatematica of Giovanni Giustino Ciampini," *British Journal for the History of Science* 8 (1975): 138–54; Maria Pia Donato, *Accademie romane: una storia sociale, 1671–1824* (Naples: Edizioni Scientifiche Italiane, 2000), 26–44; Federica Favino, "Beyond the 'Moderns'? The Accademia Fisico-matematica of Rome (1677–1698) and the Vacuum," *History of Universities* 23 (2008): 120–58; see also scattered references in Jean-Michel Gardair, *Le Giornale de' letterati de Rome, 1668–1681* (Florence: Olschki, 1984).

Giuseppe Campani.[147] The second was the demonstration of a set of improvements to Leeuwenhoek-type microscopes—a better focusing mechanism and a rotating sample holder that could accommodate seven different objects simultaneously.[148] A similar sample holder had been used earlier by Nicolas Hartsoeker (1656–1725), and so Homberg probably carried the design to Italy with him. Indeed, a microscope strikingly similar to the one illustrated by Ciampini appears in a portrait of Hartsoeker dating from this period.[149] Cellio, Campani, and Carlo Antonio Tortoni also presented new microscope designs to the Accademia around the same time, indicating that this was a serious topic of interest for the group during 1686.

What was Homberg observing with his improved microscope? The *Giornale de' letterati* gives some details about the Accademia's contemporaneous microscope studies, but the key clue appears in a letter from the prefect of the Vatican Library, Emmanuel Schelstrate, to the editor of the *Acta eruditorum*, Otto Mencke. Schelstrate recounts that the academicians were then making observations of "the *animalcula* in the semen of all animals."[150] Although the letter does not mention Homberg explicitly, the topic it cites allows a connection to be made with later references to his studies. In his 1690 *Système de philosophie*, Pierre-Sylvain Régis (1632–1707) discusses views on animal generation; among those who believe that the animal seed comes from the male, he first cites Hartsoeker, but then notes that "Mr. Homberg has communicated to us the manuscript of a treatise he has written on the generation of animals based upon the same principles, and he maintains this opinion with reasoning that is so probable that if it should still appear surprising, that is only because it is new."[151]

Homberg's claim that the "animalcula" or "worms" in semen were responsible for generation situates him as an early proponent of the "animal-

147. Giovanni Giustino Ciampini, *Nuove inventioni di tubi ottici dimostrate nell'Accademia Fisicomatematica Romana l'anno 1686* (Rome, 1687), 4–5; Silvio Bedini, "Seventeenth Century Italian Compound Microscopes," *Physis* 5 (1963): 383–422, at 397–400.

148. Ciampini, *Nuove inventioni*, 8–9.

149. *Journal des sçavans* (1678): 355–56; item no. 205 in Marjorie E. Wieseman, *Caspar Netscher and Late Seventeenth Century Dutch Painting* (Doornspijk: Devaco, 2002).

150. Schelstrate to Mencke, 15 June 1686, in *La correspondance d'Emmanuel Schelstrate*, ed. Lucien Ceyssens (Rome: Academia Belgica, 1949), 206–7; printed in *Acta eruditorum* (1686): 371–72.

151. Pierre-Sylvain Régis, *Cours entier de philosophie* [*Système de philosophie*], 3 vols. (Amsterdam, 1691), 2:642; on Régis and generation see Dennis Des Chene, "Life after Descartes: Régis on Generation," *Perspectives on Science* 11 (2003): 410–20.

culist" side of a debate about generation (against the "ovists") that would rage for the next two decades all across Europe.[152] The Batavian was again one of the first on the scene of a celebrated scientific topic, as he had been with phosphorus, but now it was a topic in the life sciences rather than in chymistry or physics. Homberg's manuscript treatise must have circulated fairly widely, for several other authors cite it. When surveying ideas about animal generation for an article written for the *Journal des sçavans*, Leibniz lists Homberg and Hartsoeker together, alongside Malebranche, Swammerdam, and Leeuwenhoek.[153] Homberg's views are mentioned again in the review of a disputation on the topic held at Caen in 1711.[154] Indeed, in 1715 Homberg's *Nachlass* included "a little treatise on generation," and Homberg's nephew Denis Dodart (1698–1775) was preparing the same work, listed as "a compendium of experiments and reasonings about the generation of animals following the system of worms" for publication in 1723. Unfortunately, the publication never materialized, and the treatise is now lost.[155] If Régis saw the completed treatise before 1690, and the Accademia Fisicomatematica Romana was looking at spermatozoa while Homberg was devising and demonstrating microscopes there, it seems very likely that this lost biological work on animal generation was a product of his studies in Rome around 1686.

Homberg's Italian microscopal investigations probably also connect with a trip to southern Italy, during which he made observations (later published) of a minute parasite he found on certain spiders in the kingdom of Naples.[156] Homberg's interest in biological generation continued for some years after; in 1699 he published a paper on the reproduction of damselflies. The paper recounts how he captured male and female insects, observed their copulation, and dissected some before and some after mating. Finding certain ducts

---

152. On this debate, see Jacques Roger, *The Life Sciences in Eighteenth-Century French Thought*, ed. Keith Benson, trans. Robert Ellrich (Stanford, CA: Stanford University Press, 1997), 236–56.

153. Published in André Robinet, *Malebranche et Leibniz: Relations personelles* (Paris: Vrin, 1955), 312.

154. Review of Pierre Ango, *Quaestio medica: An homo a vermibus*, *Journal des sçavans* (13 June 1712): 376–80, at 379.

155. Nicolas Rémond to Leibniz, 23 December 1715; NLB, LBr 768, fols. 53r–54v, at fol.54r: "un petit traité de la Generation." Dodart to Montesquieu, 28 December 1723, in *Oeuvres complètes de Montesquieu*, 18:74: "un recueil d'Experiences et de raisonnements sur la Generation des Animaux suivant Le systeme des Vers." Here the term "worms" refers to spermatazoa.

156. Homberg, "Araignées," 348, 352. Homberg later showed drawings of spiders, evidently made with the help of a microscope, to the Académie, and some of these might date from his Italian period; PV 14, fol. 132v (15 June 1695).

in the males empty after copulation, he concludes that they must be reservoirs of spermatic fluid.[157]

Trips to Paris, Provence, and into the Académie, 1687–91

Homberg reappeared in Paris in May 1687, but whether this was a brief trip after which he returned to Italy or went elsewhere, or the beginning of an extended stay in the French capital remains unclear. Whichever it was, in May 1687 Homberg made his first appearances at regular meetings of the Académie recorded in the procès-verbaux. On 7 May he demonstrated an air pump constructed according to a new design, an improvement upon his earlier model—apparently his mechanical inventiveness in Rome was not limited to microscopes. With this new apparatus he performed several experiments, which included measuring the weight of the air, firing a pistol in an evacuated space, and showing the effect of the vacuum on various phenomena: phosphorus (apparently a repeat of his 1683 demonstration), dissolutions in acid, the breaking of Prince Rupert's drops, and the action of a magnet. On the same day he also exhibited his luminous Bologna Stones, and three weeks later he revealed the secret of their calcination and carried out the process in the Académie's laboratory.[158]

Do these appearances represent a renewed bid for membership? It is unlikely that Homberg would have traveled so far to appear suddenly at the Académie without having been invited to do so. But whatever hopes he might have had for admission once again came to naught. There is little evidence regarding Homberg's activities or whereabouts during the following four years, from mid-1687 until late 1691; Fontenelle passes over these years in silence. Homberg makes no subsequent appearance in the Académie's records until after his formal admission in November 1691. It is possible that he returned to Italy and spent part of these "missing years" there. But perhaps more likely is that Homberg refers to this otherwise undocumented period when he recalls how he "saw more than two hundred cases of scrofula cured in the hospitals of Aix-en-Provence, Marseille, Toulon, and of other locales in Provence."[159]

---

157. Wilhelm Homberg, "Observations sur cette sorte d'insectes qui s'appellent ordinarement Demoiselles," *MARS* (1699): 145–51.

158. PV 12, fol. 60r–v (7 May 1687): "Mr Hombert a apporté sa Machine pneufmatique qui est differente de lordinaire, il a fait plusieurs experiences." A short description of the experiments follows, which is partly reprised in *HMARS 1666–99*, 1:365–67; see also Fontenelle, "Éloge de Homberg," 88. For the Bologna Stone, see PV 12, fols. 38r–39r (28 May 1687).

159. VMA MS 130, fol. 249r: "J'ay vû guerir par ce moyen plus de deux cens ecroüelles dans les hopitaux d'aix, de Marseille, et de Toulon et d'autres endroits de la Provence."

Witnessing the cure of more than two hundred patients and visiting these multiple hospitals implies a rather extended stay in the area, and it is difficult to fit such a sojourn into any other part of Homberg's travels. This period in Provençal hospitals would also explain the source of Homberg's medical knowledge and experience, given that he had no formal medical training or degree. Intriguingly, the reference to Provence—and in particular to Toulon—suggests that Homberg had renewed contact with Bonnin, now bishop of Toulon, and was perhaps once again receiving support from him.[160] This additional contact with Bonnin would further justify Fontenelle's otherwise puzzling reference to a "grande liaison" between the two.

Significantly, just three months after Homberg's appearance at the Académie in May 1687, he received royal permission to hold benefices in France, implying that someone in authority was working on his behalf to orchestrate a funded and permanent return to France. The link to benefices suggests that someone with high clerical status—that is, with the ability to dispense a benefice—was helping Homberg, and Bonnin is the most likely candidate from what we know at present. The *lettres patentes* from Louis XIV state that Homberg "has chosen his vocation and is resolved to embrace an ecclesiastic status in the hope he has of obtaining some benefice in order to subsist by means of it in the requisite dignity and condition."[161] Which level of holy orders Homberg was ready to take remains uncertain, but he was probably planning to become an abbé like Mariotte and Duhamel. If a permanent and ecclesiastically funded return to France was being planned for him in 1687 and 1688, it too failed to materialize, and Homberg did not take holy orders.

A letter that I discovered in an unexpected locale nevertheless indicates that Homberg continued to seek a position in the Académie and that the academicians fully supported him. Following the death of the chymist Jacques Borelly in summer 1689, Homberg apparently saw a window of opportunity to gain admission as Borelly's replacement. At this time, acting on Homberg's behalf, the physician Claude Deshayes Gendron (1663–1750), whom Homberg may have met either in Paris or in Provence while Gendron was on the

160. Bonnin was not officially consecrated as bishop until 1692, owing to disagreements between Paris and Rome, but he had nevertheless taken up leadership of the diocese before this time; see Jean-Pierre Papon, *Histoire générale de Provence*, 4 vols. (Paris, 1777), 1:380.

161. AN, X/1A/8682, fols. 284v–286r, "Congé à Guillaume Hombert d'accepter et posseder Benefices," 23 August 1687. On fol. 285r: "il a fait choix de sa vocation et resolu d'Embrasser l'Estat Ecclesiastique dans l'Esperance qu'il a d'obtenir quelque benefice pour y subcister dans l'honneur et descence requise." See also O/1/31, fol. 178r, "Permission de tenir benefices pour Guillaume Homberg . . . le xxiiie jour d'Aoust 1687," and the Chambre de Comptes registration of the *lettres patentes* at PP//151, fol. 92v (6 August 1688).

medical faculty at Montpellier, wrote to ask for assistance from the antiquarian and scholar Nicolas Toinard (1628?–1706).

> I seem to remember that I have heard you speak on other occasions of the bishop of Meaux [Jacques-Bénigne Bossuet] as one of your friends. If this is the case, Mr. Homberg would have a favor to beg of you with some urgency, which would be to obtain for him an audience with the bishop by means of a letter on your part. Tell us if you are a close enough friend to him for this. It is so that Mr. de Meaux might offer witness to Mr. de Louvois about the special merits of Mr. Homberg, whom all the academicians of the sciences ask to have as a colleague. Mr. de la Chapelle [Henri de Bessé] advised Homberg to procure some access to the minister [Louvois] who willingly accommodates himself to everything that Mr. de Meaux says. The abbé de Louvois is a friend of Mr. Homberg, but even so it would be good if a learned man like you should offer testimony of Mr. Homberg's abilities, he is assuredly the most capable man in the world to fill the place of Mr. Borelly. If you can offer him this service, you would delight the most respectable [*honneste*] man in the world and the one most dear to all the savants. Should you not be able to do it, I ask you at least to write as quickly as possible so that he may take measures elsewhere.[162]

The letter reveals that Homberg and the academicians were working together to secure his admission to the Académie. Even Henri de Bessé, Sieur de la Chapelle (c. 1629–94), whom Louvois had installed in the Académie in 1683 as his agent and spokesman, gave Homberg advice on how to promote his appointment through the cultivation of social contacts able to recommend him to Louvois. Had the academicians already directly but unsuccessfully asked Louvois to appoint Homberg, perhaps in the context of his

---

162. Claude Deshayes Gendron to Nicolas Toinard, undated (c. July/August 1689), Uppsala, Uppsala University, Carolina Rediviva, Waller MS fr-03897: "ce me semble vous avoir autrefois entendu parler de Monsieur L'evesque De Meaux comme d'un de vos amis. Si cela est, Monsieur Homberg auroit une grace a vous demander avec empressement, qui seroit de luy procurer un accés aupres de luy par une lettre de vostre part. mandés nous si vous etes assés son amy pour cela, c'est pour que Monsieur De Meaux rende tesmoignage a Monsieur De louvois du merite particulier de Monsr. Homberg que tous /1v/ les academiciens des sciences demandent pour compagnon. Monsr. de la chapelle luy a donné ce conseil de se procurer quelque accés aupres de ce ministre qui s'en rapporte volontiers a tout ce que Monsieur de Meaux dict. Monsr. L'abbé de Louvois est amy de Monsieur Homberg, mais avec tout cela il seroit bon qu'un homme scavant comme vous eust rendu tesmoignage de la capacite de Monsr. Homberg qui assurement est le plus habil homme qui soit au Monde pour remplir dignement la place de Monsr. Borelli. Si vous pouvés luy rendre ce service, vous allegrés le plus honneste homme qui soit au monde, et le plus cheri de tous les scavants. en cas que vous ne le puissiés pas, au moins ecrivés je vous prie au plustost afin qu'il prene les mesures ailleurs."

1687 visit to the Académie? The letter's sense of urgency suggests that it was written very soon after Borelly's death in mid-1689 in an attempt to secure the empty position before anyone else could be put forward or appointed. Louvois's son, Camille de Tellier, Abbé de Louvois (1675–1718), who was at the time only fourteen years old, is mentioned as Homberg's friend and potentially a supplementary avenue of support. It was known from later sources that Homberg tutored the Abbé de Louvois, but this newly discovered letter shows that their relationship had already developed by this time, perhaps engineered as a preliminary step toward securing Louvois's favor.[163] The letter leaves no doubts about the high regard in which Homberg was held at this time: "the most capable man in the world" to fill Borelly's place, "the most respectable man," and "the most dear to all the savants." Most significantly, it sheds new light on the otherwise hidden negotiations and social machinations needed in order to be appointed an academician prior to the 1699 *règlement*, a topic about which very little information has hitherto been available. Whether or not Toinard agreed to help, and whether or not Homberg ever got an audience with Bossuet, is unknown. But Louvois, who by this time had lost all interest in the Académie, appointed neither Homberg nor anyone else to replace Borelly. That position, like others that had been vacated by death or departure during Louvois's watch, would remain vacant until the minister's own death allowed for the revitalization of the Académie under his successor Pontchartrain.

During the same period, Homberg worked to enhance his profile in Parisian scientific circles. In 1690, three highly complimentary accounts of Homberg's work and expertise appeared in print. Two occur in Régis's *Système de philosophie*. One cites the now-lost treatise on animal generation (noted above), while the other praises Homberg's expertise in optics and instrument design (fruit of his Italian sojourns) and his qualities as an experimental natural philosopher: "everything we write about this subject has been confirmed by the experiments that have been done by Mr. Homberg, the German gentleman, so famous for the great knowledge he has of natural philosophy, but above all, for the attention and extreme exactitude with which he carries out all sorts of experiments."[164]

In the same year, the seventh edition of Lémery's *Cours de chymie* pub-

163. Thierry Sarmant, "L'abbé de Louvois: Bibliothécaire du roi, 1684–1718," *Revue de la BNF* 41 (2012): 76–83; Bernard de Fontenelle, "Éloge de l'abbé Louvois," *HARS* (1718): 101–4. Lest it be thought that fourteen is too young to be tutored in chymistry, it might be noted that Louvois had made the young man King's Librarian at the age of nine, and he became an abbé at eleven.

164. Régis, *Cours*, 2:642 and 3:292.

lished thirty-five pages of material from Homberg dealing with the Bologna Stone and white phosphorus. "Mr. Hombert, the German gentleman, who is very well known for the beautiful discoveries he has made in natural philosophy . . . has not only recently brought to light this stone that had been nearly forgotten, but has gone far beyond everything which has been published about it. A trip he made to Italy in order to seek it out has given occasion for several excellent observations which I have drawn from him. . . . I shall report here several observations which he was good enough to communicate to me and the experiments at which I was present."[165]

It is noteworthy that now in 1690 Régis refers to Homberg as "so famous [*si fameux*]" and Lémery calls him "very well known [*assez connu*]." These adjectives correlate well with the complementary account in Gendron's letter and the support Homberg enjoyed among the members of the Académie. Homberg was by now a full-fledged savant whose knowledge and expertise were widely acknowledged in several branches of natural philosophy. Frustratingly, such notoriety did not leave any further documentary traces. This situation points to the degree to which renown was based upon direct personal contact and oral reports—presumably the products of the poorly documented informal groupings that met in domestic or other private spaces around Paris—rather than upon the kinds of written sources that are available to historians. Curiously, Homberg seems to have been rather ambivalent about publication. In an era when issues of priority and credit were hotly contested, and at a time when Homberg himself was actively angling for a position at the Académie, he was content either not to publish at all (hence the loss of his treatise on generation) or to let others (Régis and Lémery) publish his findings in their own books. This attitude might fit with Homberg's customary economy of exchange, but it also complicates, and perhaps further illuminates, issues of knowledge exchange and the value and meaning of intellectual property that remained under negotiation in the late seventeenth century.

Only upon the death of Louvois in July 1691 and the appointment of Louis Phélypeaux, Comte de Pontchartrain (1643–1725), as his replacement as ministerial head of the Académie did Homberg finally achieve his goal. Pontchartrain appointed his twenty-nine-year-old nephew, the very capable and energetic Abbé Jean-Paul Bignon (1662–1743), as president of the Académie Royale des Sciences. Significantly, one of the very first acts of the new leadership was to make Homberg a member of the Académie on 24 Novem-

165. Nicolas Lémery, *Cours de chymie*, 7th ed. (Paris, 1690), 658; see also Principe, "Chymical Exotica."

ber 1691.[166] What has not been previously noticed is that the move to admit Homberg began months earlier, almost immediately after Louvois's death, further indicating that the minister really was the main obstacle. On 14 August 1691, less than a month after Louvois's demise, Bourdelin "paid a man for having put in order and cleaned up the laboratories left by Mr. Borelly, which were in great disarray."[167] The Académie's chymical laboratories at the Bibliothèque du Roi had been left vacant after Bourdelin began working at home in 1687 and Borelly died in 1689. Why clean them up at this point unless in preparation for the arrival of a new chymist to use and take charge of them? Given the keen desire to admit Homberg that the academicians had expressed in 1689, it seems that as soon as Louvois was out of the way, they moved quickly to secure the Batavian chymist as one of their number.

According to Fontenelle, Bignon had never met Homberg and knew of him only by reputation, a statement that occludes the long-standing support Homberg obviously had among the members of the Académie, who undoubtedly knew him quite well, as Gendron's letter witnesses.[168] Once again we catch Fontenelle silently replacing the years of behind-the-scenes jockeying with a triumphal narrative of the inevitability of recognizing Homberg's genius. Homberg responded enthusiastically to the invitation he had been seeking for nearly a decade. Four days after his formal admission, at his first meeting as an academician, he promised to reveal (at last!) the coveted secret of how to make white phosphorus. He presented this material on 12 December, demonstrated some new phenomena of phosphorus he had discovered, and demonstrated the preparation of the substance in the Académie's laboratory.[169] Homberg's first contribution was another of his customary exchanges; the much-coveted process for making phosphorus that Homberg had kept secret for so long was now finally revealed, but only *after*—and immediately after—he was admitted to the Académie, seemingly as recompense for his admission. Homberg would go on to present more than seventy papers to the weekly assemblies of the Académie during the 1690s alone, and once the

166. PV 13, fol. 69v (24 November 1691).

167. Bourdelin Expense Account, fol. 128r: "payé a un homme pour avoir rangé, netoyé durant longtems les laboratoires laissés par Monsieur Borelly qui estoit en grand desordre dont on a retiré au moins quatre tombereau dordure et verres cassés."

168. Fontenelle, "Eloge de M. Tournefort," *HARS* (1708): 143–54, at 147; "il [Bignon] ne connoissoit ni l'un [Tournefort] ni l'autre [Homberg] que par le nom qu'ils s'etoient fait."

169. PV 13, fols. 70r (28 November 1691) and 71r (12 December 1691). Part of the content of Homberg's oral memoir (of which there is no copy extant) was published as his first published paper, "Phosphore brûlant de Kunkel," in the April 1692 issue of the Académie's newly initiated *Memoires de mathematique et de physique*, 74–79, and reprinted in *HMARS 1666–99*, 10:84–90.

serial *Mémoires* began to appear regularly, it would contain more than fifty full papers by him, not counting the brief notices of his contributions summarized in the *Histoire* or the many papers he read to the assembly that remained unpublished.

A Merchant of the Marvelous

The revised and enriched account of Homberg's early life presented in this chapter offers a portrait of the "Indian gentleman" very different from that constructed by Fontenelle and widely accepted ever since. Homberg's path to the Académie was not paved with formal study, university degrees, and a succession of formal apprenticeships with the leading natural philosophical lights of the late seventeenth century. Far from either the prototype of a modern scientist or Fontenelle's imagined ideal of an eighteenth-century natural philosopher, Homberg emerges instead as a character both more colorful and more firmly situated in the intellectual and social practices of the seventeenth century: an autodidact, a traveling collector of secrets and exotica, a savant in search of patrons, a deft navigator of social networks, and a merchant of the marvelous. During his years of extensive travel from Italy to Sweden, from England to the Balkans, he did meet an impressive array of learned figures— Cellio, Kunckel, Leibniz, Boyle, and others. However, these contacts were brief, tending to last only long enough to arrange an exchange of secrets and presumably to obtain letters of introduction for his next destination. A substantial, perhaps major, part of Homberg's time was also spent observing and asking questions of countless unknown practitioners, craftsmen, artisans, and possessors of chymical or mechanical secrets. Miners, metalworkers, physicians, empirics, lens grinders, refiners, mechanics, vinegar makers, and many others supplied him with new observations, information, and hands-on (or at least eyes-on) experience of their practices. What is significant is the way Homberg combined his various sources of information and informants, drawing knowledge and materials from the broadest possible range of social and intellectual levels from all across Europe, all of which would eventually be synthesized and drawn upon for his later work at the Académie.

Homberg's wandering path to the Académie was laid open by his early contact with Otto von Guericke, his most significant early influence. When Homberg left Magdeburg in 1678, he undoubtedly filled his purse with coins, but he also took with him a more substantial treasure in the form of some of the famous Bürgermeister's prized secrets. Homberg parlayed these secrets over the succeeding years into a hoard of natural philosophical and mechanical materials and knowledge that quickly brought him to the attention of the academicians and of Colbert when he arrived in Paris in 1682. When his advancement was frustrated by the death of Colbert (followed closely by that of Mariotte) and the enmity of Louvois, Homberg found a new patron in the Abbé de Chalucet, and when that too came to end, he returned to the intellectual contacts he had made previously in Italy. His progress through the penumbra of the Académie Royale des Sciences toward its center stymied, he joined the Accademia Fisicomatematica Romana, underscoring the importance of participation in scientific societies both to Homberg himself and to the social and intellectual character of late seventeenth-century natural philosophy. Having returned to France, Homberg endeavored to make a place for himself, either in the Church or in the Académie (not that the two were mutually exclusive), drawing not only upon his scientific expertise but also upon his many personal contacts. Finally, having built an impressive reputation for himself through the accumulation of contacts and impressive displays of his knowledge and secrets, Homberg at last gained a position in the Académie as soon as that institution's reinvigoration commenced under Pontchartrain's leadership.

Homberg emerges as an insatiably curious, highly intelligent, and versatile figure whose experience and knowledge spanned a wide range of fields and practices—chymistry, mechanics, optics, microscopy, animal generation, and more. His success, particularly during his travels, probably also depended upon a winning personality, a feature that would be noted repeatedly by later acquaintances. Yet Homberg's broad learning—and his preference for the showy, the exotic, and the marvelous—was by no means dilettantish, despite the "unofficial" way in which he amassed it. He did not merely collect knowledge and materials idly, but actively investigated them further, often uncovering new features and teasing out fresh and unexpected phenomena, as he did with white phosphorus, the Bologna Stone, and other items in his possession. He maintained a special degree of interest in two particular topics: luminescent substances and pneumatics. While his work on pneumatics would quickly wane during the 1690s, his investigations of the interaction of light with matter grew to become a central part of his thought and his mature comprehensive theory of chymistry, as mentioned at the start of this chapter.

By 1691 Homberg had already come a very long way. From distant origins in the East Indies, through a very brief legal career, followed by many years of collecting and trading secrets and expertise during thousands of miles of travel, he eventually, at the age of thirty-eight, gained a coveted salaried position in the most renowned scientific institution in Europe. But much more, both intellectual and personal, lay in store for him in Paris, including his firm establishment as one of the Académie's leading lights and his entry into the highest echelons of French society as a favored member of the household of Philippe II, Duc d'Orléans, the future Regent of France. The following chapter examines Homberg's life in Paris as an academician, his first experimental projects, and the status and place of chymistry and chymists at the early Académie Royale des Sciences.

2

A Batavian in Paris

In late 1691, when Homberg was finally admitted to the Académie Royale des Sciences, the now twenty-five-year-old institution was in the early stages of a revival. The years following the death of its founder Colbert in 1683 are generally seen as a time of languor for the Académie. Scholarly explanations for this weak period vary. Blame is most commonly laid at the feet of Colbert's successor, the minister Louvois. Some studies have pointed to a preferential interest on Louvois's part for immediate practical applications over open scientific inquiry, although during Colbert's years practical services of various sorts to the crown also occupied part of the Académie's time. More significant was Louvois' general loss of interest in the Académie as he himself increasingly lost favor with the king. Some of Louvois's interventions in the Académie's affairs, such as canceling publications and expeditions due to financial constraints, undoubtedly took a toll on the academicians' morale, and the same can be said of the failure to pay their stipends in 1689 and 1690.[1] The loss of a large number of prominent academicians in the 1680s, whether by death or emigration, and the failure to fill the vacancies either at all or

---

1. Alice Stroup, *A Company of Scientists: Botany, Patronage, and Community at the Seventeenth-Century Parisian Royal Academy of Sciences* (Berkeley: University of California Press, 1990), especially 51–56. See also Alice Stroup, *Royal Funding of the Parisian Académie Royale des Sciences during the 1690s, Transactions of the American Philosophical Society* 77 (1987); Elmo S. Saunders, *The Decline and Reform of the Académie des Sciences à Paris, 1676–1699* (PhD diss., Ohio State University, 1957). A quick and informative overview is Marie-Jeanne Tits-Dieuaide, "Une institution sans statuts: l'Académie royale des sciences de 1666 à 1699," in *Histoire et mémoires de l'Académie des Sciences: Guide des recherches*, ed. Éric Brian and Christiane Demeulenaere-Douyère (Paris: Tec & Doc, 1996), 3–13.

with equivalent talent, exerted further stultifying influence on the Académie's vigor and activity.[2]

Although Louvois was in fact operating under conditions of increasing fiscal difficulty for France, the following era of Pontchartrain's leadership was equally stymied by financial difficulties, as the fiscal situation of France continued to deteriorate throughout the 1690s and into the eighteenth century. Yet the bleak economic outlook during his stewardship was more than compensated by judicious and effective leadership, best exemplified by the appointment of his nephew, the Abbé Jean-Paul Bignon (1662–1743), as the Académie's president.[3] As a minor but illustrative example, toward the end of Louvois's stewardship, Bourdelin's laboratory expenses took as much as three years to be reimbursed, while under Pontchartrain the same reimbursements took only two to three months.[4] As mentioned at the end of the previous chapter, one of Bignon's first acts was to fill two vacancies in the Académie with Tournefort and Homberg. Further initiatives soon followed. Perhaps the most important was a program of increased publication. The academicians had produced a significant number of manuscripts in previous years that remained unpublished, and these materials appeared steadily through the 1690s. The Académie also launched a new monthly publication, somewhat akin to the *Philosophical Transactions* that Henry Oldenburg had initiated in 1665 and that had become associated with the Royal Society of London. Prior to the 1690s, the Académie's work had been published predominantly as large and expensive books, such as the 1671 *Histoire des animaux*, or as short accounts inserted in the *Journal des sçavans*.[5]

2. Huygens left permanently in 1681, and Roemer in 1682; Picard died in 1682, Mariotte and Carcavi in 1684, Duclos in 1685, Blondel in 1686, Perrault in 1688, and Borelly in 1689. Writing in March 1690, Philippe de la Hire complained to Christiaan Huygens that "the company is greatly diminished"; *Oeuvres complètes de Christaan Huygens*, 22 vols. (The Hague: Nijhoff, 1888–1950), 9:378. See also the chart in Stroup, *Royal Funding*, 12–13.

3. On Bignon, see Françoise Bléchet, "L'abbé Bignon, président de l'Académie royale des sciences: un demi-siècle de direction scientifique," in *Règlement, usages et science dans la France de l'absolutisme*, ed. Christiane Demeulenaere-Douyère and Éric Brian (Paris: Lavoisier Tec & Doc, 2002), 51–69; David J. Sturdy, *Science and Social Status: The Members of the Académie Royale des Sciences* (Woodbridge, UK: Boydell Press, 1995), especially 222–26, 343–46, 367–74.

4. Bourdelin Expense Account, fols. 130r, 132r.

5. Anne-Sylvie Guenon, "Les publications de l'Académie des sciences," *Guide des recherches*, ed. Brian and Demeulenaere-Douyère, 107–12; Stroup, *Royal Funding*, 50–51; August Bernard, *Histoire de l'Imprimerie royale au Louvre* (Paris, 1867), 138–54; and Robert Halleux, James McClellan, Daniela Berariu, and Geneviève Xhayet, *Les publications de l'Académie Royale des Sciences (1666–1793)*, 2 vols. (Turnhout: Brepols, 2001), 1:33–44. On the *Histoire des animaux*,

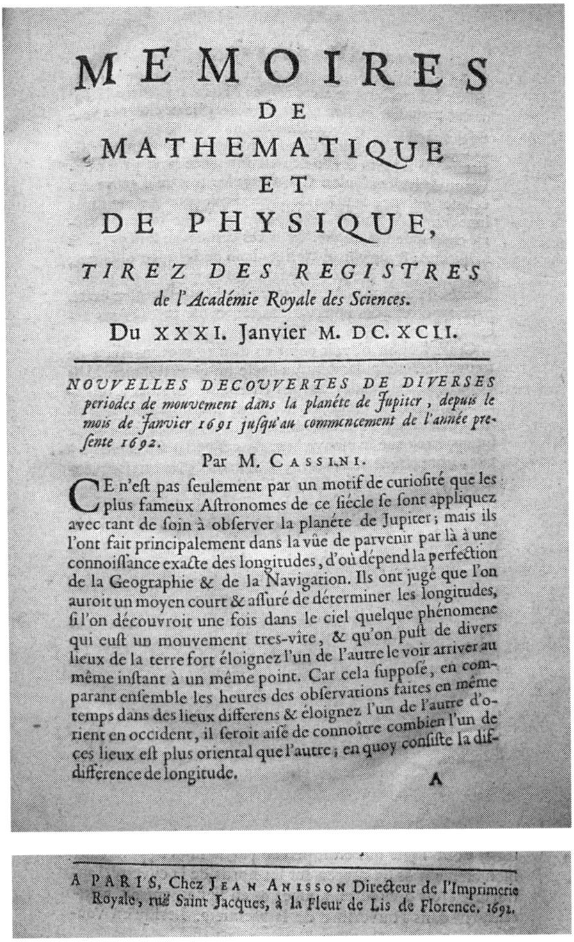

MEMOIRES
DE
MATHEMATIQUE
ET
DE PHYSIQUE,
TIREZ DES REGISTRES
de l'Académie Royale des Sciences.
Du XXXI. Janvier M. DC. XCII.

NOUVELLES DECOUVERTES DE DIVERSES periodes de mouvement dans la planéte de Jupiter, depuis le mois de Janvier 1691 jusqu'au commencement de l'année presente 1692.

Par M. CASSINI.

CE n'est pas seulement par un motif de curiosité que les plus fameux Astronomes de ce siécle se sont appliquez avec tant de soin à observer la planéte de Jupiter; mais ils l'ont fait principalement dans la vûe de parvenir par là à une connoissance exacte des longitudes, d'où dépend la perfection de la Geographie & de la Navigation. Ils ont jugé que l'on auroit un moyen court & asseuré de déterminer les longitudes, si l'on découvroit une fois dans le ciel quelque phénomene qui eust un mouvement tres-vite, & qu'on pust de divers lieux de la terre fort éloignez l'un de l'autre le voir arriver au même instant à un même point. Car cela supposé, en comparant ensemble les heures des observations faites en même temps dans des lieux differens & éloignez l'un de l'autre d'orient en occident, il seroit aisé de connoître combien l'un de ces lieux est plus oriental que l'autre; en quoy consiste la difference de longitude.

A

A PARIS, Chez JEAN ANISSON Directeur de l'Imprimerie Royale, ruë Saint Jacques, à la Fleur de Lis de Florence. 1692.

FIGURE 2.1. The title page and publisher's colophon of the Académie's first monthly publication of its *Mémoires*, in January 1692. Author's collection.

In January 1692, the Imprimerie Royale began issuing the Académie's monthly *Mémoires de mathématique et de physique* (fig. 2.1). This publication differed from most other learned periodicals of the day—such as the *Philosophical Transactions* and the *Journal des sçavans*—in that it was devoted entirely to reporting the scientific work of the Académie. As such it did not contain book reviews, news items, or letters; its content was made up exclusively of full papers by the academicians, an average of four in each thirty-two-page

see Anita Guerrini, *The Courtier's Anatomists: Animals and Humans in Louis XIV's Paris* (Chicago: University of Chicago Press, 2015).

issue.[6] The new serial could not, however, be maintained for more than two years, and it ceased production at the end of 1693 after twenty-four issues.[7] Several explanations for this short run are plausible: a lack of research materials to publish, a general lack of success for the series, or a lack of publication funds (most issues contained at least one high-quality engraving, and such illustrations were costly and time-consuming to produce). When Martin Lister (1639–1712) visited Paris in 1698, the Marquis de l'Hôpital (1661–1704) told him that the small cadre of academicians simply could not produce results at the rate necessary to keep a monthly serial supplied. Nor was the academicians' output supplemented by contributions received by correspondence, as was the case for the *Philosophical Transactions*. This explanation appears fully adequate, given that fourteen research-active academicians were expected to provide an average of four research papers every month—in other words, each academician would have had to produce a publishable paper every three to four months, a rate that is clearly unsustainable.[8] Today, the original serial issues are extremely rare; Lister complained of the "unreasonable Rate" of twenty livres he had to pay to buy a set from the printer Jean Anisson in 1698.[9] Such rarity might imply that the journal was produced in a small print run or that it was poorly disseminated.

Wilhelm Homberg was a major contributor to these monthly issues. He contributed thirteen papers during the two-year run—five in 1692 and eight in 1693. In the same period, he read four more full papers to the Académie that remained unpublished. Only Cassini and de la Hire outpublished Homberg, with sixteen and fifteen papers respectively, and their output was assisted by the frequency of astronomical events—eclipses, occultations, parhelia, and suchlike—each of which provided the basis for a publication. Homberg's output that began almost immediately upon his admission to the Académie was possible only because he drew upon the reserve of natural philosophical experience, experiments, and secrets he had accumulated in previous years. His first publication detailed the method for preparing white

6. The two contributions of Abbé Gallois—a summary of a report on echoes and extracts from a book by Jesuits in the Far East (both sent to the Académie)—represent the only items not composed of results generated at the Académie.

7. These serial issues (and materials published earlier in the *Journal des sçavans*) were reprinted in 1730 in volume 10 of *HMARS 1666–99*. A two-volume collected edition also appeared in Amsterdam in 1723, published by Pierre de Coup.

8. There were at this time eighteen academicians (not counting the secretary Duhamel), but four—Dodart, Bourdelin, Couplet, and Lefebvre—contributed nothing to the serial.

9. Martin Lister, *Journey to Paris in the Year 1698* (London, 1699; reprint, ed. Raymond Phineas Stearnes, Urbana: University of Illinois, 1967), 80, 97, 137.

phosphorus from urine—the secret he had obtained from Johann Kunckel in 1679 in exchange for that of von Guericke's Wettermännchen. The paper also gave an account of the discovery of phosphorus by Brand and its transmission through Kunckel.[10] This historical section provoked a response from Leibniz that challenged the details of Homberg's account, particularly the claim that Brand died shortly after discovering phosphorus in 1669, and that Kunckel rediscovered the process on his own. (Brand was still alive as late as 1699.) Homberg's information had presumably come from Kunckel, while Leibniz's came directly from Brand, each of whom had reasons to slant the account in his own favor.[11]

Six of Homberg's papers in the monthly *Mémoires* involve experiments with the air pump. These form a coherent series dating from December 1692 to July 1693, and although they may also include recently performed experiments, their origins lie with work Homberg carried out a decade earlier in conjunction with Edmé Mariotte. The first describes the breaking of Prince Rupert's drops in an evacuated space. These "drops" (*larmes de verre* or *lachrymae Batavicae*) were celebrated and remarkable curiosities of the late seventeenth century, and so exactly the sort of exotica that would attract Homberg's attention. Prepared by pouring molten glass into cold water, these teardrop-shaped pieces of glass have heads durable enough to be hammered without breaking, but when the tiniest tip of the thin tail is broken off, the entire drop instantly explodes into minute fragments.[12] Before Homberg ar-

10. Wilhelm Homberg, "Maniere de faire le phosphore brûlant de Kunkel," *MARS* (30 April 1692), reprinted in *HMARS 1666–99*, 10:84–90.

11. Gottfried Wilhelm Leibniz to Simon Foucher, 17 October 1692, in Leibniz, *Sämtliche Schriften und Briefe*, 2d ser., 2:609–10; Leibniz, "Historia inventionis phosphori," *Miscellanea Berolinensia* 1 (1710): 91–98. No response from Homberg to Leibniz's criticism is known, even though it is virtually certain that he knew of it; the Abbé Bignon records reading this very issue of Leibniz's *Miscellanea* in a letter to Leibniz dated 1 December 1710; NLB, LBr 68. Leibniz seems to have been rather obsessed by this topic; he mentioned it also in letters to Le Thorel (12/22 May 1699, *Sämtliche Schriften und Briefe*, 1st ser., 17:191–92) and to Rudolf Christian von Bodenhausen (3/13 January 1696 *Sämtliche Schriften und Briefe*, 3d ser., 6:607), and wrote a correction to be sent to the Académie. This correction may never have been sent, and may be the "recit seur" offered to Foucher; it is, however, published in Herbert Breger, "Notiz zur Biographie des Phosphor-Entdeckers Hennig Brand," *Studia Leibnitiana* 19 (1987): 68–73. Homberg's account remained the standard one in Paris; see, for example, *HARS* 32 (1730): 48. Curiously, Leibniz's correspondent Rudolf Christian von Bodenhausen claimed (24 October 1692) that Leibniz's own account of the discovery of phosphorus was itself wrong—according to von Bodenhausen it was discovered not by Brand but by an unnamed and secretive character who performed the operation at Brand's house. See Leibniz, *Sämtliche Schriften und Briefe*, 3d ser., 5:415–22, at 419–20.

12. For historical and technical information on the drops, see Laurel Brodsley, Charles Frank, and John W. Steeds, "Prince Rupert's Drops," *Notes and Records of the Royal Society of*

rived in Paris in 1682, Mariotte had conjectured that external air pressure was responsible for the powerful explosion of the drops, and claimed that although the drops would still shatter in a vessel evacuated of air, the explosion there would be feeble, and "the fragments ought not to be carried very far."[13] Once Homberg began his collaboration with Mariotte, they performed experiments on breaking the drops in a vacuum, and Mariotte mentioned them to the Académie in 1683. Homberg showed the experiment to the assembly himself during his trip to Paris in 1687, using the new air pump he devised while in Italy. His 1692 paper describes these experiments and shows that, contrary to Mariotte's conjecture, the drops explode with *more* violence in a vacuum, and therefore the air plays no role except to impede the motion of the exploded fragments.[14]

Homberg's early 1693 paper on the freezing of water is likewise tied to work done during his first Paris residence and linked to Mariotte's conjectures. Homberg's goal was to determine if the anomalous expansion of water upon freezing was the result of the emergence of dissolved air during congelation, an opinion Mariotte had published in 1679.[15] (Almost all substances contract upon passing from liquid to solid, but water expands—hence, glass bottles filled with water shatter when their contents freeze.) To test Mariotte's idea, Homberg designed a glass vessel equipped with a valve and simple manometer to monitor the internal pressure. He put pure water into this instrument, sealed it, and over the course of an entire year pumped air out of the vessel more than twenty separate times in order to remove all traces of air from the water. After an additional year of letting the water stand undisturbed in the evacuated vessel, Homberg exposed it to freezing weather and reported that the water did not expand as it turned to ice, and so concluded that the usual expansion seen when water freezes is indeed due to air in the water.[16] Since

---

London 41 (1986): 1–26; for a mid-eighteenth-century comment on Mariotte's and Homberg's work on them, see Claude Nicolas le Cat, "A Memoir on the *Lacrymae Batavicae*," *Philosophical Transactions* 46 (1749–50): 175–88.

13. Edmé Mariotte, "De la nature de l'air," in *Oeuvres*, 2 vols. (The Hague, 1740), 1:157–59; the essay was first published in the 1679 *Essays de phisique*.

14. Jean-Baptiste Duhamel, *Historia regiae scientiarum academiae*, 2d ed. (Paris, 1701), 222. Homberg's 1687 presentation is PV 12, fol. 60r–v (7 May 1687); Wilhelm Homberg, "Reflexions sur l'expérience des larmes de verre qui se brisent dans le vuide," *MARS* (31 December 1692): 183–87, reprinted in *HMARS 1666–99*, 10:215–20, original memoir presented 6 December 1692, PV 13, fol. 121r.

15. See *HMARS 1666–99*, 10:507–13, for Mariotte's 1672 experiments on freezing water. See chapter 1, note 118 for the (probable) notice of Homberg's 1684 experiment on this topic.

16. Wilhelm Homberg, "Experiences sur la glace dans le vuide," *MARS* (28 February 1693): 19–25, reprinted in *HMARS 1666–99*, 10:255–62; presented in PV 13, fol. 128r (28 January 1693).

this experiment required two years to complete and was reported to the Académie in January 1693, it must either have been performed in the early 1680s or started in early 1691 before Homberg's admission to the Académie.

Homberg also repeated some earlier experiments to achieve greater precision.[17] These new trials sometimes led to new discoveries. For example, the work with Mariotte for which Homberg was most noted was the determination of the relative densities of water and air. In 1693, armed with a superior pump, Homberg attempted to improve upon his now decade-old value of 630 to 1. Now he determined the value to be 1,200 to 1.[18] The large discrepancy led Homberg to recognize the difference in the density of air at different temperatures, and he conducted further experiments that measured the greater buoyancy effects of cold air.[19] (The formal relationship between the temperature and volume of a gas would be enunciated as "Charles's Law" only a century after Homberg, following the studies of Jacques Charles, 1746–1823, a later member of the Académie.) Similarly, Homberg's experiments with purging air completely from water led him to try to determine the exact quantity of air dissoluble in water. He gave up when he discovered (correctly) that the quantity varied not only with temperature but also with different samples of water.[20] Homberg's keen interest in precision and in determining exact quantities of materials forms a consistent feature of his work across multiple fields of inquiry, and reappears many times in this study.

The early association with Mariotte thus clearly played an important role in Homberg's formation—both by direct collaboration and by suggestions of ideas to be tested later. Therefore, when assessing the background to Homberg's pneumatic experiments, it is incorrect to attribute a significant formative role to an ill-defined and probably extremely brief contact with Robert Boyle.[21] Instead, the clear influences on Homberg's pneumatic work were von Guericke, Mariotte, and instrument makers such as d'Alencé and those he met in Rome who contributed to his design for the improved pump he demonstrated in 1687 and used in the 1690s.

The high profile of air pump experiments in Homberg's early publications might give the impression that such work was his major interest at the time.

17. Homberg, "Larmes de verre," 215.

18. PV 13, fol. 145v (26 August 1693); PV 14, fol. 24v (28 August 1694).

19. Wilhelm Homberg, "Observations sur le different poids d'un mesme volume d'air selon qu'il est plus ou moins dilaté par les differens dégres de Chaleur," PV 14, fols. 296v–298v (14 March 1696).

20. Homberg, "Glace," 259.

21. Marie Boas Hall, *Robert Boyle and Seventeenth-Century Chemistry* (Cambridge: Cambridge University Press, 1958), 209; on 126 Homberg is called Boyle's "disciple."

This conclusion, however, would be erroneous. Much of that work had been performed a decade earlier, and given the pressure on the academicians to fill the monthly issues of the 1692–93 *Memoires*, supplementing earlier unpublished work with a few new experiments was an efficient way for Homberg to supply the journal with a steady stream of papers. Throughout the 1690s, Homberg's main interest was unquestionably chymistry; this conclusion is made clear by a consideration of the unpublished materials in the procès-verbaux, the pochettes de séances, and other sources. To be sure, Homberg's interest in chymistry long predates the 1690s, and it is a mistake to think that his turn to chymistry was a late development, or that he was an "outsider" to chymistry as has sometimes been claimed.[22] His earlier chymical work and interests are evident in his travels to learn about metals and minerals in 1680–82, his hiring by the Abbé de Chalucet for a specifically chymical project in 1684, his notable success with very difficult chymical operations like preparing white phosphorus, and his brilliant resolution of reproducibility problems in preparing the Bologna Stone. The relative silence of the written record on the score of Homberg's pre-1691 chymistry is largely an artifact of audience and reportage—early notices about Homberg reflect the interests of the members of the academies in Paris and Rome with whom Homberg wished to associate. His 1691 admission to the Académie was undeniably as a chymist; he was explicitly seen as the replacement for either the chymist Samuel Cottereau Duclos, who had died in 1685, or the chymist Jacques Borelly, who had died in 1689.[23] Homberg was given control of the Académie's chymical laboratory, the furnaces of which had lain idle since Borelly's death (Bourdelin had already abandoned them in 1687).[24] Moreover, Bignon immediately assigned Homberg a difficult and specifically *chymical* task—the assessment

22. This assessement is made by Mi Gyung Kim in *Affinity, That Elusive Dream* (Cambridge, MA: MIT Press, 2003), for example at 89. It is true that Homberg differed in terms of his approach to the subject from the other academicians involved in chymistry in the 1690s, as will become clear by the end of this chapter, but that does not render him an "outsider" to the discipline.

23. Erich Odhelius to Urban Hjärne, 6 June 1692, in Carl Christoffer Gjörwell, *Det Swenska Biblioteket*, 2 vols. (Stockholm, 1757–62), 1:337–39; and Gendron to Toinard, Waller MS fr-03897.

24. Bourdelin Expense Account, fol. 120v (Bourdelin begins working at home). Bourdelin had already begun withdrawing from the Académie's laboratory in 1682, while Duclos was still active; see PV 10, fol. 95v (15 April 1682). Borelly worked also at the laboratory at the Jardin du Roi: see AN O/1/2124 regarding payments to Borelly in 1687 "pour l'experience de l'accademie des sciences et laboratoire" recorded under the "depence fixe" of the Jardin du Roi. Borelly lived next to the Académie's chymical laboratories at the Bibliothèque du Roi from 1685, in the apartment previously occupied by Duclos; see Philippe de la Hire to Christiaan Huygens, 3 March 1688; *Oeuvres de Huygens*, 9:263–65, at 264.

of the accumulated chymical analyses that were part of the institution's long and troubled project to compile a natural history of plants.

Homberg's work on this project showcases his characteristic approaches to chymistry, as well as his theoretical commitments, experimental acumen, and desire to discover and identify the simple substances, or "principles," from which all compound bodies are composed—an endeavor that he would pursue throughout the rest of his mature career. But getting a clear picture of what Homberg actually did with the plants project is complicated by the fact that he worked on several unrelated projects (both major and minor) at the same time, and thus reports of his work with the plants project, which stretch over a decade, are interleaved with other results from which they have to be disentangled. When that is done, however, a strikingly coherent research program emerges. Equally important, Homberg almost immediately realigned the plants project to fit his own research interests. In order to understand what he did and its significance—particularly for illustrating the transformations of chymistry in the period—it is necessary to review the earlier trajectory of the plants project. This background paints a picture of chymistry's place at the Académie before Homberg's admission and sketches a portrait of Homberg's most important and influential chymical predecessor at the Académie, Samuel Cottereau Duclos (fig. 2.2).

The Chymical Analysis of Plants: Principles and Problems

Since 1668 the Académie had been involved in the chymical analysis of plants as one facet of a huge undertaking to produce a comprehensive natural history of plants replete with engravings, descriptions, botanical and horticultural information, uses, and analyses. This project has already been the subject of several scholarly treatments, of which Alice Stroup's is the most comprehensive and reliable.[25] The research for the proposed *Histoire des plantes* was collaborative, involving most of the members of the early Académie to one degree or another. But various factors conspired to complicate, delay, and finally doom the project. From the start, its goals were too grandiose. The magnificent engravings, of which some three hundred plates were produced, were extremely expensive, and Colbert was forced to cut project funding in 1681. Although Louvois reinstated some support, most plates were never printed as part of a comprehensive work on plants. But certainly the most

---

25. Stroup, *Company*. See also Frederic L. Holmes, *Eighteenth-Century Chemistry as an Investigative Enterprise* (Berkeley, CA: Office of the History of Science and Technology, 1989), 61–83.

Agathange Cottereau du Clos
de L'Académie Royale des Sciences

FIGURE 2.2. Samuel Cottereau Duclos, here described using the name "Agathange" by which he was occasionally known. Engraving from the Österreichische Nationalbibliothek.

problematic—and likewise the most innovative—part of the project was the chymical analysis of plants. The early Académie's chief chymist and oldest member, Samuel Cottereau Duclos, was initially in charge of this aspect of the project. Duclos was one of the most active and productive academicians of the period.[26] He designed (and lived adjacent to) the Académie's laboratory

26. On Duclos, see Doru Todériciu, "Sur la vraie biographie de Samuel Duclos (Du Clos) Cotreau," *Revue d'histoire des sciences* 27 (1974): 64–67; Alice Stroup, "Censure ou querelles savantes: l'Affaire Duclos (1666–1685)," in *Règlement, usages et science*, ed. Demeulenaere-Douyère and Brian, 435–52; Sturdy, *Science and Social Status*, 107–9; Victor D. Boantza, *Matter and Method in the Long Chemical Revolution* (Burlington, VT: Ashgate, 2013), 17–92; Rémi Franckowiak, "Du Clos, un chimiste post–*Sceptical Chymist*," in *La philosophie naturelle de Robert Boyle*, ed. Miriam Dennehy and Charles Ramond (Paris: Vrin, 2009), 361–77; and Lawrence M. Principe, "Sir Kenelm Digby and His Alchemical Circle in 1650s Paris: Newly Discovered Manuscripts," *Ambix* 60 (2013): 3–24, at 17–19.

at the Bibliothèque du Roi on the rue Vivienne, and set down the basic questions and procedures for the chymical analysis of plants. Most of the laboratory operations were carried out by Claude Bourdelin; his seemingly endless series of analyses, steadfastly performed over the course of nearly thirty years, fills thousands of manuscript pages.[27] Jacques Borelly (d. 1689) was a third member of the chymical group at work on this project and, like Bourdelin, worked in the laboratory on analyses.[28] In 1674, the leadership of the project began shifting away from Duclos to the physician Denis Dodart (1634–1707). Naturally enough, conflict erupted between Duclos and Dodart that further disrupted the project—the nature of their conflict, detailed below, illuminates the shifting and disputed contours of chymistry in the period. Dodart did nevertheless succeed in producing the *Mémoires pour servir à l'histoire des plantes*, a sumptuous work in large folio that appeared in 1676 and included full-page plates and a lengthy introduction describing the project's rationale and methodology. The hopes of producing a second volume were dashed, however, when its completed manuscript was supposedly stolen from Dodart in 1681 by highway robbers.[29] No further installments ever appeared. Thus the single 1676 volume fell far short of the much grander production the academicians had envisioned.

The chymical analysis of plants provoked varied responses from later generations. In the mid-eighteenth century, the Académie thought enough of it to have all of Bourdelin's results reorganized and recopied into hefty folios, although apparently nothing further came of this labor.[30] Later eighteenth-century voices—led perhaps by the Académie's secretary Nicolas de Condorcet—condemned the project as a waste of time. In fact, Condorcet's dismissive attitude toward both the project and those who worked on it resulted in several persistent historical errors about the early chymists of

27. Bourdelin's laboratory records survive in five different copies held in various repositories; see the note on sources.

28. On Borelly, see Pierre Chabbert, "Jacques Borelly (16. .–1689): Membre de l'Académie des Sciences," *Revue d'histoire des sciences* 23 (1970): 203–27. The *Index biographique de l'Académie des Sciences* (Paris: Gauthier-Villars, 1979), 146, gives 1674 for Borelly's admission to the Académie, but he was already at work on plant analyses by May 1670, as witnessed by Bourdelin's record of laboratory expenses; BNF, MS NAF 5147, fol. 31v. In July 1670 the English diplomat Francis Vernon reported to Henry Oldenburg that Borelly, "of sharp wit & well verst in Chymistry," was a full member of the Académie and receiving the standard pension; *Oldenburg Correspondence*, 7:60–64, at 60. Huygens calls him "un de notre assemblée" in December 1671; see Chabbert, "Borelly," 207.

29. Stroup, *Company*, 103–7; see also PV 10, fol. 72r. Dodart's report of the theft may be little more than the seventeenth-century equivalent of "the dog ate my homework."

30. See note on sources.

the Académie, as if he either could not have been bothered to get the details of such unprofitable persons right, or preferred to have them written out of the institution's history. He continued the conflation, possibly begun by Fontenelle, of Jacques Borelly with the unrelated Pierre Borel, who was never an academician, (an error corrected by Chabbert in 1970 but still encountered in some recent secondary literature). In speaking of the chymical work, Condorcet asserted that it "was without object" and did not "serve to establish any theory, nor even to verify any general fact." Condorcet's evaluation of Duclos is equally dismissive, and more geared toward praising Stahl and Rouelle than memorializing Duclos.[31] Most subsequent writers, including some mid-twentieth-century historians of science, followed suit.[32] Frederic L. Holmes, however, refreshingly pointed out that in terms of contemporaneous chymistry, this first generation at the Académie "had taken an entirely reasonable approach" to plant analysis.[33] Alice Stroup's careful study and discovery of new manuscript sources at last placed this major feature of the early Académie into its proper context. The analysis of plants showcases critical issues and controversies in late seventeenth-century chymistry and provides the necessary background for Homberg's investigations in the 1690s.

The goal of the chymical analysis of plants was simple enough: to divide plant materials into their component substances in order to determine their composition and the origin of their medicinal effects. By the mid-seventeenth century, chymistry had become increasingly tied to analytical work, even to the point of being defined routinely as the art of separating compound bodies into their constituents.[34] The great looming figure in the movement to use chymical analyses to reveal the hidden nature of compound bodies is the Flemish natural philosopher and chymist Joan Baptista Van Helmont (1579–1644). Van Helmont made quantitative analysis a crucial part of his work, even if he occasionally fell short of his expressed ideal of accurate quan-

31. Marie-Jean-Antoine-Nicolas Caritat, Marquis de Condorcet, *Éloges des Académiciens de l'Académie Royale des Sciences, morts depuis 1666 jusqu'en 1699* (Paris, 1773), 66–77, 155–57; "Liste de Messieurs de l'Académie Royale des Sciences," *HMARS 1666–99*, 2:364–65; Chabbert, "Borelly."

32. For example, James R. Partington, *A History of Chemistry*, 4 vols. (London: Macmillan, 1962), 3:12; Boas Hall, *Robert Boyle*, 7. See Stroup, *Company*, 302n2.

33. Frederic L. Holmes, "Analysis by Fire and Solvent Extractions: The Metamorphosis of a Tradition," *Isis* 62 (1971): 129–48, at 135–36.

34. As a few examples, see Jean Beguin, *Les elemens de chymie* (Paris, 1620), 1; Christophle Glaser, *Traité de la chymie* (Paris, 1663), 3; Nicolas Lémery, *Cours de chymie* (Paris, 1675), 2–3. See also William R. Newman and Lawrence M. Principe, "Alchemy and the Changing Significance of Analysis," in *Wrong for the Right Reasons*, ed. Jed Z. Buchwald and Allan Franklin (Dordrecht: Springer, 2005), 73–89.

titative results. Van Helmont still remains in many ways a "hidden figure" of the seventeenth century. Although his historical importance is readily acknowledged—usually more in general terms than in specific ways—we are still far from fully appreciating how gigantic and inspirational a figure he actually was, and how pervasive and influential his ideas were across multiple disciplines. As I will show here, *Helmontian ideas are necessary for making sense of the initial terms of the analysis of plants project as conceived in the late 1660s and equally of Homberg's own research program during the 1690s.*

Helmontian Chymistry at the Académie Royale des Sciences

On New Year's Eve 1666, two years before the plant project was launched, Samuel Cottereau Duclos spoke to the assembled academicians about the chymical principles of mixed bodies and laid out a series of critical questions. It bears stressing that the very first meeting of the Académie Royale des Sciences *began* with chymistry: the first pages of the procès-verbaux record that the assembly "resolved to begin its natural philosophical studies with a consideration of natural mixts."[35] In his presentation, Duclos referred to the traditional triad of chymical principles, the *tria prima* of mercury, sulphur, and salt. This system dated back to Paracelsus (1493–1541) and was still favored by some writers and workers in the mid-seventeenth century. But Duclos also referred to the then more common pentad of spirit (mercury), oil (sulphur), salt, phlegm, and earth popularized by Étienne de Clave (1587–1645) and others.[36] This pentad was the product of experimental results from the analysis of organic materials, particularly vegetable matters, well before the Académie began its work. The standard method of analysis for such substances was by what is today called *destructive distillation*. In this process, a plant or animal substance was placed in a retort and heated. The heat drove

35. PV 1, pp. 2–16 (31 December 1666), at 2.

36. On these principles, see Principe, *Aspiring Adept*, 36–42. On Étienne de Clave, see the papers in *Dossier Étienne de Clave, Corpus: Revue de philosophie* 39 (2001): 9–99, especially Bernard Joly, "La thèorie des cinq éléments d'Étienne de Clave dans la *Nouvelle Lumière Philosophique*," 9–44. See also Joly, "De l'alchimie à la chimie: le développement des *Cours de chymie au XVIIe siècle*," in *Aspects de la tradition alchimique au XVIIe siècle*, ed. Frank Grenier, *Textes et travaux de chrysopoeia* 4 (1998): 85–94. For earlier background to the five-element system in Quercetanus and others, see Didier Kahn, "Helisaeus Röslin, Joseph Du Chesne et la doctrine des cinq éléments et principes," in *Nouveau Ciel, Nouvelle Terre: La révolution copernicienne dans l'Allemagne de la Réforme*, ed. Edouard Mehl and Miguel Granada (Paris: Belles Lettres, 2009), 339–54. See also William R. Newman, "Alchemical and Chymical Principles," in *The Idea of Principles in Early Modern Thought: Interdisciplinary Perspectives*, ed. Peter Anstey (New York: Routledge, 2017), 77–97.

off a variety of volatile materials that then condensed and were collected in the receiver. Some had a penetrating odor and/or taste, and these were called the *spirits*. Others were weak and watery, and were called the *phlegm*. Still others were oily and flammable, and these were the *sulphurs*. When nothing more could be made to distill, the retort was allowed to cool. The blackened residue, called the *caput mortuum* ("dead head") was then extracted with water to produce a lixivium that after filtration and evaporation yielded the crystalline *salts*. The insoluble residue left over from washing the caput mortuum was the *earth*, the last of the five principles.

Many ambiguities complicated this process. For example, some substances yielded a highly volatile, inflammable fraction. This fraction could be called spirit on the basis of its volatility, but could just as well be called sulphur on the basis of its flammability. Which was it? Some materials, when heated, gave salts that were volatile and passed over into the receiver rather than remaining fixed (that is, nonvolatile) as part of the caput mortuum. The order of the appearance of the various materials in the receiver varied and generally overlapped. Since the majority of those doing such distillations were interested primarily in making pharmaceuticals, the main question involved which of the separated substances to choose as the most medicinally useful. Phlegm and earth were almost universally rejected as inactive, while the other three were considered "active" as the carriers of the essential virtues or qualities of the original, mixed substance.

A comprehensive treatment of the gamut of views on the status, number, and identity of the chymical principles during the seventeenth century remains a desideratum, but the pursuit of that topic at present would lead us too far afield. It suffices here to say that seventeenth-century conceptions of these five "chymical principles" are frequently oversimplified, misunderstood, and even parodied, in some secondary literature. While it is undoubtedly true that *some* writers thought of these principles as elemental or quasi-elemental substances, this view was far from universal. Many chymists denied them this status, and still more cared little about such questions, holding the principles to be simply groups of substances encountered in laboratory operations and broadly classifiable on the basis of sensible qualities. Naturally enough, those chymists involved primarily in the production of saleable products were less worried about the intrinsic nature of the principles than were the more philosophically inclined. Indeed, I suggest that the often confusing landscape of contradictory discussions, declarations, and assumptions about the chymical principles can be seen as the result of two often overlapping but potentially conflicting chymical agendas—one more natural philosophical and the other more practical.[37] The first stems from the desire to *identify and understand* the basic constituent(s) of compound substances, while the second responds to the desire to *prepare* chymical, often specifically pharmaceutical, products. Several issues relating to chymical practices and theories of the period—including the fluctuating fortunes of the Académie's analysis of plants project—can be approached profitably by keeping these two often competing agendas in mind.

Duclos was one of the many chymists who denied elemental status to the separated chymical principles. He stated clearly enough to the Académie in 1666 that "in the final resolution of natural mixts, nothing apparently remains but water," because water "is their first matter."[38] This simple declaration that water is the primary substratum of all substances clearly demonstrates Duclos's allegiance to Van Helmont, whose arguments for the primacy of water are well known. His famous willow tree experiment—familiar in decontextualized form even today to biology students—was intended to demonstrate water's ability to be transformed into all other substances. But Duclos's Helmontianism goes much deeper. Van Helmont argued that water is transformed into all other substances by the action of *semina*, or seeds, generated from "ferments" initially implanted in creation by God. These seminal transformations are (particularly in the case of animate creatures) overseen by organizing entities he called *archei*. Following these same principles, Duclos explained to the Académie how water is transformed by "an impalpable and spiritual efficient" and "some internal agent"—that is, the Helmontian ferments and seeds—and he likewise invoked the action of the "archeus which is the principal author of the form of the kind of mixts that are vivified, and the director of the ferment."[39] When Duclos goes on to enunciate a series of twenty questions about the chymical principles for the Académie to study, *nearly every one of them* is drawn from Van Helmont's writings or based upon his chymical system.[40] In short, Duclos enunciates a fully Helmontian course of study for the Académie's first communal undertaking.

Among the Helmontian ideas that Duclos presented to the academicians were doubts about the suitability of analysis using fire. Van Helmont famously argued that many of the substances resulting from the distillation

37. A similar but not identical division (between philosophical and "vulgar" chymists) can be applied to Robert Boyle's targets in his 1661 *Sceptical Chymist*; see Principe, *Aspiring Adept*, especially 30–35.

38. PV 1, p. 2 (31 December 1666).

39. PV 1, pp. 2 and 7.

40. PV 1, pp. 14–16.

of organic materials were actually *produced* during heating rather than being preexistent components that were merely separated unchanged from the
analyzed material.[41] He pointed to the fixed alkalies in particular—defined
as the salts isolated from the caput mortuum—as productions of the fire,
arguing that they arise from the combination of a volatile salt initially present
in the plant material with the plant's sulphur (that is, its oil). Clearly in reference to this Helmontian theory, Duclos suggests that the Académie study
whether "alkalies are productions of the external fire, and if they are made by
the linkage of volatile salt with earth."[42] The remarkable fact that throughout
the seventeenth century all the academicians—not just Duclos—routinely
referred to alkalies under the odd name of *sels sulphurés* ("sulphurated salts")
witnesses how fully ingrained Helmontian chymistry was at the time. Alkalies
show no properties that an observer would associate with sulphur; only an
implicitly held Helmontian theory of alkalies—namely, that they contain a
hidden sulphur—allows the term "sulphurated salt" to make sense as a name
for alkalies. Thus, when the plants project was launched in 1668, with Duclos
as one of its chief architects, Helmontian ideas were foundational. Indeed,
the impetus for incorporating chymical analysis into a "history of plants" can
be seen as stemming directly from Helmontian concerns, in particular from
Van Helmont's powerful statement that "we read in our furnaces that there
is no more certain genus of acquiring knowledge for the understanding of
things through their root and constitutive causes than when one knows what
is contained in a thing and how much of it there is."[43]

Van Helmont's system nevertheless posed two problems for the Académie's analysis of plants. The first is a practical conundrum—where should
analysis stop? This point relates directly to the potential tension between the
natural philosophical and the practical agendas. Water may well be the uni-

41. Allen G. Debus, "Fire Analysis and the Elements in the Sixteenth and Seventeenth Centuries," *Annals of Science* 23 (1967): 128–47. For a valuable view of the problem in the early eighteenth century, see John Powers, "Fire Analysis in the Eighteenth Century: Herman Boerhaave
and Scepticism about the Elements," *Ambix* 61 (2014): 385–406.

42. PV 1, p. 13. On the theory of alkalies, see Joan Baptista Van Helmont, *Ortus medicinae*
(Amsterdam, 1648; reprint, Brussels: Culture et Civilisation, 1966), "Blas humanum," no. 38,
p. 187; and William R. Newman and Lawrence M. Principe, *Alchemy Tried in the Fire: Starkey,
Boyle, and the Fate of Helmontian Chymistry* (Chicago: University of Chicago Press, 2002), 84–
89. Note that here Duclos says earth rather than sulphur as Helmont did, but since he later
claims that sulphur itself is resolvable into salt and earth, it comes nearly to the same thing and
indicates Duclos's further development of Helmontian ideas.

43. Joan Baptista Van Helmont, *De lithiasi*, chap. 3, no. 1, p. 20, in *Opuscula medica inaudita* (Amsterdam, 1648; reprint, Brussels: Culture et Civilisation, 1966). On Van Helmont and
quantitative chymistry and analysis, see Newman and Principe, *Alchemy Tried in the Fire*, 56–91.

versal material substratum and thus the ultimate product of all analyses, but while an analysis all the way to water would enlighten (in part) the natural philosophical agenda, it would defeat the practical. Reducing diverse plant materials into water converts their valuable variety of properties and ingredients into insipid uniformity and renders them pharmaceutically useless. Even the philosophical agenda would be only half-satisfied by such an "extreme" analysis, for it would defeat the Académie's goal of identifying the active principles in plants and thereby constructing a system for rationalizing and predicting their medicinal virtues. An analysis all the way to water—produced identically from *all* plants—would tell the chymist nothing about the makeup of any *particular* plant. Thus, the place to stop is at the intermediate principles that preserve the valuable virtues of the plant. Practical chymists as early as Jean Beguin (1550–1620) recognized this fact; that progenitor of the French didactic tradition in chymistry asserted that the principles mercury, sulphur, and salt *could* be further resolved into simpler substances but *practically* there was no point in doing so because the efficacious virtues of the original substance would be lost.[44] Duclos explicitly acknowledges the same point when he notes that the "extreme analysis of plants" into water "serves only for us to have a more exact knowledge of the constitution of the subject, because there no longer remains anything which is of any utility."[45] Thus, the more productive and informative course of action was to stop at an "incomplete" analysis—only as far as the "proximate principles" of spirit (mercury), oil (sulphur), salt, phlegm (water), and earth—for, as Duclos explained to the Académie, "all the qualities and differences [of mixts] can be very well referred to these first products."[46]

A second problem arises from Van Helmont's doubts about the suitability of fire analysis. For the Flemish natural philosopher, analysis by fire was deceitful. It produced new substances and combinations, and destroyed the *semina* that transform water into diverse substances. As the fire weakened these *semina*, the identities of the substances they produced were compromised. Substances exposed heedlessly to the fire would end up being something less than they were originally. As Van Helmont warned his readers: "Whatever is distilled by fire alone retreats far from the virtues of the compound."[47]

Van Helmont provided a solution for both problems, but one that proved

44. Jean Beguin, *Elemens de chymie* (Paris, 1620), 26–30.

45. PV 4, fol. 53r (9 June 1668).

46. PV 1, p. 3.

47. Van Helmont, *Ortus*, "Spiritus vitae," no. 7, p. 196: "quidquid solo igne distillatur, procul à viribus concreti recedere."

difficult to put into practice. The solution was to use not ordinary fire, but rather "the fire of Gehenna," namely, a solvent he called the *alkahest*. Any substance treated with the alkahest would be resolved first into its proximate principles—with no loss of virtue—and then, upon further digestion and distillation, into primordial water. Unlike the heat of the fire, the alkahest could separate mixed bodies without destroying or impairing their semina—in other words, without damaging their properties, identities, and virtues. (Only upon lengthy digestion and heating would the alkahest mortify the semina, thus turning the separated principles themselves back into water.) The alkahest could perform this function because it was composed of the smallest particles of matter, able to insinuate themselves in between the heterogeneous substances composing the mixed body and thereby gently disentangle them. Unlike other solvents, such as acids, the alkahest did not combine with what it dissolved, and so could be gently distilled away, leaving the disjoined principles behind in a pure state. The alkahest thus offered the optimal means for analyzing mixed bodies. Through its use one could determine the exact quantities and identities of the proximate constituents of mixed bodies, and separate them in a form—called the *ens primum*, or first essence—that maintained their virtues and properties intact. Both knowledge and practicality would be satisfied. Van Helmont describes several pharmaceuticals he claims to have prepared using the alkahest, most notably a cure for kidney and bladder stones made from a mineral he calls *ludus*, and an "elixir of life" prepared from Lebanon cedarwood.[48] The only obstacle in this glorious royal road for chymistry was that no one knew how to prepare Van Helmont's alkahest.

For the balance of the seventeenth century and into the eighteenth, scores of chymists endeavored to discover the secret of making the alkahest—and Duclos was one of them. He noted in his 1666 address that "one of the most renowned of modern chemists" (obviously a reference to Van Helmont) claimed to know of "a most expeditious method for resolving all mixed bodies into their principles . . . he called it the alkahest."[49] In a lengthy memoir that filled three assemblies in 1668, shortly after his protocols for plant analysis were delivered, Duclos reviewed the Paracelsian and Helmontian

48. On the alkahest see Principe, *Secrets*, 134–35; Paulo A. Porto, "*Summus atque felicissimus salium*: The Medical Relevance of the *Liquor Alkahest*," *Bulletin of the History of Medicine* 76 (2002): 1–29; and Bernard Joly, "L'alkahest, dissolvant universel, ou quand la thèorie rend pensible une pratique impossible," *Revue d'histoire des sciences* 49 (1996): 308–30.

49. PV 1, p. 5.

literature for hints about how to prepare the alkahest.[50] His unpublished *Dissertation sur les sels*, written in 1677, devoted five chapters to the alkahest and similar "resolutive" solvents.[51] Undoubtedly, he felt strongly that using the alkahest would be the best way to proceed. But without a bottle of the alkahest on the laboratory shelf, the Académie's analysis of plants would remain at a standstill. Therefore, since "this mysterious and secret means is not yet well known to us," Duclos offered other methods to be used provisionally "while waiting until the time we are able to have it."[52]

In 1668, the academicians decided that Duclos's "method would be very useful for composing the history of plants."[53] His proposals included describing the appearance and cultivation of each plant, but focused specifically on three methods of chymical analysis that responded to Helmontian warnings about fire analysis. First, the fresh juice or a decoction of the plant would be tested with various chemical indicators. The different colors and/or precipitates formed with aqueous solutions of vitriol of Mars (iron sulphate), salt of Saturn (lead acetate), sublimate (mercuric chloride), and, somewhat later, extract of *tournesol* (similar in action to litmus) would indicate the kinds of substances the juice contained.[54] Second, the "essential salts"—that is, the saline

54. PV 4, fols. 127v–133v (11 August 1668); fols. 134r–166v (18 August); and fols. 167r–175r (25 August).

51. BNF, MS fr. 12309; this manuscript was first identified as a work of Duclos by Stroup, "Censure."

52. PV 1, pp. 6–7: "en attendant que nous le puissions avoir." Duclos here suggests the use of air and oils to volatilize fixed alkalies into an alkahest-like solvent, which is an arcanum of Helmontian origin uncovered by George Starkey in the mid-1650s and only hinted at by him in print; see Newman and Principe, *Alchemy Tried in the Fire*, 142–47, 187. Did Duclos independently interpret Van Helmont's enigmatic utterances in just the same way as Starkey, did he correctly interpret Starkey's own enigmatic writings, or was there some communication of the process between the two? There is evidence of the existence of now lost Starkey letters sent to France dealing with just this subject, and possibly known to Duclos. See Hellot Caen Notebooks, 3:181v, 272r; as well as William R. Newman and Lawrence M. Principe, eds., *The Alchemical Laboratory Notebooks and Correspondence of George Starkey* (Chicago: University of Chicago Press, 2004), xxviii–xxix. For earlier indirect contact between Duclos and Starkey, see Principe, "Digby," 19.

53. PV 4, fols. 48r–55r (9 June 1668). Duclos gave another memoir of analytical protocols in 1670, but the relevant PV volume is lost; a précis is preserved in Duhamel, *Historia*, 89–90; see also *HMARS 1666–99*, 1:121.

54. It is not out of place to mention here that *tournesol* is *not* a type of sunflower, nor is it heliotrope, nor even, apparently, *Chrozophora tinctoria* (the illuminator's *folium*), which seems not to act as an acid-base indicator according to Christoph Krekel, *Chemische und kulturhistorische Untersuchungen des Buchmalereifarbstoffs folium und weiterer Inhaltsstoffe aus Chrozophora tinctoria und Mercurialis perennis* (PhD diss., Munich, 1996). All these plants have been given

material crystallized from the plant juices—would be examined, again using indicators and other tests. Third, the plants would be distilled into "their separate constituent parts: distilled waters, spirits, oils, fixed and volatile salts." These distilled substances, Duclos said, should likewise be subjected to indicator tests in order to identify and classify them consistently. Duclos's instructions aim at minimizing the negative aspects of fire analysis by distilling at the lowest temperature—that of a *bain marie* (hot water bath) or *bain vaporeux* (steam bath)—as much as possible. But the alkahest remained very much on Duclos's mind. Just two weeks later he asserted that "it is these external resolutive menstrua [like the alkahest] that we ought principally to seek out in order to use them in the chymical analyses."[55]

There is little evidence that Duclos's first two methods were used to any appreciable extent. Indicator tests on juices and crystallized solids could give little information, only indicating the presence or absence of the particular component being tested for, but without any ability to truly analyze, that is, to *separate* the various components present. Accordingly, distillation quickly became the fallback method. Yet, still heeding Van Helmont's warnings, the early analyses carried out by Bourdelin show the greatest care in using the lowest effective temperatures. In stunning displays of patience and operational fortitude, Bourdelin extended each distillation over a period of *twelve to fourteen days* of continuous heating, increasing the temperature gradually from gentle warmth to an open flame. In another attempt to limit the use of elevated temperatures, some plants or their juices were "digested," or fermented, at low temperatures in order to "loosen" their principles before distillation; such work seems particularly connected with Borelly. These digestions were continued for anywhere from four to nine months, and reduced the plant matter to a stinking slime that was then gently distilled. But even with these operational cautions, the results remained problematic. Working backward from the distilled fractions to the medicinal virtues of the plant was not straightforward. Very different plants did not necessarily yield obviously different analytical results. Even identifying the various fractions was fraught with unexpected difficulty because the indicators often gave ambiguous results. As Bourdelin's experimental reports steadily accumulated through the early 1670s, Duclos probably began insisting more and more on finding the alkahest as the only way to resolve such issues. It was also around this

----

in the secondary literature as the identity of *tournesol*; it thus remains unclear at present what it actually was.

55. PV 4, fols. 63v–66r (23 June 1668), at fol. 64v: "Ce sont ces Menstrues resolutifs externes que nous devons principalement rechercher pour les employer aux Analyses Chymiques."

time that Duclos, now in his seventies, began losing control of the project to the younger Denis Dodart.[56]

Competing Visions of Chymistry: Dodart vs. Duclos

Denis Dodart (1634–1707) had been admitted to the Académie as a botanist in 1671; he was a physician by training and profession.[57] He held a medical degree from the University of Paris, was licensed by the Faculty of Medicine in 1660, and acted as physician to the Duchesse de Longueville and the Princesse de Conti. Relative to other academicians, such as Duclos (and later Homberg), Dodart was not particularly productive—his few memoirs concern medical topics: a study of transpiration after the fashion of Sanctorius, a study of bloodletting, and observations on diseases of the poor. Unlike Duclos who, despite holding a medical degree and the title *medecin ordinaire du Roi*, "was not much interested in seeing patients, but preferred to devote himself to study and chymical experiments," Dodart focused his energy on medical practice and showed no particular interest in chymistry.[58] By early 1675, it was Dodart, not Duclos, who provided analytical protocols to the assembly. In January 1675, he pointedly asked Duclos "to meditate again" on the alkahest "to see if there might be anything to be hoped for in this dissolvent," and asserted at the same time that any value lay in the "marvelous [medicinal] effects attributed to the *entia prima*" it could produce, and not, he declared confidently and without explanation, "for analysis, for which it cannot be of service."[59] For his part, Duclos redoubled his Helmontian concerns about fire

56. The procès-verbaux covering the years 1670–75 are lost, rendering it difficult to speak definitively about the timing of the Académie's activity during these years.

57. On Dodart, see Sturdy, *Science and Social Status*, 184–89; Bernard de Fontenelle, "Éloge de M. Dodart," *HARS* (1707): 182–92; and Stroup, *Company*. The *Index biographique de l'Académie des Sciences* (Paris: Gauthier-Villars, 1979), 223, relying upon Fontenelle's report, gives the date of Dodart's reception as 1673, but Stroup (*Company*, 83n57) argues for 1671 on the basis of stipends Dodart received in 1671 and 1672 for his work with the Académie. Her revised dating can be further corroborated with an entry in Bourdelin's Expense Account, fol. 54v, dated 29 December 1671, that refers to "Monsieur Doudart qui est de lassemblée." The date 1671 also accords well with the testimony of Charles Perrault, who writes that Dodart was admitted a "little while after [*peu de temps après*]" Jean-Dominique Cassini, who became an academician in 1669; see *Mémoires de ma vie*, ed. Paul Bonnefon (Paris: Renouard, 1909), 45–46.

58. *Nouvelles de la république des lettres* (October 1685): 1139–43, at 1139.

59. PV 8, fols. 2r–7v, at fol. 4r; note that this memoir is cited with the incorrect date of 1674 in Frederic L. Holmes, "Investigative and Pedagogical Styles in French Chemistry at the End of the Seventeenth Century," *Historical Studies of the Physical and Biological Sciences* 34 (2004): 277–309, at 277. It is with regret that I must point out that this posthumously published paper

analysis, launching on 1 February 1675 a painstaking analysis of the various parts of lemons, distilled separately and so slowly with such gentle heat that each distillation was drawn out over *twenty-nine days!*[60]

Just a year later, in 1676, Dodart brought out the *Histoire des plantes*, prefaced by a lengthy justification of the chymical analysis of plants that directly rejects Duclos's entire approach. Dodart first dismisses two of Duclos's three protocols for analysis, leaving only fire analysis. He then waves off Helmontian concerns that the fire produces new substances. If there is any change, he asserts confidently and without evidence, it is "slight." As for the alkahest, in a direct rebuff to Duclos, Dodart writes dismissively that "in order to know what plants are, we have not believed that we were obliged to toil at resolving them into what the chymists call their *entia prima*, to resolve them irreversibly into an apparently simple liquor containing their virtues by means of some supposed universal dissolvents described enigmatically by Paracelsus, Van Helmont, Deiconti, etc."[61]

Dodart even claims that if the alkahest were found, its products would be more difficult to analyze than the plants themselves. For Dodart, fire is a perfectly satisfactory means of analysis, seemingly because it is an available method that produces *results*. He is clearly unconcerned, however, about the quality of the results and whether or not they are meaningful. Dodart systematically replaced Duclos's concerns about separating the constituents of plants without alteration and identifying them consistently, as well as his Helmontian principles, with a pragmatic doctrine of "good enough." The impression is that Dodart was keen to "get on with it" rather than worrying about such "minor" details as whether or not the methodology was suitable. Indeed, in February 1676 the Académie agreed with Dodart's proposal "not to carry out the analyses in such great detail anymore."[62] Duclos's Helmontianism may also not have been welcome to a physician from the conservative Paris faculty like Dodart. Dodart may furthermore have been more easily swayed than Duclos by the Académie's practical need to show tangible results to its royal

---

and its companion piece in the same journal and volume ("Chemistry in the Académie Royale des Sciences," 41–68) contain numerous errors and should be approached with caution. Undoubtedly, Holmes would have corrected these points had he lived to do so.

60. AdS, Bourdelin Notebooks, 4:20r–27v; Duclos's direct supervision in this set of experiments is signaled by the citation of his name in Bourdelin's marginalia.

61. *Mémoires pour servir à l'histoire des plantes*, ed. Denis Dodart (Paris, 1676), reprinted in *HMARS 1666–99*, 4:424–572, at 453–56.

62. Bourdelin Notebooks, 5:50v (10 February 1676): "Sur ce que Mr dodart à proposé de ne faire plus les analyses dans un si grand detail, la Compagnie est demeurée daccord qu'on ne feroit plus que lanalyse des parties usuelles des plantes dans le temps le plus favorable à leur usage."

patron. Duclos envisioned a broad audience of natural philosophers for the *Histoire des plantes*, evidently not fully grasping that the Académie's publications needed primarily to impress an audience of one—Louis XIV. This latter fact is made clear by the lavish format in which both the *Histoire des plantes* and the *Histoire des animaux* were produced: richly bound, oversized folios produced in small print runs that went almost entirely for presentation copies.[63] The pragmatic emphasis of the physician Dodart preempted the natural philosophical aims and theoretical concerns of the chymist Duclos, and redirected the project.

Duclos did not accept the rebuff quietly. He wrote a vitriolic page-by-page critique of Dodart's essay.[64] Duclos forcefully reiterates his (and Van Helmont's) contention that the fire changes the identities and qualities of the plants' constituents, and claims that Dodart speaks purely out of ignorance. As for the alkahest, "the author of the project, who has neither the use, nor the knowledge, nor the experience of this sort of analysis rejects it as impossible and as less useful than the ordinary analysis made by fire . . . [he,] who speaks always in the name of the Company without having been charged to do so, should not reject universal dissolvants as vain and useless, nor should he engage the Company to share in his blame for having so misunderstood the means by which several famous chymists assure us they have had good success."[65]

Duclos's ire was to no avail. Dodart's published text shows no changes

---

63. This feature was recognized by contemporaries: "I perceive the Academy Royal at Paris goes on with their Design of publishing a general History of Plants. . . . I have several Things to object against this mighty French Work, design'd rather for the Glory of the Monarch than the Use of the Subjects"; Tancred Robinson to John Ray, 18 April 1684, in John Ray, *Philosophical Letters*, ed. William Derham (London, 1718), 153–55. My thanks to Alexander Wragge-Morley for bringing this letter to my attention. On the *Histoire des animaux* and its format and print run, see Guerrini, *The Courtiers' Anatomists*, 149–50.

64. BMHN MS 1278, "Remarques sur le Projet de l'histoire des Plantes dressé par Mr. Dodart de l'Academie Royale des sciences, et imprimé au Louvre. Par Mr. Duclos de la meme Academie." The manuscript is in the hand of the Académie's botanist, Antoine-Laurent Jussieu (1748–1836), who presumably copied it from Duclos's now lost original. For more on the Dodart-Duclos feud, see Stroup, *Company*, 83–98; Lawrence M. Principe, "The Chymist and the Physician: Rivalry and Conflict at the Académie Royale des Sciences," in *Alchemy and Medicine from Antiquity to the Enlightenment*, ed. Jennifer M. Rampling and Peter M. Jones (London: Routledge, forthcoming 2020); and Victor D. Boantza, "Alkahest and Fire: Debating Matter, Chymistry, and Natural History at the Early Parisian Academy of Sciences," in *The Body as Object and Instrument of Knowledge: Embodied Empiricism in Early Modern Science*, ed. Charles T. Wolfe and Ofer Gal (Dordrecht: Springer, 2010), 75–92.

65. BMHN, MS 1278, fols. 3v–4r; Principe, "Chymist and Physician," contains further analysis of and quotations from this document.

whatsoever to the lines specifically critiqued by Duclos, indicating either that Duclos's objections were entirely ignored or that Dodart's essay had already been published without a draft ever having been shown to Duclos, even though Dodart was supposed to be only the editor of a communally composed document. Duclos turned to other projects, while Dodart took over the lead on the history of plants. The natural philosophical agenda promoted by Duclos and the practical agenda promoted by Dodart went their separate ways.

No further results from the plants project were published, but the laboratory work carried on. Dodart continued to present lists of plants for analysis and suggestions for particular experiments for the rest of the 1670s, as did Bourdelin and Borelly.[66] During the 1680s, Dodart's participation in the project steadily evaporated, as did the sorts of innovations Bourdelin had proposed in the 1670s. Bourdelin thereafter carried on mechanically with the analyses, cranking through lists of plants, submitting each to destructive distillation and registering the results in ever multiplying notebooks. In 1682, he began to separate himself physically from the Académie, starting to vacate the laboratory at the Bibliothèque du Roi in favor of his home laboratory; in 1687, he stopped going to the Académie's laboratory altogether.[67] At the same time, the number of plants he analyzed dropped dramatically; from October 1682 until May 1684, he analyzed merely a handful, and although the pace picked up again thereafter, it remained a fraction of the pre-1680 rate.[68] Bourdelin thus came to do his work in a perfunctory and narrowly circumscribed manner—he was content to distill plants, record the practical results, and not worry at all about either interpretations or applications. He would persist in this way until his death in 1699.

66. For example, Denis Dodart, "Memoire pour le travail a faire dans le Laboratoire sur les plantes et sur les Animaux, durant cette année 1678," PV 8, fols. 172r–178r (18 May 1678); Claude Bourdelin, "Pendant le reste de cette année et la prochaine 1679," PV 8, fols 190v–191r (23 November 1678); a similar text appears as "proposition de Bourdelin" dated 16 November in Lavoisier Bound MSS, QD B76++, 6:463–64.

67. PV 10, fol. 95v (15 April 1682); Bourdelin Expense Account, fol. 120v. Although the Académie allowed Bourdelin to work from home, it refused to reimburse the salary of a laboratory operator (1 livre per day) employed there. Bourdelin resigned himself—somewhat grudgingly—to the decision: "il est a remarquer que l'on m'a permis de travailler chez moy pour L'academie, en mesme tems Monsieur de la Chapelle m'a dict par deux fois qu'on ne me payeroit point de garçon quoy que J'en aye autant de besoin comme si je conduisois le laboratoire, mais quoy qu'il soit il faut souffrir."

68. AdS, Bourdelin Notebooks, vol. 9. There was a similar slow period from August 1690 until June 1692 when only about twenty plants were analyzed in almost two years, and most of those were done in June/July 1691; AdS, Bourdelin Notebooks, 10:161v–180r.

Duclos moved in the opposite direction. In keeping with his devotion to questions of broader natural philosophical import, he pursued his own, independent study of the chymical principles and the fundamental constitution of compound substances. He applied himself first to the study of salts, and then to the study of sulphurs. In what must have been a frenzy of writing in 1677, perhaps energized by his frustration over Dodart's rather imperious actions of the previous year, Duclos composed a weighty volume on salts in the form of letters, plus a treatise on the chymical principles, three more on salts, and one on sulphurs.[69]

The first dissertation, "Sur les principes des mixtes naturels," is a draft of what would be published in 1680 as Duclos's *Dissertation sur les principes des mixtes naturels*. Stroup, who first examined the manuscript, noted how a committee composed of Nicolas-François Blondel, Jean-Baptiste Duhamel, Edmé Mariotte, and Claude Perrault reviewed the text and apparently refused permission for it to be published.[70] The committee's response does not survive, leaving their reasons open to question. Duclos later blamed Duhamel specifically, claiming he was "always opposed to it because of some opinions he could not accept."[71] Perhaps echoing some of the committee's remarks, Duclos recorded that "some delicate philosophers" would find the dissertation to have a "bad taste" because it savored of "Platonism" (meaning, apparently, in this case, a sort of Neoplatonic hermetism), a system that "seems to them no longer in vogue; they want currently fashionable sciences."[72] What was probably at least as problematic for the Académie is Duclos's negative assessment of the chymical analysis of plants project. "The search for the principles of natural mixts has exercised the most capable natural philosophers for a long time. I have also labored in various ways to learn about them. In working on the chymical analysis of plants, I busied myself in vain to reduce

69. BNF, MS fr. 12309 and MS fr. 1333, "Dissertations physiques du Sieur Du Clos." The work on salts is mentioned by Henri Justel to Henry Oldenburg in a letter of 18 July 1677: "Nous aurons bien tost un traitte des sels de M. du Clos," *Oldenburg Correspondence*, 13:325–27. MS fr. 1333 also contains an autograph version of Duclos's "Remarques sur le livre des essays physiologiques de Mr. Boyle," fols. 238r–262v. Lavoisier Bound MSS, QD B76++, 6:465–86, contains a lengthy section dated 1677 and entitled "Diferentes observations sur les sels" that is probably related to Duclos's compositions.

70. Stroup, "Censure"; the names of the reviewers appear on fol. 1r of MS fr. 1333.

71. *Nouvelles de la république des lettres* (October 1685): 1141.

72. MS fr. 1333, fol. 42v [in Duclos' autograph]: "Cette Dissertation, renduë publique, pourra estre trouvé de mauvais goust à quelques Philosophes delicats, qui ne peuvent souffrir ce que leur paroist sentir de Platonisme. La doctrine de L'anciene Academie ne leur semble plus de Saison, ils veulent des Sciences à la mode qui court." See Stroup, "Censure," especially 447–51.

these mixts into some simple materials that could be reputed primary and pass for principles."[73]

Given that Dodart's *Histoire des plantes* had appeared the previous year, and that chymical analyses were continuing in the Académie's laboratory, the publication of such a sentiment would scarcely have been politic. Duclos's dissertation also contains a significant amount of theological material and biblical exegesis, and this inclusion must also have played a major role in the reviewers' refusal to grant permission to publish. Much of what Duclos scribbled at the end of the rejected manuscript attempts to justify his inclusion of such theological material—thus suggesting that it was indeed a sticking point for the reviewers. Such material would certainly have attracted unwelcome attention from the theologians of the Sorbonne and violated the Académie's strict policy of avoiding anything theological. Taken as a whole, there is little in the dissertation that would have been particularly welcome to most academicians. Its publication came about only when Duclos handed a reworked version to Daniel Elzevier (during the latter's visit to Paris), who then published it in Amsterdam in 1680.[74]

What has not been previously noted is that the five "dissertations" of MS fr. 1333 are not five separate works, but a single coherent treatise. Indeed, the reviewers read not just the first dissertation, but explicitly "le tout."[75] Duclos's slim *Dissertation sur les principes* as published in 1680 is no more than the introductory chapter dissevered from a 500-page treatise, and represents less than one-fifth of the whole. The full work appears as such in the list of books "ready to be printed" that was presented to Louis XIV during his December 1681 visit to the Académie.[76] Thus, evaluations of Duclos's chymical thinking that rely solely upon the published fragment are incomplete. The treatise as a whole, which bears the autograph title *Dissertations physiques*, is Duclos's ambitious attempt to compose a comprehensive treatise on the origin, composition, and properties of chymical substances—in essence, the very goal he proposed to the Académie at its first meeting in December 1666 and toward which he consistently tried to direct the chymical analysis of plants. When we recognize the unity of the composition, a far clearer and more accurate view

73. MS fr. 1333, fol. 5v; published with minor rewording in Samuel Cottereau Duclos, *Dissertation sur les principes des mixtes naturels*, in *HMARS 1666–99*, 4:1–40, at 4.

74. *Nouvelles de la république des lettres* (1685): 1141.

75. MS fr. 1333, fol. 1r.

76. PV 10, fol. 84r. The inclusion of Duclos's work in this catalogue potentially supports Duclos's contention that Colbert and "a good part of the Académie" approved the work despite Duhamel's objections; see *Nouvelles* (1685): 1141. The "illegal" Dutch publication of the first *dissertation* is not included in the catalogue's list of the academician's "printed books" on fol. 83v.

of Duclos's ideas and methods emerges. Although its full exposition must await another occasion, a brief summary of his system is necessary here in order to portray properly the various threads of chymical thought present at the Académie in the seventeenth century and to understand Homberg's later relationship to them.

Duclos's treatise begins by describing how the chymical analysis of plants failed to provide the results he hoped for. While fire analysis did separate plant materials into water, oil, salt, and earth, he found these fractions to be neither homogeneous nor elemental, nor could he succeed by further chymical operations to make them so. As a result, he gave up on chymical analysis as a means of identifying the principles of compound bodies and turned to a new methodology, concluding that "the principles can be better known by observing what comes together [*concourt*] in the generation of mixts." That is, Duclos turned toward tracing out the natural *synthesis* of compound bodies rather than relying upon their chymical *analysis* alone. Through observations of the natural world, he identifies water, earth, and air as the three most simple, noninterconverting, and ubiquitous substances—coincidentally the three corporeal members of the traditional Aristotelian four elements. Since these three principles are themselves inert, a motive principle is necessary. He identifies this fourth, but incorporeal, principle as emanating with sunlight and calls it the *esprit ignée*. Received upon the Earth, this igneous spirit (seemingly akin to the last of the four Aristotelian elements) insinuates itself into water, which provokes the water to combine with earth to produce first salt and then sulphur. He notes that the sharp taste of a salt cannot be referred to either the water or the earth in its composition, and must thus be a quality arising from the incorporated *esprit ignée*. But while the *esprit ignée* allows for reactivity and the combination of the corporeal principles, it has no ability to organize matter further or to bring things to their natural ends. That is the task of a fifth principle in Duclos's system, a thing he simply calls *nature* and which is something midway between Aristotle's *physis* (the "internal principle of change") and Van Helmont's organizing archei. This *nature* enters into the composition of mixts only after the incorporation of the *esprit ignée*. Thus, Duclos posits five principles of mixed bodies: three corporeal (water, air, earth) and two incorporeal (igneous spirit and nature). What laboratory efforts toward the analysis of compound substances could not fully answer, Duclos completes by observing their synthesis. It is thus by joining analysis and synthesis that Duclos comes to his conclusions about the principles of mixed bodies.

The desire to understand the fundamental nature and character of the chymical principles (the natural philosophical agenda) drove Duclos's inquiries. He had reservations, following Van Helmont, about the reliability of fire analysis, and its ambiguous and unsatisfactory results led him finally to abandon ordinary chymical analysis. Distillation methods proved incapable of providing clear or meaningful results, and the alkahest—the only reliable way forward—could not be found. As a consequence, he turned away from seeking answers solely in the laboratory operations that had proven too weak to provide sound results, and turned instead toward a priori reasonings and the teachings of the Bible and of ancient schools (especially that of "thrice greatest Hermes [*le trois fois très grand Hermes*]") and then supported these reasonings with experimental results and observations.[77] In contrast, the desire to obtain results applicable to medicine and pharmacy (the practical agenda) led Dodart to adopt a policy of "good enough" in regard to the chymical analyses, and to dismiss the wider natural philosophical implications of the plants project. While this attitude enabled him to produce one volume of the *Histoire des plantes*, the continuing ambiguity of the results in terms of practical application (coupled with Dodart's greater devotion to his own medical practice) caused him to lose interest in the project. Bourdelin abandoned both practical application and the search for natural philosophical meaning, as he continued to carry out fire analysis and to record his raw results mechanically long after Duclos and Borelly had died and Dodart had moved on to other things. Thus, within the Académie there existed significantly divergent conceptions of chymistry's scope and goals. This was the situation when Homberg entered the Académie in late 1691, and was asked to cast a fresh eye over the by then highly problematic project of analyzing plants chymically.

An Important Clarification about Fire Analysis vs. Solvent Analysis

Before leaving this phase of the analysis of plants project it is worth taking a moment to clarify two points about the Académie's chymical analysis of plants that can be unclear or problematic in the secondary literature. Both points involve the distinction between analysis by means of distillation and analysis by means of solvents. In 1971, Holmes usefully identified and discussed the complex history of these two approaches to the analysis of organic materials: analysis by fire (distillation) and analysis by solvent extraction.[78] While distillation continued (and continues) to be used for some preparative

77. MS fr. 1333, fol. 43v; not in the published version.
78. Holmes, "Analysis by Fire."

purposes, during the eighteenth and nineteenth centuries it was gradually re-placed by solvent extraction as a technique for the analysis of organic materi-als. The latter method employs particular solvents (such as alcohol, ether, or water), sometimes used sequentially, to dissolve out particular components of an organic material that contains a range of substances with differing solu-bilities in different solvents. Brewing a cup of tea, for example, is a solvent ex-traction of the tea leaves using water as solvent. If one used alcohol instead of boiling water, the resultant brew would contain a different set of compounds extracted from the leaves. Holmes deployed the Académie's analysis of plants project as an example of these two possible practices: Dodart and Bourdelin employed distillation, while early eighteenth-century academicians, led by Simon Boulduc and Charles-Joseph Geoffroy, turned increasingly to solvent extraction. Because solvent extraction avoids any thermal decomposition of the material being analyzed, it represents a superior method and the move toward it is generally seen as "modern." This evaluation played a key role in the rejection of the Académie's earlier plant analyses by fire as a "waste of time" by both some academicians of the later eighteenth century and some historians of the twentieth.

The first point to clarify is exactly how Duclos fits into this narrative. He was clearly an advocate of the use of solvents (called *menstrua* in early modern parlance) for analysis. But the kind of solvent analysis he proposed was far from modern, for he stipulated the use not of alcohol, ether, or some other familiar solvent, but rather of the Helmontian alkahest. The alkahest differs fundamentally from the solvents ordinarily used in extraction analy-sis, because it is not an *extractive* solvent at all. It does not dissolve one part of a mixed body and leave the rest untouched; it was supposed to reduce the *entire* mixed body into one or more liquids—their *prima entia*, or first essences. Duclos was clearly aware of this distinction. He himself explicitly divided solvents into three classes—corrosives, extractives, and resolutives. The first, like acids, dissolve (or more properly speaking, *corrode*) the entirety of the materials put into them, but alter them in the process, often by com-bining with them. The extractives, like alcohol and water, do not alter the dis-solved material but dissolve out only a part of the substances put into them, leaving the rest as a residue. Only the resolutives, typified by the alkahest, dis-solve (or more properly speaking, *resolve*) substances in their entirety with-out altering them. Duclos considered the first two types of menstrua to have little importance in analysis—the first destroyed the identity of the compo-nents, and the second isolated only some components and not others. Only the resolutive solvents "are of singular esteem," in Duclos' words; they alone

dissolve mixed bodies entirely and separate all their components without altering them. They are the only truly *analytical* solvents.[79] Thus, the history
of the ascendancy of modern solvent techniques over distillation includes in
its lineage—surprisingly and ironically enough—Duclos's advocacy of Van
Helmont's mysterious alkahest.

The second point to address is how some treatments in the recent secondary literature of the dichotomy between fire analysis and solvent extractions are confused by their use of the terms "solution methods" and "solution chemistry." This confusion arises partly from an erroneous conflation
of the words "solvent" and "solution," and partly from the dual meaning of
"analysis." These errors stem both from an unfamiliarity with the laboratory
operations involved and from a misinterpretation of what Holmes (correctly)
set forth in his 1971 paper. Chemical analysis can mean *either* (1) the physical
separation of a mixed substance into its individual components (*separative
analysis*) or (2) the determination of the contents or identity of a substance
without any physical separation of its components (*diagnostic analysis*).
Fire analysis and solvent extractions both represent methods of *separative*
analysis—they begin with a mixed substance and end with its various components physically separated into different vials. Indicator tests, in contrast,
represent *diagnostic* analysis. Litmus paper is a prime modern example: a strip
of the paper is treated with a drop of an unknown material, and whether the
paper turns red or blue, or remains unchanged *indicates* whether the material
being tested is acidic, alkaline, or neutral—but no actual *separation* of components occurs. The confusion comes to a head, particularly in the writings
of Mi Gyung Kim, but also elsewhere, in the context of the Académie's plant
analyses, when indicator *tests* are conflated with solvent *extractions* under the
vague and misleading terms "solution chemistry" and "solution methods,"
as if diagnostic indicator tests could possibly be an alternative to separative
fire analysis, or are somehow equivalent to solvent extractions. Indicator tests
are an *adjunct*—not an alternative—to solvent extractions and fire analysis,
and were routinely used as such at the Académie. After plants were distilled
(separative analysis), the resulting fractions were always subjected to indicator tests (diagnostic analysis). Not understanding these practical facts sets up
a false dichotomy between practitioners of fire analysis and supposed advocates of alternative "solution methods" (whatever that means), and leads to
inaccurate conclusions that cloud our historical understanding.[80]

---

79. PV 4, fols. 63v–66r (23 June 1668).

80. Allen Debus, given his expertise as a chemist, of course used the term correctly in his
"Solution Analyses Prior to Robert Boyle," *Chymia* 8 (1962):41–61; cf. Kim, *Affinity*, 6–7, 11, 50,

Homberg's Role in the Plant Analysis Project

Given the Académie's continuing problem of what to do with the history of plants project, it is probably no coincidence that the first academicians admitted under the fresh leadership of Abbé Jean-Paul Bignon were a botanist (Tournefort) and a chymist (Homberg). Indeed, immediately upon Homberg's admittance to the Académie in 1691, Bignon asked him to review the chymical results of the plants project. In February 1692, Homberg presented his "reflections upon Mr. Dodart's memoir for the chymical projects."[81] Those reflections were not recorded, but a forty-four-page manuscript in Homberg's autograph hand that must have been compiled in preparation for the report survives at the Bibliothèque du Muséum d'Histoire Naturelle.[82] Dating from late 1691 or early 1692, it preserves a glimpse of Homberg's investigative methodology and his theoretical commitments during his earliest days as a member of the Académie.

Bourdelin had reported the quantities of separated fractions in terms of *onces*, *gros*, and *grains*; Homberg first recalculated these values as ratios relative to the starting material, clearly in order to facilitate comparisons. Homberg then endeavored to draw exact conclusions from his calculations. In the case of two species of rest-harrow (*ononis spinosa* and *non spinosa*), for example, he notes that both are recommended as diuretics, but since analysis shows that the latter species contains a higher percentage of salts (upon which diuretics depend), then one should prefer that one for medicinal use.[83] In the case of fraxinella he notes that only the root yields a volatile salt (one-two thousandth of the total weight), and that the leaves have one-third more oil,

---

52 and passim; in later chapters (especially the third), the term "solution chemistry" is used to refer vaguely to anything done in or with a liquid. There is a similarly odd notion of affinity as a substitute for chymical principles in theories of chemical composition, e.g. 7, 13, and passim.

81. PV 13, fol. 81v (27 February 1692).

82. BMHN MS 1279, "Liste des plantes analysées du premier Volume de l'histoire des plantes." The "first volume" must refer to some manuscript collation of analysis results; it cannot refer to the 1676 *Histoire des plantes* since that publication does not contain specific laboratory results, and most plants treated in Homberg's manuscript are not included there. What Homberg saw, and referred to elsewhere as "Mr. Dodart's memoir," was probably a summary taken from Bourdelin's tenth autograph notebook (at the AdS preserved in the Fonds Bourdelin)— each entry there bears the endorsement "Abrégé" which might signal the making of a summarized memoir for Homberg, while the eleventh notebook begins with the note that "le dernier ayant esté mis dans les mains de Monsieur Dodart." BMHN MS 1279 was first identified by Alice Stroup, "Wilhelm Homberg and the Search for the Constituents of Plants at the 17th-Century Académie Royale des Sciences," *Ambix* 26 (1979): 184–202.

83. BMHN MS 1279, 38.

which would "encumber its salts and weaken their action." Homberg notes that practical experience confirms this prediction, since it is the root that is used for resolving obstructions, while the leaves are useless for this purpose.[84]

Homberg's initial assessment was thus moderately optimistic. His second report, however, delivered in September 1692 after having examined "three volumes" of results, was more mixed. He praised the quantity of analyses performed and the precision of the observations. He noted the "great uniformity" in the products of analysis—most plants gave the same set of separated materials, but with potentially significant differences in their relative quantities. Nevertheless, he also observed that plants with completely opposing medicinal effects sometimes yielded very similar analytical results, and so he concluded that "one cannot therefore fully judge the [medicinal] effects of a plant by means of its analysis."[85]

Homberg's conclusion completely undermined the utilitarian medical rationale for chymical analysis that Dodart had substituted for Duclos's broader motivation to use the analyses to learn about the chymical principles. Yet Homberg's revised conclusion was not a universally negative assessment of the plants project. He went on to say that "I have found here some experiments that establish truths that have been highly contested in the past and even entirely denied by some authors, for example, [regarding] the volatile salts of certain plants, and the differences among lixivial salts, which are very considerable discoveries that should please the learned [sçavants]. But it was further hoped that the public might be able to find some particular utility for medicine in this work, in regard to which it seems one is referring to the use of simples or of plants."[86]

Having already stated that the pharmaceutical value of plants cannot be properly ascertained by chymical analysis, Homberg's comment left "the public" with what it already had before the plants project ever began—namely, herbals and lists of simples and their observed medicinal uses. Homberg was clearly much more enthusiastic about the strictly *chymical* results—such as the presence and character of volatile and lixivial salts—which he considers should be of interest to learned natural philosophers.

84. BMHN MS 1279, 27: "Cette analyse fait conjecturer que la racine doit produire des effets plus efficaces et differens de ceux de feuilles et des tiges de cette plante, parcequ'elle est aromatique et amere et contient du sel volatile . . . l'herbe contient à proportion ⅓ plus d'huile que la racine, laquelle doit embarasser ses sels et en affoiblir l'action; l'experience confirme cecy, car la racine est fort utilement employée dans touttes sortes d'obstructions, sans qu'on fasse aucun cas de l'herbe."

85. PV 13, fols. 115v–118r (3 September 1692), at fol. 116r.

86. PV 13, fol. 116r.

The exact nature of Homberg's chymical interest in these results becomes clear in his fulfillment of Bignon's order that he "consider to what purpose one might employ that which has come out of this work." Dramatically, Homberg chose to bypass entirely the *results* of the analyses and to examine instead their *products*. He set aside the thousands of pages of carefully penned experimental results and focused instead on the hundreds of bottles of the separated materials stored in a room of the Académie's laboratory. The 1689 inventory of the Académie's laboratory lists more than four hundred vials of liquids and more than five hundred vials of salts isolated from the plant analyses and arranged methodically in cabinets on lettered shelves.[87] The neat ranks of countless bottles and flasks shown at the far left of the headpiece to the *Histoire des plantes* may depict these stored materials (fig. 2.3). Homberg proposed, therefore, to make an "individual examination of each of the materials found in these analyses," and he set to work immediately with the thick, black oils or *huiles puantes* (stinking oils) "that are found at the end of all the analyses." By redistilling such oils repeatedly from quicklime, he found that more than half their weight was converted into water, and the remainder rendered light in color, more fluid, and free from its foul odor, which was replaced with the smell of the plant from which it had been originally distilled. Thus, Homberg concludes that the huiles puantes are composed of water and the plant's essential oil mixed with particles of burned vegetable matter.[88]

On its own, this experiment is neither particularly interesting nor significant, even if Homberg, with a brief nod to medical application, found that the rectified oil could be used for paralytic and rheumatic pains. Only when it is juxtaposed with other experiments scattered among the rest of Homberg's work in the 1690s does a coherent picture emerge of Homberg's first long-term research program at the Académie. The ultimate aim of his endeavor reveals much about Homberg's own view of chymistry and how he envisioned its proper goals and applications. In short, Homberg redirected the plants project once again. Dismissing as unworkable the practical, medicinal direction into which Dodart had steered it, Homberg returned the project to the more properly chymical goals of the natural philosophical agenda that Duclos had envisioned and fought for. *Homberg would now spend almost a*

87. Chabbert, "Borelly," 219, 223–24.

88. The work is first mentioned in PV 13, fol. 109v (9 August 1692), and described on fols. 116r–118r (3 September 1692). In some trials Homberg used the analogous stinking oil distilled from tartar (predominantly potassium bitartrate), perhaps because it could be produced more quickly and in greater quantity.

FIGURE 2.3. The headpiece by Sebastian Leclerc to the 1676 *Histoire des plantes*. While the overall depiction is an artistic pastiche of various locales — the Académie's laboratory was miles away, on the opposite side of the Seine from the Jardin du Roi, which here is depicted outside the window — the shelves of vials on the left side do approximate the description of such a storeroom provided by an inventory of the laboratory.

decade using the accumulated substances from the plants project as starting ma-
terials for experiments to identify the true principles of compound bodies. Homberg organized much of this work specifically to test Helmontian theories
of composition, including the idea that water is the universal substratum of
all substances. Using an orderly and exhaustive methodology, similar to that
he used in the early 1680s to solve the problem of the Bologna Stone, Homberg began to examine each class of substance separately, attempting either
to break it down into simpler substances or to convert it into one of the other
classes, thereby collapsing the list of separated "principles" to the smallest
possible number, and perhaps to just one—Van Helmont's primordial water.

The 1676 *Histoire des plantes* reported that distillation analysis gave the
standard five principles, some of which were subdivided into subclasses:
two types of oils (essential and black), three spirits (acidic, sulphureous, and
urinous), three salts (two of them fixed, the saline and the lixivial or *sulphuré*,
and one volatile, the urinous), water, and earth.[89] Thus, Homberg had a list of
ten classes of substances with which to begin. His initial work with the huiles
puantes collapsed the two varieties of oil into one—he converted the black
oils into water and essential oil. What, then, was the composition of the essential oil? To answer this question, Homberg first studied how such oil is
formed in nature and then designed a probatory experiment. He first compared immature seeds with mature ones. Immature seeds contain "a great
deal of phlegm, little oil, and much fixed salt," yet the same seeds when mature contain "much oil, little phlegm, and very little fixed salt." Olives and
nuts stored for three or four months before being pressed yield more oil than
when they are pressed fresh. Plant materials from which all the oil has been
pressed out yield new oil when submitted to the fire of distillation. Homberg
therefore concluded that oil must be produced by the combination of water,
salt, and earth, and that this synthesis of oil is promoted by heat. His study
of the huiles puantes supported this conclusion: by means of six reiterated
distillations from lime, he reduced 16 ounces of huile puante into 13 ounces
of watery phlegm and one residual ounce of oil, as the quicklime absorbed
the earthy and saline parts. Further distillations, he was confident, would
turn the residual essential oil into water, earth, and salt as well. Combining
the observed synthesis and experimental analysis of essential oil, Homberg
concluded that "in young seeds, the phlegm combines over time with the
salt and a part of the earthy material to produce the quantity of oil found in
ripe seeds, and by means of art one can separate the compound body into the

---

89. Dodart, *Histoire des plantes*, 464–65. Fourteen substances are listed there; the list above
ignores mixed fractions that were later separated into the others by Bourdelin.

same simple materials from which nature formed it."[90] Just as Duclos had done earlier, Homberg drew his conclusions from a combination of both laboratory analysis and the observation of natural synthesis. Homberg went yet further to prove his idea about the natural synthesis by mimicking it artificially through the extraction of the oil from cacao in three different ways. By simple distillation, he obtained 3 ounces and 2 gros of oil from 1 pound of cacao—that is, about *one-fifth* of its weight. By expression, he obtained 2 ounces, then boiling the residue in water provided another half-ounce, and distillation of the boiled residue gave a further 2½ ounces—thus 5 ounces total, or about *one-third* of its weight, a greater proportion. Finally, by "crushing the cacao upon a hot stone, in the way one makes chocolate" and boiling 13 ounces of the resulting paste in water, Homberg obtained 6 ounces of oil, and by distilling the residue another ounce and 3 gros, for a total of 7 ounces and 3 gros, or more than *one-half* the weight of the initial cacao.[91] Homberg argued that the increased amount of oil he obtained must have resulted from combination of fixed salt and earth in the cacao with the added water under the influence of heat to form more oil de novo. Thus, by combining analysis and synthesis, Homberg concluded that oils are compounds of water, salt, and earth, and this allowed him to strike oils off the list of principles.

The elimination of oils left spirits and salts for similar inquiry. Homberg turned first to the spirits, specifically the acid spirits. He proposed that these acidic liquids are actually nothing other than a volatile acid salt mixed with water. This claim is traceable to Van Helmont.[92] To prove this idea, Homberg needed to demonstrate the existence of such a volatile acid salt by isolating it from the water. The difficulty of doing this in practice was that distillation would be useless as a means of separation because the putative salt was as volatile as the water, and so they distill together. So Homberg devised an indirect method (fig. 2.4). He began with aquafort (nitric acid)—which according to his hypothesis should consist of water and the acid salt—and used it to dissolve 2 ounces of silver. To this solution he added common salt, which caused all the silver to precipitate out of solution as a white powder. This precipitate weighed 2½ ounces, and for Homberg the increased weight demonstrated that the silver had successfully combined with the acid salt. Although the acid salt was successfully separated from the water, it was now bound even more

90. PV 14, fols. 122r–125r (1 June 1695), "Extrait d'une Dissertation de Mr. Homberg touchant les parties qui composent les huiles."

91. PV 14, fols. 124r–125r. In the early modern French system Homberg uses, there are 8 gros in an ounce, and 16 ounces [onces] in a pound.

92. Van Helmont, *Ortus*, "Blas humanum," no. 44, p. 188, where he writes that distilled vinegar "is water impregnated with a volatile acid salt."

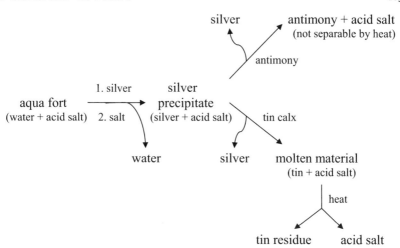

FIGURE 2.4. Homberg's scheme for isolating the dry volatile acid salt from aquafort (nitric acid). Silver dissolves in aquafort by combining with the acid salt, taking it away from the water. The addition of common salt precipitates the silver combined with the acid salt. Antimony can then displace the silver and combine with the acid salt, but since both are volatile, they cannot be separated by heat. Alternatively, tin can also take hold of the acid salt and displace the silver (which is recovered in its original form and weight). Finally, gentle heating at last isolates the volatile acid salt by subliming it away from the tin residue. (Those familiar with chemistry will recognize that the isolated material is not in fact related to anything initially present in the aquafort.)

tightly to the silver. Homberg knew this to be the case because "even hot water is not able to separate" the salt from the silver precipitate—washing it with hot water provokes no change, no loss of weight, no acidification of the wash water. So he tried to transfer the salt to a substance from which it might be more easily separable. He first tried antimony, which reacted quickly with the silver precipitate, liberating metallic silver. Noting that the silver reappeared in its original form and weight, he concluded that the antimony took up all the acid salt. Unfortunately, antimony proved as volatile as the acid salt, and so upon heating, the two sublimed together (as butter of antimony, today antimony trichloride) without separation—so Homberg was no better off than when he started. Homberg then substituted a less volatile metal—tin. By gently heating the silver precipitate with tin calx, the silver was liberated (in its original weight), and a liquid composed of the tin combined with the acid salt could be poured off. Upon cooling, this liquid solidified. Finally, with gentle heat, Homberg succeeded in subliming beautiful saline crystals out of the solidified mass. This white sublimate was volatile, weighed nearly half an ounce, was powerfully acidic, and dissolved readily in water and spirit of wine to form acidic liquids that dissolved metals, thereby proving to Homberg's satisfaction that he had successfully isolated the volatile acid salt present initially in the aquafort.[93] Homberg demonstrated his success to the assembled academicians with a dramatic flourish by throwing a solution of this separated salt onto the stone floor of their meeting room where "it boiled up just like aquafort."[94]

The thoughtful experimental design of this process reveals a good deal about Homberg's chymical thinking. First, he uses comparative weights to follow the progress of an invisible material (the acid salt) from one combination to another. Gravimetric analysis had been taking on an increasingly important role in chymistry for over a century; for Homberg it is a routine tool for monitoring experiments.[95] Second, Homberg implicitly uses principles that would later become known as *affinities*. The acid salt first leaves the water to combine with silver, to which it becomes more firmly attached; it then forsakes the silver for tin, and finally by means of heat it is sublimed away from the tin and isolated. While the famous affinity table or "Table des rapports" was published by Étienne-François Geoffroy (Homberg's student and closest associate at the Académie) only in 1718, it is clear that Homberg already had much of its content and concept in mind as implicit operational principles well before it was cast into tabular format.[96] Finally, the isolation of this salt is one of many examples where Homberg designs an experiment specifically for the purpose of hypothesis testing. It is sometimes imagined that seventeenth-century chymists did not routinely engage in explicitly theory-testing experiments; Homberg's work refutes this notion repeatedly and thoroughly. Homberg set out not only to prove that acids consist of a dry volatile acid salt in aqueous solution, but also explicitly to refute "some chymists who doubt that there is any volatile salt in minerals."[97]

Having shown that acid spirits are simply acid salts dissolved in water, Homberg could readily extend this conclusion analogically to the other spirits: sulphureous and urinous spirits are just water containing a *sulphuré* salt or a urinous salt, respectively. Now spirits could also be struck off the list of principles. This left Homberg with salts. Salts seemed more stable and simple

93. Wilhelm Homberg, "Maniere d'extraire un sel volatile acide minérale en forme séche," *MARS* (December 1692): 171–76; reprinted in *HMARS 1666–99*, 10:202–8. What Homberg actually prepared was a hydrate of tin tetrachloride.

94. Homberg, "Sel volatile acide," 171; this occurred on 26 March 1692, see PV 13, fol. 86v.

95. Newman and Principe, *Alchemy Tried in the Fire*, 46–91.

96. On the affinity table, see Frederic L. Holmes, "The Communal Context for Etienne-François Geoffroy's 'Table des rapports,'" *Science in Context* 9 (1996): 289–311; Étienne-François Geoffroy, "Table des differens rapports observés en chimie entre differentes substances," *MARS* (1718): 202–12. See chapter 3 for more on the background to Geoffroy's *rapports* at the Académie.

97. Homberg, "Sel volatile acide," 171.

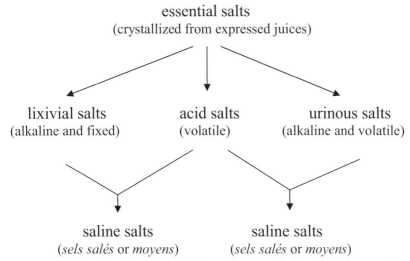

FIGURE 2.5. The family of salts. The essential salts contain some or all of the more fundamental salts in the middle row—lixivial, acid, and urinous. The combination of an acid salt with either of the two alkali salts produces saline salts (*sels salés*), later called *sels moyens*, or "middle salts," because they are produced from two opposing (acid and alkali) salts. These saline salts can be decomposed back into a volatile acid salt and either a fixed or a volatile alkali salt.

substances than either oils or spirits, and they proved a more complex and diffi-cult topic of study; Homberg spent many more years experimenting with them.

By the end of the seventeenth century, salts constituted a complicated but well-organized group of substances. Van Helmont had laid the foundations for classifying salts into classes. In England, George Starkey and then Robert Boyle further developed this scheme, which the Belgian chymist had, in his customary manner, outlined rather diffusely and obscurely.[98] Most chymists at the Académie adopted a version of this classification system, although it re-mains unclear whether their usage came directly from Van Helmont, through one of his Continental interpreters, through Starkey and Boyle, or, perhaps most likely, through some combination of all three. Figure 2.5 summarizes this classification of salts as it existed at the Académie in the seventeenth cen-tury, particularly in relation to plant and animal substances, although varia-tions could embrace also the mineral realm.[99]

98. Newman and Principe, *Tried in the Fire*, 275–86.

99. Holmes, *Eighteenth-Century Chemistry*, 33–59, very correctly pointed to the study of salts as an important area of research at the Académie, even if he did not recognize the larger context—the testing of Helmontian water theory and the search for principles—for such study in Homberg's work.

The essential salts, isolated by crystallization from plant or animal juices, contain various of the three simpler salts: lixivial, acid, and urinous. Two of these simpler salts are alkaline (they turn color indicators like tournesol blue)—the lixivial and the urinous—and one is acidic (it turns tournesol red).[100] The acid and the urinous salts are volatile, they pass over into the receiver during distillation. The lixivial is "fixed" (nonvolatile) and is isolated from the distillation residue. Acid salts can combine with either of the two alkaline salts (lixivial and urinous) to produce a new salt that is neither acidic nor alkaline, but merely "salty" and hence is called "saline." Of the five types of salts listed in figure 2.5, the three in the middle line were clearly less compounded than the others, and it is these that Homberg chose to study.

In a clear reference to Van Helmont, Homberg noted that the avidity with which acid salts dissolve in water "has given occasion for some authors of our age to suspect that they are nothing other than a certain modification of common water, and that by long digestion and many cohobations upon certain earthy or metallic bodies that they have dissolved, one can return them to their original insipidity."[101] To test this idea, Homberg sealed up aquafort, oil of vitriol, and spirit of salt (the three mineral acids, nitric, sulphuric, and hydrochloric, respectively), aqua regia, and vinegar in sets of small glass vessels. He placed one set near the fire, a second set farther away but still in the warmth, and a third where it received no warmth from the fire. He left them like this for *four years*.

Two years after beginning this experiment to "render [acid salts] insipid or similar to common water, as certain of the most enlightened chymists [that is, Helmontians] of our age claim," Homberg carried out a parallel experiment. He dissolved as much mercury as possible in 12 ounces of aquafort, and distilled the solution to dryness. The distillate was no longer acidic, and the residue weighed 3 ounces more than the original mercury; therefore, Homberg concluded that the mercury combined with all the acid salt from the aquafort. He then heated the residue with quicklime and recovered the original weight of mercury by distillation, thus leaving the acid salt combined with the quicklime. He then increased the temperature to drive out all the volatile acid salt, which "appeared as red fumes in the receiver,"

100. Van Helmont himself thought that only the fixed lixivial salt was "sulphureous"; however, by Homberg's time, the urinous salt was also being thought of as sulphureous, presumably because of its alkaline nature (akin to that of the fixed lixivial salt) as revealed by acid-base color indicators.

101. PV 14, fols. 69r–72r (9 March 1695), at fols. 70v–71r. Cohobation is an operation wherein a volatile liquid is distilled from a nonvolatile solid, the distillate poured back over the residue and distilled off again, and the process repeated multiple times.

and thereby regenerated the aquafort. He found that this recovered aquafort could dissolve barely half an ounce of mercury, whereas at the start it dissolved 15 ounces. What had happened to the corrosive acid salt? "One can plausibly conclude that by artifice a corrosive acid can be changed into an insipid material."[102]

Here Homberg is undoubtedly exploring the Helmontian concept of exantlation, according to which corrosives are "weakened" or "exhausted" by the act of dissolving. Indeed, Van Helmont's conclusion from his own experiment (which was not as well controlled gravimetrically as Homberg's) summarizes Homberg's results equally well: "dissolvants are changed even if that which is dissolved loses nothing of its material or substance." After dissolving the mercury, Homberg's aquafort distilled over as an insipid phlegm, just as Van Helmont had reported that "whatever is distilled thence is mere water." When the acid salt was recovered at the end of the operation, it was considerably weakened.[103]

While the longer term experiment of decomposing acid salts into water by simple digestion continued quietly, Homberg turned to lixivial salts "to discover the parts that compose them." He achieved the "extraordinary" result of making these extremely fixed salts volatile by cohobating them with various forms of alcohol—spirit of wine, *eau de vie*, and white wine. Beginning with the lixivial salt of wormwood—apparently a stored product from the plant analyses—Homberg caused most of it to co-distill with wine, finally giving a liquor in which no trace of the salt could be detected by indicator tests, as if the salt had been so "penetrated and divided by the sulphureous spirit . . . it was made into something completely different from the salt it had been."[104] Thus, these salts, despite their apparent stability, are not permanent entities. It is rather difficult to identify in modern chemical terms what exactly Homberg was so proud of achieving, but the process closely resembles the "alcoholization" of alkalies to render them volatile, hinted at by Van Helmont and described more fully by George Starkey in his 1658 *Pyrotechny Asserted*. The volatilization of fixed alkalies (that is, lixivial salts) was considered a great secret by Helmontians. It opened the way for preparing

102. Wilhelm Homberg, "Essay sur l'adoucissement des Acides," PV 16, fols. 40r–43r (6 March 1697).

103. On exantlation, Newman and Principe, *Tried in the Fire*, 80–83; Van Helmont, *Opuscula, De febribus*, chap. 15, no. 20, p. 57.

104. Wilhelm Homberg, "Essays sur l'analyse des sels des plantes," PV 16, fols. 194r–196r (17 July 1697); published in Stroup, "Homberg," 194–96. "Sulphureous spirit" here refers to the alcohol in the wine; it is considered sulphureous because of its inflammability. An early stage of this experiment is probably alluded to in PV 13, fol. 94r (23 April 1692).

a powerful solvent that could perform many of the feats of the alkahest.[105] Duclos had himself expressed the desire to prepare such a solvent, given the difficulty of making the alkahest, and in 1673 oversaw a process that one Theodore Aleman conducted in the Académie's laboratory that promised to do so.[106] Homberg gives no indication here that he had such a lofty goal in mind. His last memoir, however, presented in 1714 and not published until after his death, contains experiments whose origins clearly lie in these experiments to volatilize fixed lixivial salts during the 1690s. That memoir will be more closely examined in chapter 6. For now it suffices to mention that Homberg there explicitly states that he carried out this operation because "I believed that by this same operation I would be able to obtain a volatile salt from the fixed salt of tartar, which Paracelsus and Van Helmont have so greatly praised."[107] This is the sole mention of Van Helmont by name in Homberg's work, although the foundations of his 1690s experiments testify clearly to the Belgian chymist's influence.

Having made lixivial salts volatile using spirit of wine, Homberg next attempted to do the opposite: to make spirit of wine fixed using lixivial salts.[108] This reciprocal process was again something that Van Helmont had claimed to be possible: "a volatile salt . . . is materially the same as an alkalizate [that is, lixivial] salt, and just as volatile salt can be made alkalizate, an alkali can in turn be volatilized."[109] Thus, still apparently continuing to test Helmontian ideas, Homberg moistened the lixivial salts isolated from a variety of plants with spirit of wine, and heated the mixture in a sealed flask for two months, gradually increasing the temperature. After this time he added another portion of spirit of wine and heated again. After subjecting the material to intense heat for fifteen days to expel all volatile materials, he found that the salt had gained weight, and so concluded that the lixivial salt had fixed and retained a portion of the "oily and [volatile] saline parts of the spirit of wine." Van Helmont promised repeatedly that lixivial salts could fix spirit of wine: "a salt is made from spirit of wine; it is coagulated in the salt of tartar." This process

105. George Starkey, *Pyrotechny Asserted* (London, 1658), 126–27, 137–38, 142; Newman and Principe, *Tried in the Fire*, 136–38; Van Helmont, *Opuscula, De febribus*, chap. 15, no. 26, p. 58.

106. AdS, Bourdelin Notebooks, 1:353–56 (3 July 1673).

107. Wilhelm Homberg, "Memoire touchant la volatilisation des sels fixes des plantes," *MARS* (1714): 186–95, quotation on 190. On "alcoholization," see Starkey, *Pyrotechny*, 126–27, 137–38; the clearest description is on 144–45, compare this with Homberg's process on 192–93 of his 1714 memoir. See also Van Helmont, *Ortus*, "Blas humanum," no. 45, p. 188.

108. Wilhelm Homberg, "Observations sur les sels fixes des Plantes," PV 18, fols. 37v–39v (3 December 1698).

109. Van Helmont, *Ortus*, "Blas humanum," nos. 41–42, p. 188.

was a step in producing the *balsamus samech*, a celebrated but secret Paracelsian preparation.[110] Homberg's claim that spirit of wine is both sulphureous and saline was also asserted by Van Helmont.[111] Homberg then repeated the experiment with saline salts (see figure 2.5), but obtained no weight increase, because "the purely lixivial part of these salts being already sufficiently saturated with acid parts . . . they cannot saturate themselves a second time with the oily and saline parts of the spirit of wine."[112] What Homberg demonstrated experimentally was the interconversion of fixed lixivial salts and volatile spirit of wine or, more broadly speaking, the interconversion of salts and sulphurs, thus showing that they are not really two separate and distinct principles. Finally, Homberg examined the last class of salts—the urinous—and found them to be the most resistant to alteration. Nevertheless, he devised an experiment that produced a urinous salt from a lixivial one, thus showing their interconvertibility.[113] As a result of these experiments, Homberg could again condense his list of chymical principles.

Thus, by the end of 1698 Homberg had struck both sulphurs (oils) and spirits from the list of principles and had shown that the various classes of salts were interconvertible, thereby collapsing the original list of ten separate substances to merely three: water, earth, and salt. Significantly, *every one* of Homberg's experiments in this six-year project had a demonstrably Helmontian foundation or inspiration. In this regard, Homberg was part of a tradition of Helmontian chymistry stretching back to Van Helmont himself, a tradition that included also Starkey, Boyle, Duclos, and many others of, in Homberg's own words, "the most enlightened chymists of the age." Clearly, Homberg was thoroughly versed in the most prominent chymical theories and literature of the day, spent much of the 1690s putting these ideas to the test experimentally, and made use of the plants project in order to do so. All that remained was to open the flasks of acids that he had left digesting for nearly four years to determine if the acid salts had been partly or wholly reduced to water—the ultimate transformation proposed by Van Helmont. Homberg opened the flasks in early 1699. His final memoir on the topic, presented on 28 January, unambiguously cites his initial intention of testing

110. Van Helmont, *Ortus*, "Inventio tartari in morbis temeraria," no. 24, p. 242; cf. "Blas humanum," no. 39, p. 188 ("aqua vitae fixata in alkali tartari"); on *balsamus samech*, see "Progymnasia meteori," no. 27, p. 72. Compare Starkey, *Pyrotechny*, 138: "*Helmonts* Doctrine; *That volatile Spirits* (as of Wine, Vinegar, &c.) *are fixed by means of fixt Salts.*"

111. Van Helmont, *Ortus*, "Spiritus vitae," no. 3, p. 196.

112. Homberg, "Observations sur les sels fixes," fol. 39v.

113. Wilhelm Homberg, "Observations sur le sel urineux des plantes," PV 17, fols. 40r–43r (27 November 1697).

Helmontian theory, but also makes a measured retreat from his former confidence in the chymists he had held to be "the most enlightened of the age."

> Many modern philosophers have thought that all the parts of compound substances, or the principles of them that are uncovered by chymical analyses, ought to be reducible into a single material, and that the difference among the principles consists solely in the different modifications of that material, which they suppose to be common and insipid water. They base their reasons for this opinion upon particular experiments. Although these experiments do not give sure proof of their reasonings, they have not failed to give me the curiosity to make a particular study of them by means of several trials upon all the principles of mixts.[114]

Homberg goes on to describe how he tested the strength of the acids first by taste(!) and then by measurement of the weight of metals they could dissolve, and how he compared these results with those obtained at the start of the experiment four years earlier. He discovered that the mineral acids had not changed in the least, only the vinegar had degraded. Homberg concludes that if the mineral acids could ever be degraded by this means, it would take far longer than four years. This result did not make him conclude definitively against Helmontian theory, but the failure of the acid dulcification experiment may explain the rather dampened enthusiasm he expresses in the opening lines of the memoir. In any event, this memoir was nearly the last contribution Homberg made to the plants project.[115] His turn away from that project and toward other projects probably not only indicates that he had now come to the end of his exhaustive and methodical study of each of the separated fractions, but also reflects the death of Claude Bourdelin later in the same year (1699), which brought an end to the distillations of plants that had been ongoing for thirty years.

Homberg's last word on the project appeared in the *Mémoires* for 1701, almost a decade after Bignon had requested his assessment. There Homberg repeats some of the concerns he had expressed early in the 1690s: plants with very different medicinal properties—he uses the example of cabbage and deadly nightshade—give very similar analytical results by distillation, and different methods of analysis provide different results from the same plant.

114. Wilhelm Homberg, "Observations sur la addoucissement des acides," PV 18, fols. 97r–99v (28 January 1699).

115. One exception is PV 19, fols. 340v–349r, "Observations sur les huiles des plantes" (28 August 1700), but this is about how to improve the yield of essential oils rather than about analysis; following fol. 349r there is Homberg's autograph drawing of a flask for collecting essential oils.

He also notes that recombining the separated principles does not give anything resembling the original material, casting further doubt on the notion that the separated materials preexisted in the unanalyzed material. After summarizing some of his earlier findings, he added a further experiment (similar to his earlier processes on cacao) wherein he treated grape juice in three different ways. He first destructively distilled the grape juice and obtained the usual range of fractions—water, oils, earth, and salts. He then gently evaporated the juice, obtaining the essential salt by crystallization and separating a fragrant oil. Finally, he fermented some of the juice and isolated a flammable spirit (alcohol) and water by gentle distillation and an oil and salt by extracting the residue. Significantly, these three methods represent, respectively, the standard method of Dodart/Bourdelin, one of the protocols initially suggested by Duclos, and a method pursued by Borelly. The second method gave fractions that best preserved the properties of the original substance, while the standard fire analysis altered the materials beyond recognition. Homberg's final conclusion not only recapitulates what Duclos had argued long before, but simultaneously undermines Bourdelin's innumerable analyses and Dodart's bold reassurances about their adequacy: "the analyses that employ only a strong fire are not as proper for discovering the true principles and virtues of a plant as when a gentle heat and fermentation assist in the natural separation of the principles which compose the simples."[116] Duclos and Borelly were nearer to the correct path, Bourdelin and Dodart went astray. It was presumably a rather delicate matter for Homberg to present this conclusion, although he did so on a day when Dodart was absent. Fontenelle, in summarizing the paper for the *Histoire*, tried to put the best light on Homberg's conclusion: "it seems then that the Académie thus loses the fruit of the very great number of plant analyses that it has performed, but only the truth matters to the Académie."[117]

Chymists and Chymistry at the Académie in the 1690s

The end of Homberg's activity in this project did not signal the end of his interest in identifying and understanding the chymical principles comprehensively; that undertaking continued to form a central theme of his research for the rest of his career. The next stage in this endeavor, linked to Homberg's goal of writing an entirely new kind of textbook of chymistry, is covered

116. Wilhelm Homberg, "Observations sur les analyses des plantes," *MARS* (1701): 113–17, at 117; PV 20, fols. 209r–214r.

117. Fontenelle, "Sur les analises des plantes," *HARS* (1701): 68–70, at 69.

in chapters 3 and 4. As for the Académie's chymical investigation of plants, it did not come to a halt but instead progressed predominantly along new lines that avoided the use of fire analysis, a development well documented by Holmes.[118] Ironically, Homberg himself discovered in 1697 one means of avoiding at least some of the thermal decomposition that made fire analysis so problematic. Having observed that liquids evaporated more rapidly in his air pump, Homberg set up a small distillation train inside an evacuated container and succeeded in distilling spirit of wine (ethanol) without any applied heat. This experiment represents, to the best of my knowledge, the first reduced-pressure distillation, the method used routinely by organic chemists today to lower boiling points and thus avoid thermal decomposition. Homberg imme-diately recognized the potential of this technique for answering the question of whether or not the chymical principles separated by distillation preexist in the original substance or are "a new formation . . . due to the violent action of the fire."[119] However, the technique was not further deployed, probably due to technical limitations imposed by the air pumps and glassware of the day.

Significantly, new directions in the analysis of plant materials after 1700 came mainly from academicians who had been formed under Homberg's tu-telage or who followed his ideas. Their work can be seen as a continuation of projects that Homberg directly or indirectly "spun off" to them. This obser-vation brings up the question of Homberg's relations and collaborations with other academicians in the 1690s. Although Homberg was by far the most pro-ductive chymist at the institution during the last decade of the seventeenth century, he was not the only chymist present, nor did he always work alone. The presence at the Académie of "junior" chymists guided by more senior members like Homberg also points to ways in which the structure of the in-stitution transformed the means by which chymical practitioners (although in small numbers) could be formed intellectually and practically without ei-ther undergoing the extended travels that Homberg undertook or depending upon the pharmaceutically oriented chymistry lectures given at the Jardin du Roi. This training model became more formalized in 1699 when the Acadé-mie underwent its *renouvellement* (or *règlement*) in which Louis XIV gave it its first written regulations.[120] The new rules sorted the active academicians

118. Holmes, "Analysis by Fire," 139–41; Holmes, *Eighteenth-Century Chemistry*, 68–83.

119. Wilhelm Homberg, "Nouvelle maniere de distiller sans aucun chaleur," PV 16, fols. 126v–129v (15 May 1697), at 129v.

120. The contemporary account of this *renouvellement* (presumably by Fontenelle) and the full text of the *règlementation* appear in *HARS* (1699): 1–16. Historical analyses include Éric Brian, "L'Académie royale des sciences de l'absolutisme à la Révolution," in *Guide des re-cherches*, ed. Brian and Demeulenaere-Douyère, 15–32; Jeanne-Marie Tits-Dieuaide, "L''affec-

into three ranks: *pensionnaires* (salaried members), *associés* (associates), and *élèves* (students). Each pensionnaire was able to choose one élève as an assistant or apprentice, although it must be said that the system did not always work as it was intended; it soon came to be used primarily as a convenient method for obtaining positions for academicians' sons and other relatives. The 1699 regulations also established a fixed number of positions for each of the disciplines represented in the Académie. Significantly, this meant that eight positions—three pensionnaires, two associés, and three élèves—were earmarked for *chimie*, essentially the first time that chymistry had received such clearly established and permanent institutional positions that were not explicitly linked to medical applications.

Almost as soon as Homberg was admitted to the Académie, he began looking for a laboratory assistant. In 1692, he asked a visiting Swedish chymist for help in procuring a skilled operator from Germany. Whether or not he obtained such an assistant in the early 1690s is unknown, although he did have a German laboratory assistant by the first years of the eighteenth century.[121] When Simon Boulduc (1652–1729) was admitted to the Académie in 1694, it was explicitly "to work conjointly with Mr. Homberg on chymistry," indicating that the concept of an associé predates the 1699 Règlement.[122] Boulduc, an apothecary by profession, had been giving the courses of chymistry at the Jardin du Roi since 1687.[123] Over the next several years, Boulduc's presentations to the Académie almost invariably involved projects given to him by Homberg. In most cases, he followed Homberg's instructions to repeat his experiments in order to verify the results. In one instance that relates

tion' de Louis XIV pour l'Académie royale des sciences: sur les raisons d'être du règlement de 1699," in *Règlement, usages et science,* ed. Demeulenaere-Douyère and Brian, 37–50; Sturdy, *Science and Social Context,* 281–96.

121. Erich Odhelius to Urban Hjärne, 6 June 1692, in Carl Christoffer Gjörwell, *Det Swenska Biblioteket,* 2 vols. (Stockholm, 1757), 1:336–39, at 338: "Mons. Homberg har bedt mig, gifwa sig någon anledning in docimasticis och förskaffa sig någon ifrån Tyskland, som wore öfwader til Laborant, hwarutinnan jag wil widare se hwad jag wid min återkomst kan gjöra." Homberg eventually engaged Johann Gross (Jean Grosse) as an assistant; see chapter 5.

122. PV 14, fol. 23r (7 August 1694).

123. On Boulduc, see Paul Dorveaux, "Apothicaires membres de l'Académie Royale des Sciences: III. Simon Boulduc," *Revue d'histoire de la pharmacie* 1 (1930): 5–15; Christian Warolin, "La dynastie des Boulduc: Apothicaires à Paris aux XVIIe et XVIIIe siècles," *Revue d'histoire de la pharmacie* 49 (2001): 333–53, and 50 (2002): 439–50; and Sturdy, *Science and Social Status,* 244–48. Records of payments from 1687 to 1691 for his "cours de chimie" at the Jardin and for the distribution of the medicines made during the course through local parishes and the Capuchins exist in AN, O/1/2124. Boulduc mentions his continuing course and his use of furnaces at the Jardin in PV 17, fols. 230r and 345r (4 June and 28 July 1698).

to the principles separated from plants, Homberg asked Boulduc to measure the amount of volatile acid salt in vinegar. In this case as in others, Homberg kept his own prior results secret to avoid influencing Boulduc's: "Mr. Homberg having made this experiment himself, asked me to do it in order to see if the results would agree; he has not yet told me what his results were." The method Homberg gave Boulduc to use for this determination later formed the basis of one of Homberg's most significant publications—an attempt to compare and standardize acid strengths by measuring the exact amount of an alkali they could neutralize.[124] Homberg also asked Boulduc to carry out a highly labor-intensive recipe, probably acquired during his travels, for making the famous and highly sought-after remedy known as potable gold; Boulduc showed the results to the assembly.[125] It is certainly clear that Boulduc's regular contributions to the Académie during the 1690s were more often than not the result of projects or ideas provided by Homberg.

As an apothecary, Boulduc was more interested than Homberg in pharmaceutical applications arising from the analysis of plants. In 1699, when he and Homberg became equals in the Academie—each named as one of the inaugural *chimistes pensionnaires*—Boulduc set off in his own directions. He embarked on an investigation of purgatives by means of analysis in order "to know in what their virtues consist"—an undertaking very similar to the earlier chymical analysis of plants project. This work, along with the analysis of medicinal spring waters, would occupy Boulduc for the rest of his career at the Académie. He had already begun a particular study of the Brazilian root ipecacuanha, a powerful emetic, in 1697, and this topic formed the basis of his first published memoir in 1700, to be followed by more than a dozen further papers on the analysis of purgatives stretching into the 1720s.[126] Boulduc reiterated how distillation gave unhelpful results, and turned instead to extractions using solvents—not the alkahest Duclos had hoped to use, but the much more readily available water and spirit of wine.

---

124. PV 17, fols. 153v–156r ("Experience sur la quantité de sel acide que peut contenir le vinaigre distilé," 2 April 1698); other examples of Boulduc following Homberg's orders are PV 17, fols. 112r–113v (26 February 1698) and PV 18, fols. 58r–59r (17 December 1698). Wilhelm Homberg, "Observation sur la quantité exacte des sels volatiles acides contenus dans tous les differens esprits acides," *MARS* (1699): 44–51.

125. PV 14, fol. 55r (16 February 1695); Homberg's name does not appear in the PV account, but Duhamel, *Historia*, 369, attributes the process to him without mention of Boulduc. The process is first described by Homberg in VMA MS 130, fols. 151v–153v.

126. PV 18, fol. 142v (21 February 1699); PV 16, fol. 39v (6 March 1697); Simon Boulduc, "Analyse d'Ypecacuanha," *MARS* (1700): 1–5. On Homberg's own work with the material, see *HARS* (1704): 23.

A longer lasting and much closer relationship existed between Homberg and Étienne-François Geoffroy (1672–1731), who was admitted to the Académie as Homberg's élève in 1699. Homberg had previously taught Geoffroy chymistry privately.[127] Geoffroy and Homberg would share many projects and ideas over the following years, and Geoffroy would take Homberg's place as pensionnaire upon the latter's death in 1715. Geoffroy read several of Homberg's papers to the assembly; it is unclear whether he took on this role because Homberg was not feeling well or because of his German accent, although given that Homberg was in the room at these times, the latter seems more likely.[128] As Boulduc had done previously, Geoffroy would repeat or extend many of Homberg's own projects. His first paper, published in the annual *Mémoires* established by the 1699 Règlement, presented and extended observations that Homberg had made on endothermic dissolutions in the 1690s. In regard to the analysis of plants, Geoffroy's first extended project involved the examination of essential salts, the sole remaining category of salts isolated from plants that Homberg had not yet examined himself.[129] Beginning in 1704 Geoffroy would be involved in an acrimonious dispute with Louis Lémery over the presence of iron in plant ashes, a subject that brought together plant analysis and metallic transmutation, the latter a topic Homberg and Geoffroy pursued extensively both together and independently (to be treated at length in chapter 5).[130] Geoffroy also received from Homberg the same recipe for potable gold that Homberg had shared with Boulduc, and he presented a memoir based on it to the Académie in 1713. Several of Geoffroy's other publications very probably contain information gleaned from Hom-

127. On Geoffroy, see Bernard Joly, "Etienne-François Geoffroy (1672–1731), a Chemist on the Frontiers," *Osiris* 29 (2014): 117–31; Olivier Lafont, "Échevins & apothicaires sous Louis XIV: la vie de Matthieu-François Geoffroy, bourgeois de Paris" (Paris: Parmathèmes, 2008), 73–83; Sturdy, *Science and Social Status*, 324–42; Bernard de Fontenelle, "Éloge de M. Geoffroy," *HARS* (1731): 93–100.

128. Bourdelin Diary, fols. 11v (22 August 1699) and 17r (20 February 1700). A similar cause may explain why a paper by the Italian-born Gian Domenico Cassini was read at the 1699 public assembly by his French-born son Jacques; Bourdelin Diary, fol. 3r (29 April 1699).

129. Étienne-François Geoffroy, "Observations sur les dissolutions & sur les fermentations que l'on peut appeler froides," *MARS* (1700): 110–21 (Homberg's previous work is reprised at 116–17, referring to PV 14, fols. 148v–149v, 13 July 1695); PV 18, fols. 143v–144r (21 February 1699).

130. On this dispute, see Bernard Joly, "Quarrels between Etienne-François Geoffroy and Louis Lémery at the Académie Royale des Sciences in the Early Eighteenth Century: Mechanism and Alchemy," in *Chymists and Chymistry: Studies in the History of Alchemy and Early Modern Chemistry*, ed. Lawrence M. Principe (Sagamore Beach, MA: Science History Publications, 2007), 203–14.

berg that the latter had obtained on his travels, for example, details about the production of alum in Sweden.[131] One wonders also to what extent the dissertation *An hominis primordia, vermis* ("Whether a worm is the first state of a human being?"), defended by Claude du Cerf under Geoffroy's direction in 1704, is indebted to Homberg's lost treatise on animal generation, either for its inspiration or for its contents. As it was common practice at the time for the *praeses* to write his students' theses, Geoffroy himself probably wrote the text. The thesis answers the title question affirmatively, as Homberg's treatise had maintained, but without an extant copy of Homberg's text, a closer comparison is not possible. It seems unlikely, however, that Geoffroy (or his student) would have written on this topic without input from Homberg, whose 1680s treatise on generation likewise expounded "the system of worms [that is, spermatozoa]" in animal generation.[132] The examples of Boulduc and Geoffroy indicate Homberg's important role in assisting, directing, and training other chymists at the Académie.

Other less well-known figures also worked on chymical matters, either with Homberg or along lines he proposed, although for various reasons they did not go on to become productive academicians in the way that Boulduc and Geoffroy did. When Geoffroy was promoted to *associé chimiste* in late 1699, leaving the post of Homberg's *élève* vacant, Homberg asked for Claude Berger (1679–1712) to be switched into the chymical division from the botanical, where he had been working on plants under Tournefort.[133] This slightly peculiar request may have been a way for Homberg finally to free himself from the plants project by giving what remained to someone else, since his interests had by then moved to other issues. Berger turned out, however, to be more interested in pursuing medical practice and connections at court rather than chymistry or participation in the life of the Académie. He attended

131. Étienne-François Geoffroy, "Des Teintures des métaux et particuliérement des teintures d'or," PV 32 (15 March 1713); "Manière de faire l'alun de roche," PV 21, fols. 11r–19v, at 18r–19r.

132. Claude du Cerf, *An hominis primordia, vermis?* (Paris, 1704); Jacques Roger, *The Life Sciences in Eighteenth-Century French Thought*, ed. Keith Benson, trans. Robert Ellrich (Stanford, CA: Stanford University Press, 1997), 135; Fontenelle, "Éloge de M. Geoffroy," *HARS* (1731): 93–100, at 96. On early modern dissertations and their authorship, see Kevin (Ku-Ming) Chang, "From Oral Disputation to Written Text: The Transformation of the Dissertation in Early Modern Europe," *History of Universities* 19 (2004): 129–87.

133. PV 19, fol. 15r (20 January 1700). On Berger, see Sturdy, *Science and Social Status*, 314–20; prior to Sturdy, little was known of Berger beyond his strikingly short éloge by Fontenelle, *HARS* (1712): 81–82; he published no papers in *MARS* and is mentioned only once in the *HARS* (1704): 35.

Académie meetings only sporadically and made virtually no contributions. His admission to the institution was probably based more on his status as a protégé of the king's powerful physician, Guy-Crescent Fagon (1638–1718), than on any natural philosophical aptitude or interest. Homberg, therefore, benefited not at all from this nominal élève, nor did the élève advance in the institution. Equally obscure is Adrien Tuillier, a physician who was admitted initially as Bourdelin's élève in early 1699. He planned to revisit the analysis of plants using the methods of fermentation advocated and practiced by Borelly in the 1680s, and to tackle the persistent problem of "rendering the work more useful to medicine," a goal Homberg had clearly given up on many years earlier. Tuillier's project likewise came to naught. When Bourdelin died eight months later, in October 1699, Tuillier became Nicolas Lémery's élève, but in early 1702 Tuillier was sent to attend the war-wounded in a hospital in Kaiserswerth (near Düsseldorf), where he died of fever within four months. He has the distinction of receiving the shortest éloge Fontenelle ever wrote, fewer than two hundred words.[134]

To complete the array of members working on chymical matters at the Académie in the last decade of the seventeenth century (see figure 2.6), mention must be made of three more figures. The eldest of these was Moyse Charas (1619–98), admitted to the Académie in 1692, shortly after his seventy-third birthday and just after his return to France following more than a decade abroad in the Netherlands, England, and Spain.[135] The oddity of this selection has led Sturdy to suggest that it was recompense for clandestine diplomatic activities carried out for the French crown while Charas was practicing medicine abroad, particularly in Spain, where he was in contact with members of the royal family. If so, his would not be the only such case among

134. PV 18, fol. 144r (21 February 1699); Bernard de Fontenelle, "Éloge de M. Thuillier," *HARS* (1704): 139. His last appearance at the Académie was 18 February 1702 (PV 21, fol. 75r), and his death is noted on 14 June (PV 21, fol. 247r); see also fol. 279r.

135. On Charas, see Sturdy, *Science and Social Status*, 254–57; Fred W. Felix, "Moyse Charas, maître apothicaire et docteur en médecine," *Revue d'histoire de la pharmacie* 90 (2002): 63–80; Paul Dorveaux, "Apothicaires membres de l'Académie Royale des Sciences: II. Moyse Charas," *Bulletin de la Société d'Histoire de la Pharmacie* 17 (1929): 329–40, 377–90; Bruno Bonnemain, "Moyse Charas, un maître apothicaire et docteur en médecine emblématique de son époque," *Revue de l'histoire de la pharmacie* 64 (2016): 405–18; Michel Bougard, *La chimie de Nicolas Lémery* (Turnhout: Brepols, 1999), 25–26; *Mercure galant* (February 1698): 122–40. I have identified an original and previously unknown memoir ("Teinture de soufre") that Charas read to the Académie, now at Uppsala, Uppsala University, Carolina Rediviva, Waller MS fr-01794; it is endorsed "M. Charas, lu 1er de decembre 1694" in the hand of the Académie's secretary Duhamel.

members of the Académie; Nicolas Hartsoeker, who was made foreign associé in 1699, performed similar offices in the Netherlands for Louis XIV.[136]
The post may also have been a "reward" for Charas's public confirmation of
his conversion to Catholicism in 1691. Whatever the case, Charas's chymical
interests, particularly in relation to pharmacy, are clear from his earlier position as lecturer at the Jardin du Roi (1672–80) and his publications, especially
the *Pharmacopée royale, galenique et chymique* (1676). He also claimed to have
written much or all of the *Traité de la chymie* that appeared in 1663 under the
name of Christophle Glaser. Charas contributed regularly to the Académie's
meetings in the 1690s. He provided four papers for the 1692–93 *Memoires*:
two on the chymical causes of hot and cold springs, one on a preparation
of cinchona for fevers, and one on being bitten by a viper.[137] This last paper
was part of his long-term study of snakebite, in which he made vipers bite
pigeons, chickens, frogs, cats, dogs, other snakes, and evidently (although
presumably by accident?) himself, for the purpose of testing improvements
upon the classical antidote theriac; he had published on this topic already
in the 1660s. Both the anatomist Joseph-Guichard Duverney (1648–1730)
and Homberg apparently examined some of these trials for the Académie.[138]
Charas also showed interest in some famous Helmontian arcana, speaking to
the Académie about volatilized salt of tartar and *usnea*, the moss that grows
on the skulls of those who have died a violent death and which is needed to
prepare the famous weapon-salve that cures wounds at a distance.[139] Several
of his other contributions deal with salts; his dissertation on the topic shows
the influence of both Van Helmont's writings and Homberg's recent experiments, yet remains distinct from both.[140]

136. Alice Stroup, "Nicolas Hartsoeker, savant hollandais associé de l'Académie et espion
de Louis XIV," in *De la diffusion des sciences à l'espionnage industriel, XVe–XXe siècle*, ed. André
Guillerme (Lyon: ENS Éditions, 1999), 205–28.

137. *MARS* (31 May 1692): 82–86; (30 November 1692): 155–58; (31 March 1693): 47–48,
and (31 January 1693): 9–16; reprinted in *HMARS 1666–99*, 10:92–98, 183–87, 288–90, 244–45.
Dorveaux, "Charas," 379–84, usefully lists Charas's presentations to the Académie. It does not
include the newly-discovered memoir cited in note 135 since there is a gap in the PV for several
weeks in November–December 1694.

138. Moyse Charas, *Nouvelles experiences sur la vipère* (Paris, 1669); PV 13, fols. 109v–112v,
117v, 119v (13 and 20 August, 6 September, 19 November 1692).

139. PV 13, fols. 99v and 101v (21 May and 4 June 1692) and fol. 130r (28 March 1693); Charas,
"Le sel volatile de tartre," AdS, pochette de séances 1692. On usnea and the weapon-salve see
Mark Waddell, *Jesuit Science and the End of Nature's Secrets* (Burlington, VT: Ashgate, 2015),
37–51, especially 43–44.

140. PV 14, fols. 98v–99r (27 April 1695) and 266r–271v ("Sur le sel ammoniac des anciens
et sur celuy des modernes," 15 February 1696).

A further chymical worker of interest during the 1690s was Morin de Tou-lon; unfortunately, the historical record on him is not only scanty but also con-fused. He was admitted to the Académie at the end of 1693, at which time he was described as a medical doctor; nevertheless, he spoke initially about iron ore and embarked on a project about minerals.[141] Martin Lister met him in 1698, referred to him as "a Man very curious in Minerals," viewed his mineral collec-tion, and heard from him details of porcelain manufacture at Saint-Cloud.[142] Such expertise connects well with the memoir he presented in January 1694 on Chinese porcelain and the clays used for making it. In the same month he presented a paper about a blue mineral he had found in the Auvergne in 1688 and its uses for medicine and painting; he signs the autograph document as "escuyer med[ecin]."[143] He presented a lengthy memoir on salts and vitriol over three assemblies in 1697, which was apparently part of a longer treatise on minerals. The memoir's text was not entered into the procès-verbaux, and the larger treatise, if written, has apparently not survived, although it is clear that he performed experiments for it throughout the period from 1694 to 1697, since he was reimbursed 600 livres for such work.[144] Despite his clear and exclusive devotion to mineralogy—a unique topic of study in the Académie during the 1690s, and one that certainly places him among the chymists—he is almost always labeled a botanist in the secondary literature, presumably following the 1733 list of academicians in which the scanty entry for Morin fails to provide either his first name or his dates. Only Condorcet, writing in 1773, reports that Morin was admitted as a chymist and that he had particular interests in min-eralogy and "promised some chymical works on the metals."[145] Condorcet's

141. PV 14, fol. 5r–v (16 December 1693) and 8r (17 February 1694).

142. Lister, *Journey*, 79–80, 141–42; see also 244–45, where *two* Morins, one of whom is probably the one here in question, are listed as physicians to princes of the blood. Lister's ac-count at 141–42 is slightly telegraphic, making it seem that Morin was the "Ingenious Master" of the porcelain works at Saint-Cloud, but this conclusion has been variously refuted since no Morin appears in the records of the factory; his first name is sometimes given as François, but this too is uncertain.

143. Morin de Toulon, "Observations sur l'azur des Cendres bleuës," AdS, pochette de séances 1694 (formerly classed in Dossier Morin de Toulon). A condensed version of this mem-oir exists in the same pochette; both are dated 23 January 1694. The procès-verbaux for January 1694 are missing (they should fall between fols. 5v and 7r of volume 14), but both memoirs are summarized by Duhamel, *Historia*, 338–39, and the one on porcelain by Fontenelle, *HMARS 1666–99*, 2:205.

144. PV 16, fols. 173v, 201v, 219v (19 June, 24 July, 7 August 1697); Stroup, *Royal Funding*, 142, see also 36, 46, 49.

145. *Index biographique*, 386–87; "Liste de Messieurs de l'Académie Royale des Sciences," *HMARS 1666–99*, 2:373, 380; Condorcet, *Éloges*, 162.

assignment agrees far better with the scattered references to Morin's work in the procès-verbaux. Morin de Toulon's labeling as a botanist apparently stems from his assignment to the botany section at the time of the 1699 renouvellement when both he and the newly appointed Louis Morin de St. Victor (1635–1715) became the two *botanistes associés*. But only one of these Morins—Louis Morin de St. Victor—gave the obligatory report in February 1699 outlining his projects for the coming year, and only he appears thereafter in the records; Morin de Toulon appears in the procès-verbaux for the last time in August 1698.[146] It remains unclear how to resolve this puzzle. Perhaps Morin de Toulon was shifted to botany to leave the position of associé chimiste open for the even more obscure (and less productive) Langlade, who had become "associé externe" in late 1698, although it remains unclear what he had done to deserve that honor or to be placed as chimiste associé rather than Morin in the reorganized Académie (and he did nothing afterward to justify his selection). One suspects some unspecified royal recompense for "services rendered" that may have little or nothing to do with the Académie's purpose, which seems to have been a common practice under Louis XIV.[147] Equally difficult to determine is what became of Morin de Toulon after his last appearance at the Académie in 1698. His death in February 1707 is recorded in Bourdelin the younger's diary, and his place of botaniste associé was thereafter filled by Pierre-Jean-Baptiste Chomel, although there is a contemporaneous reference to "the difficulty of finding" a person appropriately devoted to botany.[148] Whatever the case may be, Morin de Toulon's keen interest in minerals and metals during the 1690s would have rendered him a natural colleague and potential collaborator for Homberg, who shared these interests, and it is possible that they worked side by side in the chymical laboratory at the Bibliothèque du Roi, but we currently have no information in this regard.

The final member to mention is the well-known Nicolas Lémery (1645–

146. Pontchartrain's letter is PV 18, fol. 115r (28 January 1699), and the botanist Morin's project report is PV 18, fol. 145r. Morin de Toulon's last recorded appearance at the Académie was 23 August 1698 (PV 17, fol. 364r) when he exhibited mineral specimens. Morin's previous appearances include showing an African marchasite and a fossilized rib found in Montmartre gypsum (PV 14, fols. 11r and 12v–13r, 17 March and 24 April 1694).

147. PV 18, fols. 13v–14r (19 November 1698). Langlade's attendance was spotty at best, and he made no contributions to the Académie. Virtually nothing is known about him save that he was physician to the Cardinal de Bouillon at the time of his election, and after 1712, first physician to the queen of Spain (PV 31, fol. 261r). He may be the same Langlade who sent two bottles of mineral water from Périgord in 1675 to be analyzed at the Académie by Duclos; see AdS, Bourdelin Notebooks, 4:123r. His death is noted at PV 36, fol. 172v (10 July 1717).

148. Bourdelin Diary, fol. 82v (23 February 1707); PV 26, fol. 85r (12 March 1707). The *Index biographique* gives Auvergne as the place of his death, but without citing a source.

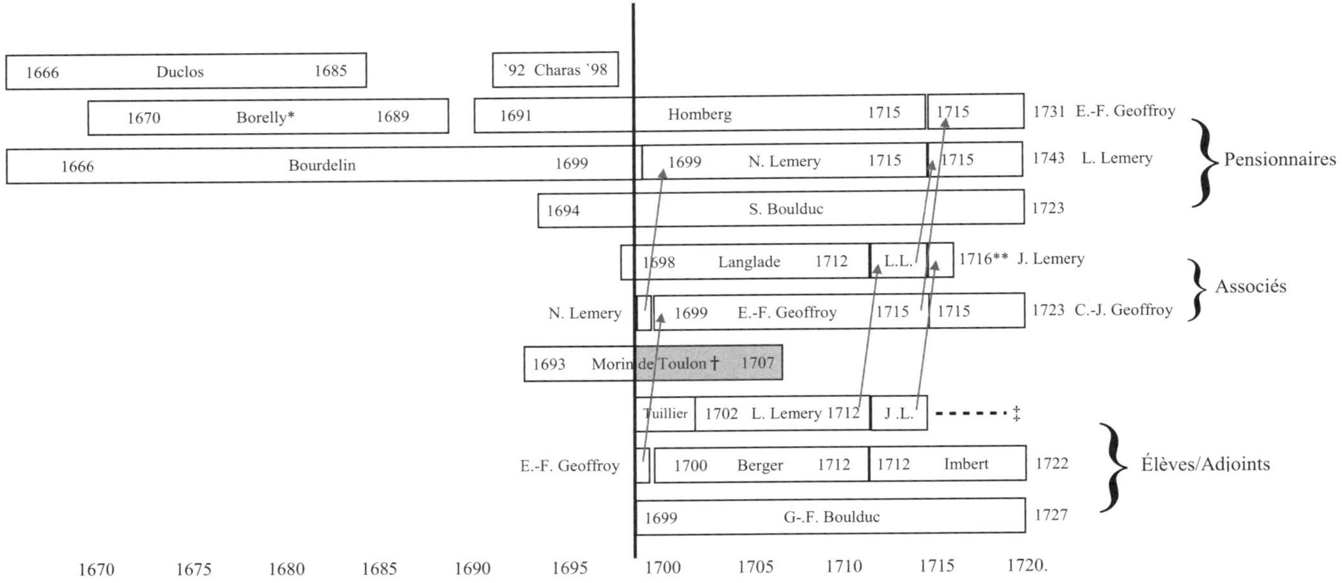

FIGURE 2.6. Chymists at the Académie Royale des Sciences, 1666–1720.

*The accepted date for Borelly's admission is 1674, but there is clear evidence that he had been admitted already in 1670 (see p. 77n28).

**After Jean Lemery died in 1716, the position was left vacant until 1721 (see p. 408).

†At the 1699 *renouvellement*, Morin de Toulon was made *botaniste associé*, although all his previous work had been in chymistry.

‡Following the *règlement* of 1716, the rank of three *élèves* was replaced with that of two *adjoints*: G.-F. Boulduc and Imbert became the first two *adjoints chimistes* (see p. 362).

1715), appointed chimiste associé in 1699, and elevated to pensionnaire in the same year, following the death of Bourdelin. Lémery, an apothecary by profession, had become famous particularly for his *Cours de chymie*, which by 1699 had reached its ninth edition in French and had been translated into multiple languages, as well as for the regular lectures on chymistry he gave at his house on rue Galande. Judging the relationship between Lémery and Homberg, who were pensionnaire chymists together for fifteen years, proves rather complicated. Prior to the entry of either into the Académie, Homberg had conveyed to Lémery a lengthy report detailing his researches on the Bologna Stone, which the latter inserted in the 1690 edition of his *Cours*. This sharing of information might suggest a fairly close relationship between the two, a notion presented as historical fact in some earlier, and actually quite fanciful, biographical treatments of Homberg and Lémery, but for which there is no evidence.[149] On the contrary, there is clear evidence of clashes between the two on multiple occasions at the Académie. Several centered on claims relating to transmutation—a topic Lémery famously derided and that Homberg supported (see chapter 5). Other clashes involved issues of chymical theory or experiment. For example, Homberg argued against those who claimed that the shape of acid microparticles is deducible from the crystal shapes of the salts from which they are distilled. While Homberg leaves such persons unnamed, it was in fact Lémery who made this claim in the second (and all following) editions of his *Cours*.[150] Likewise, Homberg pointedly expressed skepticism about explanations of chymical behavior that relied on microparticle shape, a hallmark of the "mechanical philosophy" of the late seventeenth century, and for which Lémery was (and remains) well known. For Homberg, such explanations had "but a weak semblance of truth," and he preferred to rely on experimentally measurable factors like the amount and concentration of reactive materials.[151] The diary of Académie meetings kept by Claude Bourdelin the younger, which preserves many details (especially disagreements) that were kept out of the sanitized procès-verbaux, records more arguments between Homberg and Lémery than between any other pair of academicians. They argued over the means to purify saltpeter, the utility of various cures, the composition of aquafort, and the composition of metals, and engaged in a "grande dispute" over the properties of mercury,

149. The worst of these is Paul-Antoine Cap, *Études biographiques pour servir à l'histoire des sciences*, 2 vols. (Paris, 1857–64), 1:180–226, 2:214–32.

150. Wilhelm Homberg, "Observations sur la quantité d'acides absorbées par les alcalis terreux," *MARS* (1700): 64–71, at 67; Nicolas Lémery, *Cours de chimie*, 2d ed. (Paris, 1677), 22.

151. Wilhelm Homberg, "Observations sur la quantité exacte des sels volatiles acides," *MARS* (1699): 44–51, at 50.

among many other things.[152] It is clear that whatever their personal relationship, they disagreed significantly on many points both specific and general about the content and goals of chymistry.

On a larger scale, Lémery's current place in the history of chymistry needs to be reevaluated. I have referred elsewhere to his lionization by twentieth-century historians, based on their perception of his chymical system as "Cartesian," largely in reference to his invocation of corpuscular shapes to explain chymical phenomena—for example, basing the corrosivity of acids on the pointed shape of their particles. Such explanations made Lémery seem "mechanical," and hence "modern and rational," and thus a break from a preceding chymical tradition that was supposedly obscure, lacking in sound explanatory principles, or crudely vitalistic. Fontenelle himself presented Lémery in this way, and Fontenelle's rhetoric undoubtedly played an important role in constructing the perspectives of later historians.[153] Yet this perspective has been steadily undermined by numerous recent studies. In the first place, the importance of such putative Cartesianism within chymistry was based upon a misguided attempt to make the history of science parallel that of philosophy. Indeed, the category of "Cartesian" chymistry has been largely dismantled by recent scholarship, and more plausible and immediate sources for Lémery's conceptions have been located in the writings of contemporaneous fellow apothecaries, such as Otto Tachenius and François de Saint-André, thus making a connection with Descartes not only superfluous but also unlikely.[154] On a broader scale, the notion of Lémery as either a coherent or an original chymical theorizer cannot withstand close inspection. Lémery was certainly extraordinarily successful as a lecturer and a popularizer—but predominantly of other people's ideas. As others published new chymical discoveries and concepts, Lémery spliced them into new and constantly expanding editions of his *Cours*, while carrying over earlier material without significant revision. Part of this editorial activity was provoked by the constant pirating of the *Cours*, which meant that Lémery was obliged constantly to generate new editions incorporating new material.[155] As a result,

152. Bourdelin Diary, fols. 30v, 35r, 45r, 46v.

153. Lawrence M. Principe, "A Revolution Nobody Noticed? Changes in Early Eighteenth-Century Chymistry," in *New Narratives in Eighteenth-Century Chemistry*, ed. Principe (Dordrecht: Kluwer, 2007), 1–22; Bernard de Fontenelle, "Éloge de M. Lémery," *HARS* (1715): 73–82.

154. Bernard Joly, *Descartes et la chimie* (Paris: Vrin, 2011); and Joly, "L'anti-Newtonianisme dans la chimie française au début du XVIIIe siècle," *Archives internationales d'histoire des sciences* 53 (2003): 213–24, especially 215–16; Principe, "Revolution?" 4–6.

155. I am grateful to James Voelkel for sharing results with me from his on-going study of the sequential pirating of Lémery's *Cours*.

many of the book's statements and explanations are inconsistent and sometimes directly contradictory, as is the case especially in the opening chapters
of later editions, concerning fundamental principles. A further analysis of the
changes between editions (usefully begun by Bougard) and a further identification of Lémery's sources would prove illuminating, but lie beyond the
scope of this study.[156] It suffices here to suggest that Lémery's prowess as a
chymical theorist and investigator (although not as a popularizer) has been
exaggerated, and that Fontenelle's praise of him as an innovator has been
adopted too uncritically by later historians. Evaluations by contemporaneous academicians themselves support the notion of a Lémery whose chymical contributions were relatively mediocre. For example, when Lémery read
parts of the manuscript that would become his *Traité de l'antimoine* to the
assembly, "all of his experiments were found not in the least new; Mr. Homberg has already done most of them." In response to an analysis of a mineral
water, Bourdelin *fils* remarked critically that Lémery was still ignorant of the
standard instrument for measuring the density of liquids.[157] Moyse Charas
criticized him for coopting parts of the *Pharmacopée royale* into a new edition
of the *Cours de chymie*.[158] In another instance, Lémery made an erroneous
statement regarding the behavior of mercury sublimate as an indicator, leading to an argument in which Tournefort retorted rather tartly that if Lémery
were correct, "it would overturn much of what the Académie has done in
regard to analysis."[159]

Homberg's Vision of Chymistry

From the foregoing it is clear that there were several chymists at work in the
Académie throughout the 1690s and into the first decade of the eighteenth
century, their number reaching a maximum after the 1699 renouvellement
that guaranteed eight seats for chymists (fig. 2.6). It is worth noting, however,
that the majority of them focused predominantly on chymistry's application
to pharmacy—a major part of early modern chymistry that ended up fossilized into the archaic yet persistent British usage of "chemist" to mean "phar-
macist." The elder Bourdelin, the two Boulducs (Simon and his son Gilles-
François), Charas, Geoffroy, and Lémery were all apothecaries by training

---

156. Bougard, *Nicolas Lémery*, especially 404–16.

157. Bourdelin Diary, fol. 1v (22 March 1699): "toutes ses experiences ne furent point trou-
vées nouvelles. Mr. Hombert les avoit faites les pluspart"; 18v (3 March 1700): "Il paroissoit qu'il
n'avoit pas encore l'usage de l'areometre ou Peze-liqueur."

158. Charas, "Sel volatil."

159. Bourdelin Diary, fol. 2v (11 April 1699) .

and/or profession. Charas and both Boulducs taught the *cours de chimie* at the Jardin du Roi at various times (as had Borelly), which dealt largely with pharmaceutical preparations, as does Lémery's famous *Cours de chymie*. Of these six, only Geoffroy—Homberg's associé and closest colleague—worked extensively on topics unrelated to pharmacy. The more obscure Tuillier, Langlade, and Morin de Toulon were known primarily as physicians; the first two made essentially no contributions to the Académie, and only Morin devoted himself to studies outside of medicine, but only, apparently, for a short time in the 1690s. Homberg stands in stark contrast to all these others. While he certainly did make comments about pharmaceutical applications, such deployments of chymistry constituted only one facet of his studies, and remained ancillary to his central endeavor of exploring and understanding the chymical nature and transformations of matter. Homberg's relatively minor interest in pharmaceutical chymistry appears in a remarkable statement about the potable gold process he shared with Boulduc and Geoffroy. Although the entire point of creating potable gold was to obtain a powerful medicine—perhaps even a true panacea—after Homberg actually succeeded in preparing such a fabled and sought-after material, he concludes his description of it with the stunning statement, "I do not know what effect this tincture of gold has upon the body of man; I've never tried it."[160] Homberg instead focused upon what the process taught about the chymical properties of gold—whether it could be rendered water-soluble without the use of corrosives like aqua regia—not in the possible medicinal application of this product for which it was so famed and widely pursued. Homberg thus stands in the same tradition as Duclos before him—that is, as one capable of performing the functions of a physician (unlike Homberg, Duclos held an MD degree), but choosing instead the pursuit of chymistry, and predominantly along natural philosophical lines. Their similarity is reinforced by their analogous desires to direct the plants project away from practical pharmaceutical applications and toward revealing the hidden chymical workings and composition of nature. Thus, Homberg presents a clear and early example of the pursuit of a chymistry liberated from subservience or restricted application to medicine and pharmacy. Homberg declared already in the early 1690s that "the objects of chymistry are of infinite extent."[161] Homberg not only envi-

160. Homberg, VMA MS 130, fol. 153v: "je ne scay pas quel effet cette teinture d'or fait dans le corps de lhomme je n'en ay Jamais fait d'essay." In contrast, Boerhaave copied this process specifically from Homberg's manuscript and translated it into Latin on a loose piece of paper preserved in VMA MS 130.

161. Homberg, VMA MS 130, fol. 112v: "les objets de la chimie sont d'une etendue infinie."

sioned a far broader scope and role for chymistry, but also charted a course akin to the way the field would develop later in the eighteenth century—at the Académie and elsewhere—as a key part of scientific investigations into the natural world and its transformations. Thus, for all their interest in what we might today consider the "ancient arcana" of premodern chymistry, such as the alkahest and metallic transmutation, Duclos and Homberg actually had a more "modern" view of the place and potential of the discipline than did their colleagues at the Académie (like Lémery) who rejected these arcana and shackled *la chimie* to pharmacy.[162]

The breadth and independence of chymistry were not fully appreciated at this point in the Académie's history (or, more broadly speaking, elsewhere). As noted above, most of those filling the ranks of *les chimistes* applied themselves to medicine and/or pharmacy, since they were physicians or pharmacists themselves. The same narrow perception of chymistry characterizes the contemporaneous views of Bernard de Fontenelle, who had become perpetual secretary of the institution in 1697. In his lengthy 1699 essay on the utility of the sciences, Fontenelle mentions *la chimie* only once, and even then as no more than an adjunct to medicine. More bluntly, when summarizing a highly sophisticated paper that Homberg published in 1700, Fontenelle misses—or ignores—the paper's whole point about standardizing and accurately quantifying chymical reagents and instead focuses on an ancillary comment about using ground oyster shells as an antacid, declaring pompously that "it is principally to these sorts of [medical] uses that all chymical discoveries ought to be turned."[163] For Homberg's part, his promotion of a grander view of the scope and independence of chymistry as a natural philosophical discipline would only increase throughout the following years. As Boyle had tried to do earlier in England, Homberg endeavored to liberate chymistry from a service role and promote it to the rank of an

162. I am uncertain how Jacques Borelly fits into this scheme because so little is currently known of him. I suspect that although he was trained as a physician, he nevertheless had a broader view of chymistry's domain, like Duclos and Homberg. Before his admission to the Académie, he had his own laboratory where he worked on metallic chymistry and "nostoc" (perhaps as a route to the philosophers' stone), invented a new type of furnace, and participated in the Bourdelot Academy; see Borrichius, *Itinerarium*, 3:336–67; *Oldenburg Correspondence*, 5:507–9. While he is recognized principally for his participation in the plants project, he also wrote largely overlooked memoirs on methods for metal smelting and assaying that signal his broad chymical interests; see BNF, Clairambault MS 452, fols. 544r–552r.

163. *Oeuvres diverses de M. de Fontenelle*, 3 vols. (Paris, 1724), 1:1–35 (not paginated), "Sur l'utilité des mathematiques et de la physique," at sig. Aiiiiv; *HARS* (1700): 50, summarizing Homberg, "Observations sur la quantité d'acides absorbés par les alcalis terreux," *MARS* (1700): 64–71.

independent and powerful scientific discipline. In fact, Homberg would go much further. In succeeding years he would argue for chymistry as the most powerful of all the sciences, and seize several topics traditionally classed under physics to make them integral parts of chymistry, as will become clear in chapters 3 and 4.

The "infinite extent" of chymistry, and Homberg's diverse explorations of that vast domain, are well illustrated by the sheer range of his endeavors at the Académie during the 1690s. Homberg's work with white phosphorus, the air pump, and his long-term program of testing Helmontian chymical theory and identifying the chymical principles were far from his only activities. He pursued a wide variety of projects, many of which continued interests he had acquired and developed during his *Wanderjahre*. In early 1692 he told the Académie that one of his main projects would be the study of the mineral antimony, and accordingly he presented observations and experiments about antimony to the assembly on sixteen occasions between 1692 and 1696.[164] Again unlike Lémery, who would also study antimonial preparations in the following decade but primarily for their pharmaceutical value, Homberg focused on the chymical properties of the material and only occasionally made reference to its pharmaceutical deployments. Homberg also busied himself with metallurgy, refining, and assaying, presenting reports on the production of brass, the gilding of iron, the refining of silver, the cupellation of gold, and the improvement of the color of gold alloys by surface enrichment.[165] Such work on assaying and refining was probably destined for his treatise *L'art des essayeurs*; it was one of three complete but unpublished treatises left behind by Homberg that his nephew, Denis Dodart (the younger), planned to pub-

---

164. PV 13, fol. 81v (27 February 1692). Examples of Homberg's work on antimony include PV 13, fol. 102v (ways of making regulus, 18 June 1692) and fol. 136v (making antimony cathartic, 1 July 1693); PV 13, fol. 140v (an infusion of antimony, 12 August 1693), published in *MARS* (30 November 1693):150–52, and reprinted in *HMARS 1666–99*, 10:403–6 (with an engraving in the monthly issues); "Suite des observations sur une infusion d'antimoine," AdS, pochette de séances 1694, compare PV 14, fol. 8v (27 February 1694) and fol. 204r (mercurification of antimony, 7 December 1695); and PV 15, fols. 23r–26r ("Observations sur le verre de l'antimoine," 11 April 1696).

165. PV 14, fols. 110r–112v ("Maniere de faire du Latton sans Zink et sans Calamine," 11 May 1695); PV 15, fols. 8r–11r ("Observations sur la dorure du fer," 28 March 1696), fols. 170r–171v ("De la haute Couleur que l'on peut donner à l'or bas," 13 August 1695), fols. 272r–274v ("Observations sur le raffinage de l'argent," 16 January 1697), fols. 295r–296r ("Suite des observations sur le raffinage de l'argent," 30 January 1697); and PV 16, fols. 8r–10v ("Seconde suite des observations sur le raffinage de l'argent," 20 February 1697); and PV 13, fols. 126r and 127r (21 and 28 January 1693), published as "Reflexions sur un fait extraordinaire arrivé dans une Coupelle d'or," *MARS* (15 December 1693): 172–76, reprinted in *HMARS 1666–99*, 10:427–33.

lish in 1724. Unfortunately, these promised publications never appeared, and Homberg's *L'art des essayeurs*, like his early treatise on animal generation, is now lost.[166]

Homberg also pursued his long-standing fascination with the interaction of light with matter, providing two lengthy memoirs on the Bologna Stone, demonstrating the *phosphorus smaragdinus* (the thermoluminescent mineral fluorspar), and discovering an entirely new light-emitting material (triboluminescent fused calcium chloride), soon thereafter christened "Homberg's phosphorus." He also toyed with the light sensitivity of silver salts, making a bone box look like black marble by treating it with a solution of silver nitrate and then exposing it to sunlight.[167] Two other intensively researched projects involved the production of artificial gems and the study of sympathetic inks. No publications resulted from these researches, even though it is clear that Homberg wrote, and intended to publish, a substantial paper on the latter topic. Why it was never published is unknown, and the document is now lost.[168] He also contributed a collection of memoirs on dyeing.[169] At other times he was asked to analyze some "febrifuge balls" seized from an English ship returning from the East Indies and taken by the French in the Manche, as well as to collaborate with his colleagues in addressing and remediating difficulties in casting type at the royal printing house and improving

166. Denis Dodart to Montesquieu, 28 December 1723, in *Oeuvres complètes de Montesquieu*, 19 vols. (Oxford: Voltaire Foundation, 1998), 18:73–75, at 74.

167. AdS, pochette de séances 1694, "Experiences sur la piere de bologne" (12 May 1694); PV 17, fols. 126v–129r ("Observations sur la pierre de Bologne," 19 Mars 1698). See also Lawrence M. Principe, "Chymical Exotica in the Seventeenth Century, or, How to Make the Bologna Stone," *Ambix* 63 (2016): 118–44. PV 14, fol. 12r (14 April 1694, published as Wilhelm Homberg, "Nouveau phosphore," *MARS* [31 December 1693]: 187–91, reprinted in *HMARS 1666–99*, 10: 445–48), fol. 14r (*phosphorus smaragdinus*, 8 May 1694), fol. 24v (ivory box, 4 September 1694); and PV 15, fols. 222r–225r ("Observations sur le nouveau phosphore," 5 December 1696). See also E. Newton Harvey, *A History of Luminescence* (Philadelphia: American Philosophical Society, 1957), 365–77, 381.

168. PV 15, fols. 74r–78v and 80r–82v ("Des pierres factices," 16 May 1696 and "Suite du Memoire sur les Pierres de couleur factice," 30 May 1696); PV 17, fols. 164r–168v ("Observations de l'encre sympathique," 9 April 1698), 182r–185v ("Suite des Encres de sympathie," 23 April 1698), 195v–199r ("Suite des Encres de sympathie," 7 May 1698), 214r–v ("Addition à mon dernier mémoire des Encres simpathiques," 14 May 1698), 238r ("Suite des Encres simpathique," 11 June 1698), 316r–322r ("Suite des Encres simpathiques," 16 July 1698); and PV 18, fol. 142r (21 February 1699): "Il achevera ses observations sur les Encres simpathiques afin qu'elles puissent etre imprimées a la fin de l'année."

169. PV 14, fols. 79v–80v and 88r–89v (23 March and 13 April 1695); PV 15, fols. 97v–100r (20 June 1696).

the manufacture of gunpowder.[170] He even shared with the Académie some "homely" recipes, such as a black polish to keep iron from rusting and a way of mending broken porcelain with a varnish-based glue.[171] Finally, in addition to these chymical projects, he collaborated briefly with Philippe de la Hire and Gian Domenico Cassini on studies of the magnet, and continued to display his mechanical ingenuity, making various instruments (lenses, a device for copying bas-reliefs, and one for measuring refraction) and building a mechanical eye to demonstrate that the movements of the eye required only four muscles.[172] Perhaps in the same mechanical vein, Homberg shared a payment with a painter, carpenter, and ironworker "for pieces of work supplied to the King" in 1692. What such work might have been remains unknown; perhaps a clever and "marvelous" device, perhaps stage machinery for the royal theater?[173]

Wilhelm Homberg thus covered and contributed to virtually every aspect and application of late seventeenth-century chymistry: analytical, metallurgical, commercial, artisanal, pharmaceutical, and productive. Yet amid this diversity of interests and activities, throughout the 1690s he nevertheless focused his most sustained and most inventive efforts on trying to uncover the chymical principles and to devise a coherent theory of chymistry. In this regard he stood out from his Académie colleagues who focused their efforts on the pharmaceutical applications of chymistry. Already at this still early stage of his academic career, Homberg's view of chymistry and its rightful place within natural philosophy, was far broader. In the following decade he would continue many of his diverse interests, while two overriding and not

170. PV 15, fols. 43r–46r (2 May 1696), "Analyse des Boules appellées febrifuges, dont il s'est trouvé une caisse pleine dans un vaisseau pris sur les Anglois venant des Indes Orientales"; PV 18, fols. 79r–81v (14 January 1699), "Essays pour corriger la matiere des lettres de l'imprimerie"; PV 22, fols. 295r and 300r–302r (29 August 1703), "Sur les poudres."

171. PV 14, fol. 122r (1 June 1695); PV 18, fol. 175v (18 March 1699).

172. On the collaboration see PV 13, fol. 91r–v (16–19 April 1692); see also *HMARS 1666–99*, 10:164–71. The first memoir Homberg contributed for inclusion in the monthly *MARS* described a new method for grinding lenses (PV 13, fol. 72v; 22 December 1691); it was not published, perhaps having been judged too artisanal. PV 14, fol. 20v (7 July 1694): "d'un tour a copier des bas reliefs," and fol. 130r (11 June 1695): "Mr. Homberg a fait voir un oeil artificiel avec quatre petits filets, qui donnent tous les mouvemens naturels de l'oeil, ensorte qu'il semble que les quatre muscles de l'oeil peuvent donner tous les mouvements; les deux obliques peuvent aider les autres." He also designed an instrument for more conveniently removing cataracts: PV 14, fols. 208r–209v ("Aiguille pour abattre les Cataractes de l'invention de M. Homberg," 10 December 1695).

173. Cited in Stroup, *Royal Funding*, 55.

unrelated desires would take center stage: an intensification of his endeavor to identify the chymical principles in order to create a comprehensive system of chymistry, and renewed efforts to solve the ancient problem of metallic transmutation. In that first decade of the eighteenth century, Homberg would acquire an international reputation, have his experiments followed closely and repeated elsewhere in Europe, and engage in learned disputes with a range of savants. His social position in the French society he had adopted would also change dramatically, bringing him into close contact with members of the royal family, an association that would enormously enhance his ability to conduct chymical research.

3

Essaying Chymistry

By the end of 1701, Homberg had concluded his experimental program of studying the chymical principles using the results from plant analyses. He had not neglected the mineral and metallic realm while fulfilling (in fact exceeding) Bignon's request to examine the plants project. Now, armed with a decade's worth of results, along with what he had acquired before joining the Académie, Homberg began a comprehensive synthesis of all he had learned by writing a "textbook" of chymistry. This textbook would initially take the form of a series of essays presented orally at meetings of the Académie and subsequently published in the annual *Mémoires*. Homberg presented the first two essays in early 1702, and gave sporadic sequels thereafter until 1709. His serial "Essais de chimie" represents not only Homberg's most extended publications but also his best known, having already been the subject of several scholarly treatments.[1] Nevertheless, what seems to be an orderly and sequential publication of the essays in the *Mémoires* masks their extremely turbulent evolution. Newly discovered manuscripts and other rarely cited sources now permit their evolution to be more accurately charted, revealing fundamental and previously unrecognized changes in Homberg's understanding of the principles of chymistry. Significantly, these changes can now be linked with specific experiments and their results. A more vivid picture thus emerges of the transformations of both a chymist and his chymistry under the influence of continuing experimentation, illuminating the often obscure dynamic be-

---

1. Most recently, Rémi Franckowiak and Luc Peterschmitt, "La chimie de Homberg: une vérité certaine dans une physique contestable," *Early Science and Medicine* 10 (2005): 65–90; and Mi Gyung Kim, "Chemical Analysis and the Domains of Reality: Wilhelm Homberg's *Essais de Chimie*, 1702–1709," *Studies in the History and Philosophy of Science* 31 (2000): 37–69.

tween observed laboratory phenomena and the theories generated to explain and organize them. These intellectual transformations did not, moreover, occur in isolation from the broader horizons of Homberg's life—his social position underwent equally significant changes during the same period. It was these latter changes that made possible his most provocative experiments, the ones that required the most radical revisions to his ideas about chymistry, laid new foundations for his thinking, and provided new instrumental methods for his experimentation. This chapter and the next explore these transformations.

A point that cannot be overstressed is the inherent difficulty of trying to construct a comprehensive theory of the sort Homberg wanted. The extraordinary ontological richness of chemistry (or of chymistry!) presents a major challenge to the researcher. Chemical experimentation, in one sense, provides *too much* data. Already by the late seventeenth century hundreds of distinct chemical substances had been recognized, each with its own unique set of properties. These substances reacted with one another in divergent and often unexpected ways, providing an overwhelming mass of seemingly irreducible results and phenomena. For example, aquafort (nitric acid) was known to dissolve silver but not gold, yet when some sal ammoniac (ammonium chloride) was added, it then dissolved gold but not silver. When dilute, aquafort dissolved lead, but when concentrated, it did not. With some metals it produced choking red fumes, with others a colorless and inflammable exhalation. It caused some substances to burst into flames, but others to turn cold. How could one explain, uniformly and comprehensively, such divergent and seemingly random results?

A further complication arose from the fact that in Homberg's day a fully worked-out concept of chemical purity and impurity had not yet been established. If *this* piece of iron and *that* piece of iron reacted differently, was the difference due to an impurity in one, or did the properties of the metal vary over some discrete range—in the way that different individual animals could have different markings, sizes, and behaviors, and yet all be part of the same species? How similar did two things have to be in order to be considered the same? One could not exhaustively compare *all* their corresponding properties, so which similarities would be enough to assure an experimenter of their identity? Without adequate methods of assessing quality, one batch of spirit of salt (hydrochloric acid) could give results very different from another batch acquired on a different day or from a different manufacturer—which phenomena represented the real properties of the substance? The same problem bedeviled the issue of reproducing results consistently, both from experimenter to experimenter and from one day to the next.

Homberg was keenly aware of such problems, and he tried to address them. His endeavor to document and explain uniformly the divergent re-activities of the mineral acids, as we shall see, forms one of his long-term struggles. He labored again and again to develop methods for standardizing materials in order to ensure reproducible results by removing complicating variables. As mentioned in chapter 1, already in the early 1680s he solved a problem with making the Bologna Stone luminescent by identifying and eliminating unsuspected variables, and in the same vein he tried to standard-ize even human excrement in order to get uniform results from it. At the turn of the century he developed methods for measuring and standardizing the "strengths" of acids and bases, and worked to identify, isolate, and remove unsuspected impurities in mercury that altered its properties. He likewise examined the effect of altering reaction time, concentration, temperature, sources of materials, and many other factors in order to explore and cata-logue all the possible results of reactions. As his experiments progressed and his results accumulated, he undoubtedly found himself occasionally over-whelmed, much as his predecessors at the Académie had been when they undertook the plant analyses. But Homberg's attitude differed fundamentally from both Dodart, who became satisfied with "good enough" results and ad hoc explanations, and Bourdelin, who catalogued results without worrying much about their interpretation. More like Duclos, Homberg fixed his efforts on producing a reliable, coherent, and truly explanatory system of chymistry. But unlike Duclos, Homberg never lost faith in the ability of experiment and observation to provide the answers, even if new experimental results required constant and sometimes drastic reworking of his explanatory system.

The formulation of explanatory systems is a dynamic process central to the history of science. Despite melodramatic storytelling to the contrary, ex-planatory theoretical structures do not emerge fully formed from the brains of natural philosophers, like Athena from the head of Zeus. A host of ob-servations and motivations, not to mention a broad variety of other factors, steer the course and shape the contours of scientific creativity. Theoretical structures are generally built up gradually, and modified frequently in the context of continuing experimental results. Such interplay between experi-ment and theory forms a key topic of interest for historians of science. How does a natural philosopher translate experimental results into theoretical structures? How do these theoretical systems suggest future experimenta-tion, direct the course of fresh avenues of study, and guide the evolution of increasingly developed systems? Why and how are certain experimental pathways chosen and others rejected, and how exactly are explanatory sys-tems revised, rejected, or revived? While these questions are of interest also to philosophers of science, my approach in this chapter is predominantly historical rather than philosophical—illustrated by following Homberg over a period of years, identifying the evolution of his ideas, and correlating them with practical, observational, and biographical events.

It is often difficult to trace the long-term development of a comprehensive scientific theory, especially for the early modern period. Informative documents, such as laboratory notebooks or correspondence, survive only occasionally, and even when they do, the intermediate steps of a theory's evolution are rarely transferred coherently from the researcher's brain to paper. Histories of discovery are, moreover, customarily strategically rewritten or misremembered, sometimes unintentionally, either by the natural philosopher himself or by others. They are thus made to appear more linear and obvious than they really were (sometimes for pedagogical purposes), while making the researcher appear more ingenious and far-sighted (not to say omniscient) than human beings actually tend to be. In the case of Homberg's "Essais," the multiple installments published in the *Mémoires* provide only disconnected snapshots from the middle years of Homberg's thinking about the chymical principles. Analyses based on them alone provide a misleadingly static depiction of the dynamic reality of his laboratory work and thought. The discovery of substantial new manuscript materials, coupled with a fuller deployment of previously underutilized materials, fills out a deeper, more detailed, and more surprising account of how Homberg developed his ideas and directed his experimentation. The result is a fine-grained analysis of how an influential and innovative chymical theory of the early eighteenth century developed over the course of twenty years, emerging from a convergence of prior experiences and interests, evolving laboratory experimentation, institutional contexts and collaborations, costly and cutting-edge scientific instrumentation, and lavish patronage.

The Sources for Understanding the Evolution of Homberg's Ideas

The following analysis of Homberg's evolving theory of chymistry deploys four sets of documents as its main reckoning points. First are the aforementioned and readily available papers entitled "Essais de chimie" that Homberg published serially in the *Mémoires* for the years 1702 to 1709. The earlier versions of these publications that Homberg presented orally to the Académie and that were transcribed into the procès-verbaux comprise the second set. A gap of roughly two years generally separates these initial oral presentations from their publication in the *Mémoires*—a long period of time in the context of an active research program, giving plenty of time for new experiments to be conducted, old ones to be reinterpreted, and revisions to be made prior to publication. In some cases, the procès-verbaux versions are very close in content to the published versions, but in the case of the first two essays the changes are so substantial that they signal *fundamental* changes in Homberg's view of the chymical principles. These striking differences have never before been noticed, since previous scholars have relied exclusively on the published versions and have not compared them with the manuscript procès-verbaux. Most impressively, the trigger for these changes can be identified in a series of spectacular experiments Homberg carried out in May and June 1702, only a few weeks after having delivered the first two essays to his colleagues at the Académie.

The other two documents are new to this study and were not to be found in the Académie's archives or its publications; their surprising rediscovery and provenance deserve a few words of explanation. Homberg's last explicit installment of his "Essais," presented in 1709 and published in 1711, left his promise of a complete work unfulfilled. To readers of the *Mémoires* it would appear that Homberg gave up on the project in 1709. But this was not the case. Fontenelle's éloge for Homberg mentions that a completed version of his "Essais de chimie" was found among his papers upon his death in 1715 "in good order and ready to be printed."[2] Leibniz, eager to see this culmination of Homberg's work, made inquiries about the fate of the full manuscript and when it was to be published. He was told that the manuscript of "a complete course of chymistry" (along with the "little treatise on generation") was in the hands of Homberg's collaborator Étienne-François Geoffroy and "will be published as soon as possible."[3] Yet nothing appeared. A few years later, in December 1723, Homberg's nephew Denis Dodart (the younger) reported that he was preparing his deceased uncle's works for publication as a collected volume. He listed the works then in his possession as not only a treatise "on the principles of chymistry" (presumably the completed version of the "Essais" mentioned by Fontenelle) but also the treatise on generation and another one on assaying.[4] Whether Dodart had acquired the original manuscripts from Geoffroy or had fresh copies of them made is unknown. Yet once again, this promised publication came to naught. What, then, became

---

2. Bernard de Fontenelle, "Éloge de M. Homberg," *HARS* (1715): 82–93, at 92.

3. Nicolas François Rémond to Gottfried Wilhelm Leibniz, 23 December 1715; NLB, LBr 768, fols. 53r–54v, at 54r: "les papiers de feu M. Homberg sont entre les mains de M. Geoffroy . . . on les fera imprimer le plustost qu'il le poura il y a un cours entier de Chymie et un petit traité de la Generation."

4. Denis Dodart to Montesquieu, 28 December 1723, in *Oeuvres complètes de Montesquieu*, 19 vols. (Oxford: Voltaire Foundation, 1998), 18:73–79.

of the final manuscript of the "Essais"? How did it differ from the versions published serially in the *Mémoires*? What further developments of Homberg's thinking did it contain, and what more might it have said about his views of chymistry?

Thanks to a combination of good fortune, stubborn searching, and years of negotiations to gain access to a particular archive, I at last discovered a manuscript of the final version of Homberg's "Essais." Retitled *Élémens de chimie*, the work turned up in an unexpected place: among the papers of Herman Boerhaave (1668–1738) at the archive of the Military-Medical Academy (Voenno-Meditsinskoi Akademii) in St. Petersburg.[5] The text preserved in this newly discovered manuscript differs significantly from the serial versions printed in the *Mémoires*, and as such reveals Homberg's "last word" on his comprehensive theory of chymistry. It incorporates materials presented in 1710–12, indicating that Homberg kept working on the manuscript for years after the last paper (1709) explicitly tied to the project had been presented. The recovery of this final version, lost for nearly three hundred years, came with an additional surprise of perhaps even greater significance. The *Élémens de chimie* was bound with an untitled second manuscript that also bore the format of a chymical textbook. This second text's content and style were so different from those of Homberg's "Essais" and *Élémens* that I initially concluded that the only connection between the two manuscripts was their similar subject matter. Upon closer inspection, however, I found this additional manuscript to contain several autobiographical anecdotes that unambiguously identify Homberg as its author. This manuscript turned out to be Homberg's earliest surviving attempt to write an "Elements of Chymistry." The text is datable to the early 1690s, well before he began writing his "Essais."[6]

The unexpected resting place of these two manuscripts implies that like Leibniz, Herman Boerhaave, the famous chymist, physician, and pedagogue of Leiden, made inquiries about Homberg's unpublished works and thereby acquired copies of the two manuscripts from whoever possessed the originals

5. For a fuller exposition of the tale of finding and gaining access this document, see the acknowledgments. I would like to express again here my deepest thanks to the two colleagues most instrumental in the discovery of and access to this document: John Powers and Igor Dmitriev.

6. The two manuscripts are St. Petersburg, VMA MS 130, fols. 2r–108v and 112r–291r. Fol. 1v bears Boerhaave's autograph endorsement: "Hoc libro manuscripto continentur praecipua chemica, quae ingens Hombergius fecerat cum solus, tum cum Duce Aurelianensi" ("Remarkable chymical matters are contained in this manuscript book, which the great Homberg carried out both alone and with the Duc d'Orléans").

at that time. Boerhaave showed sufficient interest in the completed *Élémens* that he began making a fresh copy of it, embellished with his own autograph Latin marginal annotations; this document also survives in the same archive.[7] The unexpected final resting place of these manuscripts is explained by the fact that upon Boerhaave's death, his papers went to his nephews Herman and Abraham Kaau-Boerhaave, who then took the manuscripts with them when they relocated to Russia in the 1740s. After their deaths, Boerhaave's papers— bearing the Homberg manuscripts along with them—passed through various hands, ending up finally in the Military-Medical Academy, where they were rediscovered only in 1929. The Homberg manuscripts, by the way, are not listed at all in the current Russian-language catalogue of the collection.[8] These two previously unknown manuscripts, together with the records of the procès-verbaux and the published *Mémoires*, now provide a portrait of Homberg's evolving authorship and thinking about the principles and status of chymistry that stretches over his entire tenure at the Académie—nearly twenty-five years. Table 3.1 briefly compares their varied contents and divisions. The close analysis of these documents presented in this chapter and the next reveals in a dramatic way how Homberg's theorizations about matter and its transformations evolved under the influence of specific experimental results and pathways of investigation.

The Chymical Didactic Tradition and Homberg's First Manuscript (Early 1690s)

Wilhelm Homberg's project of writing a comprehensive "textbook" of chymistry falls within the "didactic tradition" of seventeenth-century chymistry, a tradition already a century old when Homberg started writing in the early 1690s. This didactic tradition is exemplified in the many "courses of chymistry" published during the seventeenth century, stretching from the *Éléments* of Jean Beguin at the start of the period, through the books of Étienne de

---

7. VMA MS 128.

8. *O Nauke i Uchenykh: Arkhiv Burgave v Voenno-Meditsinskoi Akademii* (St. Petersburg: Voenno-Meditsinskoi Akademii, 2003). A more complete listing of the contents of the archive is the "Lijst van de Boerhaaviana" appended to B. P. M. Schulte, *Hermanni Boerhaave Praelectiones de morbis nervorum*, Analecta Boerhaaviana 2 (Leiden: Brill, 1959), 426–33. For more on the wanderings of the Boerhaave manuscripts, see Schulte, *Praelectiones*, 31–40; David Willemse, *António Nunes Ribeiro Sanches, Élève de Boerhaave, et son importance pour la Russie* (Leiden: Brill, 1966), 64–107; and Ernst Cohen and W. A. T. Cohen-De Meester, *Katalog der wiedergefundenen Manuskripte und Briefwechsel von Herman Boerhaave*, Verhandelingen der Nederlandsche Akademie van Wetenschappen, Afdeelig Naturkunde 40 (1941):1–45, at 3–9.

TABLE 3.1. The evolution of Homberg's "textbook" of chymistry

Early 1690s, [untitled] (VMA MS 130, fols. 112r–291r)	1702, Essais d'Élémens de Chimie (PV 21, fols. 61r–73v)	1704–9, Essais de chimie (Mémoires)	1715, Élémens de chimie (VMA MS 130, fols. 2r–108v)
1. Les principes de la chimie	1. Des principes de la chimie en général	1. Des principes de chimie en général (1704ª)	1. Des principes de la chimie en général
2. De l'alchimie	2. Du sel principe chimique	2. Du sel principe chimique (1704ª, 1708)	2. Du Soufre[b]
3. Des operations de la chimie	*Du souphre*	3. Du souphre principe (1705, 1706)	3. Du Sel
4. Des ustensiles necessaires dans la chimie	*Du mercure*	4. Du mercure (1709)	4. Du Mercure
5. De l'or	*De la terre*	*De la terre*	
6. De l'argent	*De l'eau*	*De l'eau*	
7. De l'estain	*Cours d'Operations*	*Cours d'Opérations*	
8. Du plomb			
9. Du fer			
10. Du cuivre			
11. Du mercure			
12. De l'antimoine			
Salts			
Vegetable substances			
Animal substances			

Note: Chapter titles in italics denote projected sections that were not completed.
[a]The year 1704 is the publication date of the extensively revised versions of the first two chapters; the initial versions of these chapters were presented orally in 1702. The remaining dates indicate when a new chapter or supplement was presented orally to the Académie.
[b]In addition to the previously published memoirs of 1705 and 1706 explicitly entitled as parts of the essay on sulphur, this final version also incorporates parts of other papers on the burning lens (*MARS* [1702], 147–55) and on sulphureous substances (*MARS* [1710], 226–33).

Clave, Christophle Glaser, Nicaise Lefevre, Pierre Thibault, Nicolas Lémery, and others, and thence on into the eighteenth century. The textual didactic tradition in France was identified and explored first by Hélène Metzger in the 1920s; Owen Hannaway later added to it the *Alchemia* of the Saxon pedagogue Andreas Libavius, whom he identified as a major source for Beguin, and argued that the development of such chymical pedagogy was the crucial step in the foundation of the modern discipline of chemistry.[9] Chymistry's didactic aspect expressed itself also in the regular courses of chymistry established at

9. Hélène Metzger, *Les doctrines chimiques en France du début du XVIIe siècle à la fin du XVIIIe siècle* (Paris: Les Presses Universitaires, 1923); Andrew Kent and Owen Hannaway, "Some New Considerations on Beguin and Libavius," *Annals of Science* 16 (1960): 241–50; and Hannaway, *Chemists and the Word* (Baltimore: Johns Hopkins University Press, 1975).

the Jardin du Roi in the mid-seventeenth century. These annual courses often provided the basis (or justification) for published works in the genre.[10] Most Académie *chimistes* in the years around 1700 also taught the cours de chimie at the Jardin—with the notable exception of Homberg. His teaching was far more limited. He tutored only privately, and we know of just three students he instructed—although there may well have been others. In the late 1680s he taught the young Abbé Louvois, in the 1690s he tutored Étienne-François Geoffroy, who later became his élève, associé, and closest associate at the Académie, and in 1701 Homberg gained a "tutee" and collaborator of far more exalted status, namely, Philippe II, Duc d'Orléans.

The didactic tradition and its role in the broader history of chemistry are not unproblematic. First of all, these books are often praised particularly for their degree of clarity (requisite for an avowedly teaching text), which is then set rather casually in contrast to the supposedly obscure nature of the rest of foregoing and contemporaneous chymical publications. Fontenelle, for example, casts Lémery's *Cours de chymie* in exactly this way, claiming that by means of this book Lémery "was the first who dispersed the natural or affected obscurities of chymistry." Yet Lémery's *Cours* is in fact highly dependent on earlier texts in the genre that are no less clear or straightforward. Once again, later readers have been misled by Fontenelle, believing that his statements are the unbiased observations of a contemporaneous spectator rather than the rhetorical productions of a crafter of public opinion. Fontenelle's éloge for Lémery creates a stark dichotomy within seventeenth-century chymistry between a supposedly dark, obscure, and confused past and a bright, clear, and rational present, exemplified best, of course, by members of the Académie Royale des Sciences. Yet while Fontenelle explicitly contrasts Lémery with Glaser, whom he describes as "a true chymist, full of obscure ideas, jealous of them, and very unsociable," Lémery's 1675 *Cours de chymie* is scarcely different from Glaser's 1663 *Traité de chymie* in either content or clarity.[11] In terms specifically of clarity, within the assaying and recipe literature there had long existed straightforward texts largely free of "obscurities." Even within the chrysopoetic traditions of chymistry, despite their customary and intentional "obscurity," many such texts—like those of Michael Sendivogius, Eirenaeus Philalethes, and many others—proved highly influential in spurring ideas and further re-

---

10. On the courses at the Jardin, see Jean-Paul Contant, *L'Enseignement de la chimie au Jardin Royal des Plantes* (Cahors: Coueslan, 1952).

11. Bernard de Fontenelle, "Éloge de M. Lemery," *MARS* (1715): 73–82, at 76 and 73. Fontenelle's revisionist desire to separate Lémery from his early instructor Glaser is possibly due to the latter's implication in the Affaire des Poisons and his consequent imprisonment in the Bastille; see Michel Bougard, *La chimie de Nicolas Lémery* (Turnhout: Brepols, 1999), 24–26.

search, and thus occupy an important place within the mainstream of the history of science, regardless of their degree of "clarity." The topos of a new and enlightened chymistry brought forth by academicians from an obscure and benighted foregoing tradition is an *idée fixe* for Fontenelle—and even for his mid-eighteenth-century successors as perpetual secretary—trotted out again and again on every possible occasion over the more than forty years he authored the *Histoire* and the éloges. It should not be taken at face value.

Second, the usually cited textbook writers do not represent the intellectual vanguard of seventeenth-century chymistry. There is no question that they popularized chymical operations and their utility, and trained a huge number of practitioners, but most of them were neither innovators nor contributors to new conceptions of chymistry. It is noteworthy that Robert Boyle criticized *specifically* the textbook writers in his famous 1661 *Sceptical Chymist* (and elsewhere), as a "lower order" of chymists, and he did so largely on the grounds of their having reduced the scope of chymistry to a merely practical craft, rather than recognizing and using it as an important part of natural philosophy.[12] Indeed, despite the common modern denomination of these volumes as "textbooks," an inspection of them shows that most of them would be more accurately described as recipe books, primarily of pharmaceuticals. This assessment is clear from even a cursory consideration of their tables of contents, and what follows in the body of the books supports the conclusion. Most of their bulk is devoted to preparative recipes for medicinal use, while the other dimensions of chymistry remain largely absent. General or theoretical principles are present only to the extent they directly assist in the main goal of preparing pharmaceuticals, or to the extent that such topics had become merely customary introductory material.[13]

Homberg's earliest attempt to write a textbook of chymistry continues the didactic tradition, but also departs significantly from it. It borrows the existing organizational format of the genre but provides a very different direction and set of contents. The surviving manuscript is untitled and consists of 180 folios (about 65,000 words) written in the hands of two unidentified copyists.[14] Its composition can be dated to the first half of the 1690s, based on the

12. Lawrence M. Principe, *The Aspiring Adept: Robert Boyle and His Alchemical Quest* (Princeton, NJ: Princeton University Press, 1998), 30–36, 58–61.

13. The "textbook tradition" actually incorporated two distinct strands, rigorously practical texts (like Beguin, Glaser, and Lémery), and texts that embraced a wide-ranging chymical world-system (like Croll and Davisson); see Principe, *Aspiring Adept*, 59.

14. VMA MS 130, fols. 112r–291r. Fol. 112r is written in the hand of a copyist whose native language was Germanic, as witnessed by the customary squiggles over the letter *u*; the rest of the document is in the hand of a French copyist.

nearly identical presentation of several topics both in the manuscript and in memoirs read to the Académie during this period. For example, the manuscript's description of how artisans use techniques of surface enrichment to improve the color of low-quality gold alloys reappears almost verbatim in a memoir Homberg presented to the Académie in 1695.[15] Since the procès-verbaux text shows greater development and contains further explanations, it was probably expanded from the version recorded in the St. Petersburg manuscript, implying that at least this section of the treatise was written by 1695. The same can be said of an extended treatment of the green color that copper imparts to flames, which reappears word for word, but with added introductory material, in a memoir Homberg delivered in early 1696.[16] The manuscript also contains a briefer and less-developed form of the experiment described in Homberg's 1692 paper about separating a dry volatile acid salt from aquafort using silver and tin calx (treated in chapter 2 and figure 2.4).[17] Finally, parts of the last chapter, which is devoted to antimony, contain results from Homberg's extended work on that substance that he announced to the Académie in 1692 and pursued until 1696.

Although Homberg worked on this writing project through the early 1690s, he gave up on it before it was finished. The manuscript promises several topics that were supposed be covered later but that never appear. The opening chapter describes the overall division of the book into mineral, vegetable, and animal substances—a standard division found in contemporaneous textbooks—but the sections dealing with vegetable and animal realms do not exist. Even the mineral section remained incomplete, breaking off before reaching twice-promised material about salts.[18] Homberg's extraction of several sections for oral presentation to the Académie in 1695 and 1696 may suggest that he had given up on the original scheme by that time, and was salvaging and repurposing useful materials.

15. VMA MS 130, fols. 154r–155r; PV 14, fols. 170r–171v (13 August 1695) .

16. VMA MS 130, fols. 224r–225v; PV 14, fols. 234v–236v (21 January 1696; part of the memoir "Observations de Monsieur Homberg sur la flame verte qui paroît lorsqu'on rougit du Cuivre au feu," 234r–236v).

17. VMA MS 130, fols. 175v–177r. In the manuscript, Homberg notes the mercurial nature of the distilled liquid, a topic detailed more fully in PV 14, fols. 35v–36v (5 January 1695). The PV process is clearly more developed, hence almost certainly later in time; for example, the 1695 PV text cites the need to moisten the silver calx in order to avoid the production of a dry white fume; however, the manuscript describes the production of this fume as the normal occurrence.

18. On fol. 113v, Homberg mentions topics that "we will examine at the end of the vegetable realm"; on fol. 132v, he defers a topic "to the place where we will speak of salts"; and on fol. 226r–v he promises to treat natural and artificial vitriols "in their place," presumably also in the unwritten section on salts.

The text of the St. Petersburg manuscript also preserves traces of aborted revisions that must have been present in Homberg's (presumably autograph) original. For example, the third chapter ends with the puzzling line: "the ordinary manner in which dissolutions are explained, and what acids and alkalies are, must be indicated here." This odd line must originally have been Homberg's note to himself about what he needed to add to the text in order to complete the chapter, and the copyist must have transcribed it indiscriminately as if it were part of the text.[19] Other sections contain explanations presented explicitly as alternatives to what immediately preceded them, yet the supposedly preceding explanation is not present, as if a foregoing paragraph had been crossed out in the original manuscript without being rewritten, and therefore was omitted by the copyist.[20] The most striking alteration, however, preserves evidence of the extent to which Homberg's chymical ideas really were in flux during this period. It occurs in the enumeration of the chymical principles found in plant and animal substances. The result is a confusing and self-contradictory paragraph:

> [Chymists] have found that most compound substances are composed of *four* different materials, namely, *oil, salt, and earth*, which they have called principles. But some of the moderns have reduced the number of these *five* principles and have excluded spirit from them . . . thus according to them there are only *four* principles.[21]

Here four principles are initially cited, but then only three are listed, and immediately thereafter the customary number of five principles appears as if that number had just been mentioned. The most plausible explanation for this *locus desperatus* is that the copyist was working from an original that Homberg had rendered virtually illegible by repeated alterations and/or

---

19. VMA MS 130, fol. 132v: "Il faut marquer icy la maniere ordinaire par laquelle on explique les dissolutions et ce que cest que les acides et les alkalis." The unknown copyist was often quite careless. In many cases he transcribes words incorrectly (e.g. "embarassé" for "embrasé," fol. 212v) and shows unfamiliarity with chymical terms, as witnessed by errors such as "sel gomma" (instead of *gemma*, 143r), "laine d'acier" (for *lame d'acier*, 216r), "algaral" (for *algarot*, 282r), and a reference to the "figure comique" of acid particles on 143v, later altered to the more sensible but less amusing *conique*.

20. For example, VMA MS 130, fols. 143v–144r.

21. VMA MS 130, fol. 113v: "Ils ont trouvé que la plus part des mixtes sont composés de *quatre* matieres differentes scavoir *d'huille, de sel et de terre*, ce qu'ils ont appellé principes mais quelques uns des modernes ont abregé le nombre de ces *cinq* principes et en ont exclu l'esprit . . . ainsi selon eux il n'y a que *quatre* principes" (emphasis added here and in the text to highlight the contradictions). The exclusion of spirits here accords with his 1692 declaration that acid spirits are simply volatile acid salts dissolved in water.

had abandoned midway through his changes. This tortured text witnesses the changing nature of Homberg's thinking about the chymical principles of plants during the 1690s, while his experiments to discover and identify them were ongoing. The following paragraphs most frequently settle on *four* principles—oil, water, salt, and earth—but since the projected part of the treatise on vegetable materials was never written, we cannot tell which or what number of principles Homberg finally chose, or indeed if he came to any final conclusion about the matter while he was working on this document. Certainly, his citation here of oil as one of the principles must have preceded his 1695 decomposition of oil into the simpler bodies of salt, water, and earth. The accumulating results of his study of plant components in the 1690s may have led not only to the confusion noted above, but also to his decision to abandon this document before writing the section on vegetable substances. It is also possible that Homberg wrote at least a part of the section of vegetable materials, but then discarded it when its theoretical content was undermined by the results of new experiments.

Much greater confidence surrounds Homberg's treatment of metals and minerals a few pages later. There he is clear and confident that the metals are composed of three chymical principles—the Paracelsian *tria prima* of mercury, sulphur, and salt. He is also surprisingly definitive, relative to contemporaneous texts, about the identity of these metallic principles. He claims that what he calls the three "simple minerals"—common mercury, common sulphur, and mineral salts—are themselves the three metallic principles in impure states.[22] Common mercury is the mercury principle mixed with earthy and watery impurities, a notion that dates back to the medieval Geber and that remained popular (particularly among chrysopoeians) down to and beyond Homberg's day.[23] The three simple minerals can combine with each other as well as with metals, both in nature and in the laboratory, to form compound minerals (*mineraux composés*). Cinnabar is produced when common mercury and common sulphur combine, and the vitriols result from the combination of a metal with a mineral salt. Homberg then explains the composition of each metal. The mercury principle gives rise to the various metals when it combines underground with varying quantities and qualities of metallic sulphur and metallic salt. Mixed with the proper proportion of

22. VMA MS 130, fols. 116v–117r: "les mineraux simples ne consistent chacun qu'en un seul principe metallique impur."

23. On the early background to this view of the metallic principles mercury and sulphur, see William R. Newman, "Mercury and Sulphur among the High Medieval Alchemists: From Rāzī and Avicenna to Albertus Magnus and Pseudo-Roger Bacon," *Ambix* 61 (2014): 327–44.

sulphur and salt, and cooked in a clean location underground, the mercury principle congeals into the perfect metal, gold. The presence of excess metallic salt generates silver, but if the salt is spoiled (*gâté*) by a heterogeneous salt, tin results. If both the salt and the sulphur are spoiled by impurities, the product is lead.[24] Homberg's far greater emphasis on the metals and minerals throughout this text suggests that he was either more interested in metallic chymistry, or simply more confident in his understanding of it at this time. Thus, despite initial promises to cover all three realms of nature, the vast majority of the text as Homberg left it—nearly 90 percent—deals specifically with the metals, making the document look like a work devoted to them. Indeed, one of the notebooks that the academician Jean Hellot compiled in the 1750s, which preserve excerpts from then surviving papers of earlier academicians, contains a section undoubtedly copied from the now-lost original of this Homberg manuscript, and Hellot refers to the text as "la chimie métallique de Homberg."[25]

In terms of format, Homberg's division of materials most resembles that established in Glaser's 1663 *Traité de la chymie* and later adopted with minor reordering in Lémery's 1675 *Cours de chymie*. Namely, the opening chapters cover the chymical principles, chymical operations, and chymical apparatus, and this material is followed by treatments of individual substances classified into the mineral, vegetable, and animal realms (see table 3.1). Even Homberg's specific order of chapters for the mineral realm almost exactly follows Lémery's, with coverage beginning with gold and silver, and then progressing through the remaining five metals—tin, lead, copper, iron, and mercury—and then on to antimony. Lémery's material thereafter continues through arsenic, salts, and other mineral bodies, while Homberg's ends with antimony, and he never got as far as the mineral salts, much less the vegetable and animal realms. Since Glaser's and Lémery's books were the most recent and popular examples of the genre when Homberg was writing his treatise, it is not surprising that he adopted what they had established as a virtually canonical order of materials.

The similarity to these earlier texts ends, however, with the division of topics and order of chapters. Homberg chooses an entirely different con-

24. VMA MS 130, fols. 115v–119v; on mercury as the base of metals, see also 167v and 263v.

25. Hellot Caen Notebooks, 3:268v: "Voyés la chimie métallique de Homberg." Hellot copied this material from Homberg's refutation of a supposed fixation of mercury into silver recounted in VMA MS 130, fols. 245r–246v. It is unclear whether "chimie métallique" is merely Hellot's convenient description of the document or a title Homberg himself provided that was not transcribed into the St. Petersburg manuscript. For more on Hellot's notebooks, see the note on sources.

tent and central aim for his book. Earlier textbooks focused mainly on the preparation of pharmaceuticals, even if some authors (like Glaser) praised the breadth of chymistry or were active in chymistry's other applications (as with Beguin's pursuit of transmutation). Homberg's treatise instead provides a more inclusive coverage of the breadth of chymistry as it was actually practiced in the seventeenth century. Homberg's book, which at the very outset praises chymistry's "infinite extent," actually touches upon much of that extent.[26] While it does contain many pharmaceutical recipes, they occupy only a relatively small fraction of the total, and share the stage with extensive and detailed material about the mining and natural occurrence of metals, the refining of ores, assaying, metallic transmutation, pigments, metallurgy, artisanal processes, and nonpharmaceutical chymical manufacture.

There is a certain "natural history" flavor to the text, in the sense that it brings together wide-ranging material about each of the metals in turn. While a substantial portion of the content derives from Homberg's laboratory work (and is therefore relayed in the first person singular), much other material is traceable to Homberg's years of learned travel and reflects his broad interests and curiosity. The chapter on copper covers everything from natural vitriol springs in Hungary, to observations he made at Swedish refineries and their natural philosophical explanation, the production of brass in Liège and Austria and of verdigris at Montpellier, and the utility of various copper alloys for everything from clock chimes to harpsichord strings.[27] Almost half of the chapter on iron is devoted to magnetism, ranging from Homberg's own effluvium-based explanation for magnetic attraction to the determination of longitude using the declination of the compass.[28] The wonders and secrets that Homberg so loved to collect also appear prominently: he recalls an enormous piece of native silver in the cabinet of curiosities of the Elector of Saxony in Dresden; cites the largest lodestone in Europe, owned by Louis XIV and stored at the Bibliothèque du Roi; and provides a recipe for a bizarre salve of pounded garlic, onions, and lard, used to protect the feet of fire-walkers.[29]

26. VMA MS 130, fol. 112v: "les objets de la chimie sont d'une etendue infinie . . . elle comprend toutes les substances corporelles."

27. VMA MS 130, fols. 210v–223r.

28. For this latter topic, in one of the very few cases where the text cites another author by name, Homberg refers to Henry Bond and to material in his *The Longitude Found* (London, 1676); on Bond, see Richard J. Howarth, "Fitting Geomagnetic Fields before the Invention of Least Squares: Henry Bond's Predictions (1636, 1668) of the Change in Magnetic Declination in London," *Annals of Science* 59 (2002): 391– 408; and D. J. Bryden, "Magnetic Inclinatory Needles Approved by the Royal Society?" *Notes and Records of the Royal Society of London* 47 (1993): 17–31.

29. VMA MS 130, fols. 157r, 196r–v, and 216r–217r.

But Homberg's early treatise is far from a miscellany. Homberg shows a dogged insistence throughout upon uncovering the general principles that govern and explain chymical transformations. The greatest share of words in the manuscript go toward providing *causal explanations of observed phenomena*, and this feature clearly distinguishes Homberg's text from the usual *cours de chimie* of the period. No previous textbook writer displays so extensive, orderly, and consistent a devotion to the natural philosophical agenda for chymistry. Homberg's attitude here correlates well with his contemporaneous redirection of the plant analysis project away from medical utility and toward chymical investigation. He uses many of the preparative procedures he provides primarily as jumping-off points for extended illustrations or explorations of generalized chymical theories rather than acting merely as instructions for preparative chymists. This focus departs radically from the "recipe book" character of the foregoing didactic tradition exemplified by authors like Beguin, Lefevre, Thibaut, and Glaser. Lémery has commonly been seen, following Metzger, as a more philosophically sophisticated author than his predecessors, based upon the "remarques" that often follow and comment upon the preparative processes in his *Cours*. But these expositions most frequently concern only further details of manual operations, and those that do involve a deeper investigation of causes and chymical theory are limited and clearly of secondary importance relative to the conveyance of practical recipes. Homberg's consistent emphasis on understanding over material production is of a different order, and displays his more coherent and sophisticated vision of chymistry as a key part of natural philosophy. His explanation of dissolution and precipitation provides a good illustration of the coherence of his chymical theory, and the prominence he gives to developing theoretical explanatory principles. Homberg's conclusions on the subject depart significantly from other contemporaneous ideas, including those often cited in the secondary literature as "characteristic" of late seventeenth-century chymical thought.

Dissolution and Precipitation: Homberg's Theory of *Semblance*

Homberg's theory of dissolution and precipitation is the most frequently invoked explanatory principle in the manuscript. Each of the eight chapters dealing with a specific metal proceeds in a similar sequence that always includes a discussion of its dissolution and precipitation (always preceded by its natural occurrence, refining, and calcination). This high profile reflects the fact that dissolution and precipitation are the observable phenomena most routinely encountered in chymical laboratory work, and the most in need of fundamental explanation given their puzzling "irregularities." It was well known, for example, that aquafort (nitric acid) can dissolve silver and lead, but not gold or tin. Aqua regia shows the opposite ability, while both acids dissolve copper and iron. What determines which solvents dissolve which substances? How do solvents break apart solid bodies into the invisibly small particles that then "disappear" into solution? It was also well known that the addition of certain substances to solutions caused the dissolved materials to reappear as solid precipitates. How and why did one substance cause another to reappear in a solid, insoluble form?

Homberg's explanatory system relies upon a uniform principle I will term "assimilation by similitude." The central tenet is that "the dissolvent of a material is always one of the principles that composes that material."[30] To explain his meaning, Homberg first treats vegetable and animal substances. Their two chief component principles are oil and salt. He observes that "fat, oils, and gums dissolve only in spirit of wine," and argues that this is because spirit of wine "is itself an oil," certainly in reference to the inflammability of spirit of wine (ethanol) that reveals its sulphureous (that is, its "oily") nature. Sulphureous spirit of wine therefore dissolves sulphureous components. Homberg then contrasts oily substances with salts, remarking that "fixed and volatile salts of whatever nature they might be dissolve only in common water because they are one of the most simple and first modifications of common water, whose form they resume more easily and with less difficulty than other bodies."[31] The solubility of salts in water is thus due to their being very similar to water—they are its "most simple and first modification"—clearly a reference to the Helmontian notion that water is the basic substratum of all substances, a theory Homberg was actively testing while writing this treatise, as described in the previous chapter.[32]

Having shown that a principle of similitude operates in the dissolution of plant and animal substances—oily things dissolve in oily solvents but not in watery ones, and watery things dissolve in watery solvents but not in

30. VMA MS 130, fol. 129r: "le dissolvent d'une matiere est toujours un des principes qui composent cette matiere." Homberg's principle of similitude bears a certain resemblance to the modern chemist's rough rule of thumb that "like dissolves like," though the two rules are based on different underlying matter theories.

31. VMA MS 130, fol. 132r–v: "les sels fixes ou volatils de quelque nature qu'ils soient ne se dissolvent que dans l'eau commune parcequ'ils sont une des plus simples et des premiers modifications de l'eau commune dont ils reprennent la forme plus aisement et avec moins d'embaras que les autres corps."

32. It is worth noting that Duclos also claimed that salt was the first modification of water, when it joined with earth and the igneous spirit; see chapter 2.

oily ones—Homberg then turns to the mineral realm. Vegetable substances are easy to separate into their watery and oily constituents, allowing for the solubility characteristics of each to be observed directly, thus establishing the principle of similitude. Metals and minerals, however, are not easy to separate into their component mercury, sulphur, and salt, so Homberg begins using the similitude principle *in reverse*. Instead of predicting solubility from observed composition, he now uses the same principle to predict composition from observed solubility. Silver is observed to dissolve in aquafort but not in aqua regia; therefore, there must be a similarity between silver and aquafort that does not exist between silver and aqua regia. Homberg's experiments had already shown that acids are composed of a volatile salt dissolved in water. The observation that silver dissolves in aquafort, therefore, indicates that the salt in silver must bear a similarity to the salt in aquafort. Contrariwise, silver's insolubility in aqua regia demonstrates that silver's salt must be dissimilar to aqua regia's salt. Just the opposite can be said of gold. Gold dissolves in aqua regia but not in aquafort because gold's salt is similar to the salt in aqua regia but not to the salt in aquafort. These different solubilities "make one suspect that the salts that enter into the composition of metals are not all of the same nature and that they approach the nature of their dissolvents. . . . We will undertake to prove this hypothesis in the place where we will talk about salts. It may well be that it is for this reason that . . . silver dissolves only in the spirit of saltpeter [that is, aquafort], because its salt is similar to saltpeter."[33]

Iron dissolves in both aqua regia and aquafort, as well as in other solvents; thus, using the same principle of similitude, "it must be that there is a great deal of salt in its composition because it dissolves in all sorts of liquids whether acidic or not." The inability of iron to dissolve in molten gold or silver to form a stable alloy provides additional evidence that its composition is very different from that of both gold and silver.[34] Homberg thus extends his principle of similitude, worked out first in the vegetable realm, in or-

33. VMA MS 130, fol. 132v: "Ces differents dissolvens font soubçonner que les sels qui entrent dans la composition des metaux ne sont pas tous d'une même nature et qu'ils aprochent de la nature de leurs dissolvens. . . . Nous tacherons de prover cette hipothese dans l'endroit ou nous parlerons des sels. Il peut fort bien être que par la même raison l'or ne se dissout que dans lesprit de sel marin et que l'argent ne se dissout que dans l'esprit de salpêtre que parceque son sel est semblable au salpêtre."

34. VMA MS 130, fol. 203r: "Le fer se dissout aussi bien par les eaux fortes que par les eaux regales, ce qui marque qu'il n'est ni de la nature de l'or ni de la nature de l'argent, aussi ne se mêle-t-il pas avec eux dans le fonte ni avec aucun autre metal. Il faut que dans la composition

der to explore the otherwise obscure composition of metals and to explain their differing solubilities. He even applies the same principle to magnetism. Having adopted the common contemporaneous view that the phenomena of magnetism are caused by an effluvium of invisibly small particles emitted by magnets, which Homberg calls the *matière aymantine*, he attempts to identify the composition of this "magnetic material" based on its observed effects. Since only iron is affected by magnetism, and lodestones themselves are iron-based, there must be a particular similitude between the matière aymantine and iron that allows for their interaction. Although one cannot be absolutely certain about the chymical composition of the magnetic effluvium, "if one may judge according to appearances and the known effects it produces, we would be able to say with some probability that it is a sulphureous material, in fact a metallic sulphur that exhales from the globe of the Earth. . . . The reason one can suppose it is a sulphureous and metallic material is that it does not attach itself nor come to rest anywhere save on the most sulphureous metal that we have, that is, on the material that resembles it the most, which is iron."[35]

This is not the only case where Homberg extends a chymical explanation to phenomena that other natural philosophers considered to belong entirely to physics, but as Homberg asserted at the start of the treatise, "everything that exists on the terrestrial globe is the object of chymistry."[36] In the next decade he will extend chymistry's dominion to the stars.

Homberg also explains how his principle of assimilation by similitude operates at the microscopic level. When silver is placed in aquafort, "the volatile acid salts of aquafort, being similar to the salt of silver, join together with and augment the volume of the [salt of the] silver and thereby carry away and break up the other parts of the silver." The idea here is that the acid salt unites specifically with the metalline salt that it resembles, swelling the bulk of each individual salt particle such that the fabric of the metal is split apart into

---

il y ait une fort grande quantité de sel parcequ'il se dissout dans toutes sortes de liqueurs soit acides ou non."

35. VMA MS 130, fols. 201v–202r: "Nous n'avons aucune certitude de ce qui peut être cette matiere mais s'il est permis de juger selon les apparences, et selon les effets certains qui en sont produits, nous pourrons dire avec quelque vraysemblance que cest une matiere sulphureuse et même que cest un soufre metallique qui exhale du globe de la terre etant mis en mouvement et etant poussé par son feu interne que beaucoup de philosophes ont appellé le feu central. La raison pourquoy on le peut imaginer que cest une materre sulphureuse et metallique, que d'une autre nature cest que elle ne s'attache et qu'elle ne s'arrête qu'au metail le plus sulphureux que nous ayons, comme a la matiere qui selon les apparences luy ressemble le plus, qui est le fer."

36. VMA MS 130, fol. 112v: "Tout ce qui est sur le globe terrestre étant lobjet de la chimie. . . ."

minute fragments as the salt particles grow in size.[37] This same mechanism explains how common water "dissolves everything that it touches better than anything else . . . a very little known discovery that will indeed surprise some savants." Water acts as a general solvent of everything because "its substance enters into the composition of all compound substances." When water encounters parts of a substance that are similar to itself, especially saline parts, which are the most similar, "it joins itself to those parts, swells them, and augments them," rendering the particles too large "to be lodged in among the other principles of the compound substance." Thus, this "excess material of one of the principles breaks apart the connection [*liaison*] with the other principles, carries them off one from another and breaks them into parts more or less tiny depending upon how more or less perfectly the principles were mixed together in the composition."[38] Dissolution thus takes place when the solvent assimilates itself to similar parts of the substance being dissolved, leading to a mechanical rupture of the fabric of the compound body into pieces so small they disperse into the solvent.

The same principle of assimilation by similitude governs precipitation. Homberg draws his first example from the production of fulminating gold, a highly explosive curiosity often studied in the seventeenth century (sometimes with disastrous results). Gold is first dissolved in aqua regia to provide a clear yellow solution. A solution of salt of tartar (potassium carbonate) is then added dropwise, which causes the gold to precipitate as a brownish yellow solid, which is the fulminating gold. Homberg first describes how acids

37. VMA MS 130, fol. 158r: "les sels volatils acides de l'eau forte etant semblables au sel de l'argent se joignent ensemble, augmentent le volume d'argent et ecartent par la et brisent les autres parties de l'argent." Homberg refers the reader to the section on gold for an explanation more detailed than his brief recapituation here; unfortunately, that section was not transcribed into the St. Petersburg manuscript. But compare fol. 163r–v, where the salt of the lime used in the cementation of silver "enfle et derange les petites parties en sorte que n'y ayant plus de liaison par le gonflement extraordinaire."

38. VMA MS 130, fols. 129v–130v: "cest une decouverte tres peu connue et qui surprendra bien des scavants. Elle dissout mieux que quoyque ce soit tout ce qu'elle touche . . . sa substance entre dans la composition de tous les mixtes . . . l'eau rencontrant dans un corps des parties de sa nature particulierement des salines dont elle est le dissolvent ordinaire, se joint a ces parties, les enfle, et les augmente. Les ayant augmentée il s'y en trouve une plus grande quantité qu'il n'en faut pour la composition de ce mixte, cest a dire, qu'il y en a trop pour être logé entre les autres principes de ce mixte et pour conserver la même forme que le mixte avoit avant cette augmentation. Ce trop de matieres de l'un des principes desunit la liaison des autres principes, les ecarte les uns des autres et les brise en parties d'autant plus menües que les principes du mixte ont eté plus ou moins exactement mêleés dans sa composition."

are prepared by strongly heating mineral salts; the fire separates the mineral salt into two components: a volatile salt that combines with the water present in the salt and distills over as the acid, and an earthy, fixed salt that remains behind. The volatile acid salts remain dissolved "until they encounter an earthy and fixed saline body that resembles the one from which the fire drove them, and then they resume their former places [in it]." Homberg proves this assertion with a simple demonstration: he mixes each of the mineral acids (spirit of salt, spirit of saltpeter, and spirit of vitriol [hydrochloric, nitric, and sulphuric acids, respectively]) separately with salt of tartar (an earthy, fixed saline substance), thus producing a salt that is identical (or at least very similar) to the mineral salts from which the acids were originally produced. Homberg calls this process a "revivification" of the original salt, and it may well be dependent upon Boyle's famous "redintegration" of niter.[39] Homberg then explains that when salt of tartar is added to the gold solution, the "acid spirits that kept the gold dissolved" immediately desert the gold to take up their former places in the salt of tartar. In short, the acid salt separates from the gold in order to combine with the salt of tartar and "revivify" its original form. The departure of the acid salt allows the gold's own salt particles to shrink back to their former size, and thus the separated particles of gold can regain their original *liaison* and return to the solid state.

Keen experimentalist that he was, Homberg noticed that the precipitated fulminating gold weighs more than the gold used at the beginning. This added weight, he explains, is due to the fact that the aqua regia used to prepare it was composed of aquafort mixed with sal ammoniac (ammonium chloride), but only the salt of the aquafort combines with the added salt of tartar, leaving the sal ammoniac free. Thus, as the particles of gold recoalesce into a solid, they trap a portion of the sal ammoniac and water within their structure. This trapped water and sal ammoniac cause both the weight increase and the explosive properties of fulminating gold. When fulminating gold is heated, the water and volatile sal ammoniac try to escape, but cannot do so because they are imprisoned within the fabric of the gold. As the heat increases, their expansive power finally reaches a critical point, when they violently rupture the microscopic clumps of gold with a dangerous explosion and pass away into the air. Homberg corroborates this claim using his (correct) observation that if the aqua regia is made with spirit of salt rather than

39. Robert Boyle, "A Physico-Chymical Essay . . . on Salt-Petre," in *Certain Physiological Essays* (1661), in *Works of Boyle*, ed. Michael Hunter and Edward B. Davis, 14 vols. (London: Pickering & Chatto, 1999–2000), 2:93–133.

with sal ammoniac, the precipitated gold shows neither increased weight nor explosive properties.[40]

Homberg's treatment of the precipitation of silver further develops these explanations. One part of this treatment is especially important. He notes that when a piece of copper is placed in a solution of silver, the copper dissolves and the silver precipitates. Homberg explains this observation in terms of the "very great difference in power [*force*]" of the aquafort to dissolve different metals. Copper dissolves more easily and in more dilute acid than does silver, and this greater ease of dissolution is, for Homberg, evidence of a greater similarity between copper and acid than between silver and acid. Therefore, when copper is placed in the silver solution, the acid salt leaves the silver to assimilate itself to the copper instead, thereby dissolving the copper and allowing the silver to reagglomerate into solid form and fall out of solution.[41] The same explanation describes how iron precipitates copper from vitriol springs: when iron bars are thrown into the water, "the liquid that holds the copper in solution, finding iron in its path which is easier to dissolve than copper, dissolves it, and the liquid deposits the copper that it held to the same extent that it dissolves the bars of iron."[42] Thus, Homberg's system includes *relative degrees* of similitude; things more alike combine more effectively than things less alike.

Homberg's theory of dissolution and precipitation relies on the notion of similitude, in part an extension of the ancient notion *similis simili gaudet* (like rejoices in like). A given chymical principle assimiliates itself to the principle in another substance that is most like itself. Salts assimilate to salts, sulphurs to sulphurs, mercuries to mercuries, water to water, and in so doing they disrupt the *liaisons* that hold a composite substance together, causing it to burst apart and dissolve. These assimiliations also discriminate between

40. VMA MS 130, fols. 145r–151v. On fulminating gold, see Georg Steinhauser, Jürgen Evers, Stefanie Jakob, Thomas M. Klapötke, and Gilbert Oehlinger, "A Review on Fulminating Gold (Knallgold)," *Gold Bulletin* 41 (2008): 305–17. Homberg's observations of fulminating gold were careful enough that he was the first to notice the differing explosive power (and hence differing composition) of the fulminating gold produced at different points during the precipitation process. Homberg also includes a refutation of the common claim that fulminating gold explodes downward, by noting that a coin placed atop the powder is thrown upward into the air during the explosion. This demonstration, along with another argument about fulminating gold, is taken from Thomas Willis, *Diatribe duae medico-philosophicae* (Amsterdam, 1663), 77–78, thus revealing one of Homberg's early sources.

41. VMA MS 130, fol. 162r–v.

42. VMA MS 130, fol. 210v: "la liqueur qui tenoit le cuivre dissout trouvant en son chemin du fer qui est plus aisé a dissoudre que n'est le cuivre, le dissout et a mesure qu'elle dissout les barres de fer, elle depose le cuivre qu'elle tenoit."

things that differ in *degree* of similarity. An acid salt will forsake one metal to assimilate to a different metal to which it is *more* similar. Likewise, an acid salt will assimilate to the salt within a metal only until an earthy fixed salt to which it is *more* akin presents itself. Homberg's early theory not only differs significantly from contemporaneous and presently better known ideas about chymical reactivity (see below), but also sheds striking new light on the background for the later practices and ideas of his colleagues at the Académie Royale des Sciences.

Homberg's *Semblance*, Geoffroy's *Rapports*, and Duclos's *Symbole*

Simon Boulduc's use of solvents to analyze plant materials, which eventually displaced the previously standard but inferior method of destructive distillation, is probably linked to Homberg's ideas on dissolution. In the first decade of the eighteenth century, Boulduc—who had "worked conjointly" with Homberg throughout the 1690s—used water to remove saline components, and spirit of wine to remove oily (sulphureous) components from pharmaceutical simples, such as ipecac and hellebore. Homberg's early treatise *explicitly* cites the applicability of water and spirit of wine for exactly these purposes, in the context of his explanation of dissolution. Of greater significance, however, is the probable connection of Homberg's early system to the famous and highly influential 1718 "Table des Rapports" of Homberg's protégé Étienne-François Geoffroy, a formulation that gave rise to subsequent affinity theory, a major organizing principle for eighteenth-century chemistry.[43] Much ink has been spilled in attempts to delineate the background to Geoffroy's idea of *rapports*. At one time it was thought to be related to Newtonian notions of force, but this connection has been refuted. More recent literature has linked it to artisanal and other practical experience and know-how, as well as to "general knowledge" held at the Académie.[44] But now that Homberg's early

---

43. Étienne-François Geoffroy, "Table des differens rapports observés en chymie entre differentes substances," *MARS* (1718): 202–12. On affinity theory see Michelle Goupil, *Du flou au clair? Histoire de l'affinité chimique* (Paris: CTHS, 1991); Ursula Klein, "E. F. Geoffroy's Table of Different 'Rapports' Observed between Different Chemical Substances—A Reinterpretation," *Ambix* 42 (1995): 79–100; and Mi Gyung Kim, *Affinity, That Elusive Dream* (Cambridge, MA: MIT Press, 2003).

44. Henry Guerlac, *Newton on the Continent* (Ithaca, NY: Cornell University Press, 1981), 77; William A. Smeaton, "E. F. Geoffroy Was Not a Newtonian Chemist," *Ambix* 18, 1971, 212–14; Frederic L. Holmes, "The Communal Context for Etienne-François Geoffroy's 'Table des rapports,'" *Science in Context* 9 (1996): 289–311; Ursula Klein, "The Chemical Workshop Tradition and the Experimental Practice: Discontinuites within Continuites," *Science in Context* 9 (1996):

theory of dissolution has been revealed by the discovery of the St. Petersburg manuscript, we can identify a striking and more specific resemblance between Geoffroy's rapports and Homberg's theory of selective assimiliation by similitude. Geoffroy's table indicates how a given substance or principle that is combined with another will relinquish that combination in order to combine with a different substance with which it has greater rapport. Homberg's system describes the same transferral of a material from combination with one substance into combination with another, based on relative degrees of *semblance*. Indeed, Geoffroy's own summary of the principle of his table—with the substitution of semblance for rapport—equally well summarizes Homberg's explanation of dissolutions and precipitations. "Whenever two substances that have some disposition to join with one another are found joined together, if a third substance that has more rapport with one of the two comes upon them, it unites with the one by making it release the other."[45]

Suggestively, the first columns of Geoffroy's table deal with acid salts, the same substances with which Homberg most fully works out his theory of assimilation. In fact, the very first column depicts the same sequence Homberg describes, wherein an acid salt that is combined with a *substance métallique* will leave the metal to combine with a *sel alkali fixe*, such as salt of tartar. It is certain that Geoffroy knew of Homberg's early ideas, not only because he owned the original manuscript after Homberg's death in 1715, but even more compellingly because Geoffroy was initially taught chymistry by Homberg himself exactly at the time when the Batavian was writing this treatise. It is not unreasonable to imagine that Geoffroy may even have been taught chymistry by Homberg using parts of this very treatise. This is not to say that Geoffroy's "Table" simply borrows Homberg's notions—Geoffroy had more information and different theories of composition (even Homberg's views on the latter point would change within a few years)—but the main thrust of Geoffroy's rapports shows sufficiently significant parallels with the earlier semblances of his teacher and longtime colleague Homberg to argue for a direct and substantial influence.

These comments on the history of the concept of rapports at the Académie would remain incomplete without mention of its existence *prior* to Homberg, namely, in the writings of Samuel Cottereau Duclos. Several of Duclos's chymical ideas are sufficiently similar to Homberg's to argue for a direct in-

251–87; and Bernard Joly, "Etienne-François Geoffroy (1672–1731), a Chemist on the Frontiers," *Osiris* 29 (2014):117–31, esp. 127–30. See also William R. Newman, "Elective Affinity before Geoffroy: Daniel Sennert's Atomistic Explanation of Vinous and Acetous Fermentation," in Gideon Manning, ed. *Matter and Form in Early Modern Science and Philosophy* (Leiden: Brill, 2012), 99–124.

45. Geoffroy, "Rapports," 203.

fluence. Homberg may well have had access to some of Duclos's writings, and it is further possible that he might have met and collaborated with Duclos during his residence in Paris in the early 1680s. Such a meeting and even collaboration is not at all unlikely, since Homberg arrived in Paris bearing the secret of making white phosphorus at the very time Duclos was struggling (unsuccessfully) to prepare it himself. Unfortunately, no currently known sources can shed light on the matter. In any event, Duclos refers in several unpublished writings to what he calls the *symbole* between two substances, and alludes to how such substances can interact *symboliquement*. While the exact meaning of this term initially seems murky, it is clear from Duclos's usage that he means it to express the tendency of similar things to unite together, a meaning closely tied to the original Greek συμβολή, "a bringing together." For example, Duclos uses the term to describe how sulphureous substances can assimilate and bind with other sulphureous substances.[46] He uses it also to explain differences in the results from the distillation of salts, and why sulphureous solvents dissolve only the sulphureous parts of compound substances.[47] Clearly then, *symbole* refers to a type of affinity that causes similar things to join together, and suggests (but not definitively) the additional feature of Homberg's and Geoffroy's systems, in which a substance will forsake one combination in order to combine with another substance with which it has greater similarity. The close connection of Duclos's *symbole* with later affinity ideas is strikingly expressed in the nearly contemporaneous *Dictionnaire de l'Académie françoise*. The *Dictionnaire* defines the verb *symboliser* as *avoir du rapport*, using the very same word that Geoffroy would later adopt for his "Table." It also gives, as its example of the verb's use, "les alchimistes disent que les Planètes symbolisent avec les métaux," implying that the word was already tied especially to chymistry.[48]

Homberg's Chymical System in Context: Descartes, Lémery, Régis, and Others

Homberg was, of course, not the only person to propose explanations for dissolution and precipitation in the late seventeenth century. In fact, he explic-

46. Samuel Cottereau Duclos, "Experiences de l'augmentation du poids," PV 1, 40–52, at 50–51 (January 1667).

47. PV 5, fol. 7v (12 January 1669); PV 4, fols. 325v–326v (15 December 1668); Victor D. Boantza, *Matter and Method in the Long Chemical Revolution: Laws of Another Order* (Burlington, VT: Ashgate, 2013), 54, 61 (note that the citations therein reference PV 6 for a Duclos memoir that actually appears in PV 5).

48. *Dictionnaire de l'Académie françoise* (Paris, 1694), s.v.

itly differentiates his own theory from the ideas of his contemporaries. He clearly spells out what he calls the "ordinary" alternative to his theory of assimilation by similitude.

> The dissolution [of gold] by aqua regia is ordinarily explained in another manner, namely, that the acid points of the dissolvent, which are supposed to be of a conical shape, being put into motion by the subtle matter, enter easily into the pores of gold that are supposed to have a complementary shape [*même figure en creux*], and are thus very suitable for receiving them. Having entered there with great force thanks to the activity of the subtle matter that pushes them, they carry off by the bulk of their bases the pores of gold that have received their points without resistance, and they seize little invisible parcels of gold in this way bit by bit until the entire mass has been broken into similar parcels.[49]

Homberg here describes the mechanical-chymical system ordinarily attributed to Nicolas Lémery and his *Cours de chymie*. Lémery's system has frequently been celebrated as applying Cartesian ideas to chymistry and hence represents a significant step in "rationalizing" a supposedly irrational or vitalistic chymistry by rendering it mechanical.[50] But as I and others have argued elsewhere, Descartes actually has very little to do with Lémery's ideas.[51] As Bernard Joly aptly remarks, "when one has a look at the *Cours de chymie* he

49. VMA MS 130, fols. 143v–144r: "On explique la dissolution par l'eau regale ordinairement d'une autre maniere, scavoir que les pointes acides du dissolvent que l'on suppose de figure conique, etant mis en mouvement par la matiere subtile entrent aisement dans les pores de l'or que l'on suppose de même figure en creux et par consequent fort propres pour les recevoir; y etant entrés avec grande force par lactivité de la matiere subtile qui les pousse, elles ecartent par la grosseur de leur base les pores de l'or, qui en avoient reçu les pointes sans aucune resistance et en arrachent de cette maniere peu a peu des petites parcelles invisibles jusqu'a ce que toute la masse de l'or ait eté brisée en parcelles semblables."

50. Metzger, *Doctrines chimiques*, 281–338; this viewpoint was popularized in the anglophone context by Richard S. Westfall, *The Construction of Modern Science* (Cambridge: Cambridge University Press, 1971), 68–73, among others.

51. Lawrence M. Principe, "A Revolution Nobody Noticed? Changes in Early Eighteenth Century Chymistry," in *New Narratives in Eighteenth-Century Chemistry*, ed. Lawrence M. Principe (Dordrecht: Springer, 2007), 1–22, at 3–6; and Bernard Joly, "L'anti-Newtonianisme dans la chimie française au début du XVIIIe siècle," *Archives internationales d'histoire des sciences* 53 (2003): 213–24, especially 215–16; Joly, "Could a Practicing Chemical Philosopher be a Cartesian?" in *Cartesian Empiricisms*, ed. M. Dobre and T. Nyden (Dordrecht: Springer, 2013), 125–48; and Joly, *Descartes et la chimie* (Paris: Vrin, 2011), which expounds Descartes's chymical thinking and influence, and notes that Lémery's ideas are "totalement étrangères aux textes cartésiennes" (176, see also 203–4). See also Sylvain Matton, "Cartésianisme et alchimie: à propos d'un témoignage ignoré sur les travaux alchimiques de Descartes," in *Aspects de la tradition alchimique au XVIIe siècle*, ed. Frank Greiner (Paris: SEHA, 1998), 111–84.

published in 1675, one can even wonder how [Lémery] could cast a figure as one of the prominent representatives of Cartesian chymistry."[52] Reference to shaped particles does not automatically indicate a dependence on Descartes. Whereas Descartes postulated mechanical particles bearing a wide variety of specific shapes and sizes that were intended to explain a broad range of phenomena, Lémery's own deployment of shaped particles is almost entirely limited to interactions between pointy acids and porous recipients. This narrow scope points to a more immediate and plausible source for Lémery's model— namely, the apothecary Otto Tachenius's quasi-mechanized version of Van Helmont's embryonic acid-alkali duality.[53] Tachenius (1610–80) endeavored to explain all chymical reactions as interactions between pointed acid particles and porous alkali particles, and explicitly modeled these reactions as sexual encounters between phallic acid particles (which he denominated as male) and vacuous alkali particles (which he denominated as female). He even calls the neutralization interaction between acids and alkalies an "impregnation," and parallels the vigor and driving force of such reactions with animal sexual desire. Tachenius's ideas were widely disseminated and, after being stripped of their rather coarse sexual metaphors, were developed further and more systematically by François de Saint-André, a physician at the University of Caen (and hence a colleague of Lémery), in his *Entretiens sur l'acide et l'alkali*, which appeared first in 1672, just three years before Lémery's *Cours*, and ran quickly through three successive and expanded editions.[54]

Given the absence of rigorously Cartesian ideas from Lémery's *Cours*, it is striking that Homberg mentions a *matière subtile* in his description of the alternative and more "ordinary" explanation of dissolution.[55] Homberg claims

52. Joly, "Practicing Natural Philosopher," 140.

53. On Tachenius, see James L. Partington, *A History of Chemistry*, 4 vols. (London: Macmillan, 1961), 2:291–96; Heinz-Herbert Take, *Otto Tachenius, 1610–1680: ein Wegbereiter der Chemie zwischen Herford und Venedig* (Bielefeld: Verlag für Regionalgeschichte, 2002). A more comprehensive study of Tachenius's chymistry remains a desideratum. The same is true of acid-alkali theory, for which we currently have recourse only to Marie Boas Hall, "Acid and Alkali in Seventeenth-Century Chemistry," *Archives internationales d'histoire des sciences* 9 (1956): 13–28, which is not an adequate sketch of the topic.

54. Principe, "Revolution?," 5–6; and Principe, "Revealing Analogies: The Descriptive and Deceptive Roles of Sexuality and Gender in Latin Alchemy," in *Hidden Intercourse: Eros and Sexuality in Western Esotericism*, ed. Wouter Hanegraaff and Jeffrey J. Kripal (Leiden: Brill, 2008), 209–29, at 224–26.

55. Antonio Clericuzio, *Elements, Principles and Corpuscules* (Dordrecht: Kluwer, 2000), 174 has identified one unquestionably Cartesian idea that appears in the 1687 edition of Lémery's *Cours* (at 582): the definition of fire as a rapid motion of particles about their center. But this statement does not represent a revision of Lémery's earlier ideas; his earlier reliance on discrete

that this subtle matter, the "first element" in Descartes's system, provides the motive force behind the acid points that enables them to penetrate the pores of a metal. Lémery's *Cours* does not mention the Cartesian matière subtile as the motive force behind the pointy acid particles; they move on their own to penetrate porous bodies. Thus Homberg's mention of the matière subtile suggests that he is referring not directly to Lémery, but instead to Pierre-Sylvain Régis's 1690 *Cours entier de la philosophie*. The chymical sections of Régis's *Cours* are indeed indebted to Lémery, whose course of chymistry he attended in the 1680s, but Régis's own book renders the apothecary's material more "Cartesian" throughout.[56] Régis's explanation of how aqua regia dissolves gold, unlike Lémery's, does assert that the "first element" (that is, the matière subtile) accompanies acid particles and gives them the great speed that allows them to enter the pores of gold and break its substance into minute parts.[57] Homberg and Régis certainly knew each other personally. As noted in chapter 1, Régis made reference to Homberg's expertise in optics and experimentation, and Homberg shared his unpublished treatise on generation with Régis.

Homberg himself rarely uses anything identifiable as Cartesian language in his own explanations, and then only during the 1690s. On the rare occasions when Homberg does use Cartesian vocabulary, he does not employ it with a standard Cartesian meaning. His understanding of subtle matter, which he invokes in only one other place in the early treatise, is different from that intended by Descartes himself and used by the generality of Cartesians. For Homberg, subtle matter is simply identical with the matter of fire, a topic that I will treat at greater length below. Given the paucity of anything Cartesian in Homberg's ideas and his explicit rejection of it here in his early treatise, it is clearly incorrect to claim that Homberg brought the Académie's chymistry

---

fire particles continues unchanged in the same edition (for example, 271–72). The Cartesian idea identified by Clericuzio appears in a section on white phosphorus inserted for the first time in the 1683 edition (at 561). The contradiction between this new statement and his previous claims about the nature of fire that he carried over unchanged into the new edition is further evidence of Lémery as a rather indiscriminate editor, rather than an original theorist, one who rarely bothers to harmonize the sometimes contradictory statements he culled at different times from different authors.

56. Joly, *Descartes*, 186–93; Luc Peterschmitt, "The Cartesians and Chemistry: Cordemoy, Rohault, Régis," in *Chymists and Chymistry: Studies in the History of Alchemy and Early Modern Chemistry*, ed. Lawrence M. Principe (Sagamore Beach, MA: Science History Publications, 2007), 193–202.

57. Pierre-Sylvain Régis, *Cours entier de la philosophie*, 3 vols. (Paris, 1690), 2:318; compare also his explanation of the precipitation of silver by copper, at 318 and 321–22.

closer than ever "to the reigning Cartesian discourse."[58] Unlike Descartes and the advocates of the acid-alkali theory, Homberg was reluctant to postulate shapes for the invisible microparticles of matter and then to use those imagined shapes to explain observed chymical results. He does speak of "points" in his early treatise in regard to acids, but goes no further. For Homberg, an acid salt's points do no more than *facilitate* the dissolving action, they are not its *cause*. Water is clearly free of such points (since it does not prick the tongue when tasted, unlike acids), yet for Homberg water is the most general solvent of all. Common mercury is likewise free from points, yet it dissolves all the metals (except iron). The real cause of dissolution, for Homberg, lies in the action of similitude that causes like things to join together.

Homberg's Changing Ideas at the End of the 1690s

By the last years of the 1690s, Homberg's reliance upon the principle of similitude had begun to wane, and was gradually—although fitfully and never completely—replaced by more mechanical explanations of chymical transformations. He flirted with a broader use of particle shape as a way of explaining differential solubilities, then rejected it again quite forcefully, but still later he would be obliged to incorporate some consideration of particle shape into his explanatory principles. While there are no currently known sources that explicitly address the exact causes of Homberg's sequential changes of heart, considerable insight can be gained by a close examination of the experiments he was carrying out during this wavering period toward the end of the 1690s.

At the end of the 1690s Homberg began examining the acid-alkali theory more seriously, and it is that theory which relies most heavily upon particle shape as its explanatory principle. As he had done previously with the Helmontian water theory, Homberg conducted new experiments to test the acid-alkali theory and to reveal its limits. It seems likely that his final dissatisfaction with Van Helmont's water theory in the late 1690s drove him to seek an alternative organizational theory of chymistry. Whatever the cause, in 1699 he began a series of experiments dealing specifically with acids and alkalies, in order to understand their properties and to employ them more accurately. He first devised methods for determining the exact weight of volatile acid salt contained in various acids. He created an improved *aréometre*—essentially a type of volumetric flask—to measure their densities accurately relative to

58. *Pace* Kim, *Affinity*, 6, 11–13, and 83.

water. He carefully neutralized acids with salt of tartar (potassium carbonate), evaporated the residual water, and measured the weight gain in the resultant "revivified" salts, which he took to be equivalent to the quantity of acid salt present in the original acid.[59] These experiments are of considerable interest as being among the earliest attempts to devise a quantitative analytical technique that would later be known as titration. Homberg does not describe his method in detail, but it is possible that he used color indicators like *tournesol* to detect the endpoint. He then used his results to explain the different dissolving power of different sorts of aqua regia. Rather than appealing to similitudes, he noted instead that according to his measurements the varieties of aqua regia that dissolved more gold simply contained more acid salt: "one produces double the effect of the other only because it contains twice the amount of acid salt." He explicitly preferred this new quantitative explanation to the usual "recourse to the softness of the points of one of the acids which are easily blunted, and the hardness of the points of the other that act for a longer time and more powerfully carry off the little particles of gold, or by some similar explanation, which has at best but a *feeble appearance of truth.*"[60]

Homberg briefly changed his mind the next year after conducting analogous experiments to measure the relative neutralizing power (*force*) of various "earthy" (that is, water-insoluble) alkalies such as lime. He discovered that when 1 ounce each of aquafort and spirit of salt of the same density (thus, in his opinion, containing the same weight of dissolved acid salt) were poured separately over an excess of alkali, the aquafort almost always dissolved more alkali than the spirit of salt. He initially concluded that this difference was due to the "different configurations" of the two acid salts and the "different pores of the alkalies that must receive them"—a notion seemingly more in line with typical acid-alkali explanations of the day. But he explained that an acid composed of blunter points (*cones plus obtuses*) dissolves less because each acid particle is fatter and so weighs more than a thin, sharp one; thus, equal *weights* of acid salts could contain either a *smaller number* of fat, blunt par-

59. Wilhelm Homberg, "Observation sur la quantité exacte des sels volatiles acides contenus dans tous les differens esprits acides," *MARS* (1699): 44–51; PV 18, fols. 245v–251v (29 April 1699). Homberg had asked Boulduc the year before to carry out similar experiments on vinegar in order to verify his own results; PV 17, fols. 153v–156r (2 April 1698). In the context of the analysis of plants project, a similar but less successful attempt to know "assez precisement combien il y a de sel aceteux dans une certaine quantité de liqueur acide" by measuring the weight of powdered coral a given distillate could dissolve had been briefly carried out at the Académie in early 1680; see Cornell University Library, Lavoisier Bound MSS, 6:331–41, 445–51.

60. Homberg, "Sels volatiles acides," 50; emphasis added.

ticles or a *larger number* of thin, sharp particles. Since each acid particle acts as a "cleaver [*tranchoir*]," the acid containing more cleavers can divide (that is, dissolve) more alkali.[61] Significantly, although Homberg does now appeal to particle shape, he nevertheless insists upon combining such explanations with weight analyses—the experimental probe that always remained paramount to him—so that his final answer turns out to be an essentially quantitative one. Sharper points themselves do not directly explain the greater solvent action; it is simply that there are more of them in an equal weight of acid. This conversion of a shape argument into a quantitative argument is not typical of acid-alkali explanations, but it illustrates Homberg's strong and repeated preference for concrete, measurable quantities over imagined and unobservable shapes.

Yet even with this modification, Homberg still remained uncomfortable with an explanation that seemed to venture too far from laboratory observations into mere speculations about invisible shapes. Between his oral presentation of the memoir in February 1700 and its publication in 1703, he reworked his experimental protocol and turned his explanation around so that it relied first and foremost on the different *quantities* of acid salt in the two acids, and only secondarily on the shapes of the particles. Thus, the final version of the paper makes no explicit claims about particle shapes, save to refute Lémery's claim that such microscopic particle shapes can be inferred from the shapes of the crystals produced from each acid.[62] Interestingly, Fontenelle's summary in the *Histoire* of Homberg's paper clearly refers to the 1700 procès-verbaux version rather than to the revised version that actually appeared in the *Mémoires* in 1703, which may suggest that Homberg's revisions were made very late in the publication process. In any event it underlines the peril of relying on Fontenelle's secondary accounts in the *Histoire*.[63]

A similar transformation occurs in Homberg's contemporaneous treatment of the replacement reactions that form the backbone of affinity theory. His early treatise described how an acid salt releases a metal to which it is joined in order to join to another with which it has a greater *semblance*. Greater semblance is revealed by a metal's easier solubility in the given acid,

61. PV19, fols. 61r–67r (20 February 1700).

62. Wilhelm Homberg, "Observations sur la quantité d'acides absorbées par les alcalis terreux," *MARS* (1700): 64–71; refutation of Lémery (unnamed), at 66. Lémery had made this argument in the 1677 second edition of his *Cours*, at 22, and retained it for all later editions. Homberg notes that the salts produced by dissolving different substances in the same acid crystallize in a variety of shapes; therefore, the shape of the acid's particles cannot be the determining factor in the shape of the crystals.

63. Bernard de Fontenelle, "Sur la force des alkali terreux," *HARS* (1700): 48–50.

and hence more easily dissoluble metals precipitate those that are harder to dissolve. In a 1701 paper describing a new and improved method for refining silver, Homberg revisits this phenomenon. He preserves unchanged the same sequence of replacements he had previously observed—copper replaces silver, and iron replaces copper—and continues to rationalize this order based on ease of dissolution. But now ease of dissolution is in turn based not on semblance, but on the "different texture of the metals." One metal dissolves more easily in acid because it has larger pores that allow for the easier entry of the acid particles and the metal's subsequent disaggregation. Critically, Homberg's new explanation is *not* based on some imagined notion of pore shape or postulations about the metal's invisible microstructure akin to those dreamed up by various more or less Cartesian contemporaries; it is again based strictly on a *quantitative measurement*. Metals vary widely in density, and those with a higher density must have their "texture more compressed [*tissu plus pressé*]"—that is, they must contain more solid matter and thus less porosity per unit volume. Therefore, the easily perceptible and measurable density of a metal reveals its invisible degree of microscopic porosity. Thus, density predicts, and indirectly explains, the observed replacement series. "Iron being less dense than copper, and copper less dense than silver, one is dissolved more easily than another in that order, and likewise one precipitates the other."[64] Here again, while Homberg moves toward more mechanical explanations of chymical phenomena, he does so only to the extent that claims about invisible microstructures can be plausibly and logically revealed by direct measurement and observation. One can sense him turning reluctantly toward the "more usual" mechanical explanations based on shape, and then dismissing them again when he identifies an experimentally verifiable solution that he finds more solid and satisfying. Again and again Homberg makes clear that he places a premium on the witness of the senses and of direct measurement in strong preference to speculations about the imperceptible. The first installment of his 1702 "Essais" will make this preference explicit and definitive.

Within a couple of years, Homberg decided to reject the acid-alkali the-

---

64. Wilhelm Homberg, "Observations sur le raffinage de l'argent," *MARS* (1701): 42–46, at 45; PV 20, fols. 72r–75r (26 February 1701). Homberg's observation of the relationship between relative density and the replacement sequence is completely accurate for the substances he knew, but largely coincidental. The argument that greater density in metals reveals lesser porosity dates back to the medieval Geber; see his *Summa perfectionis*, ed. William R. Newman (Leiden: Brill, 1990), 159–62, 471–75, 725–26. Homberg repeated again the "remarkable fact" that more easily dissolvable metals displace less easily dissolvable ones from their combinations in his "Essays de chimie," *MARS* (1702): 33–52, at 44.

ory and its claim that all chymical transformations depend upon the interaction of an acid with an alkali. After making more careful distinctions among the fermentations, effervescences, and ebullitions that proponents of the acid-alkali theory commonly conflated, Homberg showed that many such processes—even violent ones—do not involve the mutual interaction of acids and alkalies. In fact, building upon an observation made first by Olaus Borrichius (1626–90), Homberg showed that the most violent interactions actually occur between an acid and a sulphur; oil of vitriol (sulphuric acid) mixed with oil of turpentine can grow so hot it self-ignites.[65] He demonstrated this conclusion in his customarily dramatic way to the audience at the Académie's public assembly at the Louvre in April 1701. "In order to combine a philosophical spectacle with his reasonings," Homberg added a few drops of concentrated aquafort (nitric acid) to oil of cinnamon mixed with gunpowder. After a few moments the mixture burst spontaneously into flames. What the spectators thought of the ensuing smoke and acrid vapors remains unrecorded.[66]

Starting Over: Homberg's Curtailed Second Attempt, the "Essais d'elemens de chimie" (1702)

Sometime around 1700, Homberg embarked on a fresh attempt to write a comprehensive and explanatory textbook of chymistry. He presented the Académie with the first two chapters of a work entitled "Essais d'elemens de chimie" in February 1702.[67] This new treatise would be entirely different in form from his aborted treatise of the 1690s, and an even greater departure from the standards of the didactic tradition. We cannot be fully certain of

65. Borrichius initially used Venice turpentine and nitric acid; "Efficere, ut duo spiritus tactu frigidi, invicem confusi flammam edant," *Acta medica et philosophica Hafniensia* 1 (1671–72): 133–35. It is noteworthy that Borrichius reports that he performed this experiment in Paris in the presence of Abbé Boucaud, a close associate of Duclos, and thus some report of it may have come down to Homberg locally. Frederick Slare also published on the inflammability of essential oils with nitric acid in "An Account of Some Experiments Relating to the Production of Fire and Flame," *Philosophical Transactions* 18 (1694): 201–18.

66. *Mercure galant* (April 1701), 2 vols, 1:112–13; Wilhelm Homberg, "Observations sur quelques effets des fermentations," *MARS* (1701): 97–101; and PV 20, fols. 115v–120r, at 120r: "Il en a fait voir a l'assemblée les principales Experiences." See also PV 20, fol. 214v. One difference between the oral presentation and the published memoir is that Homberg softened his tone about how partisans of the acid-alkali theory had conflated different types of reaction; the initial "le vulgaire confond" (116r) became "on confond ordinairement" (97).

67. Wilhelm Homberg, "Essais d'elemens de chimie," PV 21, 61r–73v (15 February 1702); Homberg's reading of his text occupied four meetings of the Académie (1, 4, 8, and 15 February).

Homberg's exact vision for the scope of the work because the project was derailed just two months after its first two installments were presented and subsequently reformulated. But judging by the title and the two extant chapters, Homberg intended his new treatise to focus primarily, if not exclusively, on the chymical *principles* rather than on chymical *substances* (see table 3.1). Its aim was to establish firm foundations for understanding and explaining chymical transformations. In a clear break from tradition, the text would contain few if any descriptions of individual chymical substances and, even more strikingly, no preparative recipes.[68] The narrow scope of the topics and the exclusion of the material that normally filled nine-tenths of the standard chymical texts then available meant that this treatise was clearly aimed at a very different audience than books in the didactic tradition like Glaser's or Lémery's. Useless as a workshop manual for apothecaries, its appeal would be to the far smaller number of chymists and other readers devoted to the same natural philosophical agenda as Homberg himself.

The first chapter, on "the principles of chymistry in general," retains much of Homberg's earlier thinking about the nature of chymical explanations and chymistry's status, but also marks a significant shift in his ideas about the composition of compound substances. It opens with a rather standard definition of chymistry as the analysis and synthesis of compound bodies by means of fire, and then states that "chymistry is one of the parts of natural philosophy [*la Physique*]."[69] While it might not be immediately apparent, this statement makes an implicit claim about the goals and status of chymistry. *La physique* was defined as the "knowledge of natural causes which explain all the phenomena of heaven and earth" and a *physicien* as "one who knows nature and explains its effects."[70] Thus, if chymistry is a part of this larger field, it too must be predominantly about *explanation and causal knowledge*. In contrast, the Académie's secretary Fontenelle habitually pre-

68. Homberg does mention an intention to prepare a supplemental "Cours d'operations" as an appendix to his "Essais" (Wilhelm Homberg, "Essays de chimie," *MARS* (1702): 33–52, at 50); such a work, so far as we know, was never begun, and all mention of it was deleted from his final version of the work.

69. Homberg, "Essais d'elemens," fol. 61v.

70. Antoine Furetière, *Dictionnaire universel* (Paris, 1690), s.v. Interestingly, Furetière separates *la chymie* from *la physique* in his description of the Académie's work, contrary to what Homberg would have done (s.v. "Académie"). The breadth of *la physique* in this period is well treated by John L. Heilbron, *Electricity in the Seventeenth and Eighteenth Centuries: A Study in Early Modern Physics* (Berkeley: University of California Press, 1979), 9–13. Holmes's assertion that Académie chymists used the term differently ("Communal Context," 293–94) is not supported by evidence.

sented chymistry as little more than an adjunct to medicine and a practice useful primarily for preparing pharmaceuticals. Indeed, as outlined at the end of chapter 2, most of those who held positions within the division of *la chimie* at the Académie at this time were apothecaries—with the notable exception of Homberg. The content of the courses of chymistry taught at the Jardin du Roi and the general orientation of books in the French didactic tradition reinforced this practical orientation toward medicine. Thus, when in the following line Homberg uses the unusual term *la physique chimique*, he is referring specifically to a different style of chymistry, one that explains and understands, one that pursues a natural philosophical agenda, not one whose primary function is the production of materials pharmaceutical or otherwise.

Homberg's term likewise implies a different style of natural philosophy—namely, one improved and enlightened by chymical knowledge and results. He bluntly declares that chymistry gives information about the natural world that is inherently superior to that proffered by the ordinary *la physique*. The natural philosophers' ideas about the principles, by which he means speculations about matter at its most fundamental level, "are at best but plausible suppositions. These general principles are too vague, and the human spirit has not yet been able to say anything incontestable about the figure and arrangement of these first matters."[71] With this line he waves off as "too vague" the cogitations of Cartesians as well as other theorizations and speculations about the shapes, forms, and ultimate nature of matter. This opinion is fully in line with Homberg's repeated distrust of imagined particle shapes and his insistence on direct observations and quantitative measurements instead. Far preferable, he writes, is a "chymical natural philosophy [*physique chimique*], which consists only of experiments and the exposition of facts, and seeks solid truth." This sort of chymistry has established, Homberg continues, in place of the vague general principles proffered by the ordinary *physiciens*, another sort of principles, namely the chymical principles, which are "more material and sensible." By *sensible* Homberg means "apprehendible by the senses," a sentiment that continues his reluctance to stray too far beyond the observable and the measurable, and in close accord with the very nature of chymistry itself as a sensual science of observable qualities. Chymists know their materials and analytical products through constant applications of their own senses—sight, smell, taste, and touch. Using the chymical principles, chymistry "easily explains its own operations in its own way and thereby

71. PV 21, fol. 61v: "ce qui ne sont tout au plus que suppositions vraisemblables. Ces principes généraux étant trop vagues, et l'esprit humain n'ayant pas encore pû déterminer rien d'incontestable sur la figure et sur l'arrangement des 1res matières."

knows more distinctly the substances it examines."[72] Given the bold nature of these expressions, it is scarcely surprising that after Homberg read this section to the assembled academicians, "several objections were made."[73] The exact nature of these objections remains unknown, but in the published version Homberg did tone down his rhetoric about the superiority and independence of chymistry—but only temporarily, as we shall see.

Homberg then lists the number and identity of the chymical principles, giving an account that diverges from the inventory he presented in the earlier treatise. In minerals and metals, he now replaces the Paracelsian *tria prima* with the newer pentad of Étienne de Clave (and others)—salt, sulphur, mercury, water, and earth. For vegetable and animal materials, the confusion of the early treatise has resolved itself into a triad: salt, earth, and water. Oil is now gone, and Homberg explains that "in the analysis of oils that I have performed"—clearly a reference to his work in 1695—he decomposed them into salt, water, and earth, and therefore oil cannot be considered one of the chymical principles.[74] Homberg also classifies the principles in terms of activity or passivity. For Homberg, sulphur and salt are both active principles ("because they alone act and make the others act"), while earth and water are passive ("they never act and serve only as vehicles for the others"), and mercury is "middle" ("it never acts on its own, but becomes capable of acting when joined to sulphur or salt").[75] Grouping the chymical principles as active or passive was standard practice among the textbook writers, but Homberg's division is unique in demoting mercury to a middle state. Glaser

72. PV 21, fol. 61v: "La Physique Chymique qui ne consiste qu'en expériences et en expositions de faits et cherche la verité certaine, a établi une second sorte de principes plus materiels et plus sensibles que ne sont ces Principes généraux, par le moyen desquels elle prétend expliquer aisement et à sa mode ses propres operations, et de conoître par là plus distinctement les Corps qu'elle examine par ses Analyses." On *sensible*, see Furetière, *Dictionnaire*: "Qui fait impression sur les sens, qui en frappe les organes." Appropriately for Homberg's usage, Furetière includes as an example: "Les atomes ne sont sensibles ni à la veüe, ni au toucher." For important further background to the idea of the sensual nature of chymical principles, see Joel A. Klein, "Corporeal Elements and Principles in the Learned German Chymical Tradition," *Ambix* 61 (2014): 345–65; and in the context of later chemistry, Lissa Roberts, "The Death of the Sensous Chemist: The 'New' Chemistry and the Transformation of Sensuous Technology," *Studies in the History and Philosophy of Science* 26 (1995): 503–29.

73. PV 21, fol. 55v (4 February 1702); compare the somewhat watered-down version that was eventually published: Homberg, "Essays de chimie" (1702): 33.

74. PV 21, fols. 62r, 63r. It is slightly odd that only one year previously, Homberg still referred to the "four principles" in plants (that is, including oil); see Wilhelm Homberg, "Observations sur les analyses des plantes," *MARS* (1701): 113–17, at 117.

75. PV 21, fol. 62r.

and Lémery had both followed the usual tradition of calling mercury, sulphur, and salt active, and earth and water passive—a notion based on which principles show medicinal properties. Homberg thus implicitly changes the basis of defining activity and passivity from a pharmacological to a physical criterion. Homberg had already voiced his choice of salt and sulphur as the two active principles a year earlier in his study of fermentations, probably as a result of his experience that the most violent reactions occur between acid salts and oils.[76]

In Homberg's treatise of the 1690s, salt was very clearly the main chymical actor. Homberg invoked salt as the cause of observable chymical phenomena more than all the other principles put together. Salts also received the lion's share of Homberg's attention during his experimental study of the principles during the 1690s, especially in terms of his attempt to interconvert them and transform them into primordial water. It is again possible that there was some influence in this direction from Duclos—either by direct contact during Homberg's first years in Paris, or through Duclos's extant manuscripts. Duclos had similarly emphasized the primary role of salt in chymistry, and placed sulphur second, just as Homberg did in his initial plan for the "Essais." Duclos had also left behind a massive work on salts with which others in the Académie, and perhaps Homberg, may have been familiar.[77] Homberg's previous focus on salt explains why he chose to put the essay on salt first in his new treatise on the chymical principles, and he devotes a large part of that essay to classifying the various species of salts. Mineral salts and the essential salts of plants, since they can be divided into simpler components, cannot be counted as principles. The three classes of salts identified by Van Helmont and which had become standard by Homberg's time—acid, urinous, and lixivial—are simpler, and since "it is not yet in our power to separate them" into anything yet more simple, they can legitimately be called principles in that sense. Here Homberg takes the pragmatic route of defining as a "principle" anything that has not *yet* been successfully divided experimentally into something more simple, the same route Lavoisier would take two generations later in defining an element.[78] Nevertheless, for Homberg these salts must still be compound substances rather than the pure chymical salt principle, otherwise they would not differ from one another; we simply do not *yet* have the power to break them down further and thereby isolate the one true primordial salt principle they all contain and that is the foundation

76. Homberg, "Fermentations," 101; PV 20, fol. 120r.

77. Samuel Cottereau Duclos, *Dissertations sur le sel* (1677), BNF MS fr. 12309.

78. PV 21, fols. 63v–64r.

of their common properties. According to Homberg, that primordial salt principle, like the sulphur principle, never appears in a pure state but only in combination with another principle that acts as its vehicle.[79] The background to this slightly odd statement may lie in Homberg's inability in practice to isolate a single primordial salt principle from any of the three classes of simple salts. His researches during the 1690s did succeed in interconverting the various species of salts—he volatilized the fixed and fixed the volatile—which demonstrated that they must have a common component, the putative salt principle, but he could never isolate that substance. His conclusion here, perhaps simply a practicality, is that the salt principle just cannot be isolated experimentally.

Now, if both the salt principle and the sulphur principle must always be joined to one of the other principles as a vehicle, and given that Homberg lists *three* species of salts (acid, urinous, and lixivial) that are equally close to the true salt principle, it would seem logical to take the step of saying that each of these simple salts is composed of the salt principle combined with one of the three nonactive principles—earth, water, and mercury. He does demonstrate by experiment that lixivial salts contain earth, presumably as their vehicle, but goes no further.[80] One wonders if he considered that the earth vehicle would explain the fixity of lixivial salts, leaving the other two simple salts, acid and urinous, to have water and mercury respectively as their vehicles, both of which are volatile like the acid and urinous salts themselves. It all *seems* to work out very neatly, but Homberg does not make such connections explicit. Perhaps, although he was confident that "we almost [*à peu près*] know" the composition of the three species of salts, that was not yet sufficient for him to make the leap without fuller experimental evidence.[81]

Much of the 1702 essay on salt details the interconversions of various salts and their interactions with each other and with other substances. As such, the chymistry is strikingly—almost confusingly—dynamic rather than descriptive or taxonomic. One could derive in retrospect a classification scheme for the multiple kinds of salts Homberg cites, but he seems less interested in such an exposition (as helpful as it would have been to readers!) than in exploring observations and experiments, and what each one of them says about composition, interaction, and causation. Thus, virtually all traces of the "natural history" approach visible in the early treatise have now evaporated. Hom-

79. PV 21, fol. 62v: "ne sauroient paroître à nos yeux sans être joints à quelqu'un des autres trois principes qui les servent de vehicule."

80. PV 21, fol. 72v.

81. PV 21, fol. 64r.

berg takes up again the old problem of why different acids dissolve some substances and not others. His former principle of similitude is not entirely absent, but has been turned largely on its head and more closely coupled with mechanical explanations. The key change here is a new idea: that the acid salt in spirit of niter (aquafort, our nitric acid) contains a sulphureous component. Based upon the long-standing tradition within chymistry that color is an indicator of sulphur, Homberg argues that the red vapors of aquafort witness the presence of a sulphur. When, why, and how Homberg adopted this new view of the constitution of aquafort is unclear. The only other writer of whom I am aware who claimed a sulphureous nature to aquafort was Duclos—once again suggesting a connection to the thought of Homberg's predecessor at the Académie. Homberg's claim of aquafort's sulphureous character was criticized by Lémery as soon as it was presented.[82] In any event, this sulphureous character now provides a new explanation of why aquafort cannot dissolve gold. Gold, as demonstrated by its own beautiful color, contains a great deal of sulphur. In Homberg's thinking of the previous decade, this similarity to aquafort would have *guaranteed* the acid's ability to dissolve the precious metal, but now it *prevents* it. The aquafort instead "finds the interstices of the gold filled with a matter similar [*semblable*] to itself, which, already occupying the place which ought to receive the sulphureous acid of niter, prevents the spirit of nitre from entering there."[83] Homberg's notion now seems to be that a sulphur can occupy only a particular *locality* within a metal's composition, and if that position is already filled, no more sulphur can be added; gold's own sulphur *blocks out* the aquafort's sulphur.

In contrast, the acid salt in spirit of salt (our hydrochloric acid) can dissolve gold because it contains no sulphur—its vapors are colorless—so it can enter the gold and break it apart. The same acid cannot, however, dissolve silver because silver, being free of sulphur (it is a white metal), does not have its interstices narrowed by the presence of sulphur, hence these interstices are larger than those in gold, but the spirit of salt's points are so thin (*deliées*) they pass right through the spaces without breaking apart the fabric of the metal. In aquafort, the points are fatter wedges that can break silver apart and so dissolve it. This new system boils down to little more than the notion that aquafort's points are accommodated to silver's pores, and aqua regia's to gold's—which does not seem like much of an explanation. It retreats from the more universalizing endeavor of Homberg's earlier treatise (semblance explains everything uniformly) into more *ad hoc* explanations.

82. Bourdelin Diary, fol. 46v (8 February 1702).
83. PV 21, fol. 66r–v.

It also retreats from Homberg's former reluctance to use explanations based on particle shapes that are not directly demonstrable from observation and experiment; indeed, it even seems to undercut some of the bold statements about the way that chymistry operates, with which he began the "Essais." To make matters yet more confusing, Homberg also now expands his refutation of those who would extrapolate the invisible shape of acid particles from the observable shape of the crystals formed by substances dissolved in that acid. Instead, he says, the crystal shape is more attributable to the alkalies the acid has dissolved, "which serve them as their bases."[84] The noteworthy point here is that Homberg thus seems to be the first chymist to use the term "base" in a sense very close to more modern chemical usage. For him, the alkali is literally the *base substance* into which an acid lodges itself, the foundation into which the acid salt fixes itself, thus producing a new salt that Homberg calls a *sel moyen* (middle salt), *sel salé* (the older, Helmontian term), or *sel composé*. The credit for introducing the word "base" with this meaning is ordinarily given to Guillaume-François Rouelle (1703–70) for work published in the mid-eighteenth century.[85]

It cannot be said that Homberg's 1702 salt essay is entirely successful in terms of coherence, explanatory clarity, or fulfillment of his expectations for a chymical theory. He might have planned to rethink its structure and its theoretical foundations in the time between its oral presentation and its publication. But whatever such plans he might have had were abandoned two months later, when his first experiments with a new scientific instrument revealed unexpected phenomena that overturned fundamental parts of his developing theory of chymistry. Between the 1702 oral presentation and the 1704 publication, Homberg rewrote both essays extensively, changing his most basic ideas about the principles, deleting large sections of text, and composing a considerable amount of new text that doubled the length of the salt essay. This dramatic shift in Homberg's ideas, and the resultant divergence between the eventually published "Essais" and their original oral format, has never before been noticed. The story of this shift not only showcases how new and unexpected experimental data could redirect scientific theories and research programs, but also highlights the role played by costly scientific instruments and by the patronage and social position needed to acquire them.

The years around 1700 reveal Homberg casting about for satisfactory ex-

84. PV 21, fol. 69r: "qui leur servent de bases."

85. William B. Jensen, "The Origin of the Term 'Base,'" *Journal of Chemical Education* 83 (2006): 1130. Homberg had already used the term *baze* in the same sense once in VMA MS 130, fol. 146r.

planatory principles upon which to ground his vision for chymistry. During this period, his ideas were in an almost constant state of flux as he explored system after system, only to revise, rewrite, and reject them as they failed to live up to his expectations for a coherent explanations based on observable experimental results. We see Homberg deeply immersed in laboratory operations as he struggled to devise a comprehensive theory of chymistry that could explain satisfactorily the baffling complexity of chymical properties and reactivities. His consistent aim was to pass beyond a merely preparative approach to and deployment of chymistry. Material production, of course, remained important to him, but generating a satisfactory explanatory framework was paramount. He explored Helmontian ideas and acid-alkali theory, two of the main contenders for the role of such an explanatory system, and eventually dismissed them both. He developed a theory of selective assimilation by similitude that foreshadowed affinity theory and that seemed to work well in a wide range of cases, and he built his first attempt at composing a textbook of chymistry around it. He flirted now and then with more fully mechanical reasonings, but found them too ad hoc and generally too disconnected from the sensual experience of matter and its transformations to satisfy him. Among his requirements for a satisfactory theory of chymistry, perhaps none was more crucial than clear support and demonstration through the *sensual, observable, and measurable* experience of chymical experiments and phenomena. To his mind, other natural philosophers (*physiciens*), who were inadequately versed in chymistry and its methods, merely indulged their imaginations about the microscopic structure and activities of matter, without basing their airy speculations on the concrete realities of sensible and measurable data provided by practical experience in the chymical laboratory. The apothecaries, in contrast, cared mostly about practical preparations and stopped short of seeking out foundational chymical theories. Homberg endeavored to steer a middle course for chymistry. He sought an explanatory and organizational theory as tightly linked as possible to direct observations of material transformations. Homberg would continue this demanding endeavor throughout the following decade, pursuing fresh ideas down new avenues that he had perhaps scarcely glimpsed in previous years.

4

A New Chymical Light

The year 1702 marked a significant turning point in Homberg's life and thought. In that year he gained one of the most powerful and generous patrons in France, the most costly and impressive scientific instrument in Europe, and a new chymical laboratory that reportedly put all others to shame. These developments led to new experimental results that quickly forced Homberg to rethink his ideas about chymistry and the chymical principles, and set him on a new investigative course. The critical trigger for these changes occurred when the twenty-seven-year-old Philippe II, Duc d'Orleans (1674–1723), nephew of Louis XIV and future Regent of France, chose Homberg as his "tutor" in chymistry. How this arrangement actually came about is open to some speculation. Fontenelle claims that Homberg was recommended to Philippe in this capacity by his preceptor, the Abbé Guillaume Dubois (1656–1723), later a cardinal and Philippe's first minister during the regency. Over a decade earlier Dubois had spelled out a course of instruction for Philippe that included chymical topics, along with much else.[1] But there were many other routes through which Homberg might have been recommended to Philippe; several of Homberg's colleagues were already connected with the house of Orléans. Homberg's fellow *pensionnaire* chymist and former collaborator, Simon Boulduc, was apothecary to Philippe's mother, Elisabeth Charlotte,

---

1. Bernard de Fontenelle, "Éloge de M. Homberg," *HARS* (1715): 90. On Dubois, see Guy Chaussinand-Nogaret, *Le cardinal Dubois, 1656–1723: une certaine idée de l'Europe* (Paris: Perrin, 2000); for Dubois's schedule of instruction for Philippe, see Victor de Seilhac, *L'abbe Dubois, premier ministre de Louis XV*, 2 vols. (Paris, 1862), 6–11 and 185–205, chymical topics are mentioned at 194.

the Princesse Palatine.[2] The academician Joseph Sauveur (with whom Homberg shared an interest in mechanical contrivances) had been mathematics tutor to Philippe and was at this time discussing music and acoustics with him.[3] Claude-Jean-Baptiste Dodart (1664–1730), son of the academician Denis Dodart, was *premier médecin* to Philippe's father until the latter's death in mid-1701, and Claude Gendron—the physician who in 1689 tried to get Homberg an audience with Bossuet in order to facilitate admission to the Académie—was one of his quarterly physicians; both continued for a time as Philippe II's physicians.[4] Philippe II also attended the public *assemblée* in April 1701, where he not only heard Homberg speak, but also saw, heard, and smelled his fiery demonstration of detonating a mixture of cinnamon oil and gunpowder with a drop of nitric acid.[5] Thus, once the Duc d'Orléans developed an interest in chymistry, several possible conduits led to Homberg.

Homberg's Laboratories and His Life in the Orléans Household

Homberg accepted Philippe's invitation and the salary that came with it. The Duc d'Orléans responded immediately by building a magnificent laboratory for them at the Palais Royal, his official residence. Fontenelle praised this workspace as "the best furnished and most magnificent laboratory that chymistry had ever had."[6] The exact location of this laboratory within the Palais Royal has hitherto remained unknown. But an informative letter from Philippe's mother, Elisabeth Charlotte, indicates that it was located directly

2. On Boulduc and his family of apothecaries, see Christian Warolin, "La dynastie des Boulduc, apothicaires à Paris aux XVIIe et XVIIIe siècles," *Revue d'histoire de la pharmacie* 49 (2001): 333–54; 50 (2002): 439–50; 51 (2003): 103–10.

3. Bernard de Fontenelle, "Éloge de M. Sauveur," *HARS* (1716): 79–87, at 85–86; PV 16, fol. 107r (a letter from Philippe II to Sauveur on probabilities in dice games, 27 April 1697).

4. Jean-Jacques Peumery, "Les Dodarts, père et fils, médecins du roi," *Histoire des sciences médicales* 34 (2000): 39–46; Philippe Champault, "Les Gendron 'médecins des rois et des pauvres,'" *Transactions of the Royal Society of Canada*, 2d ser., 6 (1912): 35–120, at 90. See also *L'Etat de la France*, 2 vols. (Paris, 1702), 2:121–22. David Sturdy, *Science and Social Status: The Members of the Académie des Sciences, 1666–1750* (Woodbridge, UK: Boydell Press, 1995), 231, suggests that Denis Dodart played the major role in the choice of Homberg, but this seems unnecessary given the multiple routes cited here.

5. PV 20, fol. 109r (6 April 1701): "L'assemblée étant publique . . . et honorée par la presence de S. A. R. Mgr Le Duc de Chartres." Philippe acquired the title Duc d'Orléans only upon the death of his father three months later. Philippe was also present at the 26 April 1702 public assembly (PV 21, fol. 149r).

6. Fontenelle, "Éloge de Homberg," 90.

FIGURE 4.1. The location of Homberg's laboratory in the Palais Royal, perspective view from the north. Detail: The Jardin des Princes. Homberg's laboratory was on the ground floor, behind the tall arched-top windows with French doors opening into the garden; Philippe II's apartments were located directly above the laboratory. Engraving by Jacques Rigaud (1680–1754), "Vûë du Palais Royal," c. 1730. Author's collection.

under Philippe's own living quarters.[7] The laboratory can therefore be identified as the high-ceilinged rooms with tall French doors opening out onto the private Jardin des Princes (figs. 4.1 and 4.2). Exactly how much of the space beneath Philippe's apartments was reworked into a laboratory is open to some question, but since Elisabeth Charlotte emphasizes it was "a whole apartment," one can be fairly confident that the laboratory occupied a considerable part, and perhaps all, of the space. This ground-floor location on the south side of the Jardin des Princes provided not only easy access from

7. Elisabeth Charlotte to Sophie, 27 October 1709, in *Aus der Briefe der Herzogin Elisabeth Charlotte von Orléans an die Kurfürstin Sophie von Hannover*, ed. Eduard Bodemann, 2 vols. (Hannover, 1891), 2:231: "Mein sohn hatt im palais Royal ein gantz apartement unter dem großen apartement zum laboratoire gemacht."

Philippe's apartments, but also direct access to the garden, which was necessary for the crucial experiments that were done outdoors there (see below) and undoubtedly very helpful as well for the occasional ventilation of unpleasant or unexpected laboratory effluvia.

The new laboratory at the Palais Royal represents yet another, and very substantial, improvement to Homberg's chymical workspace and hence to his ability to carry out his researches. It was at least the fourth laboratory he had in Paris. In the early 1680s, before his admission to the Académie, he maintained a private laboratory, probably a rather small space in his lodgings equipped with furnaces and distillation equipment, but apparently sufficient for him to continue his work. In 1684, he set up another laboratory in the rented house where he worked on the Abbé de Chalucet's project to turn mercury into silver. This must have been a rather makeshift workspace, assembled rather quickly in a domestic space and used for only three months, but again, adequate for distillations and suchlike. Upon his admission to the Académie in 1691, Homberg was given control of the institution's official chymical laboratories at the Bibliothèque du Roi, which Duclos had designed in the late 1660s. Bourdelin's expense account indicates that there were actually two separate chymical laboratories at the Bibliothèque—one on the ground floor and the other on the floor above. The same source also indicates that the furnaces and other outfitting required regular renewal through the 1670s and 1680s. The inventory made of these laboratories in 1689, soon after the death of Borelly, its last occupant before Homberg, indicates that they were well stocked with equipment and materials.[8] These laboratories on the rue Vivienne must, however, have seemed isolated after the Académie transferred its seat to the Louvre at the time of the 1699 Règlement. The other spaces the Académie had been using there were given over to the library, and there was no provision for constructing a new chymical laboratory at the Louvre.

Philippe's new laboratory at the Palais Royal was ready by spring 1702, and was undoubtedly superior in size and equipment to the laboratories at the Bibliothèque. It was also more convenient to the Académie's new seat, since the Palais Royal lay just across the rue Saint-Honoré from the Louvre, rather closer than the Bibliothèque du Roi. Homberg surely gave advice on the laboratory's design and outfitting to suit both his own work and aims, and

---

8. Bourdelin Expense Account, references to the "upper" laboratory ("d'en haut") at fols. 95v and 108r, and to the "lower" laboratory ("dembas") at fol. 101r. It refers once to "three laboratories" (77r) which may refer to the upper and lower chymical laboratories plus the preparation and storage room attached to the lower one. The same spaces are mentioned in the 1689 inventory; BNF, MS NAF 5149, fols. 21r–34r.

the operations he expected to carry out with Philippe. By 1702, the Académie's laboratories were more than thirty years old; their division on two floors must have been inconvenient, and given the floor plan of the building, they were probably rather cramped.[9] Homberg probably moved his operations quickly to the new space, and rarely if ever went back to the rue Vivienne after the Palais Royal laboratory was completed. All subsequent references to his laboratory cite the Palais Royal as its location. The Académie nevertheless retained control over the spaces at the Bibliothèque that had been outfitted as chymical laboratories, and Homberg remained their official director and "occupant." It is unclear if anyone worked there regularly after 1702. The laboratory contents were dispersed, perhaps transferred to the Jardin du Roi, where the Académie maintained other workspaces used in the regular courses of chymistry the various academicians taught there. The abandonment of the original Bibliothèque laboratories is clear in the inventory made of them after Homberg's death in 1715. In contrast to the 1689 inventory, which runs for twenty pages, the list of laboratory contents in 1715 now fit on half a sheet of paper: 97 pieces of "unserviceable" glassware, 36 pieces of earthenware ("many broken"), an earthen furnace, 11 bell jars, 4 or 5 bottles, and a "cheap" armoire.[10]

Homberg and Philippe worked together in the new laboratory on chymical operations for many years. In his famous *Mémoires*, Louis de Rouvroy, Duc de Saint-Simon (1675–1755), recorded both Philippe's interest in chymistry and some details about the work he and Homberg carried out at the Palais Royal. After claiming that Philippe was driven to studies such as chymistry by solitude and loneliness, Saint-Simon writes:

> Thrown thereby into the study of the arts, he set himself to blow at coals— not to seek how to make gold, a thing he always scoffed at—but to entertain himself with the curious operations of chymistry. He built himself the best-equipped laboratory, and he engaged an artist of great reputation named Homberg who was endowed no less with probity and virtue than with ability in his craft. He watched him follow and carry out many operations, and he worked on them with him, but all very publicly. He discussed chymistry with

9. See Guy Meynell, "The Académie des Sciences at the rue Vivienne," *Archives internationales d'histoire des sciences* 44 (1994): 22–34.

10. "Etat de ce qui est de reste a la Biblioteque du Roy dans le Laboratoire qu'y occupoit feu Mr. homber," AdS, Dossiers biographiques, "Homberg." The foregoing analysis corrects the erroneous claim that Homberg continued to work at the Bibliothèque du Roi, in Frederic L. Holmes, "The Communal Context for Etienne-François Geoffroy's 'Table des rapports,'" *Science in Context* 9 (1996): 289–311, at 294. Neglecting the adjectives that describe the poor state of the equipment, Holmes cites the 1715 inventory as evidence that the laboratory was well stocked, rather than abandoned and decrepit, and was used by Homberg until his death.

all those of the profession both in the court and in the city, and occasionally he brought some of them along to see Homberg and himself at work.[11]

What few records survive of the work carried out in the great laboratory of the Palais Royal indicate that accounts of Philippe's collaboration were not mere flattering rhetoric—he did in fact work alongside Homberg on a large number of projects, suggesting that his position was considerably more than that of a "tutee," but a great many of them, *pace* Saint-Simon, did deal explicitly with the transmutation of base metals into gold. It is interesting to note how concerned Saint-Simon was to deny any involvement in transmutation, which is suggestive of how the field of chymistry, and its promises of producing precious metals, was viewed by outsiders at the time. The work of Homberg and Philippe together, particularly on transmutation, will be explored in chapter 5.

Homberg's relationship with the Duc d'Orléans and his household continued to grow in the following years. Two years later, in 1704, Philippe invited Homberg to become his first physician, succeeding Claude-Jean-Baptiste Dodart in this position. Given Homberg's relatively modest interest in medicine—a contemporary noted that "he practiced medicine only for a small number of friends"[12]—and lack of formal medical training, Philippe's choice may have had more to do with forging a closer connection with his chymical collaborator and channeling further resources and social prestige to him than with actually hiring a personal physician. Homberg's ability to accept this enhanced position, however, was complicated by the rules of the Académie. The Règlement of 1699 stipulated that academicians were not allowed to accept positions that would take them away from Paris.[13] The Duc d'Orléans was often at Saint-Cloud, Marly, or Versailles, and was likely to travel much farther afield, particularly on military campaigns, and would naturally expect his first physician to accompany him. Recognizing this difficulty, Philippe took the unusual step of dispensing Homberg from the responsibility of following him on journeys outside of Paris, and thus Homberg accepted the position and the substantial additional salary of 3,800 livres, paid partly by the house of Orléans and partly by the royal treasury, more than double the pension of 1,500 livres he already received as a pensionnaire of the Académie.[14]

11. Saint-Simon, *Mémoires*, ed. Yves Coirault, 8 vols. (Paris: Gallimard, 1983–88), 4:456.

12. Gilles Filleau des Billettes to Leibniz, 23 February 1697, in Leibniz, *Sämtliche Schriften und Briefe*, 1st ser., 13:573–78, at 575.

13. Règlement of 1699, article 4, in *HARS* (1699): 3–4.

14. *L'etat de la France*, 2 vols. (Paris, 1712), 2:131: "Un premier medecin, 2000 livres de gages par le Trésorier de la Maison, et 1800 livres de pension au Trésor Roiale—M. Guillaume Hom-

Upon inspection, however, possibly by Pontchartrain, the deal proposed by the Duc d'Orléans did not prove a satisfactory way around the regulations. Homberg thereupon declared that he would relinquish the post with Philippe in order to keep his position in the Académie.[15] Homberg's resignation did not, however, become necessary. On 24 December 1704, Louis XIV intervened and granted Homberg a special dispensation to hold both positions and both incomes simultaneously (which undoubtedly provided the Batavian chymist with a very merry Christmas that year). At the same time, the king also reworded the regulation in question "to avoid all ambiguity," such that no pensionnaire academician could hold a position in any of the royal households.[16] This "clarification" of the regulations had an immediate and unfortunate side-effect. The pensionnaire chymist Simon Boulduc had long been apothecary to Philippe's mother, Elisabeth Charlotte. Unlike Homberg, Boulduc received no royal dispensation, nor was his situation "grandfathered" as Denis Dodart's had been at the time of the 1699 Règlement.[17] Instead, he was required to choose immediately between the two appointments. Boulduc chose the Académie, and his position in the house of Orléans was thereupon transferred to his son Gilles-François (1675–1742), then an élève in the Académie. It seems, however, that this outcome did not sit well with Boulduc, and he may not actually have fully resigned as required. Elisabeth Charlotte continued to intercede with the king on Boulduc's behalf repeatedly, apparently for the next ten years, but Louis XIV absolutely refused to relax the regulation for Boulduc.[18] In light of this fact, the unique dispensa-

---

bert." The payment of the portion of Homberg's salary (for an unspecified year) from the king's accounts is recorded in AN O/1/630, fol. 43r, item 433. Note that Sturdy, *Science and Social Status*, 232–33, was not aware of this record of Homberg's substantial income from the Duc d'Orléans, and thus concluded incorrectly that his financial condition was weak. The salary as physician was possibly also in addition to the pension Homberg obtained as Philippe's tutor in chymistry, mentioned by Fontenelle, "Éloge de Homberg," 90.

15. Fontenelle, "Éloge de Homberg," 91. At the public assembly the following April, the presiding officer, Abbé Bignon, extemporaneously praised Homberg publicly for his show of devotion to the Académie by his willingness to relinquish the post with Philippe in order to remain an academician; see *Nouvelles de la république des lettres* (June 1705): 701–2.

16. PV 24, fols. 41r–42v (4 February 1705; reading of the dispensation in a letter from Pontchartrain dated 24 December 1704); see also AdS, pochette de séances 1705.

17. PV 18, fol. 115r (letter of Pontchartrain dated 27 January 1699), published in *HARS* (1699): 12–13.

18. PV 33, fol. 411r–v (22 December 1714). Denis Dodart was physician to the Princesse de Conti and an academician at the time of the 1699 règlement, and special mention was made that he could maintain both positions; see PV 18, fol. 115r.

tion Louis made for Homberg should be seen as a sign of special royal favor toward the Batavian chymist.

Fontenelle records that around the time that Philippe engaged Homberg as his first physician, an invitation "with even more considerable advantages" arrived for Homberg from the Elector Palatine, presumably Johann Wilhelm II, who held the electorate from 1690 until his death in 1716. Homberg turned down this offer.[19] I have been unable to verify this assertion; however, there is no reason to discount the possibility. In 1704, Homberg was reaching the peak of his career and his celebrity. Philippe's mother, the daughter of a former Elector Palatine, might have provided a possible avenue of communication about Homberg.

As Philippe's first physician, Homberg would have moved his residence to the Palais Royal, if he had not already done so two years earlier upon completion of the laboratory there. It is reasonably likely that he took up residence there in 1702, since practicing chymists of the early modern period ordinarily needed to reside close to their laboratories, given the need to tend to long-term experiments at all hours of the day and night, even when there was a paid operator present to tend the fires. Duclos and Borelly lived in an apartment adjacent to the Académie's chymical laboratories at the Bibliothèque du Roi, and presumably Homberg lived there as well while he worked in that locale. The exact location of Homberg's apartments within the Palais Royal is provided by clues found in the legal documents prepared immediately after his death in 1715. These documents record that Homberg's bedchamber was on the second floor and "looked out on a little court," that another of his rooms looked out "over the passage that leads to the rue de Richelieu."[20] This information localizes Homberg's apartments as shown in figures 4.2 and 4.3. Their position placed Homberg directly adjacent to the apartments of the Duc d'Orléans on the same floor, as well as close to the laboratory, which was accessible by a staircase. At least one domestic servant occupied third-floor rooms, and the ground floor housed a kitchen and a room appointed for another domestic, perhaps the operator for the laboratory.[21] A wine cellar

19. Fontenelle, "Éloge de Homberg," 90.

20. AN, Y//11647, Homberg's Scellé-Après-Décès, 24 September 1715, fol. 1r: "nous sommes passes dans une chambre ayant veue sur une petite Court qui donne sur la Corps de Logis de derriere de la porte du palais Royal," and fol. 2r: "audessus du passage qui conduit susditte Rue de Richilieu."

21. AN, Minutier-Central, LX-205, Homberg's Inventaire-après-décès, 26 September 1715, fols. 2r–3r, specifies on the ground floor "la Cuisine au rez de Chaussee," "une petite Salle attenant le passage du pallais royal" and "un Chambre audict estage ou couche les domestiques."

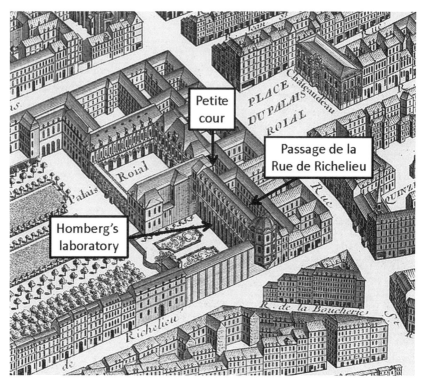

FIGURE 4.2. The location of Homberg's laboratory and the architectural features that localize his apartments in the Palais Royal, shown on the Turgot map of Paris (1734–36).

(presumably also used for exposing hygroscopic salts to cool, humid air in order to prepare *olea per deliquium*) completed the suite of Homberg's lodgings, which thus spread over four levels of the Palais Royal.[22] Homberg would reside in these apartments at the Palais Royal for the rest of his life, although he also rented an apartment at Versailles and was given rooms at Saint-Cloud, presumably for attending Philippe when he was resident in those places.[23]

Although Philippe had dispensed his new first physician from the normal obligation of accompanying him on campaigns, Homberg did follow him on one military expedition. In summer 1706, Philippe was placed in

22. At the time of Homberg's death, the cellar contained "environ une centaine de bouteilles remplie de vin de Bourgogne," Scellé-Après-Décès, 3r.

23. Record of these other apartments survives in the Inventaire-apres-décès: receipts of rent payment to Mr. Puison for an apartment at Versailles, fol. 17v; and an inventory (3 October 1715) of goods from his apartments at Versailles and Saint-Cloud, fols. 20r–21v.

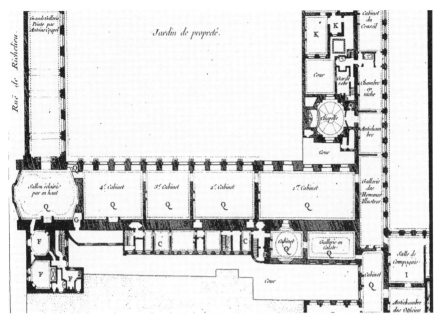

FIGURE 4.3. The location of Homberg's apartments on the second floor of the Palais Royal—the suite of rooms marked "C" as well as the two adjacent rooms marked "Cabinet Q" and "Gallerie," which were his study and bedroom. Floor plan from Jean-François Blondel, *Architecture françoise* (Paris, 1752–56). By the time this plan was made, the former apartments of the Duc d'Orléans had been converted into an art gallery, labeled here as "Cabinets"; the location of some doorways and nonbearing interior walls may also have been changed after Homberg's residence.

charge of a military campaign in Italy during hostilities with the Duke of Savoy, part of the War of the Spanish Succession. He left Paris on 1 July, led a battle at the besieged city of Turin on 7 September, and arrived back at Versailles to report to his uncle Louis on 8 November.[24] Fontenelle reports that the Duc d'Orléans "left his first physician behind in Paris" when he departed for the Italian battlefield. The attendance lists in the procès-verbaux indicate that Homberg was present at the meeting on Saturday 6 July, five days after Philippe left Paris, and so Fontenelle is right to say that Homberg was "left behind." But the further implication that Homberg stayed in Paris while Philippe was away is wrong. The later attendance lists indicate that Homberg was absent from the Académie continuously from 10 July until 17 November 1706, which closely matches the time the Duc d'Orléans was on his Italian campaign. A subsequent search of the royal *ordonnances* for 1706 uncovered a previously unnoticed entry dated 5 July that grants "per-

24. Saint-Simon, *Mémoires*, 1:759, 791.

mission to Mr. Hombert of the Académie des Sciences to absent himself for four months."[25] The diary of Claude Bourdelin the younger provides further information, recording tersely that "Mr. Homberg left Thursday 8 July."[26] Evidently, Homberg would not leave Paris (possibly jeopardizing his status in the Académie) until royal permission was officially granted. Although he was not obliged to do so, thanks to Louis XIV's and Philippe's special dispensations, Homberg nevertheless chose to follow the Duc d'Orléans on the campaign to Italy. Homberg's decision to go thus seems an act of his own accord, perhaps a sign of his devotion to Philippe, rather than the fulfillment of regulations.

Philippe was apparently unaware that Homberg was on the road a few days behind him, since one of his officers, the Comte de Nocé, concerned that Philippe was heading into battle without a physician, convinced him to summon Pierre Chirac (1650–1732), professor of medicine at Montpellier and member of the Académie Royale des Sciences de Montpellier, to attend him on the campaign.[27] At the end of the campaign Chirac returned to Paris as Philippe's physician and, upon Homberg's death in 1715, formally succeeded him as Philippe's first physician. As it turned out, Philippe was wounded by two bullets during the battle at Turin in early September—lightly in the hip and so seriously in the left forearm that some attending physicians advocated amputation. The wound was, however, healed by the application of a medicinal water, although Philippe never regained full use of three fingers on his left hand. What role (if any) Homberg played in this medical drama is unclear, but credit for the successful cure is generally attributed to Chirac, which would explain his quick promotion to become Philippe's regular physician.[28]

25. Bernard de Fontenelle, "Éloge de M. Chirac," *HARS* (1732): 120–30, at 123. Homberg's absences are documented in the registers of the meetings, PV 25, 247r–348r. The attendance lists were added to the procès-verbaux registers starting in December 1694. AN, O/1/50, fol. 73r: "5 juillet 1706 à Versailles: Permission au Sr Hombert de l'academie des sciences de sabsenter pendant quatre mois."

26. Bourdelin Diary, fol. 80v (10 July 1706): "Mr Homberg est party le jeudi 8 juillet."

27. Fontenelle, "Éloge de Chirac," 123. Fontenelle's claim that Chirac was engaged only as Philippe was en route to Italy is supported by Saint-Simon, *Mémoires*, 6:643: "[Philippe] avait toujours gardé [Chirac] auprès de lui depuis qu'il l'avait pris en Languedoc, allant commander l'armée d'Italie."

28. On Philippe's wounds, Elisabeth Charlotte to Raugräfin Louise, 16 September 1706, in *Briefe der Herzogin Elisabeth Charlotte von Orléans, 1676–1706*, ed. Wilhelm Ludwig Holland, *Bibliothek des Litterarischen Vereins in Stuttgart* 88 (1867): 477; and Bodemann, *Briefe der Herzogin*, 2:144–45 (14 September 1706). On the threat of amputation and Philippe's recovery, Elisabeth Charlotte to Dubois, 18 October 1706, in Seilhac, *Dubois*, 1:243–44. On Chirac's role,

"Thank God," Philippe's mother wrote at the time, "he has a very good doctor and field-surgeon."[29] Homberg himself was not in much of a condition to act as Philippe's physician at the time, for he had contracted dysentery. His cure may also have been overseen by Chirac, whose specific for dysentery became a major source of his renown after he used it successfully on the army of Roussillon in 1693.[30] This was the only time that Homberg followed Philippe into battle. In the following year, when the Duc d'Orléans led a campaign in Spain from April until December 1707, Homberg stayed home and Chirac went with Philippe, in short, acting as his first physician without the official title. On the return leg of the 1706 trip to Italy, Homberg may have stopped in Toulon to visit his old associate and patron Louis-Armand Bonnin, now archbishop of that city; such a stop would explain the paper Homberg offered the Académie a few months later, in 1707, that included observations of spiders he made in "a garden in Toulon."[31]

As a member of the household of the Duc d'Orléans, Homberg came into contact with Philippe's remarkable mother, Elisabeth Charlotte von der Pfalz, often known affectionately as Liselotte, or simply as Madame. Elisabeth Charlotte is a well-known figure of Louis XIV's court thanks to the thousands of letters she wrote, which provide a seemingly inexhaustible source of details about life in the court of *le roi soleil*. These letters also reveal that she and Homberg, both Germans, forged as strong a friendship as their disparate ranks could allow. Her letters mention Homberg quite frequently, and provide vivid and highly complimentary descriptions of his personality, a feature hard to gauge from his formal papers and in the absence of any significant

---

Fontenelle, "Éloge de Chirac," 123–24. The water used came from the hot springs of Balaruc in Languedoc; this water had been examined by Régis and Boulduc, *HARS* (1699): 55; see also J. Laissus, "Les eaux merveilleuses de Balaruc," *Revue d'histoire de la pharmacie* 53 (1965): 367–77.

29. Elisabeth Charlotte to Raugräfin Louise, 7 October 1706, in *Briefe 1676–1706*, ed. Holland, 480.

30. Elisabeth Charlotte to Dubois, 15 October 1706, in Seilhac, *Dubois*, 1:242–43. Homberg is probably the unnamed *pauvre* cited as being very ill along with the "pauvre Père Du Trévoux" in her letter of 10 September (238–39), since she explicitly pairs Homberg by name with the Père du Trévoux in terms of their common illness on 15 October. The relevant sentence in the 10 September letter reads as if the transcriber may have accidentally omitted the name specifying which *pauvre*. Fontenelle, "Éloge de Chirac," 121.

31. PV 26, fol. 325r (27 July 1707); and Wilhelm Homberg, "Observations sur les araignées," *MARS* (1707): 339–52. These observations might possibly have been made instead during Homberg's time in Provence in the late 1680s, and presented to the Académie only in 1707 to keep up his obligation to publish at a time when his research work must have suffered from four months away from Paris.

quantity of his personal correspondence. "One is unable," she wrote, "to get to know Homberg without esteeming him for his honest spirit; he is not at all befuddled as savants usually are, nor ponderous, but instead always merry. Everything he knows, even the most difficult arts, are for him like light banter [*seindt bey ihm wie eine badinery*], and when he jokes and jests, he laughs heartily too. He has a soft voice and speaks very slowly, but explains himself very well."[32] Fontenelle gives a similar description of Homberg's speech, writing that "his manner of explaining things was very simple, but methodical, precise, and without excess. Either because French was always a foreign language for him or because he was naturally not profuse in words, he searched for his words at almost every moment, but he found them."[33] Elisabeth Charlotte remarked also on Homberg's very sociable personality: she described him as "a gallant, learned, pleasant man who is always merry and very good company" and "a very clever man with great understanding, he is completely mild-mannered, never becomes angry, and is really very funny at times; he's not forgotten his German at all, I always speak German with him."[34] For his part, Homberg entertained Madame with stories about his travels and occasionally with chymical demonstrations; in one case, he showed her his famous pyrophorus—the material prepared from excrement fried with alum—that bursts spontaneously into flames upon exposure to air. In another case, he explained to her why the air of Paris was so unhealthy compared to that of her native Heidelberg. Wondering about this difference himself, Homberg collected and analyzed a sample of the black muck from under a paving stone in Paris and found it to be full of niter due to the constant urination in the streets. This niter, he reasoned, released acrid and unhealthy vapors into the air during the heat of summer.[35]

Other authors in contact with the Orléans household also left behind remarks about Homberg's character and personality. Saint-Simon praises Homberg as "one of the greatest chymists of Europe, and one of the most honest men who ever lived; he was the most simple and the most solidly pious."[36] Likewise, Nicolas-François Rémond (1638–1725), who like Hom-

---

32. Bodemann, *Briefe der Herzogin*, 2:311 (14 April 1712).

33. Fontenelle, "Éloge de Homberg," 92–93.

34. Bodemann, *Briefe der Herzogin*, 2:307 (19 March 1712) and 261 (21 December 1710).

35. Elisabeth Charlotte to Raugräfin Louise, 4 February 1720, in *Briefe der Herzogin Elisabeth Charlotte von Orléans aus dem Jahre 1720*, ed. Wilhelm Ludwig Holland, *Bibliothek des Litterarischen Vereins in Stuttgart* 144 (1879): 39–43, at 39–40. Homberg's familiarity with Heidelberg suggests that it too was a stop during his travels.

36. Saint-Simon, *Mémoires*, 5:742.

berg was a member of the Orléans household, remembered him in terms so closely tied to early modern French ideals of good behavior that they are difficult to render properly into modern English: "[un] tres bon homme, d'une société sure et douce, fort moderé et d'un tres grand desinteressement."[37] Besides being noteworthy in terms of unanimity, these testimonials suggest that Homberg's notable success in establishing himself quickly in new locales and with new groups of savants and others during his travels was not due only to his possession of rare and valuable secrets, but also to his possession of an agreeable and winning personality. Madame even praised Homberg's skill in wisely negotiating the often perilous waters of court culture—for example, his shrewd manner of dealing with the king's powerful and overbearing first physician Guy-Crescent Fagon (1638–1718). Delighted that thanks to Homberg's intervention her granddaughter Marie-Louise (1695–1719) would not be bled despite Fagon's orders, Elisabeth Charlotte remarked that "Homberg is no fool to have found a way to appear obedient to Fagon and still do only what he himself wants to do."[38] While Homberg would later fall afoul of palace rumor and intrigue, his lively personality, humble demeanor, and astute social observations made him a valued and respected member of the Orléans household, as well as a recipient of Louis XIV's royal favor. The high social position he achieved represents a truly extraordinary journey and level of success for the Java-born child of a war refugee.

The Tschirnhaus Burning Lens Comes to the Palais Royal

In the watershed year of 1702, Homberg's new relationship with the Duc d'Orléans and the latter's financial resources and interest in chymistry allowed for the acquisition of an extraordinary scientific instrument to which Homberg had primary access. The story begins with a visit of the Saxon mathematician and inventor Ehrenfried Walther von Tschirnhaus (1651–1708) to Paris in December 1701. Tschirnhaus, as mentioned in chapter 1, had become a (nonsalaried) member of the Académie in 1682 after he delivered a recipe for making white phosphorus. Homberg and Tschirnhaus may possibly have met in 1682 when they were both in Paris, both in regular contact with Mariotte,

---

37. Nicolas-François Rémond to Gottfried Wilhelm Leibniz, 23 December 1715; NLB, LBr 768, fols. 53r–54v, at 54r.

38. Elisabeth Charlotte to Étienne Polier de Bottens, 8 April 1705, in S. Hellman, *Aus den Briefen der Herzogin Elizabeth Charlotte von Orléans an Étienne Polier de Bottens, Bibliothek des litterarischen Vereins in Stuttgart* 231 (1903): 44; in regard to Fagon, Madame here remarks that "je le tient plus grand politique que grand medecin."

and both interested in developing closer relations with the Académie Royale, but no documentation exists to prove this. Tschirnhaus's status was normalized to that of *associé étranger* during the 1699 Règlement.[39] In that same year, he sent Homberg a description of the large convex burning lenses he was able to produce; Homberg translated the document from German and read it to the academicians.[40] In early 1700, Tschirnhaus sent the Académie a further description of how his lenses could be used as single-lens telescopes, and accompanied the report with a medallion cast from tin melted by the sunlight focused by one of his lenses. The medallion carried a Latin inscription beginning "Ecce suis radiis hunc nummum fudit Apollo . . ." ("Behold how Apollo melted this coin with his rays . . .").[41] In late 1701, Tschirnhaus visited Paris as part of his tour through the Low Countries and France to publicize Saxon products and to visit porcelain works and glassworks collecting information helpful for manufacturing processes, particularly in regard to his ongoing endeavor to replicate Chinese hard-paste porcelain.[42] While Tschirnhaus was in Paris, from early December 1701 until 19 January 1702, he spoke to the academicians on several occasions. He told them still more about his lenses and lens-grinding machines, mentioned the new glassworks he now directed in

39. *HARS* (1699): 12.

40. PV 18, fols. 190v–194v (24 March 1699); and "Effets des grands verres brulans," *HARS* (1699): 90–94. The content is close to that of the Latin article "De magnis lentibus seu vitris causticis," published in *Acta eruditorum* 9 (1697): 414–19. See also Bourdelin Diary, fol. 2r (24 March 1699).

41. *HARS* (1700): 131–34, describes a lens of 1 foot in diameter and its use as a telescope objective but makes no mention of its use as a burning lens. This latter use, the tin medallion, and its verses are recorded only in the Bourdelin Diary, fol. 16v (23 January 1700). There is no mention of these communications from Tschirnhaus in the corresponding *procès-verbaux*. An identical medallion by Tschirnhaus, dated 1698, survives in the collections of the Stadtmuseum Bautzen, and is depicted in *Ehrenfried Walter von Tschirnhaus (1651–1708)—Experimente mit dem Sonnenfeuer* (Dresden: Staatliche Kunstsammlungen, 2001), 129.

42. Tschirnhaus's original report of these travels, "Allerhand Projekte und Vorschläge," dated 16 March 1702, is preserved in the Sächsisches Staatsarchiv (Dresden), 10026 Geheimes Kabinett, Nr. Loc. 00489/01, and published in *Ehrenfried Walther von Tschirnhaus Gesamtausgabe*, ed. Eberhard Knobloch, ser. 2, part 1 (Leipzig: Sächsischen Academie der Wissenschaften, 2004), 39–48. On Tschirnhaus, see Günter Mühlpfordt, *Ehrenfried Walther von Tschirnhaus (1651–1708)* (Leipzig: Leipziger Universitätsverlag, 2008); Alfred Kunze, "Lebensbeschreibung des Ehrenfried Walther von Tschirnhaus auf Kieslingswalde und Würdigung seiner Verdienste," *Neues Lausitzisches Magazin* 43 (1866): 1–40; Eduard Winter, *Der Bahnbrecher der deutschen Frühaufklärung: E. W. v. Tschirnhaus*, in *E. W. von Tschirnhaus und die Frühaufklärung in Mittel- und Osteuropa*, ed. Winter (Berlin: Akademie Verlag, 1960), 1–82; and Hans-Joachim Böttcher, *Ehrenfried Walther von Tschirnhaus: Das bewunderte, bekämpfte, und totgeschwiegene Genie* (Dresden: Dresdener Buchverlag, 2014).

Saxony, and offered them the products of his manufacture.[43] Almost always short on cash, Tschirnhaus was seeking monetary support through the sale of his products and expertise. During this visit to Paris he succeeded in negotiating a major financial transaction: the Duc d'Orléans agreed to buy the largest lens Tschirnhaus had ever produced.

The burning lens that Philippe bought from Tschirnhaus was truly a technological wonder. More than 3 feet (1 meter) in diameter and weighing 160 pounds (75 kg), the lens was crystal-clear and free from flaws. It had been ground from a glass disk weighing 300 pounds.[44] The technological skill needed to cast a piece of high-quality glass of such size, without major inclusions and without cracking during cooling, is simply extraordinary. Much of the success in doing so was due to a talented glassworker named Constantine Fremel (1667–1748), the descendant of a family of Murano glassworkers. Fremel initially doubted that one could cast glass disks of the size and thickness Tschirnhaus wanted, but after having been shown a 2-foot-diameter lens that Tschirnhaus had already produced, and urged on by the Saxon nobleman's encouragement and financing, in April 1695 Fremel successfully produced a glass disk "of the most beautiful crystal, and both as large and thick as [Tschirnhaus] wanted." This disk, prepared by pouring molten glass into a specially prepared mold, required about three weeks to cool.[45] By late 1696, Fremel had increased the size of the blanks he could prepare yet further to nearly 3½ feet in diameter, 5 inches thick, and 300 pounds in weight. Tschirnhaus acquired three glass blanks of approximately that size from Fremel in

43. Ehrenfried Walther von Tschirnhaus, "Avis à l'Académie touchant certaines inventions," PV 20, 395r–397r (10 December 1701); Tschirnhaus attended assemblies until 14 January 1702 (PV 20, fols. 412v, 420r, 437v; PV 21, fols. 1r, 11r). On Tschirnhaus's Saxon glassworks, see Gisela Haase, "Tschirnhaus und die sächsischen Glashütten in Pretzsch, Dresden und Glücksberg," in *Experimente mit dem Sonnenfeuer*, 55–67.

44. Bernard de Fontenelle, "Éloge de M. Tschirnaus," *HARS* (1709): 114–24, at 121. Fontenelle claims that the blank from which the lens was ground weighed 700 pounds, but this is certainly incorrect. Tschirnhaus does not claim to have made a blank of such extreme weight, and since his lenses preserved the diameter of the glass blanks from which they were shaped, only rather less than half the weight would have been ground away. A supposed loss of 77 percent (a 700-pound blank ground into a 160-pound lens) would have been enormously wasteful and time-consuming. For the more reasonable value of 300 pounds, see the sources in note 46.

45. Ehrenfried Walther von Tschirnhaus, "Intimatio singularis novaeque emendationis artis vitriariae," *Acta eruditorum* (August 1696): 345–67; and Fremel to Tschirnhaus, 6 April 1695, Wrocław University Library, MS Akc 1948/0562, fol. 278r–v: "von den schönsten Cristal und gröste, wie auch dicke, wie Monsieur verlanget hat, inerhalb drey wochen auffs längste wird er abgekühlet sein." This letter is also quoted by Gisela Haase, "Sächsischen Glashütten," 57, using Eduard Winter's notes, but his transcription is imprecise.

December 1696, and it is probably from one of these enormous discs that Tschirnhaus ground the lens eventually bought by the Duc d'Orléans.[46] The process of grinding the huge slab into a focusing lens was no less remarkable than the casting of the blank. In order to do so, Tschirnhaus devised a huge, water-powered grinding and polishing machine that used various grits first to shape and then to polish the glass blanks into biconvex lenses.[47]

A letter from Elisabeth Charlotte attests not only to the exorbitant cost of the lens but also to Philippe's active participation in the laboratory with Homberg: "My son has a burning glass or burning mirror that cost him two thousand Thaler, with which he performs many experiments with his doctor, who is a German named Homberg."[48] To put this price into perspective, 2,000 Saxon Thaler equaled 6,000 French livres, the equivalent of four years' salary for an Académie pensionnaire like Homberg. While much has been made of the expensive nature of the air pump created by Robert Hooke and used by Robert Boyle, Tschirnhaus's lens cost Philippe almost twenty times as much.[49] This extraordinary, costly, huge, and technologically advanced instrument that Homberg would use extensively throughout the first decade of the eighteenth century was very much the early modern equivalent of today's "Big Science."

46. Ehrenfried Walther von Tschirnhaus, "Additio ad D. T. intimationem de emendatione artis vitriariae," *Acta eruditorum* (December 1696): 554. Compare Tschirnhaus to Leibniz, 22 October 1696, in Leibniz, *Sämtliche Schriften und Briefe*, 3d ser., 7:163–65, at 164. The *Acta* cites "nearly four feet" [*quatuor fere pedum*] for the diameter of the blanks, and the letter to Leibniz gives a measurement of "nearly 2 Leipzig ells" [*fast 2 leipziger ellen*], which would be about 44 inches (109 cm), given that the Leipzig *Fuß* was a little over 11 inches in length and there are 2 feet in an ell. A 1748 letter from Fremel to René-Antoine de Réaumur, recently discovered by Christine Lehman, refers to the three largest glass blanks that Fremel prepared, and claims that the lenses made from them were sold to the Elector of Saxony, the Duc d'Orléans, and the Landgrave of Hessen-Kassel; see Christine Lehman, in *Revue d'histoire des sciences*, forthcoming.

47. Klaus Schillinger, "Die Herstellung von Brennspiegeln und Brenngläsern durch Ehrenfried Walther von Tschirnhaus und ihre Widerspiegelung in ausgewählten Briefen," in *Europa in frühen Neuzeit*, ed. Erich Donnert, 6 vols. (Weimar: Böhlau, 1997), 4:97–114; and Werner Loibl, "Ehrenfried Walther von Tschirnhaus und der frühneuzeitliche Glasguss in Sachsen," *Neues Lausitzisches Magazin* 135 (2013): 65–96.

48. Bodemann, *Briefe der Herzogin*, 2:217 (20 June 1709). Leibniz to Johann Bernouilli, April 1702, NLM, LBr 57II, fol. 73r, gives the considerably lower price of 1,000 écus, equivalent to 3,000 livres, although the cost of 2,000 Reichsthaler is confirmed in Zacharias Conrad von Uffenbach, *Merkwürdige Reisen durch Niedersachsen, Holland, und Engelland*, 3 vols. (Frankfurt and Leipzig, 1753–54), 1:39.

49. Steven Shapin and Simon Schaffer, *Leviathan and the Air-Pump* (Princeton, NJ: Princeton Unversity Press, 1984), 38n29, estimates £25 as the cost of the air pump in 1660; the 2,000 Thaler paid for the burning lens in 1702 is equivalent to approximately £450.

During Tschirnhaus's stay in Paris, Homberg acted as intermediary between the Saxon nobleman and the Académie, thanks in part to their shared native language. More important, Homberg set up a meeting between Tschirnhaus and the Duc d'Orléans. Tschirnhaus's account of this meeting, which took place in early January 1702, provides important new information about Homberg's role and how Philippe came to purchase the lens.

> Through Mr. Homberg I had the favor of offering my most humble respects to the Duc d'Orléans. I believed at first that he would be a lover of precious stones, and I showed him several. . . . He examined them well, but remarked that these were not his intent, and asked me instead about my burning lenses, about which I had sent a report to the Académie. I showed him the tests [*Proben*] that I had also brought with me, which so pleased him that he immediately resolved to have one, and that he ought to have a yet larger one fabricated which he wanted to offer to the King as a present. It is well known in France that [Philippe] has no equal among the great lords in terms of his interest in the sciences.[50]

The report indicates clearly that Philippe met Tschirnhaus with the intention of asking about his burning lenses, and that he was obviously ready to purchase one. Homberg and Philippe must have discussed the acquisition of a burning lens in advance, and the audience Homberg arranged for Tschirnhaus was designed specifically to arrange the purchase. Such planning implies that Homberg's association with the Duc d'Orléans was already well established prior to this meeting in January 1702. Tschirnhaus suggests the same when he mentions Homberg in his report as a fellow German who "stands in high favor with the Duc d'Orléans [*welcher beym Herzog von Orleans in großer Grace stehet*]."[51] There is no hard evidence about when the relationship between Homberg and Philippe began. The date customarily cited, 1702, comes from Fontenelle, and he may have drawn it only from the start of Homberg's work with the lens at the Palais Royal in mid-1702.[52] In fact, Fontenelle's claim that Homberg's role was to teach Philippe chymistry may obscure their real relationship, which seems to have been more as collaborators from the very beginning, perhaps with Homberg in a position akin to that of a "court chymist" of the sort common in Germany through the seventeenth century. The construction of an extraordinary laboratory under Philippe's apartments, and the purchase of a Tschirnhaus lens seems far too great an investment if Homberg was simply another of the many tutors of the Duc d'Orléans. How-

50. *Tschirnhaus Gesamtausgabe*, ser. 2, 1:45.
51. *Tschirnhaus Gesamtausgabe*, ser. 2, 1:45.
52. Fontenelle, "Éloge de Homberg," 90.

ever this question might be resolved, the relationship between Homberg and Philippe must already have been well established in 1701, given that Homberg could convince Philippe to make so costly a purchase in the first weeks of 1702. Certainly, Homberg's love of the spectacular and his long-standing interest in light fired his desire to experiment with the massive lens. His interest would only have increased after having read Tschirnhaus's descriptions of its effects on metals and other substances.

During Tschirnhaus's visit, Homberg also indulged in his customary practice of trading secrets. Tschirnhaus confided to Homberg his knowledge about making porcelain, a valuable secret the Saxon nobleman had long endeavored to discover. Although the fully successful production of a white hard-paste porcelain comparable to the Chinese material occurred just before his death in 1708, while he was assisted by Johann Friedrich Böttger (1682–1719), Tschirnhaus had already made considerable progress earlier in the decade.[53] Homberg was sworn not to reveal the secret as long as Tschirnhaus lived, and indeed Homberg never revealed the secret at all, provoking the annoyance of René-Antoine de Réaumur (1683–1757). Réaumur probably had nothing to lament, since Tschirnhaus's fuller success in 1708 presumably rendered obsolete the information he had relayed to Homberg six years earlier.[54] In return, Homberg gave Tschirnhaus several chymical secrets, but it may also be the case that the porcelain secret was a commission of sorts to Homberg for having arranged the sale of the lens.[55] A lengthy digression about glass manufacture and its practical difficulties in Homberg's 1702 essay on salt may also derive from then recent exchanges with Tschirnhaus.[56]

53. On Tschirnhaus and porcelain see Curt Reinhardt, "Tschirnhaus oder Böttger? Eine urkundliche Geschichte der Erfindung des Meissener Porzellans," *Neues Lausitzisches Magazin* 88 (1912); Martin Schönfeld, "Was There a Western Inventor of Porcelain?" *Technology and Culture* 39 (1998): 716–27; and Ulrich Pietsch, "Tschirnhaus und das europäische Porzellan," in *Experimente mit dem Sonnenfeuer*, 68–74. On Böttger, see Klaus Hoffman, *Johann Friedrich Böttger: Von Alchemistengold zum weißen Porzellan* (Berlin: Verlag Neues Leben, 1985).

54. Reaumur, "Idée generale des differentes manières dont on peut faire la Porcelaine," *MARS* (1727): 185–203, at 186.

55. Fontenelle, "Éloge de Tschirnhaus," 122–23; Winter, "Bahnbrecher," 53, claims that Homberg gave Tschirnhaus "the recipe for the production of borax." This claim is not supported with evidence, and Winter is confusing borax (a naturally occurring substance) with the boric acid that Homberg first prepared *from* borax, the process of which he published in 1704 ("Essays de chimie," *MARS* [1702]: 33–52, at 50–52) and which later became known as "Homberg's sedative salt."

56. PV 21, fols. 69v–70v; Homberg edited this section down to half its original length for the published version, "Essays de chimie," *MARS* (1702): 46.

As soon as Tschirnhaus returned to Leipzig, on 5 February 1702, he set about fulfilling Philippe's order, and prepared the gigantic lens for shipment to Paris.[57] It is unclear how much further work on the lens was required, but three months elapsed before it arrived at the Palais Royal. On 15 May 1702, Geoffroy reported that the lens "has arrived safely from Germany in this city [Paris] a few days ago. It has to be mounted, after which one will set to work to make many experiments with it."[58] Whether the mounting was shipped from Germany along with the lens or prepared in Paris is unknown, but given that Homberg was already experimenting with the completed apparatus just two weeks after the arrival of the lens, it seems likely that Tschirnhaus designed and shipped an appropriate mounting that allowed the whole apparatus to be assembled in short order upon its arrival. Judging from the reported focal length of the lens, the completed instrument must have stood about 12 feet (4 meters) tall when fully vertical.[59] The sunlight collected and focused by the large lens passed into a second, smaller lens of about 1 foot in diameter mounted 8 feet down the optical axis that served to concentrate the light further, bringing it into a tighter focus, thus creating a circle of astonishing brightness and temperature only eight lignes (about 3/4 of an inch, or 1.8 cm) in diameter where samples could be placed.[60] The angle of the

57. *Tschirnhaus Gesamtausgabe*, ser. 2, 1:47.

58. Geoffroy to Hans Sloane, 15 May 1702, British Library, Sloane MS 2038, fol. 344v: "M. Le Duc d'Orleans a acheté de Mr. Tshirnhaus une de ses grandes Lentilles, de trois pieds de Diametre. Elle est arrivée a bon part d'allemagne en cette ville depuis quelques jours, on en après a la monter, apres quoy on travaillera a y faire plusieurs Experiences." Tschirnhaus's son informed his father of the safe arrival of the lens in Paris (perhaps the son escorted the lens from Saxony to France?); see Tschirnhaus to Leibniz, undated, in Carl Immanuel Gerhardt, *Der Briefwechsel von Gottfried Wilhelm Leibniz mit Mathematikern* (Berlin, 1899), 1:515–16. Gerhardt dates this letter to the "end of 1701," but this is impossible since Tschirnhaus was in the Low Countries and France from October 1701 until mid-January 1702, yet the letter states that he is at present "almost always in Dresden." Geoffroy's letter makes it clear that the letter from Tschirnhaus's son cannot date before May 1702.

59. The apparatus, which arrived in the Académie's *cabinet* after passing through other hands (see chapter 7), was set up again in 1772, at which time Lavoisier remarked that its mounting was "very large and cumbersome." Unfortunately, his manuscript breaks off before reaching his promised description of the apparatus. See Lavoisier's memoir, transcribed in Henry Guerlac, *Lavoisier: The Crucial Year* (Ithaca, NY: Cornell University Press, 1961), 204–7, at 206.

60. The details of the mounting are most clearly presented in *Journal des sçavans* (24 August 1705): 560–61. The focal length of the larger lens was 12 feet (4 meters), reduced to 9 feet (3 meters) by the use of the smaller lens. Homberg built a scale model of the whole apparatus, which he displayed at the public assembly in November 1702 when he presented a paper on his lens experiments; see *Journal de Trévoux* (January 1703): 135.

FIGURE 4.4. One of Tschirnhaus's double-lens instruments. These lenses date from before 1694 and are some of Tschirnhaus's earliest productions; the current blue and gold carriage (about 8 feet, or 2.6 meters, tall) was assembled for display purposes around 1740, and does not reproduce the original mounting. These lenses have diameters of 49.5 and 26.0 cm (19 ½ inches and 10 ¼ inches), and thus the larger of the two is only about half the diameter of the one Homberg used. Accordingly, the assembled carriage of the "grand verre du Palais Royal" must likewise have been substantially larger than this one, as well as more complicated, since its angle was adjustable with gear wheels. Courtesy of the Staatliche Kunstsammlung, Mathematisch-Physikalischer Salon, Dresden; photos by the author.

carriage could be adjusted to match the altitude of the sun using gearwheels (*roues*) and swiveled to track the sun as it moved across the sky. Surviving examples of Tschirnhaus's workmanship—although all smaller in size than the one bought for the Palais Royal—still have the power to awe the modern viewer (fig. 4.4).[61] But the sight of the size and impressive workmanship of these objects must have paled in comparison to seeing the apparatus in action. On a sunny day, any metal, including gold, when placed in the focus, melted within a few moments and then vaporized. A piece of wood merely moved through the focus burst instantly into flames. A piece of pine exposed

61. Surviving examples of Tschirnhaus lenses exist in the Mathematisch-Physikalischer Salon in Dresden, the Deutsches Museum in Munich, the Astronomisch-Physikalische Kabinett of the Museumlandschaft Hessen Kassel, and the Kunstkammer of the Academy of Sciences in St. Petersburg. The largest of these is the one at Kassel, which is nearly the diameter of Homberg's; all the others are considerably smaller.

to the focus while submerged underwater seemed to remain undamaged, but when taken out and split, its interior was found to have been converted to charcoal. Bricks, earthenware, porcelain, and even roofing slates could all be melted and vitrified with the intense heat the lens produced.[62] Given the enormous power of the lens, it is surprising that Homberg never seriously injured himself while using it; at least we have no record of such a mishap.

Although it is clear that the lens was used in the garden of the Palais Royal—specifically in the Jardin des Princes, the private garden of the Duc d'Orléans and his family, which lay just outside the laboratory—contemporaneous descriptions of the practical details of its use are lacking. It is possible that the whole device was attached to a sturdy carriage, presumably on wheels, and transported into the garden by a team of assistants or servants on sunny days for use, and otherwise kept safely indoors. The large door openings of the laboratory would have allowed for its transport back and forth from the garden. Alternatively, some sort of a shelter might have been built in the garden to house it when not in use.[63] Homberg does mention that the brilliance of the light from the lens was so great that he observed his samples through smoked glass (*verre enfumé*).[64] He also notes the difficulty of finding suitable supports for the samples to be exposed to the lens. He used a variety of materials, including charcoal (he claimed that charcoal made from green wood was better) and unglazed porcelain.[65] Although Homberg was obviously the most frequent user of the Tschirnhaus lens, it was also employed for related experimentation by his associate Étienne-François Geoffroy, and at least once by Louis Lémery.[66]

62. Tschirnhaus, "Grands verres brulans," 91–94. For more on Tschirnhaus's lenses, see *Experimente mit dem Sonnenfeuer*, especially Klaus Schillinger, "Herstellung und Anwendung von Brennspiegeln und Brenngläsern durch Ehrenfried Walther von Tschirnhaus," 43–54; and Schillinger, *Solare Brenngeräte* (Dresden: Staatliche Mathematisch-Physikalischer Salon, 1992).

63. When the lens, still in its mountings, was set up again in 1772 in the Jardin de l'Infante (between the east end of the Palais du Louvre and the Seine), a wooden shed was built there to offer it protection from the weather; see Guerlac, *Lavoisier*, 207.

64. *HARS* (1711): 16; compare PV 30, fol. 112v (11 March 1711).

65. PV 24, fols. 33v–34r (24 January 1705).

66. Louis Lémery's use of the lens is mentioned in his paper "Que les plantes contiennent reellement du fer," *MARS* (1706): 411–18, at 412, which repeated Geoffroy's experiment with the lens carried out a year earlier; PV 24, fol. 394v (16 December 1705). Geoffroy used the lens more extensively, for example, in his paper "Experiences sur les metaux faites avec le verre ardent du Palais Royal," *MARS* (1709): 162–76; and PV 28, fols. 171r–183r (18 May 1709; begun 8 May, fol. 165r).

Homberg's First Lens Experiments: Surprises about Gold and Silver

Homberg was obviously eager to start experimenting with the lens as soon as it was mounted—less than ten days after Geoffroy announced its arrival in Paris, Homberg was performing his first experiments with the new instrument. He presumably took advantage of the first sunny day after its construction to see what the device could do. Unsurprisingly, his experiments gathered an audience of spectators—certainly Philippe and possibly some relations from court, but also perhaps Homberg's associé Geoffroy and other academicians.[67] The range of materials Homberg could expose to the focus of the lens was nearly endless. While various sources report that he eventually examined its effects on minerals like marble, chalk, asbestos, limestone, rock crystal, and pebbles, as well as on gems like diamonds and emeralds (no doubt supplied by Philippe) and laboratory substances like borax and saltpeter, he did not run haphazardly through everything in the laboratory.[68] Rather than indulging in a miscellany of experiments, Homberg immediately set to work specifically on the precious metals gold and silver. He poured out his surprising results in a breathless fashion in three memoirs dating from 3 June, 9–12 August, and 15 November 1702. The single paper that appeared in the 1702 *Mémoires* (published 1704) as "Observations Made by Means of the Burning Lens" is a patchwork of these three memoirs with numerous revisions. Thus, the original sequence of observations and experiments—and what it can tell us about Homberg's experimental choices and evolving thoughts—is lost in the published version. Therefore, in order to recapture the original sequence of ideas, it is again necessary to turn to the procès-verbaux.[69]

Homberg's memoir of 3 June recounts his surprise that an alloy of gold and silver, after having been melted by the lens and cooled, did not produce the bubbles normally seen when dissolving in aqua regia. This observation drove him "to learn how the fire of the sun could have produced an effect so different from that of the fire of our laboratories."[70] Ordinary flame, he

---

67. PV 21, fol. 231r (3 June 1702).

68. Homberg himself wrote up nothing of these miscellaneous trials—in most cases they were mentioned by one of his colleagues; see Bourdelin Diary, fol. 75r (marble and pebbles, 29 August 1705); Nicolas Lémery, *Traité universelle des drogues simples* (Paris, 1723), 13, 32, 36, 136, 277–78; and PV 22, fol. 247r (14 July 1703).

69. PV 21, fols. 227r–231r , 339v, 341r–343r, and 399v–401v; Wilhelm Homberg, "Observations faites par le moyen du Verre ardant," *MARS* (1702): 141–49. Note that the page range varies between editions; see the note on sSources.

70. Wilhelm Homberg, "Observations faites par le miroir ardent du Palais Royal," PV 21, fols. 227r–231r (3 June 1702), at 228v.

argued, is "nothing other than a liquid composed of the matter of light and the oil of wood or of charcoal" or of whatever is burning. This statement marks a development over Homberg's 1690s statement that flame is merely a "sulphureous or oily matter."[71] Now he explicitly incorporates light as a key component of flame. Flame obtains its activity from the surrounding air; since the "fire fluid" of flame is less dense than air, it is pushed upward according to the laws of hydrostatics, and made to strike against other bodies, thereby heating them. This is why, he continued, flame heats more intensely when it is driven by a rapid current of air, such as from a pair of bellows. For the same reason fire cannot subsist in a vacuum, where there is no air to put it into motion. When a fusible substance is heated, the rapidly moving particles of the flame-fluid mingle with its smallest parts, disrupting its texture and thus causing it to liquefy. Flame therefore behaves like a solvent; it causes solids to become liquid by mixing with and separating their particles. As a molten material cools, the particles of flame-fluid gradually dissipate, allowing the substance to regain its former solidity. (Again, like a solution from which the solvent evaporates, leaving a solid residue.) Because solidification begins before all the particles of flame have dissipated (that is, before the material is fully cooled), some solidification occurs around these particles so that when they do at last depart, they leave behind empty spaces "molded [*moulé*]" around the particles' size and shape. These spaces become filled with air, and it is this air that is seen escaping when the cooled body—in this case, gold melted in ordinary fire—is dissolved in acid. The "fire of the Sun," however, is the pure matter of light. Rather than being moved by the air, it is impelled by the Sun itself—hence it can traverse and act in a vacuum. Free of "gross admixtures" from a burning substance, its particles are "infinitely smaller than those of a flame."[72] Therefore, when a material melted by the focused light of the lens begins to solidify, the spaces molded around the dissipating particles of light are far smaller than those produced in a metal melted with ordinary fire. With far less empty space for the air to fill, very few air bubbles escape when such gold dissolves. Homberg's important conclusions here are that (1) flame is an impure form of light; (2) light is a material substance composed of exceedingly minute particles; and (3) both behave like solvents. Dissolution and fusion are the same phenomenon; the first is brought about by ordinary liquids like water or alcohol, the second by the

---

71. VMA MS 130, fol. 211v: "la flame n'est autre chose qu'une matiere sulphureuse ou huilleuse du bois."

72. PV 21, fols. 229v–230r.

more subtle liquids flame and light, but both depend upon the insinuation of tiny particles in between the particles of another material.

Homberg's second memoir, presented in August, describes the effects of the lens on pure gold. The behavior of gold varied considerably, depending on where it was placed relative to the focus. At the focus itself, the molten gold boiled, displayed a prickly surface, and threw tiny droplets of molten gold a considerable distance. By spreading paper under the apparatus, Homberg collected a fine deposit of minute globules of gold, proving that "the whole substance of the gold evaporates [*se perd*] thereby without suffering any change . . . the true focus of our lens is too violent to keep any metal there in fusion." But when Homberg placed gold a small distance away from the focus, he obtained startling results—the gold melted, emitted smoke, and turned partly into a dark vitreous material. Generations of chymists had experienced, and been frustrated by, gold's stubborn resistance to lasting change. It had become their motto that "it is easier to make gold than to destroy it."[73] Yet using this new instrument, by placing gold in what he calls the "vitrifying region [*la place vitrifiante*]" of the lens, Homberg had now destroyed gold, transforming it irreversibly into smoke and a glassy substance. Beyond this "vitrifying region" the gold exhales smoke but produces no glass. Homberg ends his report by confessing that "I do not yet know what this smoke is."[74]

His third memoir, presented on 15 November, deals with experiments on silver. In the vitrifying region, silver smokes more than gold, boils at a lower temperature, evaporates faster, and emits a white powder that does not vitrify. Homberg then explains his strange results. The light concentrated by the lens acts not only as a solvent, but as a solvent whose particles "are tiny enough to introduce themselves into the composition of the metal and *disunite its principles*."[75] In other words, the particles of light are so small they can not only separate the individual particles of gold and silver from one another, but also penetrate inside the individual metallic particles and separate the more fundamental principles of which they are composed. Homberg thus claims success in doing what legions of chymists before him had failed

73. Robert Boyle, for example, quotes this line in the *Sceptical Chymist*, attributing it to Roger Bacon, and as a "receiv'd Axiom" in *Origine of Formes and Qualities*; see *Works of Robert Boyle*, ed. Michael Hunter and Edward B. Davis, 14 vols (London: Pickering & Chatto, 1999–2000), 2:281, 5:422.

74. Wilhelm Homberg, "Suite des observations faites au Miroir ardent," PV 21, fols. 341r–343r (12 August 1702).

75. Wilhelm Homberg, "Suite des experiences du miroir ardent," PV 21, fols. 399v–401v, at 401r: "les parties de la matière de la Lumière ou des rayons du Soleil, sont d'une petitesse capable de s'introduire dans le composé du métail pour en desunir les principes" (emphasis added).

to do, namely, separating gold and silver into their constituent principles.[76] The concentrated sunlight acts *analytically*. What no fire analysis could do to the precious metals, now light—acting like a solvent—succeeds finally in accomplishing. Strikingly, Homberg's description of light as a solvent composed of the tiniest particles of matter and therefore able to analyze the most difficult-to-decompound bodies forcibly recalls the traditional explanations of how the Helmontian alkahest works as a "resolutive solvent."[77] Did Homberg have this comparison in mind? Did he believe that he had finally discovered in light itself the resolutive solvent that Duclos and others had so fervently hoped to prepare as the replacement for fire analysis? Homberg does not provide any explicit indication on that score, but it is hard to imagine that he did not appreciate (or intend) the obvious parallel.

Homberg then explains how his observations prove that he has decomposed gold and silver. Supposing that the precious metals are composed of "mercury, metallic sulphur, and some earthy material," when these principles are disunited, the volatile mercury would be dissipated; the smoke seen rising from the gold and silver is this expelled mercury. The gold's remaining metallic sulphur and earth then combine to form a glass. Silver has less sulphur than gold, a quantity insufficient to form a glass with the earth, so the silver's separated earth remains as a white powder. But if the silver is first refined with antimony, it retains some of antimony's abundant sulphur, and in that case a glass forms. Homberg concludes triumphantly that the notion of the fixity and stability of gold or silver can no longer stand. He has volatilized and decomposed them both. The new and costly instrument of the Duc d'Orléans thus promises "not only great progress in clarifying the principles of chymistry, but to be an open gateway to a new natural philosophy, as microscopes and the air pump were in their own day."[78]

These surprising results provoked some immediate changes in Homberg's

---

76. A contemporaneously well-known demonstration of the inability of gold to be decomposed by fire involved leaving it molten in a glassmaker's furnace for two months, during which time it lost no weight; Gaston Duclo, *Apologia argyropoeiae et chrysopoeiae*, in *Theatrum chemicum*, 6 vols (Strasbourg, 1659–61), 2:1–80, at 19. Among the many authors who cite this experiment, see Boyle, *Sceptical Chymist*, in *Works*, 2:236–37.

77. On the alkahest and its method of action, see Bernard Joly, "L'alkahest, dissolvant universel, ou quand la théorie rend pensible une pratique impossible," *Revue de l'histoire des sciences* 49 (1996):308–30; and William R. Newman, *Gehennical Fire: The Lives of George Starkey, An American Alchemist in the Scientific Revolution* (Cambridge, MA: Harvard University Press, 1994), 150–51.

78. Wilhelm Homberg, "Suite des Expériences du Miroir ardent," PV 21, fols. 399r–401v (15 November 1702).

ideas. He returned to his "Essais," presented seven months earlier but not yet published, and inserted the burning lens as a means of analyzing metals. Fontenelle likewise cited the lens as an analytical instrument, calling it "a new key for entering into the inner composition of bodies" and capable of delivering "almost a new natural philosophy."[79] Subsequent writers have emphasized the hitherto unavailable high temperatures the lens could generate, and have connected its analytical abilities to this high temperature. Such readings do have legitimate origins in the primary sources, but they miss the most influential results of Homberg's first lens experiments. The high temperature, far from being key to Homberg's research, was actually a problem to be overcome. At the focus, metals were simply volatilized unchanged by the intense heat. The interesting effects, like the vitrification of gold and its emission of mercurial fumes, occurred *away* from the focus, at lower temperatures in the "vitrifying region," and this lower temperature region is where Homberg conducted his experiments on metals. The high temperature produced by the lens was crucial for Tschirnhaus, but for Homberg it was not heat that was of interest, but rather *light in concentrated form.* The concentrated light produced novel analytical effects not because of the temperatures generated but because of the extreme fineness and rapid motion of its particles, which allowed them to penetrate more deeply into the fabric of the metal and break apart its structure at the level of the principles. In order for them to act analytically, the target metal had to remain exposed to their effects for an extended time, which could not happen if it boiled away before its principles could be disentangled.

But analysis by means of light is only part of the story. A much greater impact, I argue, came from experiments that used the lens neither for high temperatures nor for analysis, but rather as an instrument of *synthesis*. This deployment of the lens has not previously been noticed, in part because of reliance exclusively on published materials. Nevertheless, it appears clearly in the very first experiments Homberg performed with the lens. The results of these experiments were so ground-shaking that he shared them only orally with his colleagues until he finished working out their full implications. By the time he finished exploring these implications, over a year later, he was forced entirely to reconceptualize his general theory of chymistry. These experiments, and the reconceptualizations they initiated, required Homberg to make extensive revisions to his 1702 "Essais," and led eventually in his daring declaration in 1705 that he had identified light itself as the true sulphur principle.

79. *HARS* (1702): 33–38, at 38.

The Chymistry of Light: From Experimental
Results to a Comprehensive Theory

Homberg's first communication to the Académie about his experiments with the lens actually occurred on 24 May 1702, ten days earlier than the first memoir included in the procès-verbaux. Only Bourdelin's diary preserves a brief record of the content of this first report: "with the burning lens . . . Mr. Homberg has melted the white metals like silver and lead into the color of gold."[80] This is a highly meaningful statement, for the usual chymical theory of metallic composition—which Homberg unambiguously affirmed elsewhere in this very year—attributed the color of nonwhite metals like gold to the large quantity of the sulphur principle in their composition.[81] Therefore, if the burning lens can convert white metals into golden-colored metals, then it must somehow be adding sulphur to them. Where did this sulphur come from? This result must have puzzled Homberg, and its possible explanation excited him.

Interpreting Homberg's first experimental result as the addition of the sulphur principle to silver explains an otherwise puzzling feature of his next choice of experiment. Homberg's next experiment was to expose a gold-silver alloy to the lens. Why would Homberg work with an alloy? If he were simply testing the effect of the lens on different metals, he would have used them pure in order to obtain straightforward results. Homberg constantly insisted on the purity and standardization of materials in order to guarantee reliable and reproducible results. But in this case he intentionally prepared an alloy of one part silver and eight parts gold to expose to the lens. Why? The most plausible explanation is that, after he discovered that exposure to the lens gave a golden color to silver, he wanted to see the effects of longer exposure. But, as he mentions elsewhere, the lens causes silver to vaporize too quickly. How could he make the silver resist the heat and remain in the concentrated sunbeams longer? A logical and practical way would be to alloy it with a more refractory metal—in other words, with gold. If this is the correct interpretation of his experimental design, then Homberg's goal at this point was to use the lens to incorporate more sulphur into the silver. Now since the dif-

80. Bourdelin Diary, fol. 50r (24 May 1702): "Mr. Homberg a fondu avec le miroir ardent de Monsieur . . . des metaux blancs comme l'argent et le plomb en couleur d'or." The procès-verbaux for 24 May 1702 (PV 21, fol. 212r) records only that "Mr. Homberg spoke of some experiments that he performed with a large mirror that belongs to Monseigneur, the Duc d'Orléans. He will give a memoir about them." He submitted a written memoir on 3 June, but omitted the experiment described by Bourdelin on 24 May.

81. PV 21, fol. 66r.

ference between gold and silver is only that silver contains less sulphur than gold, then if the experiment were successful, the silver in the alloy would be partly or wholly converted into gold. Thus, the lens would act as an agent of synthesis, forcing more sulphur to combine with silver; in short, the Tschirnhaus lens would be an *instrument of transmutation.* Homberg's next step, putting the exposed alloy into aqua regia, further supports this interpretation. Dissolving the alloy in aqua regia is an obvious next step only if Homberg wanted to assay the metal, separating the gold and silver in order to detect an increase in the amount of gold and/or a decrease in the amount of silver.

Homberg's published paper—unlike the procès-verbaux—never mentions using a gold-silver alloy, but instead describes using pure gold. Nor does he mention anywhere the tingeing of white metals with a golden color, or the final results of his first experiments (presumably the silver stubbornly failed to transmute). But I argue that these first experiments with the lens, whatever their final outcome, were the crucial observations that sparked Homberg's idea that sulphur and light were fundamentally linked, and set him on a new course to prove the identity of the true sulphur principle. In fact, Homberg never forgot about this result; he later reported an analogous experiment on iron, and explicitly linked it to transmutation. Again, only Bourdelin's diary—not the procès-verbaux—records the results. In a series of experiments that exposed iron to the lens, Homberg reported that "the iron prepared in a certain way and plunged into lemon juice acquires the color of copper on its surface," and questioned "whether this might be the start of a transmutation."[82] Once again, the color of copper, like that of gold, was attributed to its sulphur—what could be the source of this additional sulphur except the light itself?

The three years that separate Homberg's first work with the burning lens in May 1702 and his announcement in April 1705 that the true sulphur principle is actually the matter of light, mark a low point in his otherwise copious publication record (see figure 4.5.) During this time he even failed to deliver on his promise to speak about the many other observations and strange results that he achieved by exposing other materials to the lens. Indeed, it is odd that Homberg fell entirely silent about the lens experiments after his public presentation in November 1702—surely there were other exciting and surpris-

---

82. Bourdelin diary, fol. 74v (5 August 1705): "Mr. Homberg a dit que le fer preparé d'une certaine façon & plongé dans le jus de citron, y acquiert une couleur de cuivre à sa surface; savoir si ce seroit un commencement de Transmutation." Many of the experiments mentioned in this entry were published as Wilhelm Homberg, "Observations sur le fer au verre ardent," *MARS* (1706): 158–65; compare PV 25, fols. 168r–172r. There is no mention of this topic in the procès-verbaux for 5 August 1705.

ing results with which to regale his colleagues and public audiences.[83] I suggest that this low point is because he spent those years accumulating evidence for his suspicions about the identity of the sulphur principle. Nearly his only publication from 1703 and 1704 is an analysis of common sulphur, which he carried out, he says, because "before risking a definition [of the sulphur principle] I proposed to myself to examine the main substances which chymists call sulphurs."[84] Thus, Homberg attacked with renewed vigor the search for the true principles that had so occupied him since the early 1690s, but instead of focusing on salts as he had done then, his attention was now riveted on sulphurs. He had already tried to analyze common sulphur in 1694, at the time when he expressed the belief that the "simple minerals"—common sulphur, common mercury, and mineral salts—were only impure versions of the chymical principles.[85] In 1703–4 he tackled the problem anew, using chymical methods to isolate from common sulphur an acid salt, an earth, and an inflammable gum. For this analysis, Homberg used the ideas about dissolution he had put forward in his early treatise of chymistry. Specifically, when trying to isolate the sulphureous—that is, the oily—component of common sulphur, he depended upon extraction techniques using sulphureous solvents. He used essential oils at first and later turpentine, the idea being that solvents have to be similar to the materials they dissolve: "these distilled oils, being flammable solvents, are principally the dissolvent of only the inflammable part of the [common] sulphur" and are thus able to combine more tightly with its sulphur principle, and upon strong distillation the oils carry over the sulphur principle with themselves, leaving the saline and earthy components behind as a nonflammable residue.[86] Homberg then very gently redistilled the deeply colored, oily distillate, drawing off as much of the colorless oily solvent as possible, leaving behind an inflammable gum that he considered to contain the sulphur principle extracted from common sulphur. But he is

83. The promise appears in PV 21, fol. 400v: "une grande quantité d'observations que j'ay faites sur d'autres matières, dont je parleray une autre fois qui paroissent aussi étranges." In the published version (Homberg, "Verre ardent," 155) the line ends, "qui paroîtront aussi extraordinaires que celles qui viennent d'être raportées."

84. Wilhelm Homberg, "Analise du souffre commun," PV 22, fols. 85r–88r (21 March 1703) and 100v–102v (18 April), at 85v; compare "Essay de l'analyse du souffre commun," *MARS* (1703): 31–40. His only other publication between the 1702 burning lens memoir and the 1705 essay on the sulphur principle was a minor paper on an anatomical observation: "Observation sur un battement de veines semblable au battement des arteres," *MARS* (1704): 159–73.

85. Wilhelm Homberg, "Observations sur l'acide du souffre," AdS, pochette de séances 1694; VMA MS 130, fols. 116v–117r.

86. Homberg, "Analise du souffre commun," fol. 87v.

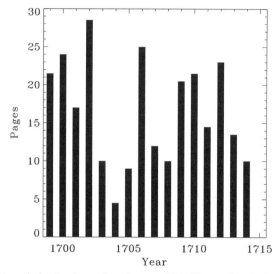

FIGURE 4.5. A graph showing the number of pages published by Homberg in the annual *Mémoires* from 1699 until 1715. The low rate of publication in 1703–5 corresponds with his focused experimental efforts in these years to identify the true sulphur principle. The more shallow minimum in 1706–7 is probably due to his extended absence from the laboratory for several months in the second half of 1706, when he followed the Duc d'Orléans on his military campaign to Italy. For the fall-off beginning in 1713, see chapter 6.

quick to say that the gum is *not* the sulphur principle in a pure state. By using his standard method of weight comparisons—comparing the weight of the oil used with the weight of the oil recovered—he recognized that the gum must retain a portion of the oily solvent. His subsequent attempts to isolate the sulphur principle from this gum failed, leading him to conclude that the pure sulphur principle cannot be contained when it is in its free state—it must incorporated in some vehicle.

When Homberg presented his analysis of common sulphur to the Académie in March 1703, and in fuller form to a public assembly in April, he still had not come to a firm conclusion about the identity of the sulphur principle. It remained "not yet known," and thus he did "not yet wish to say what the sulphur principle is."[87] The continuing modifications he made to the paper before its publication in 1705 indicate that he kept experimenting and revising his ideas accordingly throughout this period. Following his analysis of common sulphur, Homberg suggested to Geoffroy that he under-

87. Bourdelin diary, fol. 58r (18 April 1703): "Soufre est un nom equivoque. Principe de chymie non encore connu . . . Mr. Homb. ne veut pas dire encore ce qui c'est que le soufre principe." There are significant changes between the PV texts and those published in the *Mémoires*.

take the *synthesis* of sulphur from the components into which Homberg had divided it. Resynthesis would serve to verify the analytical results. Geoffroy presented his apparently successful resynthesis in 1704, but without saying anything about the identity of the sulphur principle.[88]

When the crucial experiment—the coloring of white metals yellow in the light of the lens—is seen in the context of Homberg's previous experiences and observations, the sequence of his thinking and his experimental pathway come together in a striking manner. A whole series of earlier results, ideas, and interests that had previously seemed unrelated all began to point in the same direction. As a result, Homberg reinterpreted various chymical phenomena he had previously observed and began to see them all as evidence that light was the true sulphur principle he had been seeking. The gradual development of Homberg's grand conclusion can thus be traced through a series of experiments and interests that otherwise seem unrelated. The reconnection of these disparate projects and experiences provides a more illuminating view of Homberg's thought and work, revealing them as more coherent than a cursory examination would indicate. On a larger scale, it illustrates also the remarkable adaptability of experimental results to multiple interpretations and reinterpretations, and showcases the dynamic and evolving interplay between theory and practice, and how a single experimental result can point the way toward unifying diverse and seemingly unconnected previous results into a comprehensive theory.

Homberg had been fascinated by the interaction of light with matter from the start of his career. His earliest chymical interests revolved around luminescent substances—the Bologna Stone, Balduin's phosphor, white phosphorus, von Guericke's glowing sulphur ball, and other such substances. These light-absorbing and/or light-emitting substances formed a key part of his fame through the 1680s and 1690s. He must have pondered the origin of the light emitted by these substances, and his groundbreaking work with the Bologna Stone is especially revealing. Homberg initially attributed the luminescence of the stone to ignited steams of sulphur exhaled by the calcined mineral. In 1694, he recounted how he had casually laid some calcined stones on a polished copper basin, and the next day observed a halo of corrosion spreading several inches in all directions away from the stones. He attributed this corrosion to emitted sulphur, which he could smell near the surface of the stones. Recognizing that the copper of the basins had trapped and combined with some the emitted sulphur, he put powdered copper on

---

88. Étienne-François Geoffroy, "Maniere de recomposer le Souffre commun par la réunion de ses principes," *MARS* (1704): 278–86.

the surface of the stones, and found that the stones then no longer glowed. He concluded that rather than igniting to produce a glow, the efflux of sulphur now combined immediately with the copper. Thus, already at this stage, Homberg made a weak connection between sulphur and light. At this point, however, light and sulphur were not identical; the prepared Bologna Stone merely emitted a highly inflammable sulphur that ignited upon exposure to light, in turn producing the observed glow.[89]

Homberg's study of quicklime forms another step in the sequence. Chymists had long recognized that quicklime is more caustic than slaked lime—for example, quicklime can burn skin and flesh, while slaked lime does not. Quicklime thus seemed to be a more powerful alkali. But when in 1700 Homberg determined the relative "forces" of different alkalies by measuring the amount of acid needed to neutralize equal weights of them, he found that quicklime and slaked lime had *equal* alkaline force; the same quantity of acid neutralized equal weights of each. Why, then, was quicklime more corrosive, and why did it produce much greater heat during neutralization? Homberg argued that quicklime's additional properties must indicate the presence of some additional substance capable of causing the motion we sense as heat, but which is not alkaline. Since quicklime is produced by heating limestone, oystershells, or other calcareous substances to red heat, he suggested that the product retains some *corpuscules ignées* from its manufacture. These fiery corpuscles are released when the quicklime dissolves in acid and cause the greater heat observed during neutralization, the heat sensed during slaking, and the greater corrosivity. Their activity makes quicklime *seem* more alkaline. Homberg also calls these fire corpuscles the *matière du feu*.[90]

The idea that quicklime contains fire corpuscles is not original to Homberg. Lémery had cited it previously, and in fact the notion of fire's incorporation into quicklime can be traced back to antiquity.[91] But Homberg's admission of a particulate matière du feu in 1700 does represent a change in his thinking, presumably as a result of his neutralization experiments; he had previously *denied* the existence of such particles. When describing the calcination of tin in the 1690s, he notes that 1 pound of tin gives 17 ounces of tin

89. Wilhelm Homberg, "Experiences sur la piere de bologne" (12 May 1694), AdS, pochette de séances 1694. See also PV 14, fol. 14v; "Observations sur la pierre de Bologne," PV 17, fols. 126v–129r (19 March 1698). On the Bologna Stone and Homberg's work with it, see Lawrence M. Principe, "Chymical Exotica in the Seventeenth Century, or, How to Make the Bologna Stone," *Ambix* 63 (2016): 118–44.

90. Homberg, "Quantité d'acides," 68–70.

91. Lémery, *Cours de chymie* (1675), 232; St. Augustine, *De civitate dei*, book 21, chap. 4, sect. 3.

calx, a weight increase of 1 ounce. This is, of course, an example of the crucial observation that metals increase in weight when calcined (that is, roasted in air), recognized for well over a century before its true cause—combination with atmospheric oxygen—was identified by Lavoisier in the late eighteenth century. Homberg then cites the "ordinary explanations" (the expression he typically uses for any explanation with which he disagrees) for this increase as the addition of either *particules du feu* or of an acid in the air. Boyle had studied the same phenomenon in the 1670s and attributed the weight increase to absorbed fire particles.[92] But Homberg doubted the plausibility of fire particles and notes that the air, and any acid it might contain, is so light that a vast volume of it would have to be absorbed by a very small volume of tin in order to account for the weight increase observed, and this seems highly implausible. He concludes that the observed weight increase comes instead from the iron pot ordinarily used for calcining tin: the constant stirring of the molten tin dissolves or scrapes off particles of iron. "I believe that this is the sole cause for our increase [of weight] and that there is no need to search for it so distantly as in particles of fire or in an acid of the air."[93] Homberg asked Boulduc to check the claim that antimony and antimony regulus increase in weight during calcination. Boulduc performed the experiment as requested, but only with antimony (our antimony sulphide) and found instead a weight *decrease.* He thus concluded that the previous claims of other chymists arose from "a defect in their manipulation"—namely, they had performed the calcination in a metal pan (rather than Boulduc's earthenware one), and particles dissolved or scraped from the vessel caused the weight increase—a conclusion in line with Homberg's statement about tin.[94]

Just a few years later, when discussing quicklime in 1700, Homberg cites an experiment that Duclos carried out in 1667. Duclos had calcined antimony regulus using a burning lens and found that its weight increased by one-tenth. He found a similar weight increase with both tin and lead, and recorded the same increase whether the calcination was carried out using or-

---

92. Robert Boyle, "New Experiments to make the parts of Fire and Flame Stable and Ponderable," in *Essays of Effluviums,* in *Works of Boyle,* 7:299–333.

93. VMA MS 130, fols. 175v–176r: "Je croy que cest la uniquement la cause de notre augmentation et qu'on ait pas besoin de la chercher si loin dans les particules du feu et dans l'acide de l'air."

94. Simon Boulduc, "De la calcination de l'antimoine," PV 17, fols. 112r–113v (26 February 1698). Had Boulduc tried regulus (that is, elemental antimony), he would have found a weight increase, since upon calcination elemental antimony oxidizes to antimony trioxide. The *mineral* antimony, or stibnite (Sb_2S_3), *does* lose weight during calcination—about 14 percent—as its sulphur is replaced by oxygen.

FIGURE 4.6. The calcination of antimony using focused sunlight, from Nicaise Lefebvre, *Traicte de la chymie*, 2 vols. (Paris, 1660). Courtesy of the Roy G. Neville Historical Chemical Library, Science History Institute, Philadelphia.

dinary fire as the ignition source or the concentrated solar rays from a burning lens (fig. 4.6). Duclos attributed the extra weight to volatile sulphurs in the air that assimilated themselves to the sulphureous parts of the metals.[95]

95. Fontenelle, "Experiences de l'augmentation du poids de certains matieres par la calcination," *HMARS 1666–99*, 1:21–22. The weight increase, if the oxidation goes to completion, should be about one-fifth rather than one-tenth, but the volatility of the antimony trioxide causes a considerable loss of product. Duclos invokes his idea of *symbole* (see chapter 3, section "Homberg's *Semblance*, Geoffroy's *Rapports*, and Duclos's *Symbole*") to explain the assimilation of the aerial sulphurs to the sulphureous parts of the calcined materials; Duclos, "Experiences de l'augmentation du poids," PV 1, 40–52 (January 1667). A more direct witness of Duclos's experiment, copied from Duclos's laboratory notebooks, indicates various differences from the PV version; see Hellot Caen Notebooks, 3:272v–273r. For example, the PV refers to a "miroir ardent," thus suggesting the use of a parabolic mirror like that of La Villette, while the excerpt from Duclos's notebooks clearly refers to the use of a *lens* that Duclos made by gluing together two convex pieces of glass along their rims and filling the intermediate space with water. (The

Although Homberg had denied the reality of the reported weight increase in the 1690s, by 1700 he had become convinced of it and so used Duclos's experiment as evidence of the ability of the matière du feu not only to introduce itself into other bodies—as it does in quicklime—but also to increase their weight measurably.[96] Thus, Homberg has now gotten to the point of explicitly admitting the existence of ponderable particles of fire that can incorporate with other matter, and also—since Duclos's experiment was done first with a burning lens—the existence of these same particles in light.

Just a few months later, Homberg used his new understanding of quicklime to solve a seemingly unrelated problem. In June 1700, Johann Bernoulli (1667–1748) wrote to the Académie about observations of a light emitted by the mercury in some barometers when they are shaken.[97] The academicians chose Homberg, given his long experience with luminescent phenomena, to examine Bernoulli's claims. Homberg's responses led to an extended debate—documented both in the Académie's records and in letters Bernoulli wrote to Leibniz and Varignon—that occasionally became rather insulting on Bernoulli's part.[98] Bernoulli appealed to Cartesian principles to explain the origin of the light, claiming that it was produced by *matière subtile* exiting the mercury and colliding with the Cartesian "second element" contained in the space over the mercury. He attributed the failure of most barometers to glow to impurities that form a skin over the surface of the mercury and thus impede the motion of the matière subtile. Homberg tried to replicate the phenomenon according to Bernoulli's directions, with only limited success.[99] He dismissed Bernoulli's Cartesian reasoning (as Homberg tended to

---

use of the term *miroir* was ambiguous at the Académie; even the Tschirnhaus lens was sometimes called a *miroir* even though it was obviously a *lentille*.)

96. Homberg, "Quantité d'acides," 69.

97. Johann Bernoulli, "Nouvelle manière de rendre les Baromètres lumineux," *MARS* (1700): 178–90. His letter was read to the academicians on 30 June 1700; see PV 19, fols. 243r–253r.

98. Wilhelm Homberg, "Sur la lumière du Mercure dans le vuide," PV 20, fols. 28r–34r (26 January 1701), at 28r: "La Compagnie m'a chargé de verifier les experiences que Mr. Bernoulli de Groningue a fait." Homberg's original autograph memoir, sent to Bernoulli on 26 May 1701 by Varignon, is preserved in the Universitätsbibliothek Basel, L Ia 670, Nr 45. Homberg's original differs considerably from the version in the PV, despite Varignon's remark that it is a copy of what Homberg "gave to the secretary." The Bernoulli-Leibniz correspondence on the matter is in NLB, LBr 57II.

99. Homberg tried the experiment for the academicians on 3 July 1700 (PV 19, fol. 255v). His response went to Bernoulli in a letter by Varignon dated 4 September 1700, and Bernoulli's response of 6 November 1700 is "Nouveau phosphore," *MARS* (1701): 1–9. Homberg followed up with "Sur la lumière," which Varignon sent to Bernoulli on 26 May 1701. Homberg's memoir

do with Cartesian notions), and noted that the description Bernoulli gave of his own mercury indicated that it was not pure, while the very pure mercury used at the Académie did not produce a luminous effect. Nevertheless, several academicians had seen the same barometric light previously. Jean Picard (1620–82) witnessed it first in 1676 when transporting a barometer from the Observatoire to the Porte Saint-Michel at night, and Philippe de la Hire, Cassini, and others had also owned barometers that glowed when shaken.[100] Homberg concluded that the light's erratic appearance was due to "some particular accident that occurred to the mercury that can render it capable of glowing in a place empty of air." He compared it to the Bologna Stone that had to be "disposed by a certain calcination" in order to become luminescent. Homberg then noted that a common method of purifying mercury involved distilling it from quicklime. He suggested that during this process some of the *particelles ignées* lodged in the quicklime could be transferred to the mercury and retained in its interstices in just the way he had shown that mercury could invisibly retain particles of white phosphorus that caused it to glow. Indeed, a barometer Homberg made for Jean-Paul Bignon using mercury

---

is partly summarized in "Sur le phosphore du baromètre," *HARS* (1701): 1–8. Bernoulli's subsequent response, written to Varignon on 5 July 1701, was read on 27 July (PV 20, fols. 265r–275v), and published as "Lettre de M. Bernoulli," *MARS* (1701): 135–46. The published version of the letter omits not only Bernoulli's more acrid statements, but also omits Homberg's name, replacing it with an impersonal construction ("on a fait" in place of "Homberg a fait"). Amusingly, Varignon, well aware of Bernoulli's tart personality, complimented him on the tone of his previous letter, and hoped that any further response to Homberg would be similarly polite, since the letter would be read to the Académie and since Homberg "est un parfaitement honnête homme & qui vous estime aussi beaucoup" (Universitätsbibliothek Basel, L Ia 670, Nr 45). Varignon was at least entitled to hope. As the astronomer David Gregory wrote, "Mr. Bernoulli of Groningnen is a very rough, rude man, speaking well of no mortall, and giving names to everybody"; see W. G. Hiscock, *David Gregory, Isaac Newton and their Circle: Extracts from David Gregory's Memoranda 1677–1708* (Oxford: Printed for the editor, 1937), 20. On this quarrel, see W. E. Knowles Middleton, *The History of the Barometer* (Baltimore: Johns Hopkins University Press, 1964), 354–60. Middleton is mistaken, however, to say that the examination at the Académie was carried out by Cassini, Couplet, and Varignon. Homberg did nearly all the experiments, yet Middleton never mentions him, apparently because he did not examine the manuscript materials that reveal Homberg's predominant role.

100. "Observation faite à l'Observatoire," *Journal des sçavans* (25 May 1676): 112–13. De la Hire's presentation given on 5 May 1694 (PV 14, fols. 13v–14r) is summarized by Fontenelle as "Sur la lumière du baromètre," *HMARS 1666–99*, 2:202–3; I have discovered de la Hire's original memoir in the Uppsala University Library, Waller MS fr-05046. Bernoulli learned of the phenomenon from Joachim d'Alencé's account of Picard's observations in his *Traittez du baromètres, thermomètres, et notiomètres* (Amsterdam, 1688), 50–51.

distilled from quicklime glowed when shaken.[101] When such mercury is put into a barometer—that is, into a place devoid of air—the fiery corpuscles could be drawn out and observed as light. Bernoulli was not satisfied with Homberg's response, but Homberg chose not to continue the argument. The important point is that these experiences with quicklime convinced Homberg that the "matter of fire" could incorporate with matter and reemerge as either heat (slaking quicklime) or light (shaking barometers).

Thus, when in May 1702 Homberg first observed white metals turning yellow in the concentrated light of the Tschirnhaus lens, he had already long been interested in examples of the interaction of light with matter and had already accepted the incorporation of fire particles with ordinary matter, their ability to augment its weight, and their close relationship with light. He must have viewed the new phenomena revealed by the lens in the context of his earlier results and ideas, and this unexpected result made him revise his earlier conclusions in order to incorporate the new result coherently. Therefore, it is not at all surprising that Homberg thereafter buried himself in new experiments at the Palais Royal laboratory, endeavoring to verify his new ideas and suspicions about the identity of the sulphur principle. When he finally emerged, he had completely revised his two 1702 essays and was ready to make his daring announcement at the public assembly of April 1705.

The Third Try (1704–9): Revising the "Essais" and the Central Role of Sulphur

The publication of the Académie's annual *Histoire et mémoires* was slow to get off the ground. Although the Règlement of 1699 stipulated publication at the end of every year, three years elapsed before the first volumes appeared. Undoubtedly, several factors, many of them familiar to modern editors and authors, provoked this delay: organizing a format for the new publication, obtaining finished manuscripts from authors, negotiating with printers, and so on. Additionally, a Dutch printer obtained a license from the States General in late 1700 to issue a pirated edition of the *Histoire* before the authorized work could be printed in Paris; it is possible that part of the then unfinished

---

101. Homberg, "Sur la lumière," fols. 32r–33r. The phenomenon described is an example of electroluminescence; see H. Newton Harvey, *A History of Luminescence* (Philadelphia:American Philosophical Society, 1957), 271–78. Although his explanation based on fire particles is wrong, Homberg's observations are essentially correct. In order to exhibit electroluminescence, the mercury must be scrupulously free from water; this can be effected only by distilling the mercury from quicklime (or other strong desiccants) and not by the other methods of purifying mercury used in Homberg's day.

manuscript for the 1699 edition had somehow fallen into his hands.[102] Fontenelle, who as *sécretaire perpétuel* was responsible for composing the text of the *Histoire*, editing and compiling the academicians' papers, and shepherding the manuscript through the press, found himself constantly behind schedule and having to push back the promised publication dates. In late 1700, he claimed that the volume for 1699 would appear in early 1701, but by the end of April still only about half the sheets had been printed, and the complete volume did not emerge until well into 1702. He promised that the volume for 1700 would be published during Easter 1701 but it was not available until more than two years later, in 1703.[103] Similarly, the issue for 1701 appeared in mid-1704, although the 1702 volume—bearing Homberg's first two chapters of his "Essais de chimie"—also cleared the press in 1704, probably near the end of the year.[104] These delays proved a boon for Homberg. In the two and a half years that passed between the private oral presentation of his first two "Essais" in February 1702 and their publication in the *Mémoires* in 1704, he completed much of the new research provoked by the Tschirnhaus lens, and so, rather than having to retract or contradict these first two essays, Homberg was able simply to rewrite them before they were published.

Most of the changes Homberg made to his essays involve the sulphur principle. In the first, on "chymical principles in general," Homberg demoted salt from the status of an active principle to that of a "middle" principle, thus giving sulphur new preeminence as the sole active principle: "it alone acts and causes the others to act." Water, for reasons that remain un-

102. NLB, LBr 57II, fol. 33r–v, extract (in Leibniz's hand) of a letter from Pierre Varignon to Johann Bernoulli, 30 December 1700: "un libraire de Hollande pour contrefaire seul l'Histoire de lAcademie par M. de Fontenelle a obtenu un privilege des estats comme en ayant le Ms. cependant elle n'est pas encore faite. Cette impudence a jetté M. de Fontenelle dans la derniere surprise. Car ce libraire a même osé se soutenir au correspondant du libraire de l'Academie qui nous a envoyé la lettre de ce fourbe là."

103. NLB, LBr 275, fols. 2r–3v, Fontenelle to Leibniz, 8 December 1700, and 4r–5v, 30 April 1701; see also Arthur Birembaut, Pierre Costabel, and Suzanne Delorme, "La correspondance Leibniz-Fontenelle et les relations de Leibniz avec l'Académie Royale des Sciences en 1700–1701," *Revue d'histoire des sciences et de leurs applications* 19 (1966): 115–32. Far be it from me to criticize Fontenelle on this score, given that I promised this book would appear more than ten years ago; *mea culpa*.

104. The printing of the 1701 volume could not have begun before 15 February 1704; see "Approbation," *HARS* (1701). For 1702, see *Journal des sçavans* (15 June 1705): 383. The 1701 volume was reviewed in the 26 May 1704 issue (321), although a letter from Paris published in the *Nouvelles de la république des lettres* (August 1704): 234, says that the 1701 volume appeared only "a little while ago [*depuis peu*]." The *Journal des sçavans* did not publish a review of the 1704 volume until 15 June and 24 August 1705 (383–96, 559–74).

clear, he promoted from passive to middle, leaving earth as the only passive principle. Since sulphur, now the only active principle, is required for any sort of reactivity in matter, Homberg was obliged to include it in every kind of compound substance, thus restoring it to the vegetable and animal substances from which he had previously exiled it. He also returned salt as a constituent of metals. His original 1702 memoir mentioned two methods for analyzing metals into their principles, both of them involving celebrated chymical *arcana maiora* avidly sought by chrysopoeians. One used "a prepared mercury," that is, the coveted philosophical mercury able to resolve metals radically into their principles. The other employed "resuscitating salts," that is, salts able to combine with a metal's sulphur, thereby setting its mercury free. While these two methods appear in various chrysopoetic treatises of the seventeenth century, they were perhaps most treated by Johann Joachim Becher in his *Physica subterranea*, which is almost certainly one of Homberg's sources.[105] In the revised version, Homberg added a third method for analyzing metals: "by means of a burning lens." He reports that the first method of analysis is easy, if one has the philosophical mercury (rather like how easy plant analyses would have been if Duclos had owned a bottle of the alkahest), the second is troublesome and difficult, but the third "is not difficult," Homberg explains rather nonchalantly, "provided that one has a large burning lens of three or four feet in diameter."[106] Of course, there were just three such lenses in existence, and only Homberg and the Duc d'Orléans were using theirs for chymical research.[107]

The second chapter, on the salt principle, underwent equally significant changes. Previously, Homberg had listed the three species of salt—acid, urinous, and lixivial—as the "nearly" pure salt principle, and then described their properties and mutual interconversions. As recounted in chapter 3, he had long been struggling to explain the differing solvent abilities of the acid

---

105. Johann Joachim Becher, *Physica subterranea* (Leipzig, 1703), especially 814 and 820–22. These methods appeared first in the 1675 second supplement to the work.

106. Wilhelm Homberg, "Essays de chimie: Article premier," *MARS* (1702):35.

107. The other two Tschirnhaus lenses of approximately comparable size were sold to the Elector of Saxony in Dresden and to the Landgrave of Hessen-Kassel. The latter was somewhat smaller than that in the Palais Royal, but would be used within a few years to attempt to replicate Homberg's claims; see below, chapter 7. This lens (and its smaller companion) still exist at Kassel; see Ludolf von Mackensen, *Die naturwissenschaftlich-technische Sammlung in Kassel* (Kassel: Georg Wenderoth Verlag, 1991), 128. The chymist Friedrich Hoffmann had obtained a much smaller burning lens (and a microscope) from Tschirnhaus in 1696, as mentioned in his letter to Tschirnhaus (8 April 1696, Wrocław University Library, MS Akc 1948/0562, fols. 156r–157v), but it is unclear what, if any, chymical experiments Hoffmann conducted with it, save perhaps those briefly mentioned in his *Demonstrationes physicae curiosae* (Halle, 1700), 11–12 , 17–18.

salts. In the 1690s he appealed to a *semblance* between dissolvent and dissolved as the critical factor. In 1702, he hybridized this notion with the idea of more or less sharp acid particles to say that because spirit of niter was sulphureous, it could not dissolve gold whose pores were already clogged with sulphur. Gold could, however, be dissolved with a spirit of marine salt because it was free from sulphur and composed of pointier particles that could penetrate the close texture of gold and break it apart. Now, in 1704, Homberg deletes the entire original section on differing solvent effects, for—having concluded that sulphur is necessary for *all* reactivity—*all* acid salts must contain the sulphur principle. The activating sulphur coats the otherwise inert particles of acid salts, and differences in that coating of sulphur give rise to the various acids and determine their properties. Acids from plants and animals (and this includes spirit of niter—our nitric acid—since niter is produced from putrefying animal waste) are covered with a coarse, low-density *animal or vegetable* sulphur that "augments considerably the volume of the points," disabling them from penetrating compact bodies like gold. The acids of vitriol, common sulphur, and alum (all forms of sulphuric and/or sulphurous acid) have their salt principle covered with a *bituminous* sulphur that, because it is further mixed with an inert earthy material, is less active, and thus such acids do not dissolve metals well. The acids from marine salts, however, have their particles covered with a fine-grained *mineral* sulphur, which gives the salt particles a sufficiently thin coating that it does not blunt their points, thereby allowing them to penetrate even compact metals.[108]

Taken together, these changes made in 1704 to the earlier memoirs depose salt from its former place as the primary chymical principle and replace it with sulphur. For the final manuscript of his textbook, Homberg took the next step and moved the sulphur chapter forward, making it the first section of the work after the general introduction (see table 3.1). It would seem that by this point in 1704, Homberg was largely convinced that light was nothing other than the sulphur principle in the free state, although he was apparently not yet ready to say so definitively. By the time he was ready to announce his conclusions a few months later, at the Easter public assembly of 1705 (and continued in 1706), sulphur/light had become the centerpiece not only of a chymical system but of a fully cosmological system of nature.[109]

At the public assembly on 22 April 1705, Homberg began to expound the

---

108. Wilhelm Homberg, "Essays de chimie: Article second," *MARS* (1702): 37–39, 42–43.

109. The follow-up paper is Wilhelm Homberg, "Suite de l'article trois des essays de chimie," *MARS* (1706): 260–72.

identity of the sulphur principle by describing what substances were *not* the true sulphur principle. He first recalled his 1695 experiments that decomposed vegetable sulphurs (that is, oils) into earth, salt, and water. Since oil could be broken down into simpler substances, it was clearly not a chymical principle. But now Homberg asserts that his analysis of oils must have been incomplete, because "the true sulphur principle that bound [earth, salt, and water] together into an oil is entirely lost during the analysis. . . . It will always escape whomever wishes to free it from all heterogeneous matter." This elusive sulphur principle is nonetheless present in all substances that fall into the category of "sulphureous"—from spirit of wine and oils to bitumens and metals. Only a substance "entirely pure and without any admixture" can be rightly called the true sulphur principle, "leaving the name of sulphurs and sulphureous materials to its first mixtures."[110] How, then, to identify a substance that cannot be isolated, a substance that "the more the artist struggles to isolate it, the less he finds it"? Here Homberg turns his experimental method on its head, in a move similar to what Duclos did a generation earlier with his own endeavor to uncover the true principles of compound substances. Homberg now supplements analysis with synthesis. Analysis gives important information, but cannot provide all the answers. It is only with the help of *synthesis* that a satisfactory resolution can be reached. Every attempt at analysis transforms the embodied sulphur principle into a substance that cannot be contained; therefore, proofs of its nature and identity must be carried out in the opposite direction, by trapping it through synthesis.

For more than a century, chymistry had been routinely defined as the art of analysis and synthesis. The famous spagyric method of the Paracelsians depended both practically and etymologically upon the twin processes of dividing a mixed substance through analysis (*Scheidung*) and then reconstituting it in improved form through resynthesis. But such analysis and synthesis had been conceived of largely (though not entirely) in the context of preparative processes, primarily the making of pharmaceuticals. Promising materials were divided by analysis to isolate their most efficacious components and to eliminate deleterious ones. The useful components, the separated "essences," could then be combined to synthesize potentially more powerful materials. In the seventeenth century, Van Helmont and his followers championed the use of analysis for natural philosophical purposes, as ways of knowing the true nature of materials by identifying their components and how much of

110. Wilhelm Homberg, "Suite des essays de chimie: Article troisième: Du souphre principe," *MARS* (1705): 88–96, at 88–89.

each they contain. Yet Van Helmont also paired synthetic methods with analytical ones to gain fuller knowledge about material composition. His famous
willow tree experiment was essentially a synthetic endeavor, allowing the
natural growth of a tree to synthesize a range of materials from water through
the action of its *semina*. Likewise, he paired analysis and synthesis when he
observed how sand and alkali can be synthesized into glass, and then the glass
analyzed — through fusion with more alkali followed by treatment with nitric
acid — to reisolate the sand in its original weight.[111] Homberg likewise pairs
synthesis with analysis as a way — indeed, the only workable way — of gaining
full knowledge about chymical substances and their composition. Analytical
methods had succeeded in showing what things are *not* the sulphur principle;
now only synthetic methods could demonstrate what it really *is*.

   In turning to this methodology, Homberg may well have had in mind
those first experiments with the lens when sunlight turned white metals yellow, but he does not mention them explicitly here. Instead he returns again
to Duclos's 1667 experiment showing how metallic antimony increases in
weight when calcined, and repeats the experiment using the Tschirnhaus
lens. Four ounces of antimony exposed for an hour to light from the lens
gained 3 gros of weight, or about 10 percent, despite releasing copious clouds
of smoke. In the 1690s Homberg had denied that metals truly gain weight
upon calcination; in 1700, he accepted the gravimetric result and attributed
the weight gain to the incorporation of fire particles; now in 1705 he states
definitively that "this augmentation comes only from the rays of light, or the
matter of light which binds itself [*s'est engagée*] within the regulus . . . for
no other substance was able to touch it during its calcination." Not content
with a single proof, he provides other examples of how light combines with
material substances. He refers to the increase of weight when making minium (our lead oxide) by the calcination of lead, and alludes to the production of quicklime as well, surely in reference to his study of that substance
in 1700.[112] His favored piece of experimental evidence, however, and the one

111. William R. Newman and Lawrence M. Principe, *Alchemy Tried in the Fire: Starkey,
Boyle, and the Fate of Helmontian Chymistry* (Chicago: University of Chicago Press, 2004),
50–91.

   112. Homberg, "Souphre principe," 94–95; Homberg promises that he will describe these
latter operations more fully at a later time, but never did so. In his oral presentation Homberg
incorrectly cited Borelly as the source of the regulus experiment; this was changed to Duclos
in the published version, see PV 24, fol. 129v. He may well have heard an oral report of the
experiment from a colleague rather than having read the procès-verbaux himself, which would
explain the mistake of Borelly for Duclos, and the fact that he slightly misrepresents Duclos's
experiment. Duclos used focused light only to ignite the powdered regulus, and then removed

he expounds in the greatest detail, comes from the lengthy digestion of a specially prepared mercury in a sealed flask. Homberg would return again and again to this experiment, which *became a centerpiece of his mature chymical system*. After many weeks of heating over a fire, he reports, this mercury gradually turns into a red powder that weighs more than the original mercury. When exposed to a stronger fire, the red powder releases the original mercury, but leaves behind a small quantity of a solid, malleable metal, about one-two hundredth of the total weight of the mercury. While Homberg reserves a point-by-point exposition of this experiment for a later date, in the 1705 paper he uses it to demonstrate three points. First, since the glass was sealed, "no other material save that of light was able to pass through the pores of the glass to join itself to the mercury." Flame, as Homberg had stated previously, is nothing other than the oil of the burning substance mixed with the matter of light. In this experiment, the glass walls of the flask act as a strainer, keeping out the grosser particles of oil and allowing only the light, "the smallest of all sensible material," to penetrate and act upon the enclosed mercury. Second, the addition of light changes the nature of the mercury, turning it from a silvery liquid into a red solid, and increases its weight. Third, upon being submitted to a stronger fire, much of the attached light/sulphur is expelled, thereby regenerating the original mercury, but some of this light/sulphur has become so firmly and deeply combined with the mercury that their union produces a fixed and stable metal, albeit in very small quantity.[113] I have already pointed out elsewhere the chrysopoetic import and origin of this experiment (only later did Homberg reveal that the initially unspecified "malleable metal" this experiment produced was in fact gold), particularly in the writings of Eirenaeus Philalethes (George Starkey) and Johann Joachim Becher.[114] I will return to it in the following chapter, which deals in detail with the long-term gold-making endeavors of Homberg, his Académie colleagues, and the Duc d'Orléans. For now it suffices to notice the close analogy between this experiment with mercury and that first lens experiment with silver: in both cases a white metal was turned yellow by the addition of light, and in the case of mercury, Homberg was able to isolate the newly formed yellow substance and identify it as gold.

---

it, allowing the regulus to burn on its own without further heating; Duclos, "Augmentation du poids," 40.

113. Homberg, "Souphre principe," 92–93; Homberg repeats his ideas on the composition of flame at 90–91.

114. Lawrence M. Principe, "Wilhelm Homberg: Chymical Corpuscularianism and Chrysopoeia in the Early Eighteenth Century," in *Late Medieval and Early Modern Corpuscular Matter Theories*, ed. C. Luthy, J. E. Murdoch, and W. R. Newman (Leiden: Brill, 2001), 535–56.

Having logically and experimentally identified light as the true sulphur principle, Homberg gives it a central role not only in his chymical system but in all the workings of the natural world. Indeed, the role Homberg gives to sulphur will touch even upon the theological. Light can "introduce itself into the other principles, change their shapes, augment them in weight and volume." In so doing, it produces different classes of sulphureous materials: plant and animal sulphurs (in oils and fats), mineral sulphurs (in bitumens), and metallic sulphurs (in metals). In some cases, the light/sulphur is only superficially attached, and is thus easy to expel, resulting in the decomposition of the mixed substance; this is the case with animal and vegetable materials. In other cases, the light/sulphur can "enter into the substance itself," particularly when incorporated into mineral bodies, producing minerals and metals that are difficult to decompose. The most intimate combinations of light/sulphur and mercury generate the most stable metals: gold and silver. Less complete and less pure combinations produce the base metals.[115] Homberg here continues long-standing ideas about the composition of metals — dating back to the Latin and Arabic Middle Ages — into his new theory. Gold and silver are composed of pure mercury and fixed sulphur, gold contains more sulphur than silver, the base metals contain additional earthy and saline components and unfixed sulphur. Interestingly, Homberg draws the corollary that since gold and silver contain only two components, while the base metals have more, gold and silver should be easier to synthesize artificially than the base metals.[116]

Homberg's system presents a remarkable "economy of sulphur," whereby the embodiments of light in matter and its successive liberations and reembodiments give rise to all phenomena of chymical change in the universe. All the forms that the light/sulphur principle takes when incorporated into mixed substances are potentially interconvertible because it always remains the same substance, only embodied as larger or smaller, more or less dense particles. For example, iron, copper, tin, and lead contain a metallic sulphur whose particles are larger and less dense than the sulphur principle as light. Therefore, when calcined in the fire or in the powerful light of the burning lens, these larger sulphur particles are blown away by the minute and fast-moving particles of light/sulphur, like feathers or straw blown away in the wind. The empty spaces they leave behind are then filled with finer grained sulphur newly incorporated from the light. Composed of smaller particles, it packs the vacated spaces more tightly, leaving less empty space, and as a result

---

115. Homberg, "Suite de l'article trois," 261, 264–65.
116. Homberg, "Suite de l'article trois," 267–68.

the calx weighs more than the original metal, as experiment demonstrates. But if this calx is then mixed with a sulphureous substance from the plant or animal kingdom, such as an oil, fat, or charcoal, and heated, the plant or animal sulphur regenerates the original metal by restoring to it a large-grained sulphur. Since this large-grained sulphur cannot pack as efficiently as the fine-grained sulphur captured from the flame or lens, less can be stored in the pores, and thus the revived metal returns to its initial weight, and weighs less than the calx. The poorly fixed, inflammable sulphur in iron can even be made visible. Upon warming iron filings in a solution of oil of vitriol, Homberg notes, the metal "exhales a vapor which burns like spirit of wine when a lighted candle is brought near."[117] Since light/sulphur is a principle, it is always conserved. If it is expelled entirely from a composite, it regains its free state, and "it will reenter the great mass of the matter of light which fills the whole space of the universe."[118] This is what happens whenever a combustible material burns; the light previously incorporated as sulphur reemerges again as light. Thus "sulphureous materials change their state indifferently, and pass from one species of sulphur into another species, as circumstances furnish them the opportunities."[119]

Obviously enthusiastic about the explanatory power of his central concept of light as the sulphur principle, Homberg began using it to explain an increasingly wide range of puzzling phenomena. He had noticed, for example, that when distilling spirit of wine, sometimes the drops falling into the receiver do not immediately mix with the liquid already there, but instead roll around on its surface "like peas on a table." Since this happens most when the drops are warm, he suggests the phenomenon is due to particles of the matter of light clinging to the surface of the drops and rendering them so prickly (*herissées*) that they cannot make direct contact with the surface of the liquid. The same phenomenon does not happen when water distills because it is not sulphureous like spirit of wine, and so does not accumulate the mat-

---

117. Homberg, "Matieres sulphureuses," 233. What Homberg describes here is the inflammability of the hydrogen produced by the action of dilute sulphuric acid on iron. Geoffroy ("Maniere de recomposer le Souffre," 285) had earlier made the same observation and drawn the same conclusion—possibly under Homberg's guidance—that the burning "steam" from iron dissolving in acid was evidence of the metal's highly sulphureous character. None of the other metals recognized at the time generates hydrogen with sulphuric acid, so iron would certainly have seemed different from the others, in this case "more highly sulphureous."

118. Wilhelm Homberg, "Suite des essais de chimie," *MARS* (1709): 106–17, at 111.

119. Homberg, "Suite de l'article trois," 265–66; Homberg, "Matieres sulphureuses," 226–67; and Homberg, "Observations sur les matieres sulphureuses & sur la facilité de les changer d'une espece de souffre en une autre," *MARS* (1710): 225–34, at 233.

ter of light so readily.[120] Here one can detect the persistence of his earlier idea of *semblance*: the spirit of wine as a sulphureous composite readily picks up some of the sulphureous matter of light in the air through which it passes. Such semblance appears again when he explains his observation that in the winter glasses full of water or mercury shatter if left too long near a fire, while those filled with spirit of wine remain unbroken. He suggests that the matter of light emanating from the fire easily passes into the spirit of wine because it is an inherently sulphureous liquid, and thus "more homogeneous" with the matter of light than is either water or mercury. When a vessel is filled with water or mercury, however, the matter of light is forced to tarry longer in the glass wall of the container, thereby more greatly dilating its pores and causing it to crack from the unequal expansion.[121] Similarly, in a remarkable experiment, Homberg observed that the focus of the lens is considerably less powerful when the beam of light is made to pass first through the "sulphureous" exhalations emitted by burning charcoal—the sulphur in this "vapor" absorbs some of the light because they are of the same nature.[122] Hence it is clear that Homberg continued to employ ideas of semblance as explanations of chymical combinations, thus making the connection of his semblance with Geoffroy's later rapports yet more probable.

Homberg applied his system still further, well beyond what his colleagues would have considered to be chymical phenomena. He suggested, for example, that ice melted faster in his air pump in the summer than in the open air because without the interfering air, the matter of light could beat more strongly upon the surface of the ice, causing it to melt faster.[123] In 1708, he suggested that some wind and ocean currents could be affected, and even caused, by the pressure of the matter of light from the Sun beating upon them obliquely at sunrise and sunset. In this way he explained an observation made during his travels in Hungary: in the morning the watermills on the Danube stop turning and then around noon they begin again. The matter of light from the Sun, raining down westward in the morning, he suggests, opposes the sluggish eastward flow of the water, and sufficently dampens the river's

120. PV 24, fol. 247v (22 July 1705).

121. *HARS* (1706): 5.

122. *HARS* (1705): 39–40. While the reduced intensity might have been due to the rising hot gases disrupting by refraction the proper focusing of the light, it is also possible that Homberg unknowingly observed the absorption of infrared light by carbon dioxide, having performed what was, in essence, the first experiment in absorptive spectroscopy.

123. PV 27, fol. 135r (2 May 1708).

current that not enough movement is left to turn the millwheels.[124] This paper, "which could pass for a natural history" according to the report in the *Mercure galant*, was read to a public assembly by Lémery *fils* and greeted with enormous applause. Abbé de Louvois, who presided at that meeting, "after having praised Mr. Homberg for the erudition of his discourse and his untiring application towards discoveries in natural philosophy, said that the nature of winds could not have been better defined than [Homberg] had done . . . but that he must take care lest a gust of wind should carry him back to his homeland from which France had drawn him (he meant Batavia . . . where Mr. Homberg was born) because the Académie would thereby lose too much."[125]

Homberg's attempts to universalize the explanatory power of light—now fully a *chymical* entity and substance—reflects again his broad vision of chymistry in terms of both its domain and its ability to explain natural philosophical phenomena. His initially chymical theory became a more broadly natural philosophical one. In all cases, it is light, a chymical principle, that is the sole source of change and activity in the world. Fire itself, the chymist's primary tool, has power to alter bodies only because of the light it contains. According to Homberg, therefore, "the whole Universe is filled with the matter of light . . . the Sun and the fixed stars that are spread out through the infinite space of the Universe are so many flames whose principal function is continuously to move and push forth this matter of light, which thereby collides with and penetrates every porous body it encounters in all that immense space which it fills."[126]

The Sun and stars are thus for Homberg the *engines that power the universe*, including, ultimately, every chymical process and transformation. The varieties of light/sulphur's combinations with otherwise inert matter ensure the stupendous range of substances in existence. Light's power to change the shape, weight, and volume of other matter and its ability to transfer from one substance to another account for all chymical changes. And yet Homberg goes further, concluding his treatment of the sulphur principle with the only reference to the Deity in all his writings. The variety of the combinations of light with matter, he writes, "produces infinite variety, such that if one wanted

124. Wilhelm Homberg, "Observations et conjectures touchant les causes des vents," PV 27, fols. 113r–121v (18 April 1708). A highly abbreviated version appears in *HARS* (1708): 21.

125. *Mercure galant* (April 1708), 275–76. The *Mercure*'s reporter, however, thought that Batavia was in America.

126. Homberg, "Article trois," 91–92.

to compare the variety of materials that do exist with the number that could exist by means of all the possible combinations, we would be obliged to say that the known universe is a very small thing compared to that which it might be, and even if there are other worlds like ours, they could all be furnished with objects without changing either the matter or the manner whereby these objects are composed—which points to a richness and an infinite power of the Being that produced the universe."[127]

Light has, of course, extremely powerful connections with Christian theology. Creation was called forth out of nothing with the divine command, "Let there be light." Light was the first created thing, and the origin of corporeal being. This theological connection was made explicit immediately after Homberg's presentation at the 1705 public assembly. Both the *Journal de Trévoux* and the *Nouvelles de la république des lettres* report that after Homberg had finished his presentation, the Académie's president, Abbé Jean-Paul Bignon, remarked that Homberg's new chymical theory resolves a difficulty in the interpretation of the opening chapter of Genesis, where God is said to have created light before the Sun and stars. Homberg replied that there is no difficulty in that passage, because God created a certain quantity of light, of which the Sun and stars are but portions, the rest of it remaining dispersed throughout the universe. Bignon concluded by remarking that "it is not absolutely necessary that a system of natural philosophy conform to the words of Scripture, but it seems that such a system has more likelihood when not only does it not depart from them, but also, like that of Mr. Homberg, can serve to explain them."[128]

Mercury and the Metals

The essay on sulphur begun in 1705 and continued in later installments is clearly the linchpin of Homberg's "Essais," and the sulphur principle is the central player in his chymical system. The "Essais," however, were not yet complete: three more principles—mercury, water, and earth—remained to be covered. In 1709, Homberg added the promised fourth chapter on mercury. He began the chapter by stating clearly, "in order to avoid all equivocation," that he uses the word *mercury* to refer to the substance commonly known as quicksilver (that is, the liquid metal we today call mercury). This statement was necessary since contemporaneous chymical literature, especially

127. Homberg, "Suite de l'article trois," 272.

128. *Nouvelles de la république des lettres* (June 1705):703; *Journal de Trévoux* (August 1705): 1415–26, at 1425–26.

the chrysopoetic, was often equivocal or intentionally obscure about what "mercury" actually meant—a practice that frustrated and annoyed many readers. Homberg states that he calls mercury a principle only by default, because he believes it to be a compound rather than a simple substance. In fact, he has found a way whereby mercury is "destroyed" and transformed into another substance that cannot return to the original form of mercury (this topic will be covered in chapter 5). The ability of mercury to be permanently converted into a very different form—"which never occurs with a simple body"—provokes Homberg to insist on the provisional nature of his naming mercury as a chymical principle. "Although I am persuaded that mercury does not have the character of a principle, meaning that its substance cannot be reduced into more simple materials by any analysis, I nevertheless place it among the number of my chymical principles because this analysis has not yet been found, even though there is reason to believe that it can be found in the future; mercury will then be rejected from the list, all appearances being that mercury is a compound."[129]

Nevertheless, he concludes, "it ought to find a place among my chymical principles until the parts that compose it have been discovered."[130] Since mercury is a component only of metallic substances, Homberg's paper on mercury focuses on metals. He once again revisits the keystone experiment for his system: the long-term heating of a specially prepared mercury that converts it into a red powder, most of which can be reduced back into mercury, but part of which remains as newly formed gold. The production of the small quantity of gold provides evidence that some of the matter of light from the fire entered into a new and permanent combination with the mercury to produce a stable, solid metal.

The two noble metals, gold and silver, differ only in the amount of sulphur—that is, the amount of the matter of light fixed as metallic sulphur—they contain. Since gold contains more sulphur than silver, the addition of more sulphur should transform silver into gold. This possibility seems to have been confirmed by the first experiments Homberg carried out with the burning lens in May 1702 when light turned silver yellow. Therefore, upon considering the gradual combination of sulphur/light with mercury to produce a metal, Homberg concludes that "all gold must indeed have been silver before it could attain its proper completion." In that case, there must also be

129. Wilhelm Homberg, "Suite des essais de chimie: Art. IV. du Mercure," *MARS* (1709): 106–17, at 106; compare "Suite de l'article trois," (1706): 264.

130. Homberg, "Du Mercure," 106–7; oral presentation: PV 28, fols. 117r–128v (23 March 1709). There is very little change between the PV version and the published memoir.

an "intermediate state" between the two metals, where the composition has more sulphur than true silver, but not enough to be true gold. This is the only case of which I am aware in which a chymist explicitly acknowledged this possibility, even though it was implicit in the long-standing mercury-sulphur theory of the metals.[131] Homberg then asserts that, in fact, a "middle metal [*métal mitoïen*]" is sometimes found in mines as "pale gold"—pale because it has not yet combined with sufficient sulphur to achieve the full yellow color of gold. By means of repeated fusions, however, this incomplete gold absorbs sufficient sulphur/light from the fire to become good gold. Experiment shows that even the common silver of commerce and coinage ordinarily contains some of this middle metal. Homberg recounts that if one takes silver, purifies it scrupulously of all gold by quartation, and then keeps the purified silver in fusion for at least one hour, repeats the process one hundred times, and then performs a quartation again, a small quantity of gold, evidently newly formed, will be found. This gold is produced while the silver is in fusion, at which time those of its particles "which are closest to the perfection of gold" absorb sufficient sulphur from the fire to become gold.[132] Thus, Homberg not only continued the long-standing belief that lesser metals could "mature" into gold, but also provided a new and explicit mechanism for this transmutation. His comment about the intermediate metal found in mines suggests that he also considered that such maturation was occurring naturally but slowly underground, an idea that had been part of chymistry for centuries before him.

Homberg's mercury chapter deals entirely with gold and silver, but he ends the chapter by stating that "I shall examine the lesser metals in another memoir, and I will add there the rest of my observations on mercury." This promised memoir never appeared, nor does further material on this subject exist in his final 1715 *Éléments*.[133] The reason he did not fulfill this promise is not entirely clear. It might have to do with the misfortune—detailed in chapter 6—that Homberg suffered in 1712 that forced him to curtail his experimentation. Still, it is clear that Homberg had begun to work on the lesser metals some years earlier. In 1706, he described the effects of the lens

131. There is a possible connection to the *luna fixa* described by some chrysopoetic authors, a metal white like silver but with the density and some of the chymical properties of gold; however, those who mentioned this anomalous metal did not insert it so clearly into a gradient of sulphur concentration stretching from silver to gold. On *luna fixa*, see Principe, *Secrets of Alchemy*, 114.

132. Homberg, "Du mercure," 112–15.

133. Homberg, "Du mercure," 117. The closing line promising this further treatment is absent from the 1715 *Éléments*; VMA MS 130, fol. 108v.

on iron both by itself and mixed with other metals, remarked briefly upon how other base metals were affected by the lens, and drew several conclusions therefrom about the composition of these lesser metals. Significantly, the published memoir contains a long section—absent from the procès-verbaux version—that details how "it has been discovered to us by chance" that plant ashes contain a quantity of iron.[134] That observation forms part of a lengthy argument about the possibility of producing iron de novo that was carried on between Homberg's student and collaborator Étienne-François Geoffroy and Louis Lémery from 1704 until 1708.[135] Indeed, Geoffroy's collaboration in Homberg's experiments with the lesser metals is probably indicated by the constant use of the first person plural *nous* in the added section of Homberg's paper on iron, a pronoun that occurs nowhere else in the published memoir, and only rarely in Homberg's other memoirs. I suggest that Homberg turned over much of the research on the composition of the base metals to Geoffroy. Accordingly, Geoffroy published thoughts and experiments about the composition of all the base metals in the context of his debate with Lémery in 1707, and in 1709 he read a memoir describing the effects of the Tschirnhaus lens on the five base metals: iron, copper, tin, lead, and mercury.[136] This division of labor further indicates the level and duration of the collaborative efforts between Homberg and Geoffroy. For a time, however, Geoffroy interpreted some of his results in a different framework, such that they seemed to conflict with Homberg's larger ideas about composition—an important topic covered in chapter 5.

The Final Version: *Élémens de chimie*

As mentioned in chapter 3, the manuscript of Homberg's completed version of his textbook of chymistry was described as "ready for the press" shortly

---

134. Wilhelm Homberg, "Observations sur le fer au verre ardent," PV 25, fols 168r–172r (8 May 1706); *MARS* (1706): 158–65, added section on 160–61.

135. On this debate, see Bernard Joly, "Quarrels between Etienne-François Geoffroy and Louis Lémery," in *Chymists and Chymistry*, ed. Lawrence M. Principe (Sagamore Beach, MA: Science History Publications, 2007), 203–14; Joly, "Étienne-François Geoffroy (1672–1731): A Chemist on the Frontiers," *Osiris* 29 (2014):117–31; and Joly, "Le mécanisme et la chimie dans la nouvelle Académie royale des sciences: les débats entre Louis Lémery et Étienne-François Geoffroy," *Methodos* 8 (2008), special issue: *Chimie et mécanisme à l'âge classique*, http://methodos .revues.org/1403. See also chapter 5, below.

136. Étienne-François Geoffroy, "Eclaircissemens sur la production artificielle du Fer, & sur la composition des autres Métaux," *MARS* (1707): 176–88, especially 185–88; and Geoffroy, "Experiences sur les metaux faites avec le verre ardent du Palais Royal," *MARS* (1709): 162–76; PV 28, fols. 171r–183r (18 May 1709; begun 8 May, fol. 165r).

after his death in 1715. It passed through several hands that planned to publish it, but vanished along with his other unpublished treatises in the 1720s and was thereafter entirely forgotten or presumed lost. Only in 2005 was a copy recovered from among the Boerhaave papers in St. Petersburg. This final version shows many significant changes relative to the previous version published serially in the *Mémoires*. These changes underscore important features of Homberg's ideas about chymistry as they continued to evolve with further experimentation and reflection.

To prepare the final manuscript, Homberg melded his serial essays together into a book format clearly intended as a freestanding publication. The published *articles* or *essais* became four *chapitres*, and he modified the title to *Élémens de chimie*. The originally promised chapters on water and earth, however, are nowhere to be seen (see table 3.1). Homberg must finally have concluded that they were unnecessary or not of sufficient interest to be written. The same is true of the "cours d'Operations" that was initially supposed to complete the volume. In fact, the final manuscript moved in quite the opposite direction: Homberg deleted *all* of the recipes for specific and illustrative products that had been present in the essays as published in the *Mémoires*, thus making the whole more uniform in its exclusive presentation of a comprehensive theory of chymistry and its principles.

One surprise offered by the completed version is that it incorporates more of Homberg's publications than expected. Only five publications dating from 1702 to 1709 are clearly marked in the *Mémoires* as being serial parts of his "Essais," but the St. Petersburg manuscript contains extended text from *eight* of Homberg's publications. The additional three papers are his 1702 paper on the action of the burning lens on gold and silver, a 1709 paper on acids and alkalies, and a 1710 paper on sulphureous substances.[137] Some even later materials are also inserted in the *Élémens*, such as a mention of the spontaneously inflammable substance (later known as Homberg's pyrophorus) about which he presented two papers in 1711 and 1712.[138] These unexpected inclusions show that even more of Homberg's mature work than previously thought was geared toward his central endeavor of establishing a solid and coherent system of chymistry based on experimental results. There was thus

137. Homberg, "Observations faites par le moyen du verre ardent"; Homberg, "Observations touchant l'effet de certains acides sur les alkalis volatils," *MARS* (1709): 354–63; and Homberg, "Observations sur les matieres sulphureuses."

138. VMA MS 130, fol. 41r; Wilhelm Homberg, "Observations sur la matiere fecale" *MARS* (1711): 39–46; and Homberg, "Phosphore nouveau, ou suite des observations sur la matiere fecale," *MARS* (1711): 238–45.

a greater degree of underlying coherence to his experimental endeavors and investigative pathways than has been thought, showing that he was at work on his comprehensive theory, and the book in which he presented it, for longer than previously known.

The most substantial rewriting appears in the first chapter of the book, "Of the principles of chymistry in general." The final 1715 version departs significantly from both the 1702 and the 1704 versions. One unexpected change appears in the list of chymical principles. Whereas Homberg had previously listed the five principles "salt, sulphur, mercury, water, and earth," in the final version he now lists *six*: "sulphur, salt, mercury, air, water, and earth."[139] On the one hand, the inclusion of air makes Homberg's list appear more like a juxtaposition of the Paracelsian *tria prima* with the Aristotelian four elements (lacking, of course, fire, which in Homberg's system is implicitly included under the sulphur principle). On the other, it brings together yet more of Homberg's life work by incorporating his extensive early work on pneumatics into the grand sweep of his labors. Furthermore, while Aristotelians did include air among their "elements," it is very unusual for it to be included by a chymist among the chymical principles. There are no examples of air being included among the principles by the authors within the French didactic tradition. Étienne de Clave does discuss air, but concludes that it does not enter into the composition of bodies and therefore cannot be considered one of the principles.[140] In contrast, Homberg claims that air *does* enter into combinations with the other principles. This claim is potentially extremely significant, since it was the realization that gaseous materials could become "fixed" in solid matter that opened the way for significant new discoveries in the eighteenth century. Unfortunately, Homberg does not explain exactly what he means by the combination of air with the other principles, and does not provide any examples or experiments that led him to that conclusion, so we cannot ascertain what he really had in mind.

In the final version, the sulphur principle has completed its subjugation of all the other principles. Homberg moves the chapter on sulphur into first place, displacing salt, and the sulphur chapter is by far the longest—making up nearly half of the entire text. Sulphur remains the sole active principle, but now Homberg demotes all the other principles—salt, mercury, water, air, and earth—to the status of passive, thus abolishing the category of "middle"

139. Compare Homberg, "Essays," 34; and VMA MS 130, fol. 3r.

140. Étienne de Clave, *Nouvelle lumiere chymique* (1641), ed. Bernard Joly (Fayard: Paris, 2000), 39–41.

that had previously included salt, water, and mercury. Once again, this concentration of all chymical activity in a single chymical principle is unprecedented in the earlier chymical didactic tradition.

Homberg also revisits his experimental methodology for studying the composition of compound substances by a combination of chymical analysis and synthesis. The 1702 procès-verbaux version had listed just two methods of decomposing metals into their components: using philosophical mercury or resuscitating salts. In the 1704 published version, Homberg added his new scientific instrument—the Tschirnhaus apparatus—to the list. In the final version, the first two, both traditional methods of the chrysopoeians, disappear, and only the use of "a burning lens of three or four feet in diameter" remains. But Homberg then immediately qualifies the success of this method using the lens, saying that even if one has access to such a rare and costly scientific instrument, "even then there is quite some difficulty in learning something from it." This qualification is due to the inherent limitations of analysis itself, limitations that Homberg ran up against in his experimental endeavors: "I have not been able by any analytical method—that is, by decomposition—to discover the materials of which the metals are composed. But by employing in a certain way the simple materials whose combination has produced the metals, as we shall see shortly, I have had occasion to make judgments with some reasonable probability [*vraisemblance*] about the arrangement of the materials that enter into their composition, and the difference between the perfect metals and those we call the lesser metals."[141] Synthesis stands as the necessary supplement to analysis. The main experiment in synthesis that permits Homberg to draw "reasonably probable" conclusions about the composition of metals is again his process of long-term heating of a specially prepared mercury, whereby most of it is converted into a red powder and part is converted into gold.

Homberg's lifelong praise of chymistry as the clearest, most powerful, and most expansive of the sciences continues unabated in his final version of the *Élémens de chimie*. He maintains his opening salvo about the superiority of the chymical principles to those of other natural philosophers, whose speculative principles he now calls simply "too problematic" and requiring too long a discussion. In the 1702 and 1704 versions, Homberg preferred the

141. VMA MS 130, fol. 5v: "Je n'ay pû par aucune voÿe analitique cest a dire par la decomposition decouvrir les matieres de quoy sont composés les metaux mais en employant de certaine façon des matieres simples dont lassemblage a produit des metaux comme nous le verrons dans la suitte, Jay û occasion de juger avec quelque vraisemblance de l'arrangement des matieres qui entrent dans leur composition et de la difference qu'il y a entre les metaux parfaits et ceux que nous appellons les moindres metaux."

chymical principles because they are more *sensible*—that is, directly perceptible by human senses—an outlook that squares well with his unwavering insistence on demonstrative experiments rather than imaginative speculations. But now Homberg intensifies his emphasis on the senses as the crucial way of engaging with the *sensible* chymical principles. In a series of entirely new paragraphs he *links specific senses to specific chymical principles*, and describes under what conditions each principle is sensually apprehendible.

> In its pure state and considered as a principle, sulphur cannot be perceived by any of our senses while at rest, and being in movement it is only touch and sight that receive its impressions. But being joined to the other principles it becomes sensible to all the senses.
>
> Air is always invisible to us howsoever it is modified by sulphur and howsoever it is united to the other principles; nevertheless, while the sense of sight does not perceive it in any way, the other senses do not fail to be sensibly moved by it, and principally the sense of hearing, of which it seems to be the unique object.
>
> In its pure state and considered as a principle, salt cannot appear to our eyes even being joined to sulphur, but becomes sensible when joined to mercury, to water, or to earth.[142]

These new paragraphs have several important implications. First of all, Homberg insists that true principles must be perceptible by the senses in particular and often distinctive ways, even if union with another principle is necessary for such perceptibility. The "problematic" principles of other natural philosophers—such as the Cartesian *matières*—are very different, lying as they do forever below the level of direct human sense perception. For Homberg, in contrast, even the sulphur principle, which can never be isolated, is still sensible in distinctive ways—to touch as heat and to sight as light. Air, although always imperceptible to sight, is perceived directly by other senses, and distinctively by hearing, which depends upon the air to carry sound, and thus the sense of hearing perceives nothing other than air

---

142. VMA MS 130, fols. 3v–4r. "Le soufre dans son etat pur et consideré comme principe pendant quil est en repos ne scauroit etre apperçû d'aucun de Nos sens et etant en mouvement Il ny a que le tact et la veüe seuls qui en recoivent des impressions mais etant jointes aux autres principes il devient sensible a tous les sens. L'air nous est Toujours invisible de quelque[s] manieres quil soit modifié par le soufre et de quelque maniere quil soit uni aux autres principes cependant quoyque la sensation de la veüe ne len apperçoive en aucune maniere les autres sens ne laissent pas d'en etre sensiblement agités et principalement l'oüie dont il paroit etre lunique objet. Le sel dans son etat pur et consideré comme principe ne scauroit paroitre a nos yeux meme etant joint au souphre mais il devient sensible quand il sest joint au mercure, a l'eau, ou a la Terre."

as its "unique object" of sensation. When joined to three of the other principles, salt also activates the senses, and although Homberg does not specify a particular sense explicitly here, given contemporaneous chymical laboratory practice, it would be easy to fix especially on the sense of taste whereby salts were routinely distinguished in practice from other substances. In the same vein, Homberg defines and identifies acid salts entirely by their sour taste, implicitly demoting the use of color indicators that had seen such extensive use previously at the Académie.

It is reasonable to see in Homberg's final invocation of the senses yet a further extension of the boundaries of chymistry. This extension is completely in line with his frequent statements about the independence of chymistry, about its superior methods for understanding the natural world, and about its "infinite extent." All the senses are activated immediately by the chymical principles—sight and touch by sulphur, hearing by air, taste by salt—thus whatever is sensed is by that very fact *chymical*. Human beings can interact with the physical world only through the senses, and thus the very act of human perception is inherently and unavoidably *chymical*. Having concluded that the sulphur principle is light itself, and that light is a material substance that can be used as a chymical reagent in synthesis, Homberg seems thereby to imply that subjects such as optics and vision that rely upon light, now a chymical principle, should now be considered part of chymistry. Homberg had been skilled in optics: he ground telescope lenses, made observations of prismatic colors, and built improved microscopes and an instrument to measure the refraction of light in a vacuum.[143] Might he have come to see such endeavors as somehow *chymical* insofar as these optical instruments manipulate light just as retorts and alembics manipulate other chymical substances? The Tschirnhaus lens collected and concentrated a chymical principle and reagent that could combine with other chymical substances and provoke chymical change. His new inclusion of air among the principles and its special link to hearing suggests an analogous claim. Just as the act of seeing depends upon the sulphur principle, the act of hearing depends upon the air principle. Thus pneumatics, and Homberg's extensive work with it, is thereby brought under the broad umbrella of chymistry—and might he also have thought that such subjects as acoustics then also fell ultimately within the domain of the chymist?

Homberg's vision of chymistry, or perhaps *for* chymistry, was far beyond anything ever presented by the *cours de chymie* tradition and far broader than the narrow confines that his colleagues at the Académie marked out

---

143. PV 19, fols. 198v–199r (15 May 1700); see Bourdelin Diary, fol. 21v, for a sketch.

for it. Chymistry was not only the source of more certain knowledge about the world, but was truly, as he said already in the 1690s, a subject "of infinite extent that embraces all corporeal substances." In his system, the activity, even the very purpose of the Sun and stars is inherently chymical, as they continuously cast forth, by that primordial divine command, *Fiat lux!*, the vital sulphur principle that fills the universe, gives activity and reactivity to the rest of matter, and generates all chymical change everywhere.

Shortly after Homberg's death, one of his associates in Philippe II's household wrote a letter to Leibniz that included a brief but eloquent summary of Homberg's outlook on knowledge and natural philosophy—one that accords well with the scientific worldview and personality of the Batavian chymist that the previous chapters have endeavored to put forth, particularly his unswerving devotion to chymistry and his insistence on the primacy of practical experiment and sensual observation. Nicolas-François Rémond, *chef des conseils* to the Duc d'Orléans, began by informing Leibniz about the manuscript of "a whole course of chymistry" left behind by Homberg and now in the hands of Geoffroy, and then added:

> [Homberg] didn't know how to write well but he was exact and true in his operations. All philosophy according to him lay in the use of the fire-tongs, and thus he didn't care much about either the ancients or the moderns. I often laughed about it with him, for he didn't esteem any but the chymists, and among the chymists he didn't esteem any but himself.[144]

Rémond may have intended this statement as slightly critical, but Homberg would probably have embraced it with a satisfied laugh.

The foregoing chapters have followed Homberg's endeavor to formulate a comprehensive theory of chymistry over the course of more than twenty years, from the early 1690s until 1715. In doing so, they shed light on some of the transformations going on within chymistry—in terms of theories, practices, aims, and content—during the late seventeenth and early eighteenth centuries. Central to Homberg's aim was a reliable, experimentally based identification of the chymical principles, the building blocks of compound substances. New experimental results constantly forced him to revise his assumptions, conclusions, and explanatory principles, and sent his investigative pathways down new avenues of inquiry over the course of two decades.

---

144. Rémond to Leibniz, 23 December 1715, fol. 54r: "Il ne savoit pas ecrire mais il estoit exact et vrai dans ses operations. toute la philosophie selon lui etoit dans l'usage de la pinçette et ainsi il faisoit peu de cas des anciens et des modernes. j'en riois souvent avec lui car il n'estimoit que les Chymistes et entre les Chymistes il n'estimoit que lui mesme."

The coherence of his endeavor and dynamic character of his ideas, now revealed by newly uncovered archival sources, had hitherto been obscured by a reliance on only the published serial versions of his "Essais de chimie." Significantly, many of the important changes to Homberg's ideas can be linked to specific experiments that he carried out, thus documenting some of the complex interactions between theory and practice in chymistry. Homberg began by testing received chymical theories experimentally—Helmontian water theory, acid-alkali theory, and versions of contemporaneous mechanical chymistry. When results failed to support these systems, or the systems failed to live up to his standards of measurable, observational demonstration, he modified or abandoned them, finally embarking upon the creation of a novel explanatory system of his own. By the end, Homberg had crafted an experimentally based system of chymistry and chymical principles, based upon the results of both analysis and synthesis, which embraced an ambitiously broad range of natural phenomena.

In writing up his ideas, Homberg increasingly abandoned the format of the chymical textbook established by earlier authors such as Glaser and Lémery. Departing more and more from a didactic tradition that favored the presentation of preparative (usually pharmaceutical) recipes, the successive versions of Homberg's textbook placed increasing emphasis on expounding the fundamental governing principles of chymical reactivity at the expense of presenting preparative recipes. At the same time, Homberg enlarged the borders of chymistry's domain as an independent and natural philosophical discipline, and promoted chymistry as the best means of gaining a deeper and more certain knowledge of the workings of the natural world. For him, chymistry's inherent superiority came from the directness of its contact with matter, and therefore its foundations should lie in phenomena and results perceptible by human senses. He advocated not only new theories, but also a new status, aim, and content for chymistry, in ways similar to the vision of the subject held by Boyle and Duclos before him, but differing from the ideas of many other *chimistes* both within the Académie and in the wider world beyond its walls.

The key feature of Homberg's system—the identity of the chymical sulphur principle with the matter of light—emerged from work with the massive Tschirnhaus burning lens. Yet the lens experiments did not simply provide new results. Instead, they directed Homberg toward giving new and more coherent interpretations for a miscellany of results that had been accumulating since the start of his career, indicating the degree of interpretational malleability ordinarily present in experimental results. Homberg could not have had exclusive access to the Tschirnhaus lens without the patronage and collaboration that came with his relationship with Philippe II, Duc d'Orléans. Philippe provided Homberg with a new and magnificent laboratory at the Palais Royal, equipped that laboratory with the costly Tschirnhaus lens apparatus, and funded the workspace's undoubtedly considerable operating expenses. Philippe also gave Homberg a substantial salary, fine apartments in the Palais Royal, and an entrée through the house of Orléans to the French court and royal family. The social position and contacts Homberg thus gained had significant and lasting influence on his work and thought.

While the previous chapters have necessarily focused on Homberg's long-term project of developing broad theoretical and explanatory principles for chymistry, this emphasis should not be taken to imply that he neglected chymistry's daily preparative aspects or the wide miscellany of topics, materials, and processes connected with chymistry. At the fundamental level, the chymical theory that he developed had to be *drawn from* preparative processes and was intended ultimately to *explain and guide* such processes. As I have said elsewhere, a main characteristic throughout the history of chemistry has been its uniquely close linkage of head and hand, of ideas and matter, of theory and practice. Homberg's insistence on observation and the senses entailed the preparation and transformation of material substances in a fully practical way—Homberg's philosophy truly did lie in "the use of the firetongs." His theory of chymistry thus took shape side by side with the creation of an antacid here, a sedative salt there, a better means of refining silver one day, a balsam from common sulphur the next, and ways of turning a little mercury and some silver into gold. While developing his theoretical and explanatory structures, Homberg never neglected material production, because they were in essence two sides of the same coin. He kept busy with both of contemporaneous chymistry's chief productive realms: pharmaceuticals and metals. The latter occupied more of his time than the former, and operated both by smelting and refining and, one hoped, by transmutation. The next chapter examines Homberg's endeavor—with the assistance of Philippe II and others—to solve the age-old problem of metallic transmutation, and reveals more about the goings-on in the laboratory of the Palais Royal.

5

Chrysopoeia at the Académie and the Palais Royal

Chrysopoeia formed a major aspect of the chymical tradition from its origins in late classical Greco-Roman Egypt down to Homberg's day. Seventeenth-century chymists (and many others) of every intellectual and social class, to an extent we are only now coming to realize, pursued multiple pathways to achieve metallic transmutation. The most coveted method involved discovering how to prepare the philosophers' stone, a tiny quantity of which could quickly transmute a large mass of any base metal into gold. Other methods aimed less high, endeavoring to prepare weaker transmuting agents known as *particularia*, some of which might advance a base metal only to silver rather than all the way to gold, or whose gold-producing ability was relatively feeble compared to that of the philosophers' stone. Other methods involved more laborious treatments of base metals over longer periods of time, although the cost of some such processes was sometimes claimed to exceed the value of any noble metals produced. Homberg himself described just such a process in his early 1690s textbook, remarking that "this game is not worth the candle . . . for after three days of continuous work I obtained no more than nine or ten grains [of gold, about 0.5 gram] from one mark [of silver, about 244 grams]." Yet, he concludes, "it is enough to have verified the possibility" of transmuting silver into gold.[1]

It has long been routine to claim that the search for metallic transmutation vanished from the practice of chymistry around the start of the eighteenth century, that is to say, around Homberg's time. It is undeniably true that

---

1. VMA MS 130, fols. 165v–166r: "le jeu ne vaut pas la chandelle . . . car d'un marc on m'en tire pas plus de neuf ou dix grains, apres trois Jours de travail continuel; il suffit d'en avoir verifié la possibilité."

around this time, at least in France, the words *alchemy* and *chemistry*, which had previously been used largely interchangeably, began to take on distinct meanings akin to their modern connotations—the former denoting a super-annuated and largely (if not entirely) discredited quest for making gold, and the latter a modern and productive scientific discipline and enterprise. When enumerating the various transmutations of chymistry that took place around the turn of the eighteenth century, the apparent loss of chrysopoeia—a major focus of chymistry since its origins fifteen hundred years earlier—must be considered one of the most significant. Therefore, this chapter turns on several questions about chrysopoeia: What was the status of transmutational chymistry at the early Académie? Did Homberg and his colleagues support the possibility of transmutation, and did they actively pursue it? How did new theoretical structures—such as Homberg's own—interact with traditional chrysopoeia? There is much to be said about Homberg's involvement in chrysopoeia, both on his own and with the Duc d'Orléans.

Chrysopoeia at the Early Académie Royale des Sciences

Transmutational alchemy had a strikingly bifurcated existence at the Académie Royale des Sciences. At the very start of the institution in 1666, its founder Jean-Baptiste Colbert explicitly forbade the academicians to carry out any work relating to the philosophers' stone. Although Colbert specified that the Académie should devote itself to chymistry along with the other sciences, such research must "nevertheless exclude the secrets of the philosophers' stone, to which [the Académie] is forbidden to apply itself while carrying out the other operations of chymistry."[2] Charles Perrault (1628–1703), younger brother to the institution's first leader, Claude Perrault (1613–88), repeated Colbert's injunction that the academicians must not consider the philosophers' stone "either close up or from a distance," a qualification that was presumably meant to exclude all chrysopoetic or otherwise transmutational endeavors, whether or not they aimed toward the philosophers' stone in particular.[3] The only other natural philosophical topic that Colbert specifically forbade was judicial astrology. These twin prohibitions might too easily be seen today as forward-thinking rejections of subjects that would come to

---

2. "Notes et desseins de Claude Perrault, August 1667," in *Lettres, instructions, et mémoires de Colbert*, ed. Pierre Clément, 8 vols. (Paris, 1861–70), 5:515–16, at 515. It is worth noting that Colbert's own library nevertheless contained works on metallic transmutation, some of which are now in the Ossolineum Library, Wrocław; see Iryna Kachur, "La bibliothèque de Dortous de Mairan et ses livres retrouvés," *Revue d'histoire des sciences* 68 (2015): 405–18.

3. Charles Perrault, *Mémoires de ma vie*, ed. Paul Bonnefon (Paris: Renouard, 1909), 48.

be called "pseudosciences," but such was probably not the basis of Colbert's thinking. Both transmutational chymistry and astrological prognostication were highly controversial topics both intellectually and morally, as well as potentially subversive politically, whether or not they actually worked. Prognostications about the king's health or a coming war or famine could threaten political and social stability. Indeed, just thirty-five years earlier, in 1631, Pope Urban VIII had issued the bull *Contra astrologos judiciarios* specifically to prohibit predictions about the death of popes and princes, not because he rejected astrology—which he did not—but because such predictions provoked public unrest and undermined political power.[4] In a similar way, metallic transmutation—whether or not its practitioners were actually successful— threatened economic stability by raising anxieties about the continuing value of gold and the possible debasement of the currency with artificial metal. Both transmutation and prognostication provoked their share of vigorous debate, and abuses of both were the targets of critics and of public ridicule. Significantly, these two prohibitions follow immediately upon Colbert's similar exclusion of any discussion at the Académie of political matters, since that, in his estimation, would easily lead to "perilous consequences."[5] For all these reasons, controversial subjects were just the sort of things in which a royally founded and funded institution like the Académie should not be involved.

Colbert's successor Louvois repeated the prohibition against chrysopoeia in starker terms. In January 1686, Louvois sent his spokesman, Henri de Bessé, Sieur de la Chapelle (1625–92), to deliver a message directly and specifically to the chymists of the Académie. Louvois ordered them in no uncertain terms to stay away from "the Great Work [that is, the philosophers' stone], which includes also the extraction of the mercuries of all the metals, their transmutation, or their multiplication, about which Monsieur de

4. Urban VIII, *Constitutio contra astrologos judiciarios* (Rome, 1631); on the background and results, see Brendan Dooley, *Morandi's Last Prophecy and the End of Renaissance Politics* (Princeton, NJ: Princeton University Press, 2002). See also Monica Azzolini, "The Political Uses of Astrology: Predicting the Illness and Death of Princes, Kings and Popes in the Italian Renaissance," *Studies in History and Philosophy of Biological and Biomedical Sciences* 41 (2010): 136–45; and Hervé Drévillon, *Lire et écrire l'avenir: l'astrologie dans la France du Grand Siècle (1610-1715)* (Seyssel: Champ Vallon, 1996).

5. "Notes et desseins," 515. The document also states that the Académie should not meddle in theological matters. Colbert had originally intended to include theology in the Académie, but objections from the Sorbonne caused him to abandon the idea; see Perrault, *Mémoires*, 47–48; David S. Lux, "Colbert's Plan for the *Grande Académie*: Royal Policy Towards Science, 1663–67," *Seventeenth-Century French Studies* 12 (1990): 177–88.

Louvois does not want to hear anything spoken."[6] The minister went on to urge the chymists to apply themselves instead to things useful to the king and the state, such as improving gunpowder and incendiaries for warfare, making seawater potable, examining mines and minerals, and creating new pharmaceuticals. While this ministerial intervention has sometimes been seen as Louvois's attempt to turn the Académie's work toward more strictly utilitarian ends, it is clear from the text that its *primary* purpose was to abolish transmutational talk and endeavors from the institution. Coming as it did just three months after the death of Samuel Cottereau Duclos, Louvois's injunction was undoubtedly a response to the late chymist's transmutational researches (which had just become publicly known) and concerns that such investigations might be continuing at the Académie.

In 1692, when the Académie was under the direction of Louvois's successor, Pontchartrain, a more explicit reason behind this repeated prohibition was shared with a visitor from Sweden. The chymist Erich Odhelius (1661–1704) wrote a letter from Paris to Urban Hjärne in Stockholm that described his recent meeting with Homberg, reported on the institution's reinvigoration thanks to Pontchartrain and Bignon, and mentioned ongoing work with mineral substances. Most interestingly, Odhelius also noted that the academicians were "expressly forbidden by the directors of the Académie and by the King to work on alchemical matters, for the King does not wish it to be thought that his money is produced by goldmaking [*per aurifactionem*]."[7] This remarkably clear statement indicates that the concern over transmutational chymistry, at least on the part of the Académie's royal patron, was based on economic and political worries, not on scientific objections or skepticism. The word Odhelius uses for "money," *penningar*, like its English equivalent, could mean either "financial resources" or, more literally, "coins." If Odhelius intended the first meaning, the king's prohibition could be read as anxiety over the foundations of his power and *fama*, namely, suggesting that the wealth of France was based not on the Sun King's wise governance and power, but upon the mechanical output of laborers in smoky laboratories. If Odhelius meant *penningar* in the more literal sense of "coins," then Louis XIV's concerns become even more palpable. Namely, if the official

6. PV 11, fols. 157r–158r, at 157r (30 January 1686).

7. Erich Odhelius to Urban Hjärne, 6 June 1692, in Carl Christoffer Gjörwell, *Det Swenska Biblioteket*, 2 vols. (Stockholm, 1757–62), 1:337–39: "warande expres förbudna af *Directoribus Academiae* och Konungen at laborera *in Alchymicis*, at icke Konungen må säjas hafwa sina penningar *per aurifactionem*." For more on Odhelius's travels and his relations with Urban Hjärne, see Hjalmar Fors, *The Limits of Matter: Chemistry, Mining, and Enlightenment* (Chicago: University of Chicago Press, 2015).

scientific body of France—in short, the king's chymists—was known to be at work on transmutation, some observers might begin to suspect that the French coinage, particularly the precious louis d'or, contained artificially prepared metal. While chrysopoeians insisted that chymically prepared gold was as good as (if not better than) natural gold, not everyone else acquiesced in this notion. (Such concerns have familiar counterparts in the modern world; for example, that an artificially prepared diamond is not a "real" diamond, or that the "natural" vitamin C in an orange is somehow different from the "artificial" vitamin C in a tablet.)[8] Those who did not consider alchemical gold to be fully equivalent to natural gold might then consider the French coinage to be corrupt, and refuse to accept the louis d'or at its face value, thus devaluing the currency with devastating economic consequences.

The fact that the official prohibition against transmutational chymistry was expressed to the academicians multiple times not only witnesses the desire to stamp out such endeavors, but also suggests that the pursuit of chrysopoeia continued at the Académie despite administrative warnings. In short, the academicians behaved the way all academics should toward administrators: they ignored them. The repeated warnings thus reveal a split at the Académie between administrative overseers and the chymists themselves. Odhelius explicitly states that the prohibitions were made "by the directors of the Académie and by the King," rather than by the chymists or some broader group of academicians themselves. Louvois's intervention was certainly a direct response to the activities of the recently deceased Duclos who had been at work on transmutational chymistry throughout his long career. Duclos is undoubtedly one of the pro-transmutation academicians referred to by Christiaan Huygens when he remarked in 1667 that among "our chymists are some who believe in the philosophers' stone and some who do not."[9] Curiously, the longest contemporaneous biographical account we have of Duclos, published shortly after his death, focuses not only on his transmutational endeavors but also on his supposed deathbed repudiation of them.[10] The source for this information was Nicolas Clément, the keeper of the Bibliothèque du Roi where the Académie met and where both he and Duclos had their lodgings. Clément notes that Duclos, although a physician, preferred to spend his time "on chymical experiments and the search for the philosophers'

---

8. For a study of the artificial-natural debate particularly within alchemy, see William R. Newman, *Promethean Ambitions: Alchemy and the Quest to Perfect Nature* (Chicago: University of Chicago Press, 2004).

9. Christiaan Huygens to Lodewijk Huygens, 14 January 1667, in *Oeuvres complètes de Christiaan Huygens*, 22 vols. (The Hague: Nijhoff, 1888–1950), 6:99–100, at 99.

10. *Nouvelles de la république des lettres* (October 1685):1139–43.

stone." Asked by Clément about his writings, the dying chymist mentioned his publications and the opposition he faced from some academicians, and then claimed that he had burned everything else. Asked why he had done so, Duclos replied that since many of his papers were about transmutation, he feared that his nephew, a successful painter, might try to continue such research to the detriment of his career.[11] Clément then advised Duclos that "it would surely be advantageous for the public and even for the service of the king" for him to disavow chrysopoeia in order to "restrain those who would too easily engage themselves with the unfortunate passion of blowing at coals [*la malheureuse passion de soufflerie*]." According to Clément, Duclos then asserted that "there was nothing more vain or more useless than the flattering hope that one can arrive at the transmutation of metals."[12]

It is possible that this is the way things really happened. But it is prudent to be skeptical of deathbed conversions. It is odd that Clément would have thought of explicitly asking the dying Duclos to make a statement against chrysopoeia "for the service of the king." The terms of his request redirects attention toward royal concerns about the potential political and economic subversiveness of chrysopoeia, and what academicians should and should not be doing with royal funding. At nearly the same time as this reported disavowal, Duclos, a Protestant of slightly questionable orthodoxy, also converted to Catholicism very publicly in the aftermath of the Revocation of the Edict of Nantes.[13] Perhaps Duclos saw an opportunity to safeguard and even benefit his heirs and legacy by publicly abjuring in the last months of his life the "twin errors" of Protestantism and chrysopoeia. Both repudiations could readily be seen as "services to the king," and might not necessarily reflect his real opinions on the matters.

Duclos's extensive endeavors in transmutational alchemy, including the search for the philosophers' stone, both before and during his tenure as a member of the Académie Royale des Sciences, are well documented. Recently discovered manuscripts owned by Sir Kenelm Digby (1603–65), who during his time of exile in Paris associated with Duclos, preserve substantial sections of the French chymist's laboratory notebooks dating from the 1650s that describe a wide range of chrysopoetic experimentation. These documents

11. Jacques Fricquet, a favorite student of Sébastien Bourdon (who painted Duclos's portrait, now lost), married Duclos's niece Louise Hollier on 24 September 1669. Duclos supplied a considerable dowry in the form of a house in the faubourg Saint-Antoine; see Jacques Thuillier, *Sébastien Bourdon 1616–1671: Catalogue critique et chronologique* (Paris: Réunion des Musées Nationaux, 2000), 438.

12. *Nouvelles* (October 1685):1141–42.

13. *Mercure galant* (August 1685): 136–37.

further indicate that Duclos was an active member of a Parisian circle that included both chymists and artisans (particularly goldsmiths) who shared information, texts, and experiences about metallic transmutation. The group's correspondence network reached into the Netherlands and England.[14] Even if Duclos really did burn *some* of his manuscripts, he most certainly did *not* burn them all, as Clément's account claims. Several originals bearing his handwriting survive to this day, as well as many copies made from originals that were undeniably in existence well after his death. The Académie chymist Jean Hellot (1685–1766), for example, made dozens of pages of transcriptions from a substantial cache of original Duclos manuscripts in the 1750s and 1760s. Hellot evidently had at his disposal a range of Duclos material, including laboratory notebooks (giving first-person accounts of experiments dated to the 1660s and 1670s), commonplace books, and original correspondence. Some of the material he transcribed deals with pharmaceutical and productive processes (like the making of pigments, alloys, and artificial gems), but the *majority* involves metallic transmutation and other *arcana maiora*. Hellot cites, for example, several processes for the mercurification of gold as "found in the papers of Mr. Duclos," one of which he described specifically as being "in the handwriting of Mr. Duclos." Other materials include *particularia* for turning silver into gold, practical interpretations of allegorical alchemical texts, and Helmontian arcana like the alkahest.[15] Where and by whom this *Nachlass* was preserved for at least seventy years after Duclos's death is unknown, as is the subsequent fate of such papers. Their existence until at least the mid-eighteenth century clearly calls Clément's claims into question. Beyond the rich Hellot material, a further compilation of *Extraits des recherches curieuses de physique et de chymie de M. Duclos* survived until at least the 1960s, after which the manuscript was "lost" by the science library at the Université de Bordeaux.[16] Another manuscript containing a commentary by Du-

14. Lawrence M. Principe, "Sir Kenelm Digby and His Alchemical Circle in 1650s Paris: Newly Discovered Manuscripts," *Ambix* 60 (2013): 3–24, especially 17–19; available in French as "Sir Kenelm Digby et son cercle alchimique parisien des années 1650," *Textes et Travaux de Chrysopoeia* 16 (2015): 155–82. See also Lawrence M. Principe, "Goldsmiths and Chymists: The Activity of Artisans in Alchemical Circles," in *Laboratories of Art: Alchemy and Art Technology from Antiquity to the Eighteenth Century*, ed. Sven Dupré (Dordrecht: Springer, 2014), 157–79, especially 172–74.

15. For example, Hellot Caen Notebooks, 1:39r–47r; 3:25r–27v, 148r–149v, 174r–182v, 267r–300v.

16. Originally Université de Bordeaux MS 10. When the central library was divided into separate faculties in the middle of the twentieth century, this manuscript and all the other fourteen manuscripts of the *ancien fonds* transferred to the library for "Sciences et Techniques" were lost. This loss includes other alchemical texts as well as works on astronomy, botany, and mathemat-

clos on a chrysopoetic text is extant at Orléans.[17] Multiple copies of a *Recueil de Mr. Duclos sur la transmutation des metaux* exist in at least three other libraries.[18] This last text contains not only additional first-person accounts of Duclos's transmutational experiments, but also an entry dated 1678, unambiguously indicating that such transmutational work was ongoing while Duclos was a member of the Académie, meaning that he carried out such experiments in the Académie's laboratory he had designed, and next to which he lodged, at the Bibliothèque du Roi. Multiple entries in Bourdelin's expense account for the laboratory indicate various materials he supplied to Duclos, many of them suggestive of chrysopoetic undertakings. Such work was therefore presumably well known to the other academicians, and certainly to the other chymists, Jacques Borelly and Claude Bourdelin, who shared the laboratory with Duclos. Louvois did have cause for concern.

Whether Homberg met Duclos or participated in his endeavors during his 1683 contact with the Académie and collaboration with Mariotte is unknown. Yet Homberg had already become interested in metals and their transmutation before he first arrived in Paris. As mentioned in chapter 1, he very likely obtained a recipe for Starkey's philosophical mercury from Boyle in 1680, and thereafter undertook observations of mining and metallurgy in Central Europe and Sweden. It is telling that Homberg's new patron after Colbert's death in 1683, Louis-Armand de Bonnin, Abbé de Chalucet, hired him specifically to work on a transmutational experiment. Was the abbé's choice of Homberg for this project based only on his general chymical abilities, or on a more specific interest and experience Homberg was known to have in transmutation? Given Homberg's constant pursuit of secrets and his love of the wondrous and dramatic, it is inconceivable that he would not have been attracted by the marvel of transmutation and the secret of the philosophers' stone.

Homberg's interest in transmutation appears clearly in his early 1690s treatise. Already in the first chapter, when expressing his idea that common

---

ics. Whoever had the stupid idea to divide up the University's *ancien fonds* arbitrarily rather than leaving the manuscripts safely preserved in "Lettres" (where other manuscripts from the same original cache remain well cared for) and whoever "lost" (or discarded, sold, or simply stole) a substantial part of it has a lot to answer for. Inquiries I made at the science library at Bordeaux did not locate the missing manuscripts but resulted in the phrase "Manuscrit constaté disparu en 2007" being added to each listing in Calames, the online inventory of manuscripts in France, of the fourteen lost documents.

17. Orléans, Médiathèque, MS 1037, fols. 129r–164r: "La pratique et maniere de faire la pierre philosophale par le sr. Philipon et commenté par le sr. Duclos aussi grand philosophe."

18. BNF, Arsenal MS 2517; Université de Bordeaux, Lettres, MS 3; and Bibliothèque interuniversitaire de la Sorbonne, MS 1884.

mercury is an impure version of the chymical principle mercury, he notes that when properly treated, common mercury can be turned into a solid metal: "Mercury remains a simple mineral in liquid form, as one sees it in the druggists' shops, until it has been freed from its watery and earthy feces and some metallic salt and sulphur have been introduced to it. Then, by means of an appropriate cooking, it can artificially become a metal or a composite mineral, depending upon the intent and ability of the operator." Thus, the conversion of mercury into any metal is possible, not in this case by means of the philosophers' stone or other transmuting agent, but rather by simply compounding the metal out of its ingredients.[19] The notion that common mercury is corrupted by watery and earthy impurities that prevent it from being turned into a solid metal can be traced back in chrysopoetic circles at least to the thirteenth-century Geber.[20]

Homberg devotes the entire second chapter of the book to *l'alchimie*, which he defines straightforwardly as "a particular science within chymistry that teaches the transmutation of metals . . . it teaches not only the ways of perfecting common mercury, but also how to correct the imperfect metals to turn them into perfect metal, and to do this by means of the philosophers' stone."[21] Homberg thus limits the meaning of *l'alchimie* to metallic transmutation, but also views it as a particular sort of knowledge fully within *la chimie*. Borrowing an idea found in the French didactic tradition and traceable back to Martin Ruland and Jean Beguin, Homberg claims that alchemy is "chymistry par excellence."[22] What follows is a remarkable historical account of the experiments, observations, and ideas that supposedly led chrysopoeians to produce the philosophers' stone. Homberg writes that the earliest alchemists

19. VMA MS 130, fol. 119r–v: "le mercure reste mineral simple en forme coulante comme on le voit dans les boutiques des droguistes jusqu'a ce qu'on l'ait debarassé de ces feces aqueuses et terrestres et qu'on luy ait introduit du sel et du soufre metallique, alors il pourra par une cuisson convenable devenir artificiellement metail ou mineral composé selon l'intention ou l'habileté de l'artiste."

20. William R. Newman, ed. and trans., *The Summa perfectionis of the pseudo-Geber* (Leiden: Brill, 1991), 385–88.

21. VMA MS 130, fols. 119v–120r: "chap. 2e. de L'alchimie. C'est une science particuliere dans la chimie qui enseigne la transmutation des metaux. on l'appelle alchimie ou la chimie par excellence. elle enseigne non seulement les moïens de perfectionner le mercure commun mais aussi de corriger les metaux imparfaits pour les rendre en metail parfait et ce par le moien de la pierre philosophalle."

22. William R. Newman and Lawrence M. Principe, "Alchemy vs. Chemistry: The Etymological Origins of a Historiographic Mistake," *Early Science and Medicine* 3 (1998): 32–65, at 47–50, 56, 59–60.

noticed that all metals, when molten, resemble mercury, leading them to conclude that the metals resulted from the coagulation of mercury underground.
Some of that subterranean mercury, however, failed to coagulate into a metal
because of contamination by superfluous humidity, and was found in mines
still in its liquid state. Consequently, the early alchemists tried to dry up this
excess humidity by heating liquid mercury for a long time, hoping thereby
to turn it into gold. These efforts failed. Concluding that the external heat of
a fire by itself was inadequate for the purpose, they sought a substance endowed with "internal heat" that could be added to the mercury and cause it to
coagulate into a perfect metal. They looked for such a substance in minerals,
salts, acids, and elsewhere, but found nothing that worked. Then, with "reasoning that seems very plausible," they realized that if the metals were themselves mercury that had been coagulated by some internal principle, then the
metals themselves should contain this sought-after principle of coagulation.
They called this principle "metallic sulphur." But heating common mercury
with metals did not achieve their goal because the metals held on to their
internal sulphur too tightly for it be released to the mercury. Then they theo-
rized that they needed an "incomplete" metal that could accept the metallic
sulphur from a true metal and transfer it to mercury. They finally found what
they were looking for in antimony ore (stibnite) and iron. By fusing these
two materials together, they produced a shining, metal-like substance, which
they called martial regulus. This martial regulus displayed a starlike crystalline pattern on its surface, which these alchemists saw as an allusion to the
Star of Bethlehem, meaning that "the savior of the metals had been born."[23]
The idea here was that the antimony ore could accept iron's metallic sulphur,
thus becoming "almost" a metal—the regulus is shiny like a metal but brittle
like a mineral—and then the regulus could transfer that metallic sulphur to
common mercury, causing it to coagulate into a true metal, as it should have
done underground.

After repeatedly distilling common mercury from this regulus, Homberg
continues, the alchemists were astonished to find that the mercury now coagulated by itself into a red powder upon heating. This red powder, when
cooked with more mercury, produced a small amount of gold. They then
tried dissolving gold in the prepared mercury and heating the mixture in a
sealed flask. They watched the mixture pass through several colors and then
turn into a red powder. When fused, this red powder provided a new kind
of gold much deeper in color than ordinary gold. When melted with a base

23. VMA MS 130, fols. 120r–126v; "que le sauveur de metaux etoit né" (124r).

metal, this special gold transformed ten times its weight of the base metal into normal gold. They repeated the process, amalgamating this "more perfect" gold with more of their mercury and then digesting the mixture again into a red powder; the result now transmuted a *hundred* times its weight of base metal into gold. With one more repetition of the process, they obtained a material of which "one grain converted a thousand grains of any imperfect metal into natural gold," and they called "this highly exalted material their elixir and the philosophers' stone."[24]

Homberg concludes his "historical" account of the discovery of the philosophers' stone by stating that "whether this work has ever been carried out is not something that I can decide."[25] He notes that many persons worthy of belief assert that they have seen the stone and its effects, but also that many others have bankrupted themselves trying to produce it. Therefore, he advises, it would be imprudent to set off in search of the stone without having solid ideas about how to do so.

Strikingly, Homberg's account can be identified as based almost entirely on the chapter "On the Discovery of the Perfect Magistery" in the *Open Entrance to the Shut Palace of the King* by Eirenaeus Philalethes, the pseudonym used by George Starkey.[26] That text lays out a similar sequence of trials and ideas that supposedly led alchemists to discover the philosophers' stone. But Homberg does not simply crib from Starkey, even though he reproduces a few phrases almost verbatim. Homberg's exposition is far more linear, emphasizing specific experimental results and reasonings that drove the aspiring chrysopoeians to work in particular directions. This retailoring reflects Homberg's own persistent emphasis on observable experimental results and the drawing of theoretical conclusions from them, as illustrated in previous chapters. The lofty and metaphorical language, occasional Scholastic terminology, and frequent *Decknamen* that characterize the Philalethes text are completely absent from Homberg's reworking. For example, while Starkey writes that the chymists finally found what they were looking for "hidden in the House of Aries" and in "the offspring of Saturn," Homberg writes plainly that "they

---

24. VMA MS 130, fols. 120r–126v; "un grain convertissoit mille grains de quelque metail imparfait en or naturel" (126v); "Ils appellent leur elixir et la pierre philosophalle cette matiere si fort exaltée" (126v).

25. VMA MS 130, fols. 126v–127r:"Si cet ouvrage a Jamais été executé ou non c'est ce que je ne scaurois decider."

26. Eirenaeus Philalethes [George Starkey], *Introitus apertus ad occlusum regis palatium*, in *Museum hermeticum* (Frankfurt, 1678; reprint, Graz: Akademische Druck, 1970), 647–99, especially chap. 11: *De inventione perfecti magisterii*, 662–65; *Secrets Reveal'd, or An Open Entrance to the Shut Palace of the King* (London,1669), 23–30.

were very happy to find what they sought in iron and antimony."[27] Homberg then inserts two paragraphs that have no counterpart in the *Open Entrance*, explaining in clear and sober terms—and completely consistently with his ideas about metallic composition expressed elsewhere in the manuscript— exactly why iron and antimony can perform the desired function: "Iron is a metal that abounds extraordinarily in sulphur and has very little mercury, such that this great quantity of sulphur being very little detained by so small a quantity of mercury, is more easily detached from iron than from any other metal. Of all minerals, antimony most approaches a metal. It contains a beautiful mercury and abounds in mineral and flammable sulphur, and since it has no metallic sulphur it is very appropriate for receiving it."[28]

This chapter on alchemy is the earliest, but far from the only, indication of the importance to Homberg of Philalethes and of this specific process with mercury. A version of this process was the secret Homberg may have obtained from Boyle in 1680 in exchange for the secret of making white phosphorus. Crucially, it is *this same process* that produced the "specially prepared mercury" whose long-term cooking produced a small quantity of gold that Homberg used as the central experimental illustration of "light as the sulphur principle" in his 1705, 1706, and 1708 "Essais de chimie" and in his completed 1715 *Élémens de chimie*. Indeed, Philalethes and variations on his philosophical mercury reappear constantly throughout Homberg's work. It is particularly important to stress that, rather than merely repeating the process mechanically, Homberg engaged dynamically with it. He repeatedly modified the procedure, often extensively, in response to his laboratory results. Likewise, Homberg modified the explanations he gave of his results as he developed his general theory of chymistry. Recognizing these two points— modifications of practice and modifications of explanatory theory—is crucial since there remains (in some quarters) a lingering notion that "alchemy" was somehow caught in a "loop of failure," or that it was intellectually or theoretically static, even retrogressive and contrarian in relation to "more progressive chemistry." In reality, the practices and theoretical frameworks

27. VMA MS 130, fol. 123r: "Ils ont été assés heureux pour trouver ce qu'ils cherchoient dans le fer et dans l'antimoine."

28. VMA MS 130, fol. 123r–v: "Le fer est un metail qui abonde extraordinarement en soufre et qui a tres peu de mercure en sorte que cette grande quantité de soufre etant fort peu arrestée en une si petite quantité de mercure est plus aisée d'en être detachée que de pas un des autres metaux. L'antimoine est de tous les autres mineraux celuy qui approche le plus du metail il contient de tres beau mercure et abonde en soufre mineral et brulant et quoyqu'il n'ait aucun soufre metallique il est fort propre a le recevoir." On iron's superabundance of sulphur, see Homberg's remarks on magnetism, discussed in chapter 3.

regarding metallic transmutation—at least among its many intellectually so-phisticated exponents—constantly evolved in dynamic response to both new ideas and new experimental results, and Homberg's own work provides an excellent example.

The Evolution of Theory and Practice in Chrysopoeia

When Homberg wrote about the Philalethes process in the early 1690s, his explanation of the process was similar to what Starkey had proposed in the 1650s. Common mercury first required purification from its impurities. When this was done, the addition of metallic sulphur from the martial regu-lus provided a coagulating agent, enabling the mercury to become a red solid. When gold was dissolved in the purified mercury and heated, a somewhat different red solid was produced, which, when fused, gave a "gold" that was "more perfect" than ordinary gold and could transmute ten times its weight of base metal into ordinary gold. Repetitions of the process using the "more perfect gold" gave increasingly powerful transmuting agents, in short, the philosophers' stone. Starkey's own explanation of how the process works relies on multiplying the "seed of gold" through reiterated digestions with his philosophical mercury, an idea with roots in the Helmontian notion of *semina* capable of transforming matter radically into new substances. Each repetition of the digestion with fresh philosophical mercury augments the transformative power of the aurific *semina*, allowing for increasingly power-ful transmutations.

Homberg did not invoke the idea of *semina*. Although he continued Star-key's general operational ideas about purifying common mercury and unit-ing it with a metallic sulphur, his early writings left the microlevel mechanism of transmutation unexplained. This lack of explanation accords well with Homberg's customary reluctance to provide explanations that are not clearly and closely tied to observable phenomena. Homberg had not seen success-ful transmutation using the Philalethes method, and so did not speculate on how that transformation *might* happen. Only after he became convinced that he had himself transmuted mercury into gold did he venture to provide an explanation of this result, which he did using his chymical theory of light. Al-though one cannot be sure whether the observed transmutation or the theory of light came first, yet given that Homberg was at work with the philosophical mercury long before his theory of light as sulphur coalesced around 1704–5, it is probable that he produced his transmuted gold first. If this is the case, then this apparent transmutation would count as another of the foregoing observations (enumerated in chapter 4) that fed into Homberg's formulation of his mature theory of the chymistry of light. Although Homberg worked with the Philalethes process extensively both before and after 1705, he did not reveal his transmutational result until he could also explain it coherently, and he never linked it explicitly (in public at least) to traditional approaches to chrysopoeia. Using antimony regulus to transform common mercury into a "philosophical mercury" for making the philosophers' stone represents a major avenue of chrysopoetic research in the seventeenth century and beyond.[29] It was pursued by scores, probably hundreds, of hopeful chrysopoeians during the seventeenth century and well into the eighteenth. What is truly remarkable is that this well-trodden route toward making the philosophers' stone silently became *the central illustrative example of Homberg's new theory of chymistry.*

In his landmark 1705 essay on light as the true sulphur principle, as soon as Homberg finished expounding his ideas about the nature of light, he shifted very abruptly to experimental illustrations. Without any explanation, he declared—seemingly out of nowhere—that "common mercury having been purified sufficiently by means of antimony and iron, becomes more lively and more liquid than before."[30] Many contemporaneous readers—like most modern readers—might not have found this statement noteworthy, but it would have caused seekers after transmutation immediately to sit up and take note. They would have instantly recognized the basic formulation for preparing philosophical mercury as described by Philalethes. Homberg then stated that when this special mercury is sealed in a flask with a long neck and heated, it will gradually thicken and then turn into a powder, first black, then white, and finally red. Homberg did not bother to point out that these three colors are the traditional and well-known guideposts for making the philosophers' stone, but surely some of his audience recognized the allusion.

29. For an overview of the process and its significance, see Lawrence M. Principe, *The Secrets of Alchemy* (Chicago: University of Chicago Press, 2013), 161–66. For its pursuit by Boyle, Locke, and others, see Lawrence M. Principe, *The Aspiring Adept: Robert Boyle and His Alchemical Quest* (Princeton, NJ: Princeton University Press, 1998), 153–79. For its origins in von Suchten and Starkey, see William R. Newman, *Gehennical Fire: The Lives of George Starkey, An American Alchemist in the Scientific Revolution* (Cambridge, MA: Harvard University Press, 1994), especially 125–41; and William R. Newman and Lawrence M. Principe, *Alchemy Tried in the Fire: Starkey, Boyle, and the Fate of Helmontian Chymistry* (Chicago: University of Chicago Press, 2002), 119–28. For an example of its pursuit in Germany, see Martin Muslow, "Philalethes in Deutschland: Alchemische Experimente am Gothaer Hof 1679–1683," in *Goldenes Wissen: Die Alchemie, Substanzen, Synthesen, Symbolik,* ed. Petra Feuerstein-Herz and Stefan Laube (Wiesbaden: Harrassowitz, 2014), 139–54.

30. Wilhelm Homberg, "Suite des essays de chimie: Article troisième, du souphre principe," *MARS* (1705): 92.

He noted how the resultant red powder weighs more than the initial mercury due to its absorption of the light/sulphur from the fire, and how, when this powder is heated more strongly, most of it evaporates as mercury but a small quantity remains behind as a solid metal. When addressing the public assembly in April 1705, Homberg did not state *which* metal, but those in the audience who recognized the chrysopoetic import of the process would have understood instantly that it was gold. It seems that Homberg was here engaging in a carefully measured revelation of his chrysopoetic activity, essentially hiding it in plain sight.

Only in subsequent installments of the "Essais," presented in 1706 and 1709, did Homberg identify the metal as gold and propose a complete micro-level explanation of its formation. He noted first that mercury's liquidity and high mobility strongly suggest the shape of its minute particles: they are likely to be smooth spheres that roll easily upon one another. When placed in the fire, these mercury particles are struck by the rapidly moving particles of the matter of light contained in the fire, and these impacts scratch the surface of the spheres. The minute particles of light/sulphur soon begin to stick in these scratches "by means of their natural glue [*par son gluten naturel*]." These fine particles, arrested from the current of the matter of light in the flame, eventually coat the mercury particles like a rough and prickly shell, like smooth "chestnuts covered with their green and prickly husks." As a result, the mercury particles can no longer roll upon one another, and the silvery liquid mercury becomes a dry red powder.[31]

Homberg insisted that despite the obvious change in appearance, this transformation is not yet a significant one because most of the red powder can be turned back into its original fluid mercury. When exposed to a stronger fire where the matter of light is in greater agitation, the more violent stream of light/sulphur erodes away the shells—one might visualize this today in terms of sandblasting—liberating the mercurial kernels, which then distill over as liquid quicksilver. Only a very small portion of the red powder does not return to its original liquid mercurial state, but rather remains as a solid metal. How is this metal produced? When the matter of light is attached only "superficially" to the mercury, it is easily removed; indeed, most of the red powder formed by the digestion of the mercury contains only such superficial sulphur, and is therefore readily returned to the state of running mercury by a stronger fire. A small fraction of the mercury particles, however—about one-two hundredth of the whole by Homberg's measurement—are so deeply

---

31. Wilhelm Homberg, "Suite de l'article trois," *MARS* (1706): 260–72, at 262–63; and Homberg, "Suite des essais de chimie, article IV," *MARS* (1709): 106–17, at 111.

gouged by their exposure to the hail of light/sulphur particles that some of these particles become sufficiently deeply lodged in the mercury particles that a stronger fire cannot reach them and drive them out. In this case the light/sulphur "has entered the substance itself of the mercury," and this stricter union provides a stable solid metal. The matter of light is now a *fixed* metallic sulphur. When "the matter of light . . . has become a fixed metallic sulphur which will never leave the mercury again regardless of how great a fire is given to it, holding it always in the form of metal, and according to the quantity of fixed sulphur which is detained there, the metal is more or less dense, that is to say, it is gold or silver."[32]

In 1705, Homberg further suggested that if his mercury "had been left in digestion for several years," more of the fixed metal would have been produced. By 1706, he had changed his mind, stating instead that further heating does not produce more gold.[33] Presumably, he had continued heating the red powder during the intervening year and found, no doubt to his chagrin, no increase in the quantity of gold. He explained this disappointing experimental result by arguing that the prickly coats of attached light/sulphur protect the enclosed mercurial particles from any further gouging and subsequent combination with the matter of light. But if, he added, the powder is ground with more of the prepared mercury, the "envelopes" of attached sulphur are broken open, exposing the mercury particles once again to the agitated light/sulphur from the fire. In this case, the already scratched surfaces of the mercury are cut more deeply, producing more fissures deep enough to engulf the light/sulphur particles and hold them fast, thereby transforming more mercury into gold.[34] Again, Homberg did not explicitly connect this result to any foregoing traditions. Nevertheless, what Homberg described is the *minera perpetua* (perpetual mine) cited as possible by several chrysopoeians, mostly notably by Johann Joachim Becher, an author who appears to have been an important source for Homberg. In the last supplement to his *Physica subterranea*, Becher described several "perpetual mines"—namely, long-term chymical operations that provide the chymist with regular yields of gold or silver. One of these, whose origins Becher attributed to Gaston Duclo, is strikingly similar to what Homberg described. Both digest a spe-

32. Homberg, "Suite de l'article trois," 267.

33. Homberg, "Souphre principe," 93; and Homberg, "Suite de l'article trois," 263.

34. Homberg, "Suite de l'article trois," 263–64. If, as it seems, this method of grinding the red powder with fresh philosophical mercury and reheating the mixture gave Homberg more gold, that would suggest that the gold came from the philosophical mercury, perhaps as an impurity in the common mercury with which he started, or perhaps more likely carried forward from traces of gold in the antimony ore.

cially prepared, or "animated," mercury into a red powder; both obtain gold from the red powder; both mix fresh mercury with the red powder in order to produce more gold.[35]

Using his new theory of the chymistry of light, Homberg thus provided a novel microlevel explanation of metallic transmutation—the newest chymical theory applied to a long-standing chrysopoetic practice. But Homberg did more than follow an existing process and provide a novel explanatory framework upon it. The fuller story of Homberg's engagement with this process indicates further the tight connection between chrysopoetic practice and theory, as well as between tradition and innovation. It also highlights the perennial problem of deciphering allegorical chrysopoetic texts and recounts Homberg's dogged pursuit of transmutation, often in collaboration with the Duc d'Orléans. To understand Homberg's innovative engagement with this transmutational research pathway, it is necessary first to review where the process stood when Homberg began his exploration of it.

The Philalethes Process and Its Evolution

The origins of this Mercurialist route to the philosophers' stone lie in Alexander von Suchten's sixteenth-century *Mysteria antimonii gemina*, but the process was most widely connected with the Philalethes treatises, now known to be the productions of George Starkey, which express the method in various metaphorical and allusive guises.[36] As Homberg would do later, Starkey himself continually modified his method for preparing the philosophical mercury, to make the process cheaper, faster, and ultimately (he hoped) more successful transmutationally. Initially, Starkey fused "antimony" (that is, stibnite, the native ore of antimony) with iron to produce "martial regulus of antimony." He then fused this regulus with two parts of silver, an addition he called the "Doves of Diana." He amalgamated the resulting silver-antimony alloy with mercury and then heated, ground, and washed the amalgam, and finally distilled off the mercury. The *Introitus apertus* calls each distillation an "eagle" (presumably because it makes the mercury "fly"). Starkey then amalgamated the distilled mercury with more silver-antimony alloy, and repeated the process. After seven to ten repetitions ("eagles"), the mercury was sufficiently "animated" or "cleansed" to be the *philosophical* mercury needed to make the philosophers' stone.

---

35. Johann Joachim Becher, *Physica subterranea* (Leipzig, 1703), 881.

36. On the Mercurialist school of chrysopoeia, see Principe, *Aspiring Adept*, 153–56; and Principe, *Secrets of Alchemy*, 161–63.

By 1654, Starkey had replaced silver with the less expensive copper.[37] The purpose of either metal is very practical. While mercury amalgamates fairly readily with most metals (iron excepted), it does not normally do so with antimony. Alloying antimony first with either silver or copper allows the amalgamation to occur. One of Starkey's laboratory notebooks indicates that he later tried to use martial regulus without either silver or copper, hoping to combine it directly with mercury at a higher temperature, but his recorded attempts ended with a laboratory accident when the flask broke. It is unclear if he pursued that method further. A still later notebook records that in 1656 he became dissatisfied with the use of iron in the process, probably because his previous philosophical mercuries failed to generate the philosophers' stone, and decided to use antimony metal prepared without using iron.[38] This final change does not appear in any of the Philalethes treatises—all of which were written before 1656—and in fact it undercuts their repeated insistence upon iron as a crucial ingredient. After Starkey's death in 1665, the notebook containing this final innovation ended up in the possession of Robert Boyle, and was rediscovered among his *Nachlass* only in 1995.[39]

Homberg was aware of *all* the ways "Philalethes" prepared his philosophical mercury—including the unpublished ones—and made his own further adjustments to the process. Indeed, the process keeps cropping up either explicitly or allusively throughout Homberg's life. His 1690s textbook describes a modified version of the Philalethes process, wherein Homberg combines antimony regulus (made without iron) with mercury by the rather frightening method of pouring 1 pound of molten antimony into 3 pounds of mercury heated near its boiling point. He remarks that "I have also done this operation with regulus of silver and with regulus of copper because I doubted that an amalgam could be made without the intermediacy of some metal. But I saw by the means given here that mercury combines with pure regulus of anti-

---

37. Eirenaeus Philoponus Philalethes [George Starkey], *The Marrow of Alchemy* (London, 1654–55), pt. 2, 15–17. Starkey adds to the obfuscation here by using the *Deckname* "Venus" to refer alternatingly to *both* antimony ore and copper.

38. George Starkey, *Alchemical Laboratory Notebooks and Correspondence*, ed. William R. Newman and Lawrence M. Principe (Chicago: University of Chicago Press, 2004), 189–97, 238–54; Newman and Principe, *Alchemy Tried in the Fire*, 190–94. In fact, the martial regulus contains only a small percentage of iron; most of the iron employed ends up in the slag as a sulphide. Nevertheless, Starkey believed that the sulphureous part of the iron was incorporated into the martial regulus, and was crucial for the following steps; compare the description of discovering the stone according to Starkey, as recapitulated by Homberg, above.

39. Lawrence M. Principe, "Newly-Discovered Boyle Documents in the Royal Society Archive: Alchemical Tracts and His Student Notebook," *Notes and Records of the Royal Society of London* 49 (1995): 57–70, at 59.

mony as well as if [the regulus] had been mixed with some metal."[40] It seems plausible that Homberg learned about Starkey's last innovation (the use of ordinary regulus alone) from Boyle, who owned the relevant notebook—the only place known that Starkey recorded this fundamental change to his procedure. The technique of pouring molten antimony into nearly boiling mercury in an iron mortar seems, however, to be Homberg's own innovation, although it may be a practical adaptation of Starkey's method of combining antimony and mercury directly at elevated temperatures. Starkey's first method—using silver (the "Doves of Diana")—was relayed to Boyle in 1651, where the American chymist referred to the process as "a Key into antimony." Accordingly, a Latin version of part of Starkey's letter that circulated in manuscript was explicitly entitled "Clavis," or "Key." Homberg may be hinting that he possessed a copy of such a document when he writes that "Philalethes described this operation but hid it so thoroughly under riddles that it is impossible to understand him without a key [*clef*]." Boyle might also have shared a copy of Starkey's letter with Homberg, although it is also possible that the Batavian chymist obtained one of the Latin versions that circulated privately, as did, for example, Isaac Newton.[41]

Homberg wrote candidly in the 1690s that "I have carried out this operation many times, not only with the intention to follow the entire work of Philalethes, but also to experiment with [the prepared mercury] in the known operations upon mercury, since I saw that being purified in this way the mercury is indeed different from common mercury, and more active." Homberg reported that he witnessed "many effects that I have not observed with common mercury," and that when used in place of common mercury to prepare known mercurial pharmaceuticals, the drugs acted more powerfully.[42] In terms specifically of chrysopoeia, however, "I have not been able to discover the excellencies that Philalethes attributes to it, nor to achieve the

---

40. VMA MS 130, fol. 231r: "J'ay fait aussi cette operation avec le regule d'argent et avec le regule de cuivre parce que Je doutois que l'on peut faire un amalgame avec lantimoine sans l'intermede de quelque metail. Mais J'ay vû dans cellecy que le mercure se lie aussi bien avec le regule d'antimoine pur que si il y avoit quelque metail mêlé."

41. William R. Newman, "Newton's *Clavis* as Starkey's *Key*," *Isis* 78 (1987): 564–74; VMA MS 130, fol. 232v: "Philolethe a decrit a peu prés cette operation mais il la si fort cachée sous des enigmes, qu'il n'est pas possible de l'entendre sans une clef."

42. VMA MS 130, fols. 232v–233r: "J'ay fait cette operation plusieurs fois non dans l'intention seulle de poursuivre tout l'ouvrage de Philalethe mais aussi pour en faire des experiences dans les operations connuës sur le mercure parce que j'ay vû qu'etant purifié de cette maniere il est effectivement different du mercure commun et plus agissant"; fol. 231v: "plusieurs effets que je n'ay pas observé dans le mercure commun."

whole procedure that he gives us for it. I have been able to attain only the first red powder without ever being able to succeed by any of the means I have tried to make an amalgam of this powder with fresh mercury."[43] Thus, Homberg clearly acknowledged his repeated attempts to make the philosophers' stone by the Philalethes process. Despite the roadblocks he encountered, he kept returning to the process and modifying it in the hopes of better success.

Philalethes and his philosophical mercury make several easily missed appearances in Homberg's work prior to taking center stage in his mature chymical theory. The first published instance occurs in 1692, just a year after Homberg was admitted to the Académie, thereby indicating how Homberg's pursuit of chrysopoeia was advanced even early in his career. One of the strange effects of this philosophical mercury is its ability to cause metals, under the right conditions, to "grow" into astonishing treelike shapes. This power was mentioned by Starkey and Boyle, and witnessed by me as well when I reproduced the original process.[44] Homberg also witnessed this striking phenomenon and mentioned it in passing in a paper on "metallic vegetations." Most of the "vegetations" Homberg cited are minor variations on the rather trivial "Tree of Diana," a crystalline growth produced in aqueous solutions of silver salts, or the "silicate gardens" popularized by Johann Rudolf Glauber in mid-century, both of which had by 1691 become very well known. Amid these common tricks, Homberg mentioned one vegetation "of a different sort," made in a sealed vessel from an amalgam of silver or gold with a mercury "well purified by six or seven different sublimations," but without identifying it as Philalethes's philosophical mercury.[45] This "vegetation" stands out in the accompanying illustration as the only one in a sealed, long-necked flask—a vessel known as the "philosophical egg" used particularly for making the philosophers' stone (fig. 5.1).

In 1700, Homberg published a process for making the philosophical

43. VMA MS 130, fol. 233r: "je n'en ay pas pû decouvrir les excellences que Philalethe luy attribüe, ni achever tout le procedé qu'il nous en a donné. J'en ay seulement pû atteindre la premiere poudre rouge sans jamais pouvoir reussir de quelque maniere que je m'y sois pris pour faire un amalgame de cette poudre avec du nouveau mercure."

44. Lawrence M. Principe, "Apparatus and Reproducibility in Alchemy," in *Instruments and Experimentation in the History of Chemistry*, ed. Frederic L. Holmes and Trevor Levere (Cambridge, MA: MIT Press, 2000), 55–74; Principe, *Secrets of Alchemy*, 164–66 and pls. 4–6. Another curious property is its exothermic amalgamation with gold, which allowed me to identify it with Boyle's famed "incalescent mercury"; see Principe, *Aspiring Adept*, 160–65.

45. Wilhelm Homberg, "Reflexions sur differentes vegetations metalliques," *Mémoires de mathematique et de physique* (30 November 1692): 145–52; reprinted in *HMARS 1666–99*, 10: 171–77; the paper was presented 12 November 1692, PV 13, fol. 119r.

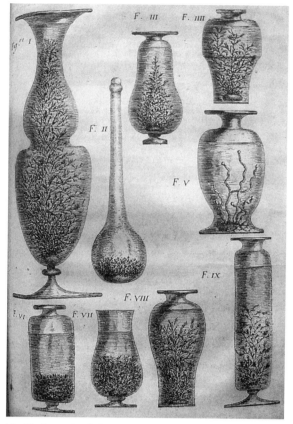

FIGURE 5.1. Metallic "vegetations" from Homberg, "Reflexions sur differentes vegetations metalliques," *Mémoires de mathematique et de physique* (30 November 1692): 145–62. "F. II" shows the "vegetation" produced using the Philalethes/Starkey philosophical mercury; the others are variations on the "tree of Diana." Author's collection.

mercury—but yet again without indicating its nature or origin. Having observed that common mercury left behind an insoluble black residue when dissolved in a certain way using spirit of salt, Homberg suggested that common mercury was actually heterogeneous. He then provided a supplemental paper that explored this heterogeneity further, using a preparation of philosophical mercury in the course of which he separated a nonmercurial gray powder from the mercury.[46] Significantly, Starkey's 1651 letter to Boyle points

46. Wilhelm Homberg, "Observations sur les dissolvans du mercure," *MARS* (1700): 190*–96*; and Homberg, "Suite des observations sur les dissolvans du mercure," 190–96. The first paper is missing from early printings of the first edition, and is not listed in the table of contents;

in just the same way to the "heterogeneities of mercury" revealed by his own preparation of philosophical mercury.[47] Homberg's paper lifts many lines verbatim from his aborted textbook, and repeats his method of pouring molten metal into boiling mercury in an iron mortar. Nevertheless, it presents what was at that time for Homberg a largely superseded process. Rather than using pure antimony, Homberg's published process employs the copper-antimony alloy made from a martial regulus. The use of copper and iron had been abandoned not only by Starkey himself, but moreover by Homberg. Perhaps Homberg was willing to publish this recipe *because* it was superseded, even though using pure antimony (rather than an alloy containing iron and copper) would have strengthened his main argument that the gray powder he isolated came from the mercury and not from anything added to it. It is also possible that this publication was in effect an advertisement to other chrysopoeians of what Homberg was working on, perhaps with the implicit goal of making contact with others working on a similar process. The same explanation, namely, to interest "divers *Philalethists*," has been shown to lie behind Boyle's 1676 publication of his "incalescent mercury"—the same substance as the philosophical mercury under a different name—in the *Philosophical Transactions*.[48]

As in the paper on metallic vegetations, and strikingly unlike his unpublished textbook, Homberg's 1700 publication never connects the process to Philalethes or to chrysopoeia. Homberg probably "neglected to mention" these connections due to the administrative warnings forbidding academicians to indulge in chrysopoeia "either close up or from a distance." Yet other sources make it clear that Homberg did not hide his chrysopoetic activities from his fellow academicians during their private meetings. In the same year, Bourdelin *fils* recorded in his diary that after Simon Boulduc read a paper on mercury as a chymical principle, Homberg remarked that "mercury has been easily extracted from lead, about which all lead-workers agree, [and] that it is drawn from antimony by a method that [Homberg] knows, even though Mr. Boulduc failed to extract it by this method which [Homberg] had

---

later printings do insert the missing paper, duplicating the pagination (as indicated here by the asterisks); see note on sources.

47. See Newman and Principe, *Alchemy Tried in the Fire*, 301–11, for more details on Homberg's reliance on Starkey for this paper.

48. [Robert Boyle], "Of the Incalescence of Quicksilver with Gold," *Philosophical Transactions* (1676): 515–33, at 529; on the background and reception of this paper, see Principe, *Aspiring Adept*, 155–65.

communicated to him."[49] Shortly thereafter, Homberg reiterated that "he has collected about 12 ounces of mercury of antimony, which is much more sulphureous than common mercury, and it identifies itself by this mark, that upon rubbing a silver spoon with this mercury, it gilds it instantly. He has resolved to make several experiments with this mercury." The reference to gilding a silver spoon as proof of the "philosophical" nature of the mercury actually comes from von Suchten (and was mentioned by Starkey and Boyle).[50] Not everyone agreed with Homberg's claims, particularly Nicolas Lémery. Lémery had advertised his anti-transmutation credentials—which included a denial of the mercuries of metals—in the 1679 edition of his *Cours de chymie*. He would continue to be a critic of transmutation and metallic mercuries, leading to several skirmishes with Homberg.[51] The following week, for example, Lémery stated that "he doubts that one can extract mercury from antimony even though Mr. Homberg said three days ago in the assembly that he has already extracted 12 ounces of it."[52] A year later, in 1702, a meeting erupted in a "grande dispute" between Homberg and Lémery about mercury and its properties. Bourdelin again recorded some of the oral exchanges that are never recorded in the procès-verbaux: "common mercury does not calcine by itself except after a long time, about a year or eighteen months. Mr. Homberg calcined some mercury purified with antimony by itself in the manner of Philalethes in three months." Here Homberg was completely frank about following Philalethes, and went on to mention explicitly the transformation of this Philalethean mercury into a *poudre rouge*.[53] There was thus a divide between what Homberg would claim (or *could* claim) in publications versus

49. Bourdelin Diary, fol. 27v (1 December 1700): "a cette occasion M. Homberg dit qu'on tiroit aisément du mercure du plomb, que tout les Plombiers en conviennent; qu'on en tire de l'antimoine par une methode qu'il scait, quoyque Mr. Boulduc ait manqué d'en tirer par cette methode qu'il luy avoit communiqué."

50. Bourdelin Diary, fol. 29r (29 January 1701): "Il dit qu'il avoit amassé environ 12 onces de mercure d'antimoine qui est beaucoup plus sulphureux que le commun & qui se reconnoit à cette marque qu'en frotant une cueilliere d'argent de ce Mercure il la dore sur le champ. Il a resolu de faire quelques experiences avec le Mercure." For the use of the test by von Suchten and Boyle, see Principe, *Aspiring Adept*, 171.

51. On Lémery's sudden turn against chrysopoeia in 1679, see Newman and Principe, "Alchemy vs. Chemistry," 59–61.

52. Bourdelin Diary, fol. 30v (1 February 1701): "Il doute qu'on tire du mercure d'antimoine quoyque M. homberg ait dit il y a 3 jours à l'assemblée qu'il en avoit desja tiré 12 onces."

53. Bourdelin Diary, fol 45r (11 January 1702): "Le mercure commun ne se calcine par luy même qu'aprez un tres long tems comme d'une année ou 18 mois. M. Homberg calcine par luy même du mercure purifié par l'antimoine à la maniere de Philalethes en 3. mois."

what he said in private meetings of the Académie—and this difference possibly relates to concerns about how the academicians' activities were perceived publicly, recalling Louis XIV's anxiety about the French coinage. This divide would not only persist but become even deeper in the succeeding decades, as will be documented in chapter 7.

Philippe II and Chrysopoeia at the Palais Royal

It was very shortly after these disputes that Homberg moved into the Palais Royal laboratory built for him by the Duc d'Orléans. There he continued his work not only on the Philalethes method, but also on many other routes that promised metallic transmutation. Homberg's chymical endeavors at the Palais Royal were promoted not only by Philippe II's generous financial patronage, but also by the prince's hands-on collaboration in the laboratory and his acquisition of new texts and recipes. The most extensive practical modifications of the Philalethes/Starkey method also date from Homberg's time in the Palais Royal, as the Batavian chymist continued to strive for the transmutational results Philalethes had promised.

Following Fontenelle's 1715 "Éloge" for Homberg, it has been claimed that Homberg and Philippe worked together—even side by side—in the Palais Royal laboratory. Unfortunately, there has hitherto been almost no documentation of their collaboration or any clear indication of what sorts of work they carried out together.[54] In this regard, the notes taken by the mid-eighteenth-century academician Jean Hellot once again open a window onto these activities. Hellot had access not only to original Homberg manuscripts, but also to substantial laboratory records, which he calls the "manuscripts of the Palais Royal" and the "Practice of the Palais Royal." The latter of these documents contained chymical processes described as "memoirs of Monseigneur the Duc d'Orléans, Regent."[55] Hellot frequently references these documents, and even prefaces one extended excerpt from them with a description of their provenance: "Operations carried out at the Palais Royal from 1705 until 1711 by Mr. Homberg under the Duc d'Orléans. I have drawn all the details from Mr. Homberg's original manuscripts, sold by his widow to Mr. Geoffroy the

54. Bernard de Fontenelle, "Éloge de M. Homberg," *HARS* (1715): 82–93, at 90. The only published account of collaboration between Homberg and Philippe involves producing glass copies of some engraved gems in Elisabeth Charlotte's collection; see Wilhelm Homberg, "Maniere de copier sur le verre coloré les pierres gravées," *MARS* (1712): 189–97, at 191.

55. Hellot Caen Notebooks, for example, 3:147r: "tiré des MSS du Palais Royal, article datté, fait en 1707"; and 8:18v (recipes for artificial gems, linked to Homberg and Philippe).

elder, and from several letters and notes in the hand of the late Duc d'Orléans giving his orders and advice to Mr. Homberg."[56]

Hellot's notes must preserve only a tiny sliver of the documentary material that once existed, and that was initially—like the manuscripts of Homberg's three unpublished treatises—in the hands of Homberg's protégé Étienne-François Geoffroy. Hellot's notes nevertheless provide otherwise unknown and unobtainable information about the laboratory work that Homberg carried out at the Palais Royal, sometimes with Philippe and sometimes in conjunction with various assistants and collaborators. None of this information appears in the records of the Académie, whether the unpublished procès-verbaux or the published *Histoire et mémoires*. Indeed, perhaps surprisingly, Homberg seems to have reported very little of the work he performed at the Palais Royal to the Académie. This situation underlines how incomplete—even misleading—the standard Académie sources can be when one is trying to assess the entirety of an academician's activities.

Hellot's transcriptions demonstrate that while Homberg carried out a wide range of experiments at the Palais Royal, he directed a considerable amount of effort toward chrysopoeia, often with Philippe's close participation. Without the benefit of the Hellot manuscripts, historians had been left with conflicting testimony about Philippe's thoughts about transmutation. The Duc de Saint-Simon was at pains to deny that the Duc d'Orléans had any interest or belief in transmutation, describing chrysopoeia as something "he always scoffed at." In contrast, Philippe's mother, Elisabeth Charlotte, wrote quite casually that "my son can indeed make gold without gold, because he is very learned [*curieux*] in chymistry; but he says that there is no profit in it. He who once sticks his nose into this curiosity can get himself out of it again only with difficulty."[57] One of the most striking pieces of testimony in the Hellot notebooks confirms Elisabeth Charlotte's claim rather than Saint-Simon's.

56. Hellot Caen Notebooks, 5:143v–144r: "Operations faites au Palais Royal depuis 1705 jusquen 1711 par Mr Homberg sous Mr le duc d'Orléans. J'en ai tiré tous les details du MSS original [*sic*] de Mr Homberg vendus par sa veuve a Mr Geoffroy l'aine et de quelques Lettres et notes de la main de feu le Duc d'Orléans, donnant ses ordres et ses avis a Mr Homberg." Since Hellot made these notes in the 1750s and 1760s, the manuscripts had by that time passed out of the hands of the elder Geoffroy, who had died in 1731, probably passing to his younger brother Claude-Joseph (*le Cadet*) with whom Hellot was very close.

57. Elisabeth Charlotte to Sophie of Hannover, 27 October 1709, in *Aus der Briefe der Herzogin Elisabeth Charlotte von Orléans an die Kurfürstin Sophie von Hannover*, ed. Eduard Bodemann, 2 vols. (Hannover, 1891), 2:231. The same letter reports that "the father of the Countess of Gramont," namely, Sir George Hamilton (c. 1607–79), "entirely ruined himself" in searching for the philosophers' stone, and that "the good doctor Tack [presumably Otto Tachenius]" was equally unsuccessful.

The Comte de Nocé and the Chevalier de Béthune have said that they saw the transmutation of mercury into gold performed once in the interior laboratory of the Palais Royal. It was done by projecting a certain quantity of a red powder which had been prepared as much by the prince as by his artists. But having assured himself of the possibility [of transmutation], he did not wish that any research in this regard be continued. . . . Mr. Grosse who was an assistant under Mr. Homberg certified this fact to me, but he never entered into the secret place of the principal operation.[58]

This astonishing account contains several noteworthy points. First, Philippe demonstrated his transmutation to several friends, including Charles de Nocé (1664–1739) and Louis-Marie Victoire de Béthune (1671–1744), who apparently later gave testimony about the event. (The Comte de Nocé accompanied Philippe on his 1706 military campaign in Italy, and it was he who recommended Chirac as a field physician.) Second, the transmuting powder was prepared collaboratively by Philippe and his "artists"—presumably Homberg and his assistants—indicating that the Duc d'Orléans had a high degree of participation in the laboratory. Third, and perhaps most strikingly, the experiment was carried out in the "interior laboratory," also called "the secret place." This is the only indication we have that the Palais Royal laboratory was divided into sections, and that it had a special room for secret operations that laboratory assistants were not permitted to enter. Presumably only the Duc d'Orléans, his guests, and Homberg were allowed in this part of the laboratory. It is plausible that this "secret place" within the Palais Royal laboratory was used regularly for research into metallic transmutation, and was perhaps even designed specifically for such operations. The idealized *domus chymica* devised by Andreas Libavius (1555–1616) a century earlier describes a similar hierarchy of chymical workspaces. While specific open spaces in the *domus* were assigned to particular chymical operations, there was also a special interior room specifically for making the philosophers' stone and to which only the master and highly advanced persons had access.[59] Although Hellot's note indicates that Philippe did not pursue transmutation any fur-

58. Hellot Caen Notebooks, 5:144r: "Mr le Comte de noce et le Chevalier de Bethune, ont dit avoir vu faire une fois la transmutation du mercure en Or dans le laboratoire interieure du Palais Royal en projettant une certaine quantité de poudre rouge, que le prince avoit préparée, tant lui meme que par ses artistes, mais que s'etant assuré du fait il n'avoit pas voulu qu'on continuât aucune recherche à cet égard. . . . Mr Grosse qui estoit artiste sous Mr Homberg ma certifié ce fait, mais il na jamais entré dans le lieu secret de l'operation principale."

59. See William R. Newman, "Alchemical Symbolism and Concealment: The Chemical House of Libavius," in *The Architecture of Science*, ed. Peter Galison and Emily Thompson (Cambridge, MA: MIT Press, 1999), 59–77, at 70–72.

ther after this one supposedly successful event, that might not entirely be the case, as we shall see shortly.

Jean Grosse, Little-Known Academician and Homberg's Laboratory Assistant

Hellot's report also names Jean Grosse—born Johann Gross, a German physician who later became a member of the Académie—as Homberg's laboratory assistant. Almost nothing has previously been known about Grosse before his admission to the Académie in 1731.[60] This lack of information does not appear to be the fault of historians but rather goes back to Grosse himself. Jacques-François Demachy (1728–1803), while praising Grosse's profound knowledge of chymistry, recalled that the German chymist's "whole ambition seems to have consisted in making himself unknown."[61] It is known that he was close to Gilles-François Boulduc (1675–1742), and lived with him from about 1713 until Boulduc's death in 1742, and that Boulduc proposed Grosse to the Académie in 1731 to fill a recently vacated position of *associé chimiste*. That proposal failed—Louis-Claude Bourdelin (1696–1777), already *adjoint chimiste*, was chosen instead. A few months later Boulduc proposed Grosse again, this time for the entry rank of *adjoint chimiste* left open by Bourdelin's advancement, which Grosse obtained and held until his death in 1744. In the 1730s, Grosse read several papers to the Académie and published five memoirs from 1732 to 1736, three of them co-authored with Henri-Louis Duhamel de Monceau (1700–1782).[62] He devised his own version of a "Table des Rapports" building upon and modifying Geoffroy's.[63] Paul-Jacques Malouin (1701–77) drew upon Grosse's experience and laboratory results for a series of papers on zinc, as did Hellot for his own studies of zinc. Judging by his frequent mention as an authority and collaborator in papers by other academi-

---

60. The only extended biographical study is Paul Dorveaux, "Jean Grosse, médecin Allemand, et l'invention de l'éther sulfurique," *Bulletin de la Société d'Histoire de la Pharmacie* 17 (1929): 182–87. See also Jean-Jacques Dortous de Mairan, "Éloge de M. Boulduc," *HARS* (1742): 167–71, at 171.

61. Jacques-François Demachy, *Receuil des dissertations* (Amsterdam, 1774), 94–100, at 94.

62. Henri-Louis Duhamel du Monceau and Jean Grosse, "Des differentes manieres de rendre le tartre soluble," *MARS* (1732): 323–42; "Sur les differentes manieres de rendre le tartre soluble: Seconde partie," *MARS* (1733): 260–72; "Recherche chimique sur la composition d'une liqueur tres-volatile, connuë sous le nom d'Ether," *MARS* (1734): 41–54; Jean Grosse, "Recherche sur le plomb," *MARS* (1733): 313–28.

63. Alistair Duncan, *Laws and Order in Eighteenth-Century Chemistry* (Oxford: Oxford University Press, 1996), 119–21.

cians, it seems that Grosse was the person to whom they all regularly turned for advice regarding their own chymical projects. Voltaire also sent chymical inquiries to Grosse in 1737, calling him "a very intelligent and laborious chymist" as well as a "gnome"—a curious label that might refer to the same reclusive nature that caused Demachy call him "the celebrated anachorite of the chymists."[64]

Statements scattered through Hellot's notebooks provide much new information about Grosse, drawn both orally from the chymist himself and from his surviving manuscripts. After Grosse's death, Hellot paid "Mr. Leberecht, his nephew" for access to Grosse's papers. Some sixty pages in one notebook alone come from Grosse's *Nachlass*.[65] These notes indicate that Grosse, like many of the Académie's chymists, occasionally taught the *cours de chimie* at the Jardin du Roi. They cite Grosse's many analyses of mineral waters carried out from at least the mid-1720s until 1735, and report that Étienne-François Geoffroy called upon him to analyze a particular water in 1730. Geoffroy also turned to the German chymist for an explanation of an opaque passage in Sir Kenelm Digby's *Chymical Secrets*.[66] Grosse was asked to do other chymical analyses—in one case he provided certification that a certain white powder did not contain arsenic, scrupulously left unnamed as anything other than "the suspected material [*la matière soupçonnée*]." He may therefore have been involved in the forensic testing of suspected poisons—a service that Homberg himself occasionally carried out for the Paris police.[67]

64. Especially Paul-Jacques Malouin, "Sur le zinc: Seconde mémoire," *MARS* (1743): 70–86, at 81–83; Voltaire to Bonaventure Moussinot, D1347 (29 June 1737), D1349 (6 July), D1351 (8 July), and D1353 (13 July), in *Complete Works*, ed. Theodore Bestermann (Geneva: Institut et Musée Voltaire, 1969), 88:337–39, 346–47, 349–50; and Demachy, *Dissertations*, 95.

65. Hellot Caen Notebooks, 1:57r, 8:214v–243v. That the papers went to his nephew suggests that Grosse had no direct heirs, and was therefore unmarried or a widower without children—which makes sense if he lived with Boulduc from 1713 until 1742 and then died shortly afterward. Demachy (*Dissertations*, 95) also mentions Grosse's nephew, whom he calls Brek, and notes that he served Boulduc as *garçon de laboratoire* and lived near him on the rue des Boucheries.

66. Hellot Caen Notebooks, 8:231v–232r (report to Boulduc on operations performed at the Jardin on 10 August 1730), 242r (on fulminating gold made at the Jardin in 1729); on mineral waters, 8:218v–219v (report dated 29 March 1730) and 214v–216v, 221v–226r (dated 1726–35); on Digby, 5:30v.

67. Hellot Caen Notebooks 8:226r (20 November 1730); the affidavit is signed by both Grosse and "Hammerer," whom I have not identified. Homberg's analyses of suspected poisons appear in two autograph letters preserved in the investigation files of one Abbé Blanche; see BNF, Arsenal MS 10588, fols. 115r (21 July 1710, to d'Argenson), 116r (7 September 1710, to Camuset), and 117r–119r (report of analyses); fol.1r–v is the envelope for the letter to d'Argenson, addressed in Homberg's autograph. Part of the text of the letters is published in *Archives de la Bastille*, ed. François Ravaisson, 19 vols. (Paris, 1866–1904), 12:3–4.

Demachy reports that Grosse was "in correspondence with all the chymists" of Germany. Accordingly, Hellot's notes record his correspondence with Caspar Neumann (1683–1737), who attended one or more of Grosse's lectures at the Jardin in 1718. Grosse sent a transmutational text (again relating to zinc) to Johann Friedrich Henckel (1678–1744), who translated and published it. Grosse also wrote a treatise (left unpublished) on the mineral wealth of France, particularly in the Auvergne, and possessed various chymical secrets.[68] Some were technical manipulations or pharmaceutical preparations: a means of isolating gold from "sulphur auratum" and of recovering silver and gold from metallic fabrics, an improved way to amalgamate copper, and how to produce a "sperm of mercury . . . that he held as a great mystery" and used as a "good remedy." Others clearly involved metallic transmutation: a particular for turning silver into gold, a means of isolating the mercury of lead, and so forth.[69]

It was certainly not known previously that at least twenty-five years before Grosse's admission to the Académie, he had been Homberg's laboratory assistant at the Palais Royal. Grosse presumably lived there as well, since he would have been obliged to tend the fires at all hours. A ground-floor room directly adjacent to the laboratory, and described as "for a domestic," may have been his lodging. Besides the testimony he gave Hellot regarding the transmutation performed in the laboratory's "secret place," Grosse reported on other events involving Homberg that took place at the Palais Royal laboratory, such as how the Batavian chymist accidentally poisoned himself with arsenic fumes in 1711.[70] Grosse's German nationality undoubtedly suited him linguistically for working with Homberg, who (as mentioned in chapter 2) had sought a specifically German assistant as early as 1692. Germans were renowned in this period for their skill as chymical laboratory operators; several of the most successful of Robert Boyle's many operators, for example, were

68. Demachy, *Dissertations*, 94, on his correspondence with Germany; Hellot Caen Notebooks, 8:236v–238r (letter from Neumann, 28 February 1718); 238r–241v (minerals of Auvergne, 1735); P. M. Respour, *Besondere Versuche vom Mineralgeist* (Dresden and Leipzig, 1742), "Neue Vorrede," sig. 2r.

69. Hellot Caen Notebooks, 5:30v (dated October 1725), 5:31v, 5:62v, 8:19v, and 1:57r (compare 8:21r and 243r–v).

70. Hellot Caen Notebooks, 8:228r. Homberg met with dangerous results from his experiments more than once. He suffered powerful evacuations from accidentally inhaling the dust produced when grinding ipecac (PV 23, fol. 170r, 14 June 1704), and nearly poisoned himself with another dust produced by grinding in 1710: Elisabeth Charlotte to Étienne Polier de Bottens, 16 November 1710, in *Aus den Briefen der Herzogin Elizabeth Charlotte von Orléans an Étienne Polier de Bottens*, ed. S. Hellman, *Bibliothek des litterarischen Vereins in Stuttgart* 231 (1903): 94–95.

also Germans.[71] It might even have been Homberg who first brought Grosse to France. In any event, during his tenure with Homberg in roughly the first decade of the eighteenth century, Grosse was another member of the Académie's "penumbra" mentioned in chapter 1. Significantly, his better known connection and residence with Boulduc began only after, and immediately after, his association with Homberg at the Palais Royal was forced to come to an abrupt end around 1713 (for reasons to be discussed in chapter 6).

Try, Try Again: Reinterpreting the Allegories

Hellot's notes report that Homberg expended enormous effort on transmutational endeavors, frequently with the assistance of the Duc d'Orléans. Homberg continued to experiment, for example, with the tantalizing (but frustrating) method of Philalethes. While Homberg did believe that he had successfully produced a small quantity of gold by heating the philosophical mercury for an extended period, he was not (apparently!) successful in turning this mercury into the philosophers' stone as Philalethes had described, nor did the *minera perpetua* he had attempted fulfill its initial promise. The Philalethes texts thus provided a further source for Homberg to consider in addition to his own experimental results and theory, but Homberg needed to decipher Philalethes's allegorical writings *correctly*. Unable to produce the philosophers' stone from the various mercuries he had prepared, Homberg must have concluded that the preparative methods he was using—which came ultimately from Starkey—were actually incorrect interpretations of Philalethes's texts. (Homberg was unaware, as were almost all readers at the time, that Starkey himself had written the Philalethes treatises.) Thus, Homberg embarked, around 1707, on a sequence of new experiments to find the true but still undiscovered method of preparing and using the philosophical mercury encrypted by Philalethes. In a lengthy section of the "manuscript of the Palais Royal" Homberg reported six new and sequential variations he made to the process.[72] He first describes the familiar process of using antimony to "cleanse" mercury in order to prepare what he now labels "the mercury commonly but *falsely* called the 'mercury of Philalethes.'" Despite

71. For example, Friedrich Slare (Schloer) and Ambrose Godfrey Hanckwitz, on whom see Marie Boas Hall, "Frederick Slare, F.R.S.," *Notes and Records of the Royal Society of London* 46 (1992): 23–41; R. E. W. Maddison, "Studies in the Life of Robert Boyle, F.R.S.: Part V, Boyle's Operator Ambrose Godfrey Hanckwitz," *Notes and Records of the Royal Society of London* 11 (1955): 159–88. For similar views of Germans in Sweden, see Fors, *Limits of Matter*, especially 54–60, 103–7.

72. Hellot Caen Notebooks, 5:144r–148r.

all the painstaking work Homberg had already carried out, probably since the 1680s, using antimony both alone and alloyed with silver or copper, he now decided—based on theoretical considerations—that an alloy of antimony with gold would be better, since the precious metal contains more "fixed metallic sulphur."[73] While this would obviously be far more expensive, Homberg now had the financial backing of Philippe, and so cost was no longer a consideration. When this change still proved unsatisfactory, he turned to a "second mercury of Philalethes," based on a recipe Philippe sent to him from Madrid in 1708. Even though Philippe was then in Spain on military campaigns, he apparently still found time to seek out transmutational recipes and send them back for Homberg to try out in Paris. But Homberg decided that the recipe as written was not "according to the sense of Philalethes," and so devised a variation on it that seemed "more conformable to the idea of the author." These alterations involved a fundamental reinterpretation of the "Doves of Diana" and the "eagles."[74]

Homberg now suggested that the two "Doves of Diana" were not two parts of silver (as Starkey unambiguously said they were) but instead niter and common salt (potassium nitrate and sodium chloride). Treating common mercury with these two salts produced mercury sublimate (today, mercuric chloride) by a process well known in Homberg's day. Continuing this reinterpretation, Homberg decided that the "eagles" were not sequential distillations of mercury from antimony regulus, but rather sequential combinations of mercury sublimate with the regulus. For each "eagle," Homberg heated mercury sublimate with antimony regulus, causing "butter of antimony" (antimony trichloride) to distill over, followed by "revivified" mercury. Homberg then treated this revivified mercury with more niter and salt into turn it again into mercury sublimate, which he then combined once again with fresh antimony regulus, and repeated the whole operation ten times to accomplish the ten "eagles." Homberg recorded his observations during this lengthy process in great detail, and even analyzed the distillation residues by multiple methods to see what they contained. By the end, the mercury smelled different and had increased slightly in density. Yet he found the operation not fully to his liking and "very tiresome [fort penible]." Still "persuaded that common

73. Hellot Caen Notebooks, 5:144r (emphasis added): "Pour faire le mercure qu'on appelle communement mais faussement le mercure de Philalethe."

74. Hellot Caen Notebooks, 5:145r: "N'ayant pas trouvé la preparation du mercure cy dessus décrite selon le sens de Philalethe. . . . Nous nous sommes imaginés la 2° methode qui suit, la croyant plus conforme à l'idée de l'auteur." Between the sentences, Hellot added the explanatory parenthetical "Mr. Homberg parle icy d'un procedé que Mr. Le Duc d'Orléans avoit envoyé de Madrid en 1708. J'ai vu l'ecrit original de la main de ce Prince."

salt and niter are the two Doves of Diana, we tried to use them in a different manner," which led to a "third mercury of Philalethes."[75]

This third preparation avoided the laborious preparation of mercury sublimate, and substituted far easier steps that would have a similar effect. Homberg now dissolved mercury in *spirit of* niter (nitric acid) and then added common salt, which produced a white precipitate. This white precipitate he mixed with antimony regulus, which upon heating caused a butter of antimony (but "black as ink") to distill over, followed by the revivified mercury. He dissolved this revivified mercury anew in spirit of niter, reprecipitated with salt, and heated the new precipitate with antimony. After four repetitions, his initial 2 pounds of mercury had dwindled to a mere 5 ounces; 85 percent of the mercury had vanished! This dramatic loss of mercury made Homberg give up on this process, but it also "gave us occasion to make some extremely interesting observations on the dissolution and precipitation of mercury."[76] Puzzled by the loss of mercury, Homberg went to search for it. He thus discovered that salt precipitated only a fraction of the mercury, leaving the rest invisibly in solution, and so he had unknowingly discarded it with the filtrate. Never one to let a curious observation go unpursued, Homberg carried out a battery of further experiments. He found that when he used less spirit of niter to dissolve the mercury, the salt precipitated a larger proportion of the dissolved mercury. Thus, although this process failed to deliver the philosophical mercury, it did provoke Homberg to identify for the first time a phenomenon that would not be explicable for more than a century after. Namely, the action of nitric acid on mercury can produce two different products—mercurous and mercuric nitrate—and common salt precipitates the former but not the latter; the less acid used, the greater the proportion of mercurous over mercuric. Such careful and alert observations bear out contemporaneous references to Homberg's "great exactness in making experiments." Homberg's inherent curiosity, close observations, and constant monitoring of weights made him set aside temporarily his search for the philosophers' stone in order to learn something new and "fort curieuse" about mercury.

Undaunted, Homberg moved on to a *fourth* process that involved yet another reinterpretation of Philalethes. "In one part of his treatise, Philalethes

75. Hellot Caen Notebooks, 5:146r–v: "La sublimation du 2ᵉ mercure de Philalethe nous ayant paru fort penible, et estant persuadés que Le sel et le nitre sont des deux colombes de Diane, nous avons tenté de les employer d'une autre maniere."

76. Hellot Caen Notebooks, 5:146v: "Cette operation nous a donné occasion de faire des observations fort curieuses touchant les dissolutions et les precipitations du mercure."

describes an operation to purify mercury that, by changing very little in it, we believed might well be his grand operation of the eagles." Here Homberg implicitly assumed that Philalethes employed the tactic known as *dispersion*— that is, placing different parts of a process disconnectedly through a text, with the expectation that worthy readers could identify and reconnect the dispersed fragments.[77] Homberg referred then to Philalethes's direction to grind mercury with salt and vinegar before treating it with antimony regulus. Homberg then collapsed the two operations into one, and ground mercury with salt, vinegar, and antimony all at once, then submitted the ground material to distillation. He then reground the distilled mercury with the three ingredients and redistilled, repeating the operation ten times. The first results were promising; the mercury increased slightly in weight with each distillation—note here, yet again, Homberg's routine reliance on weight comparisons—which Homberg took as evidence that the "metallic sulphur of the iron which is in the regulus" was successfully being incorporated with the mercury and increasing its weight.[78] But once again final results were not satisfactory, and so Homberg moved on to two "other attempts at the mercury of Philalethes." The next method provided vapors "so stinking and insupportable" that Homberg had to leave the laboratory until the next day (one wonders what the effect of such experiments was on Philippe's chambers directly above the laboratory). Rejecting this fifth method, Homberg tried yet another interpretation. This sixth method gave results interesting enough that the Batavian chymist noted that he "wants to repeat it," but Hellot's transcription ends at this point.[79]

This lengthy sequence of attempts displays another aspect of Homberg's exhaustive experimental methodology—changing variables one at a time in controlled ways and monitoring the changing results by weight and other means. The same pattern appeared in his experiments using mortars of different metals when preparing the Bologna Stone, which allowed him to identify the crucial role of trace copper and the deleterious role of trace iron.[80] Another notable feature is how these various attempts toward the philoso-

77. On the technique of dispersion, which dates back to Arabic alchemical texts, see Principe, *Secrets*, 44–45, and references therein.

78. Hellot Caen Notebooks, 5:147r: "Philalethe décrit dans un endroit de son traité une operation pour purifier le mercure, laquelle nous avons cru pouvoir bien etre sa grande operation de Aigles, en y changeant tres peu de choses." Note that now Homberg has gone back to using martial regulus, in accordance with the directions of the *Introitus apertus*.

79. Hellot Caen Notebooks, 5:147v–148r.

80. See chapter 1; also Lawrence M. Principe, "Chymical Exotica in the Seventeenth Century, or, How to Make the Bologna Stone," *Ambix* 63 (2016): 118–44, at 130–32.

phers' stone are occasionally interspersed with other research that Homberg must have been working on simultaneously. For example, the text also contains a close study of various methods of preparing the "volatile salt of vitriol," a substance soon after popularized as "Homberg's sedative salt." (This *sal sedativum Hombergi* was actually boric acid, a substance prepared for the first time by Homberg; it persisted under that name in pharmacopiae for two centuries afterward.)[81] Thus, one can imagine a busy Homberg at work on multiple projects simultaneously in the Palais Royal laboratory. Perhaps—given the apparent size of the locale—he maintained different areas of the workspace for different projects, and "spun off" some projects to Grosse, Geoffroy, and perhaps others to be pursued simultaneously. At two points in the sequence of "Philalethean" mercuries, Homberg switches from first person singular to first person plural before switching back. This very selective use of the plural "we" might indicate that those interpretations and practical work were collaborations between Homberg and the Duc d'Orleans. If so, it would further indicate that what Fontenelle wrote in 1715 about Philippe's laboratory work was not mere flattery to the recently installed Regent. "The Prince Philosopher went [to the laboratory] nearly every day, he eagerly received instructions from his chymist and often even anticipated them right away. He engaged with all the details of the operations, carried them out himself, and thought up new ones, and I often saw the master surprised by his student. [Homberg], who was nearly the only one to witness [Philippe's] talents, used to tell me in these very words: 'People don't really know him; he is a hard worker.'"[82]

While there are many examples of early modern rulers patronizing transmutational endeavors, there are very few cases indeed in which the patron got his hands dirty with the coals alongside his chymist in the laboratory, and perhaps no others where the level of direct participation was as significant as Philippe's. He would certainly make full use of his abilities as a "hard worker" several years later when as Regent he labored mightily to repair the weaknesses of the kingdom left behind by Louis XIV.

Besides the process he sent from Spain, the Duc d'Orléans acquired many

81. Homberg provided the preparation of this material (there called *sel volatile narcotique de vitriol*) in the published version of Wilhelm Homberg, "Essays de chimie," *MARS* (1702): 33–52, at 50–52. Since it is not present in the earlier procès-verbaux version, PV 21 (1702), fols. 61r–73v, his discovery of boric acid presumably dates between the oral presentation in 1702 and its publication in 1704.

82. Fontenelle, "Éloge de Homberg," 90. Fontenelle's comments are undoubtedly also aimed at countering contemporaneous negative and unfounded gossip (which persist even to the present) about the habits and capabilities of the regent.

other recipes that he brought to Homberg. In 1707 he obtained a process for transmuting silver into gold, and in 1708 he bought a recipe "from one named d'Irman" that involved heating congealed mercury, verdigris, salts, and an odd mash of figs and raisins in a crucible. The goal of the latter process is not explicitly stated, and the text ends with a note copied from the margin of the original document, saying that "the use of this whole operation is a mystery that cannot be revealed."[83] In the same year, Philippe received a method for a "profitable" augmentation of gold from one de la Villardiere. Another chrysopoetic recipe—which mentions Philalethes and his "eagles" before diverting to another pathway—was sent to Philippe from a chymist named de la Baume; the process ended up among Homberg's papers, no doubt because it was given to him to try out, whether alone or with the collaboration of the Duc d'Orléans.[84] A lengthy process for a material able to transmute twenty times its bulk of silver or mercury into gold arrived in a letter to Philippe from "Mr. Joly of Lyon." Hellot adds that "it seems to me that he sold this process to the prince through the intermediacy of Mr. Homberg." Other letters to Philippe from the same Mr. Joly provide lengthy instructions for the "great work," that is, making the philosophers' stone, and for its use as both a transmuting agent and a universal medicine.[85] The relationship between Joly and Homberg was of long duration, since Hellot also refers to a letter from "Joly of Lyon" to Homberg dated 1 March 1701.[86] Little more is known at present about this chymist and his relationship with Homberg, although the intermediacy of Homberg in acquiring a recipe bought by Philippe recalls the Batavian chymist's similar intermediacy in obtaining the Tschirnhaus lens. Philippe also acquired pharmaceutical recipes, for example, a "universal medicine of Mr. Bourguignon, a secret bought in 1712 by Monsieur le Duc d'Orléans, and carried out at the Palais Royal by Mr Homberg." The collaboration between Homberg and Philippe appears yet again in the recipe

83. Hellot Arsenal Notebooks, MS 3006, 125–56: "Teindre argent en or par l'antimoine et le fer, procedé donné par Mr. Dupuis a Mgr. le Duc d'Orleans, 1707"; Hellot Caen Notebooks, 3:271r.

84. Hellot Arsenal MSS, MS 3006, 3–4: "Augmentation d'or a 60 pour 100 de profit par les vitrifications donné à Mr. le Duc d'Orleans par J. de la Villardiere 1708"; and 4–7: "Procedé donné à Mr le Duc d'Orleans par Mr. de la Baume, MSC de Mr. Homberg."

85. Hellot Caen Notebooks, 5:33v: "tiré de la lettre originale écrite à Mgr le Duc d'orleans regent par Mr Jolly de Lyon. Il m'a paru qu'il avoit vendu le procedé au Prince par l'entremise de Mr Homberg"; of course, Philippe was not yet regent when this letter arrived, the addition of that title is part of Hellot's description. Hellot Arsenal Notebooks, MS 3008, 21–23: "Grand Oeuvre et Medicine Universelle tirée des lettres de Mr Jolly de Lyon a Mgr. le Duc d'Orleans."

86. Hellot Arsenal Notebooks, MS 3008, 21–23; see also 45.

for a mercury-based "universal purgative," which Hellot notes he found "in the handwriting of Mr Homberg, who adds that the Regent had several trials made of it."[87]

Homberg, of course, obtained plenty of information on his own about transmutation and other chymical processes from his experiments, his reading, and his own informants. Hellot's notes preserve *dozens* of such processes from Homberg's papers. Chrysopoetic entries include a way to augment gold by a third, and a fixation of silver into gold by means of a tin preparation. The latter process ends with the advice to "assay [the result] by quartation to see what the profit of this operation is."[88] Other transmutational processes include several "tinctures of gold" (one that "multiplies itself infinitely" and is dated 1712), two fixations of mercury into gold from a Mr. Berterot, a liquid able to tinge eighteen times its weight of mercury into gold, an oil of "fixed sulphur" reputed to transmute fifty times its bulk of mercury into gold, and a "graduating solvent" able to turn silver into gold with "considerable profit."[89]

Homberg and Geoffroy: Collaboration, Disagreement, Phlogiston, and Transmutation

Homberg and Philippe were not the only collaborators at the Palais Royal laboratory, nor the only ones who pursued transmutation there. Homberg's protégé Étienne-François Geoffroy was also involved in similar endeavors. Geoffroy had been privately tutored in chymistry by Homberg in the 1690s, was admitted to the Académie in 1699 as Homberg's élève, promoted to associé a few months later, and finally succeeded to his mentor's pensionnaire position upon Homberg's death in 1715. While Geoffroy certainly pursued some of his own projects, much of his early work was done in concert or in parallel with Homberg. Geoffroy often worked on projects, probably given to him by the elder chymist, that tested, complemented, refined, or gave further evidence for Homberg's findings and ideas. (Homberg had had the same relationship with the elder Boulduc prior to Geoffroy's admission to the Académie; see chapter 2.) For example, after Homberg completed his analysis of common

---

87. Hellot Caen Notebooks, 3:24v–25r: "medecine universelle de Mr. Bourguignon, secret acheté en 1712 par Mr le Duc d'Orléans, exécuté au Palais royal par Mr Homberg"; and 1:87r: "de la main de Mr Homberg qu'ajoute que Mr le Regent en a fait faire plusieurs epreuves."

88. Hellot Arsenal Notebooks, MS 3008, 173, "Huile d'Emeril: Mss Homberg"; and 158–59, "Fixation d'argent en or par l'etain Msc Homberg . . . faites le depart pour voir quel est le profit de cette operation."

89. Hellot Arsenal Notebooks, MS 3006, 11–15, 24–26, 126, 136–38. On graduating solvents, see Principe, *Secrets of Alchemy*, 113.

sulphur, Geoffroy carried out its synthesis from the components Homberg had identified, and presented his own work as continuous with Homberg's.[90] In the course of this endeavor, Geoffroy obtained results that appeared to indicate the production of iron de novo from nonferrous ingredients—clay and linseed oil—in short, the synthesis of a metal, a feat that would go a long way toward elucidating metallic composition, a topic that Homberg avidly pursued. Those present at the public assembly when Geoffroy announced these results immediately recognized their implications for chrysopoeia. One attendee reported that "the alchemists who found themselves at this assembly were somewhat consoled regarding their expenditures for charcoal, for upon learning that Mr. Geoffroy has made artificial iron, they did not despair of finding the means for making gold." The Abbé Bignon, in the remarks that he customarily made at public assemblies to summarize the academicians' presentations, pointed to the same ramifications but tinged them with the economic concerns about gold-making that led to its prohibition at the Académie. Geoffroy's "conjecture about the metals . . . can awaken the hopes regarding more precious metals that so greatly delight the hearts and minds of the majority of people," Bignon remarked, but he would not "dare to say whether such success was more to be feared than desired."[91]

In order to verify his initial results, Geoffroy tried to find an earthy substance perfectly free of iron. He thus discovered that all vegetable ashes contain iron, apparently produced once again de novo during the incineration of the plant. In pursuing these experiments, he became embroiled in what ended as a rather acrimonious dispute with Louis Lémery about the nature of metals and the proper theoretical foundations of chymistry. Bernard Joly insightfully observed that their dispute terminated with the young Lémery associating Geoffroy with Johann Joachim Becher (the source for the clay and oil experiment) while associating himself with the mechanistic principles of Cartesianism, in an attempt to create a division (and a self-serving one) between an "older" transmutational chymistry exemplified by Geoffroy and Becher, and a "newer" approach represented by Lémery *fils*.[92] To what ex-

90. Wilhelm Homberg, "Essay de l'analyse du souffre commun." *MARS* (1703): 31–40; Étienne-François Geoffroy, "Maniere de recomposer le Souffre commun par la réunion de ses principes," *MARS* (1704): 278–86.

91. *Nouvelles de la république de lettres* (January 1705): 91, and (May 1705): 505–6; see also *Journal de Trévoux* (1705): 168.

92. The dispute and its interpretation are lucidly set forth by Bernard Joly, "Quarrels between Etienne-François Geoffroy and Louis Lémery at the Académie Royale des Sciences in the Early Eighteenth Century: Mechanism and Alchemy," in *Chymists and Chymistry: Studies*

tent this dispute carried out by the two *chimistes associés* was a proxy battle based on smoldering disagreements between their respective pensionnaires, the pro-chrysopoeia Homberg and the anti-chrysopoeian Nicolas Lémery, remains open to conjecture.

Much of Geoffroy's work at this period involved iron, and such experiments fit neatly with other less well-known research he carried out with the metals. Homberg and Geoffroy appear to have divided the project of studying metallic composition: Homberg focused on the noble metals gold and silver, while Geoffroy worked on the base metals. Geoffroy was convinced that he had produced iron from its components, and Homberg believed he had produced gold from its components.[93] Geoffroy's results with iron, however, soon led him to ask fundamental questions about mercury's status as a chymical principle of the metals, since there was no obvious source of mercury in clay, oil, or plants, yet their combination still produced iron. Was mercury really a component of iron? Was it a chymical principle at all or just a peculiar metal of its own that happened to be fluid? If metals do contain mercury, are the mercuries of all metals the same or different? Geoffroy voiced these doubts orally to his fellow academicians at the end of 1705, but he deleted them before the paper was published in the *Mémoires*, perhaps because they were too open-ended, or perhaps because they conflicted with the chymical theory that Homberg was then putting forward.[94]

Geoffroy's questions, and some of the answers he gave to them, clearly reflect an influence from Becher (as the younger Lémery had noted) and possibly from Georg Ernst Stahl (1659–1734) as well. In fact, Geoffroy, well before Lémery's criticism, freely noted that his synthesis of iron was drawn from Becher's *Physica subterranea*—a book also of importance to Homberg. Equally significantly, Geoffroy's library catalogue indicates that the copy of

---

in the History of Alchemy and Early Modern Chemistry, ed. Lawrence M. Principe (Sagamore Beach, MA: Science History Publications, 2007), 203–14; see also Bernard Joly, "Le mécanisme et la chimie dans la nouvelle Académie royale des sciences: les débats entre Louis Lémery et Etienne-François Geoffroy," *Methodos* 8 (2008), http://methodos.revues.org/1403; Joly, "Etienne-François Geoffroy (1672–1731), a Chemist on the Frontiers." *Osiris* 29 (2014): 117–31, especially 123–27; and Joly, "À propos d'une querelle concernant la production artificielle du fer. Les divergences entre Nicolas Lémery et son fils Louis," *Revue de l'histoire de la pharmacie* 64 (2016): 375–84.

93. Homberg, "Suite de l'article trois," 268.

94. Étienne-François Geoffroy, "Problème de chimie: Trouver des cendres qui ne contiennent aucunes parcelles de fer," PV 24, fols. 393r–395r (16 December 1705), on 394v–395r; published with significant alterations in *MARS* (1705): 362–63.

Becher's *Physica subterranea* he owned was the 1703 edition, which contains also the *Specimen Beccherianum*, Stahl's commentary on the work.[95] Geoffroy would later reference the *Specimen* explicitly, and in 1713 he became the first chymist at the Académie to refer to Stahl by name.[96] The influence of these two German chymists becomes clearer in Geoffroy's papers of 1707 and 1709 that describe his experiments to determine the composition of iron, copper, lead, tin, and quicksilver. Neither publication cites mercury as a constituent of the metals, indicating that the skepticism he expressed tentatively in 1705 persisted in subsequent years.[97] These papers repeatedly assert that the "basis" of the metals is not mercury at all, but an "earth capable of vitrification [*terre susceptible de vitrification*]," an obvious reference to Becher's vitrifiable earth [*terra vitrescibilis*], which the German chymist similarly calls the "basis" of metals.[98] According to Geoffroy, the metals are produced when such earth combines with "the principle of inflammability [*principe d'inflammabilité*]," which seems the same as a sulphur principle. When strongly heated, whether by ordinary fire or the "fire of the sun" from the Tschirnhaus lens, the principle of inflammability is expelled from ignoble metals, leaving the vitrifiable earth behind. This residual earth, Geoffroy continues, can then be heated with any inflammable substance, such as charcoal or oil, in order to restore its lost inflammable principle, whereupon the original metal is regenerated. This regeneration indicates that the inflammable principle is the same for all metallic, mineral, plant, and animal substances, although the earths that form the basis of the metals are unique to each one. Quicksilver, for example, has a "red earth" as its basis; this red earth appears when mercury is heated for a long time in order to drive off its volatile principle of inflammability.[99]

95. Georg Ernst Stahl, *Specimina Beccherianum* (Leipzig, 1703); *Catalogus librorum Stephani-Francisci Geoffroy* (Paris, 1731), 91, item 1325.

96. Kevin (Ku-Ming) Chang, "Communications of Chemical Knowledge: Georg Ernst Stahl and the Chemists at the French Academy of Sciences in the First Half of the Eighteenth Century," *Osiris* 29 (2014): 135–57 (see table on 142); the reference to the *Specimen* is Étienne-François Geoffroy, "Eclaircissements sur la Table inserée dans les *Memoires* de 1718," *MARS* (1720): 20–34, at 26.

97. Étienne-François Geoffroy, "Eclaircissemens sur la production artificielle du Fer, & sur la composition des autres Métaux," *MARS* (1707): 176–88; and "Experiences sur les metaux faites avec le Verre ardent du Palais Royal," *MARS* (1709): 162–76.

98. Becher, *Physica subterranea*, 119–25. On Becher's principles, see Antonio Clericuzio, *Elements, Principles, and Corpuscles: A Study of Atomism and Chemistry in the Seventeenth Century* (Dordrecht: Springer, 2000), 194–96; and J. R. Partington, *A History of Chemistry*, 4 vols. (London: Macmillan, 1961), 2:643–48. For a mercifully brief summary in Becher's own words, see his *Alphabetum minerale* in *Tripus hermeticus fatidicus* (Frankfurt, 1689), 105–7.

99. Geoffroy, "Experiences sur les metaux," 174–75.

Geoffroy's remarks are noteworthy and rather surprising. In the first place, they conflict with Homberg's system. Geoffroy acknowledges as much when he remarks that his new theory of metal composition "appears very much opposed to the idea that has been formed up to this point concerning the formation of metals whose base is thought to be mercury."[100] Second, Geoffroy's "principle of inflammability," although it might initially *seem* to be a version of the sulphur principle, is not Homberg's matter of light/sulphur; it is instead essentially *identical* to Stahl's famous phlogiston. The difference between Homberg's sulphur principle and Stahl's phlogiston (and Geoffroy's inflammable principle) needs to be stressed. Geoffroy's inflammable principle, like Stahl's phlogiston, is *expelled* upon heating or calcination, whereupon the metal collapses into a dusty ash or calx. It is restored to the calx by heating it with any flammable substance—whether from the animal, vegetable, or mineral realm—thus regenerating the original metal. Strikingly, Geoffroy is seemingly unaware of (or at least unconcerned by) the "weight problem"— namely, that metal calxes weigh *more* than the original metal and not *less*, indicating that something has been *absorbed* rather than *expelled* during heating. Homberg, always obsessive about carefully measuring weights, used his measurement of the increased weight of calxes as evidence that his light/sulphur was being incorporated into the metal. It is impossible that Geoffroy was unaware of this obvious fact so crucial to his mentor, yet still he asserts that the red earth (which we now know to be mercuric oxide) obtained by heating quicksilver is a *component* of the quicksilver, the leftover residue after its principle of inflammability has been expelled. This is exactly the *opposite* of Homberg's view, which explained the transformation of liquid mercury into a red powder as the *incorporation* of light/sulphur with the mercury—as verified by an increase in its weight. Even Geoffroy's vocabulary implies that he rejected the central feature of Homberg's chymistry of light, that is, light's material character as the sulphur principle and its ability to incorporate with ordinary matter and increase its weight. Unlike Homberg, who refers to the "matter of light" offered by the lens, Geoffroy cites simply the "fire of the sun [*feu du soleil*]," which for him acts merely as an instrument of division, merely a stronger fire, not as a chymical reagent. Finally, Geoffroy's contention that each metal is composed of a unique vitrifiable earth means that each metal requires one particular earth for its composition. This demand would seem to rule out metallic transmutation, since the earth found in mercury, for example, could not serve as the earth needed to make gold or silver.

Extant sources do not report what Homberg thought of this apparent

100. Geoffroy, "Eclaircissemens sur la la production du fer," 188.

turn away from his ideas and toward Becherian/Stahlian ideas by his student and collaborator. It is clear, however, that Homberg must have been aware of Geoffroy's thinking—and the experimental results upon which it was based—as it developed. Geoffroy performed his experiments using the Tschirnhaus lens at the Palais Royal, and so it is impossible that Homberg was not party to them. In fact, he probably collaborated on them to one degree or another, given his greater expertise with operating the lens and his guardianship of the costly instrument. It might be significant that Fontenelle later referred to Geoffroy's 1709 experiments on the base metals as *Homberg's*, perhaps he knew something (otherwise unrecorded) about Homberg's role in them.[101] Weight problem aside, Geoffroy's results showing the interchangeability of metallic, vegetable, and animal sulphurs actually served to demonstrate experimentally the claims made by Homberg in 1706 (and upon which he expanded in 1710) regarding how his sulphur principle passes readily from combination in one body into new combinations in others. It is therefore plausible that Geoffroy's work was initially intended to provide experimental proofs of Homberg's theory of the circulation of sulphur through various combinations. Alternatively, they may represent the exploration, in concert with Homberg, of another and potentially rival chymical theory of composition. Homberg had done the same previously with his extensive testing— and eventual rejection—of both Helmontian water theory and acid-alkali theory.

However the origins of Geoffroy's thinking and experiments may be resolved, in the end his ideas turned out to be partly conformable to Homberg's chymical theory, and in fact led to a further refinement of them. In the same year (1709) that Geoffroy seemed to deny metallic mercuries, Homberg presented the mercury chapter for his "Essais." There he described his "destruction" of mercury. To do so, he first "introduced a sufficient quantity of the matter of light into its substance" by means of a "very long operation," such that the mercury "became a metal"—clearly in reference to his synthesis of gold by the long-term heating of philosophical mercury. He then turned to analysis using the Tschirnhaus lens. When gold or silver (that is, mercury deeply penetrated by a greater or lesser quantity of light/sulphur) is exposed to the lens, the hail of pure and enormously agitated light/sulphur can drive out even the deeply combined sulphur "as one nail drives out another." The slower, less concentrated light in ordinary fire could not, of course, perform this feat. When treated with light from the lens, however, "nearly the whole

___

101. *HARS* (1710): 46; the reference to Homberg remains constant even in later editions of the *Histoire*.

substance [of the metal] vanishes in smoke, and only an earthy and light powder remains if silver was exposed to the burning lens, or a little earth which in the end becomes an earthy and friable material if gold has been exposed."[102] Homberg identifies these residues as the "ruined parts of the mercury." The matter of light in the fire first gouged the smooth spheres of mercury and lodged in the fissures it created, thereby converting the mercury into gold. The more severe assault of the matter of light from the burning lens then drove out those deeply lodged particles, leaving void spaces behind, and thus the microparticles of mercury remained like sponges or pumice stones, which appear to the eye like an earth.[103]

Here Homberg silently reinterprets his first experiments with the lens, incorporating a part of Geoffroy's more recent results. In 1702, Homberg attributed the transformation of gold into a glassy substance (here described in 1709 as an "earthy and friable material") to the combination of gold's residual sulphur with its earth, based upon the notion that gold is composed of mercury, sulphur, and earth. That claim contradicted the traditional idea that the noble metals, gold and silver, were composed of pure mercury and sulphur, without any earthy or other admixture. In 1702, this dissonance did not seem to bother Homberg. After all, he, like Tschirnhaus, had seen a glass produced from gold, and the formation of a glass requires an earthy component; therefore, gold *must* contain earth, regardless of more traditional claims about gold's composition. His subsequent synthesis of gold from mercury and light/sulphur alone, reported in 1705, might have troubled him a bit, especially when he asserted that gold and silver are produced from mercury and sulphur alone, while only the base metals have more complex compositions that could include earth.[104] Accordingly, in 1707, when redescribing the vitrification of gold by the lens, Homberg stated that the earth needed to produce it was not an essential ingredient of the metal, but only an adventitious admixture: "in all compound substances . . . one always finds a certain portion of an earthy material during their analyses." Thus, there was sufficient earthy material even in the perfect metals to explain the appearance of the glass.[105] By 1709, however, Homberg had resolved the problem coherently and experimentally: mercury and metallic earth are really the *same thing*. Depend-

102. Wilhelm Homberg, "Suite des essais de chimie: Art. IV. du Mercure," 107. Compare Homberg, "Suite de l'article trois," (1706): 264.

103. Homberg, "Du Mercure," 107–9.

104. Homberg, "Suite de l'article trois," 267–68. Homberg does not fail to draw the implication that gold and silver should therefore be easier to make than the base metals.

105. Wilhelm Homberg, "Eclaircissements touchant la vitrification de l'or au verre ardent," *MARS* (1707): 40–48, at 43.

ing on the condition of its particles—smooth microspheres, gouged spheres filled with light/sulphur, or the same gouged spheres devoid of light/sulphur particles—mercury can be a liquid metal, a component of solid metals, or an earth.[106] Mercury thus remained for Homberg the constituent base of the metals, and Geoffroy could claim simultaneously and without contradicting Homberg that the basis of the metals is earth. The difference depended merely on whether one looked at the composition of metals from an analytic or from a synthetic point of view. Mercury does go into the production of a noble metal synthetically, but the mechanical transformation its microparticles undergo in the process means that it cannot be recovered analytically from the metal as liquid mercury but only as a solid earth. Even Geoffroy's claim that there is a characteristic earth for each metal can fit with this revised theory of composition. In gold and silver the mercury particles are gouged in different ways; therefore, the lens reveals the differently damaged mercury as two different earths. In the base metals, given their more complex composition, one would expect the residues after treatment with the lens to differ from one another. Yet in every case, the mercury from which the various metals were synthesized started out the same and can rightly be called their "basis." Thus, *both* Homberg's insistence on mercury as common basis of all metals *and* Geoffroy's observation that different metals leave behind different earths can be accounted for simultaneously. The only remaining dissonance is the fact that with the base metals—unlike with gold and silver—the residual earth weighs more than the starting metal, not less.

Soon after Geoffroy's presentation argued for the universality of the sulphur principle, Homberg added a further supplement to his "Essais" that underscored and extended his own "economy of sulphur." At the same time Homberg also very specifically and explicitly revisited and reasserted the "weight problem," no doubt as a follow-up to the implications of Geoffroy's ideas. This 1710 supplement—not previously recognized as part of the "Essais" but woven seamlessly into the final *Élémens*—addressed at length why calxes weigh more than the metals from which they are prepared and how inflammable substances are able to turn calxes back into metals. Expanding upon what he had already proposed in 1706, Homberg writes that the light/sulphur principle can incorporate with other principles to produce different kinds of sulphurs that exist as particles of different sizes and densities. Incorporated solely with mercurial particles, as in gold and silver, the sulphur/light particles remain tiny and dense—true "metallic sulphur," in Homberg's terminology. Because of their small size and tight connection to the mercury

106. Homberg, "Du Mercure," 109.

particles, only the purest and fastest moving particles of light from the lens can dislodge them and thereby destroy the composition of gold and silver. The coarser light/sulphur particles in ordinary fire cannot do this, and hence gold and silver remain unharmed by fire alone. In base metals, which have a more heterogeneous composition, their light/sulphur is not incorporated directly with their mercury but instead with an earthy or watery vehicle, thus producing an oily material (a "bituminous sulphur") whose particles are large and less dense. Both Geoffroy and Homberg report having observed this oily substance emerging from iron subjected to the Tschirnhaus lens. A base metal's bituminous sulphur, because of its larger size and lesser density, can be driven off even by the particles of light/sulphur in fire (that is, the lens is not required). The smaller, denser light/sulphur particles from the fire can then more completely fill the voids left by the departure of the coarser bituminous sulphur, and thus the resultant calx weighs more than the original metal. When heated with any inflammable substance—that is, a substance containing light/sulphur in an earthy or watery vehicle—the calx reacquires the type of coarse-grained sulphur it originally had, expelling the finer sulphur absorbed from the fire, thus turning back into a metal and resuming its former weight.[107] Homberg then takes a further illustrative step. He describes an experiment he designed in order not only to separate but also to isolate and collect the bituminous sulphur from iron, tin, and zinc. He claims to have successfully isolated it as an inflammable oil, thus directly demonstrating the presence of the sulphur principle in the form of an oily material in the base metals.[108]

The details of exactly how influence flowed between Geoffroy and Homberg, on the one hand, and Stahl, on the other, cannot be thoroughly explored here. For the present, it is clear that Geoffroy was a careful (and early) reader of Stahl, and equally clear that the ideas he expressed in 1707 and 1709

107. Wilhelm Homberg, "Observations sur les matieres sulphureuses & sur la facilité de les changer d'une espece de souffre en une autre," *MARS* (1710): 225–34; compare Homberg, "Suite de l'article trois," (1706): 265–66.

108. Homberg, "Observations sur les matieres sulphureuses & sur la facilité de les changer d'une espece de souffre en une autre," 230–34. In this experiment, Homberg vigorously heated an iron-tin alloy in a crucible, generating a fluffy (*cottoneuse*) white sublimate. He then digested this cottony material with vinegar, filtered the solution, and distilled it to provide, after the removal of the phlegm, an inflammable oil. He then did the same with zinc and obtained similar results. Although the thermal decomposition of metallic acetates can produce a volatile and an inflammable liquid, it is not clear what Homberg could have obtained when he repeated the process using dilute sulphuric acid in place of vinegar. My own replication of the process on zinc using vinegar did produce a reddish, oily, inflammable distillate, as Homberg described.

were influenced by the same chymical theory of Becher that also influenced Stahl. Geoffroy's explanation of how his "principle of inflammability" works in the calcination and reduction of metals, which shares so many key features with Stahl's phlogiston, nevertheless appears to have preceded by nearly a decade Stahl's fully developed system, which appeared only in 1718 in Stahl's *Zufällige Gedanken*. Geoffroy's doubts about the mercury principle voiced in 1705 also reappear in the same 1718 publication by Stahl.[109] Homberg's "economy of sulphur"—presented first in 1705–6 and repeated at greater length in 1710—describing how the sulphur principle circulates readily (in most cases) between the mineral, animal, and vegetable realms, successively incorporated and liberated from various chymical combinations—is so strikingly similar to the analogous circulation of phlogiston that Stahl described in 1718 that it is difficult to imagine that they were arrived at independently. Stahl was certainly familiar with Homberg's work—he spends many pages of the *Zufällige Gedanken* examining Homberg's analysis of sulphur—and he certainly read the Académie's annual *Mémoires*.[110] Homberg's sulphur principle would therefore appear to be a source for Stahl's formulation of his inflammable *Grund-Wesen*, better known as phlogiston. The key difference, however, is Stahl's dismissal of the weight problem that was crucial to Homberg's experiments and formulations.[111] Working out the exact direction, timing, and nature of such transfers of ideas will, however, require further work, which is beyond the scope of this study. Such work would be highly enlightening for the history of chemistry generally, especially in terms of further clarifying the intellectual contacts and exchanges between Paris and German lands.[112]

109. On the development of Stahl's ideas about phlogiston, plus valuable comments on the chymical principles, see Kevin (Ku-Ming) Chang, "Phlogiston and Chemical Principles: The Development and Formulation of Georg Ernst Stahl's Principle of Inflammability," in *Bridging Traditions: Alchemy, Chemistry and Paracelsian Practices in the Early Modern Era: Essays in Honor of Allen G. Debus*, ed. Karen Hunger Parshall, Michael T. Walton, and Bruce T. Moran (Kirksville, MO: Truman State University Press, 2015), 101–30. For early developments of the idea, see Kevin (Ku-Ming) Chang, "Fermentation, Phlogiston and Matter Theory: Chemistry and Natural Philosophy in Georg Ernst Stahl's *Zymotechnia fundamentalis*," *Early Science and Medicine* 7 (2002): 53–57. Georg Ernst Stahl, *Zufällige Gedanken und nützliche Bedencken über den Streit von dem so genannten Sulphure* (Halle, 1718).

110. Chang, "Phlogiston," especially 113–14, 118–19; Stahl's comments on Homberg and sulphur are found in *Zufällige Bedancken*, 358–69.

111. In the 1703 *Specimina* (137) Stahl allows that calxes weigh more than the metals from which they were produced, but denies that the metal can be recovered in its original weight—instead there is always considerable loss in the form of glassy scoria. He therefore dismisses the significance of such weight determinations.

112. On this topic, see Chang, "Communications of Chemical Knowledge."

It is worth pondering whether Homberg's German origins and the remarkable internationalism he developed in his youth played a role in facilitating or promoting such exchanges, perhaps through correspondence that has not survived.

The apparent disagreement between Geoffroy and Homberg shows again how new experimental results provoked refinements to Homberg's comprehensive theory. In short, an alternative chymical theory, in this case deriving from Becher and possibly Stahl, was examined experimentally. Its unsatisfactory parts (like the weight problem) were rejected, while the useful results and observations were incorporated into extensions or revisions of Homberg's theory. As for Geoffroy's contrary conclusions, he eventually repudiated them—seemingly on the basis of the weight problem—and readopted, with slight adaptation, Homberg's principles. Geoffroy later asserted that the calcination of lead causes the expulsion of an "oily principle [*principe huileux*]," rather than a "principle of inflammability," a term more in line with Homberg's oily bituminous sulphur. Again in harmony with Homberg's principles, Geoffroy then went on to say that the voids left behind by the departure of this *principe huileux* become filled with the element of fire (*parties d'élément de feu*), which cause the weight of the calx to increase.[113]

There is yet more to say about Geoffroy, metallic mercuries, and transmutation. Hellot's notes suggest that Geoffroy engaged in a particular study of metallic mercuries, perhaps as a means for resolving his doubts about the mercury principle, although since the entries are not dated, we cannot say whether these experiments occurred before or after 1709. Homberg's first "Essai" (as presented in 1702) promised the method of isolating mercury from the metals using "resuscitating salts"—a method and term that derives (again) from Becher's *Physica subterranea*. While the published version that appeared in 1704 continued to list this method as an option, Homberg de-

---

113. Geoffroy, "Eclaircissements sur la Table," 27. Geoffroy did not have another paper published in the *Mémoires* until 1717, although he did deliver some oral presentations recorded in the procès-verbaux; this sudden drop-off of his productivity is probably the result of his being named professor of medicine at the Collège Royal in 1709. The portrayal here of Homberg and Geoffroy as collaborators usually (but not universally) in agreement contrasts with their depiction by Mi Gyung Kim, who prefers to place them into separate camps—Homberg as a "natural philosopher" and Geoffroy as an "apothecary." While it is true that Homberg had broader natural philosophical interests (as emphasized also by the present study), the distancing of Geoffroy from such interests and from Homberg is not borne out by the sources and is contradicted by the long and close relationship between the two and the evidence of their collaboration presented here. See Kim, *Affinity, That Elusive Dream* (Cambridge, MA: MIT Press, 2003), 96–103 (note that "Geoffroy's confident statement" on 97 is actually Fontenelle's), and recapitulated in Kim, "The 'Instrumental' Reality of Phlogiston," *Hyle* 14 (2008): 27–51.

leted his promise to teach it, and focused instead on his newfound method of analysis using the burning lens.[114] Hellot's transcriptions indicate that Geoffroy worked on the topic of resuscitating salts, perhaps initially to help Homberg fulfill his original promise, or perhaps as a "spin-off" project while Homberg focused on the lens experiments. In one excerpt, Geoffroy summarizes that "the strongest resuscitating salts are potash, willow ashes, salt of tartar, salt of vine twigs, ash of bracken," and records his work to isolate the mercury of lead using plant ashes. Possibly to help explain how he could have made iron seemingly without mercury just by incinerating plants, Geoffroy adds "NB: the mercury of plants is also extracted with great labor, in particular from hay that is still green."[115] An entry marked "Mr. Geoffroy's problem" (reminiscent of the title for his presentation on finding iron-free plant ashes) asks, "can't one arrive at the mercurification of the metals" using a generalized process employing salts.[116] Elsewhere "a memoir in the elder Geoffroy's hand" describes his combination of two salts in an attempt to produce a "mercurifying spirit" that can liberate the mercurial component from metals. Hellot also transcribes a procedure he found among Geoffroy's papers for isolating the mercury of antimony, again using resuscitating salts and written in Homberg's autograph hand—perhaps a procedure given to the younger chymist to try out. Geoffroy also described the use of potash (*potasse*), a salt he listed among the best resuscitating salts, to extract mercury from antimony ore.[117]

Despite any doubts about transmutation that might have for a time stemmed from the notion that each metal required a specific earth for its composition, Geoffroy worked on a number of expressly transmutational endeavors. Like Homberg, and certainly in concert with him, Geoffroy expended efforts on the Philalethes process. Working from a manuscript "in the hand of Mr. Geoffroy the physician," Hellot transcribed a method for making philosophical mercury according to Homberg's original method and turning it into a red powder that "yields a little gold" and that can be brought onward to "The Great Work."[118] This mercury is used in the immediately following entry for making "oil of gold," which Hellot cites as "drawn from

114. Compare Homberg, "Essais d'Elemens de Chimie," PV 21, fol. 63v (15 February 1702); and Homberg, "Essays de chimie" (1702), 35.

115. Hellot Caen Notebooks, 3:270r: "Les sels ressuscitatifs le plus forts sont la Potasse, les Cendres de Saule, le sel de tartre, le sel de sarmens, la cendre de fougère"; "N.B. on tire aussi du mercure des Plantes avec un grand travail, et particulierment du foin qui est encore verd."

116. Hellot Caen Notebooks, 3:270r: "ne pourroit on pas parvenir à la mercurification des métaux, en prenant. . . ."

117. Hellot Arsenal Notebooks, MS 3008, 185–87.

118. Hellot Caen Notebooks, 3:146v–147r.

the manuscripts of the Palais Royal, article dated 'done in 1707,'" showing that the younger chymist was at work on such processes in the palace's laboratory alongside Homberg. Geoffroy also prepared an "animated mercury" (an alternate term for philosophical mercury) by yet another modification of the Philalethes method, using bismuth in place of antimony, and which eventually produced a "solar tincture."[119] He noted further that he had discovered that a few drops of oil of vitriol would allow mercury and antimony regulus to amalgamate—presumably a way of getting around using either alloys with silver or copper, or Homberg's perilous method of pouring molten regulus into boiling mercury.[120] Geoffroy later recorded an augmentation of gold using iron, and in 1730, just a year before his death, gave Hellot a recipe for preparing a particular for producing gold. A process for making another particular that converts copper into silver is marked as "proven true [*eprouvé veritable*]." In 1722, he wrote about the "mercurial seed" in lead, continuing at length about how the starting materials for the philosophers' stone are to be found in lead, specifically in unrefined lead ore; Hellot summarized this section with the marginal note "lead is the material for the philosophers' stone, Geoffroy the elder."[121] Geoffroy's original notes even recorded his attempts to decipher the allegorical language of chrysopoetic texts in order to uncover the hidden means of preparing the philosophers' stone.[122] He likewise made translations of entire chrysopoetic treatises (a less celebrated endeavor than his work to translate parts of Isaac Newton's *Opticks*), such as *Alchymia denudata*.[123]

Some of this transmutational work dates prior to Geoffroy's 1709 thoughts about metallic composition and their potentially (but not necessarily) negative consequences for transmutation. But some entries are clearly dated *after* 1709, indicating that Geoffroy either considered that his remarks about metals did not rule out metallic transmutation (Becher himself was in the

119. Hellot Arsenal Notebooks, MS 3008: 53–54: "Sublimé rouge d'or par le mercure animé par le Bismuth, pap[iers] de feu Mr. Geoffroy."

120. Hellot Arsenal Notebooks, MS 3008: 187: "Paucae guttulae olei vitrioli regulo antimonii, facillimè mercurium associant teste experientia. Note MS de feu Mr. Geoffroy."

121. Hellot Caen Notebooks, 3:276v–278v: "le plomb est la matière de la P. Pphale. G[eo]ff[ro]y L['aine]."

122. Hellot Caen Notebooks, 2:97–98; 3: 270r–271r, "axiomes abregés des sages"; 3:276v–278r; Hellot Arsenal Notebooks, MS 3006, 126–27; I incorrectly stated in an earlier publication (in *Osiris*) that this note read "épreuve veritable."

123. This manuscript is listed in *Catalogue des livres de feu M. Hellot* (Paris, 1766), 101, no. 1805. The book was first published in 1708, and reissued in extended form in 1716, 1723, and 1728. Unfortunately, since the MS is now lost, we cannot determine which edition Geoffroy translated.

same state, having proposed that each metal held a unique *terra vitrescibilis,* yet still supporting chrysopoeia), or that he later rejected the negative consequences and did not give up belief in its possibility. More strikingly, some of this transmutational material is dated to the last decade of Geoffroy's life, specifically, contemporaneous with or after he presented his celebrated 1722 paper "Some Cheats concerning the Philosophers' Stone," which has often been taken as an indicator of the "end of alchemy."[124] Chapter 7 reexamines this famous paper and its context.

This closer study of what Geoffroy was doing as an associé, much of it recorded only in Hellot's notebooks, gives a fuller idea of what Homberg himself was doing and reveals more about their collaborative work. While much of Geoffroy's work was seemingly directed by Homberg to support his own research program, Geoffroy was an independent thinker whose interpretation of his experimental results sometimes contradicted those offered by Homberg's own, constantly evolving theory of chymical composition. Geoffroy also apparently looked for ideas far beyond the walls of the Palais Royal and the Académie to consider experiments and concepts described by Becher and adopted (with modifications) by Stahl. In the end, further experimentation was able to reunite the ideas and theorizations of Geoffroy and Homberg in regard to the problems of metallic composition. It nevertheless remains tantalizingly unclear how concepts of an inflammable principle expelled during burning and calcination, for a time supported by Geoffroy and later a central principle for Stahl, may have moved between Paris and Halle. It is clear, however, that in Paris, it was weight determinations that soon swung the balance in favor of Homberg's explanation of the formation and reduction of metallic calxes as exchanges of particles of the light/sulphur principle of differing size and density.

This chapter has demonstrated that the leading chymists of the Académie during its first fifty years—Duclos, Homberg, and Geoffroy—were all involved in the search for metallic transmutation. The institution's administrators, in contrast—Colbert, Louvois, Pontchartrain, and Louis XIV himself—forbade such investigations on economic and political grounds. The two Lémerys also criticized transmutational chymistry. There thus existed a tension within the Académie regarding chrysopoeia—forbidden and

---

124. Étienne-François Geoffroy, "Des supercheries concernant la pierre philosophale," *MARS* (1722): 61–70. For more on the interpretation and context for this publication, see chapter 7; also Lawrence M. Principe, "The End of Alchemy? The Repudiation and Persistence of Chrysopoeia at the Académie Royale des Sciences in the Eighteenth Century," *Osiris* 29 (2014): 96–116.

criticized by one group, yet pursued and promoted by the other. The Académie's most prominent chymist, Homberg, not only claimed success in metallic transmutation but also made it the centerpiece of the new theory of chymistry expounded in his *Élémens de chimie*. The academicians were well aware of Homberg's chrysopoetic endeavors, but such work never appeared explicitly as such in the Académie's publications. Perhaps due to the administrative prohibitions, it was left to readers to recognize the allusions Homberg made to traditional chrysopoetic processes, and it is impossible that these went unnoticed. Homberg worked assiduously on Philalethean processes for many years, as well as on a host of others, hoping to produce not a minute quantity of gold from mercury, but rather the philosophers' stone, capable of transmuting vast quantities of base metals into gold. Significantly, the work he directed toward metallic transmutation was not a side project independent of his other chymical endeavors, but was instead a key route toward revealing and understanding the hidden composition of metals, a question central to his ongoing desire to create a comprehensive explanatory theory of chymistry based on a true understanding of the chymical principles.

The Duc d'Orléans supported Homberg's transmutational work, not only by building, equipping, and financing a magnificent laboratory for him, but also by acquiring processes and working side by side with Homberg at the furnaces. The laboratory even had a secret room where Philippe and Homberg reportedly converted mercury into gold before witnesses using a transmuting agent they had prepared collaboratively. There remains much more to say about Homberg's work and life at the Palais Royal, and about what daily life in the laboratory was like during its heyday. The next chapter completes Homberg's biography, beginning with his unusual marriage in 1708, proceeding through a perilous crisis in 1712 and its lasting impact upon his research, and finally to his long illness and death in 1715. It also examines further the crucial role of hands-on practical experience in chymical matters, and the way that such knowledge could (and could not) be communicated between individuals. In terms of such transfer of knowledge, it also reveals the surprising chrysopoetic and other contacts that Homberg and Philippe cultivated outside the walls of the palace, both in Paris and farther afield, and how Louis XIV and his ministers began cautiously reconsidering their earlier prohibitions on transmutational endeavors in the desperate hope that chrysopoeia might prove the solution to France's increasingly dire financial situation.

6

Chymistry in Homberg's Later Years:
Practices, Promises, Poisons, and Prisons

By the middle of the first decade of the eighteenth century, Homberg had achieved the most enviable position imaginable for a chymist of his day. He had achieved high social status as the foremost *pensionnaire chimiste* of the Académie Royale des Sciences and *premier médecin* to Philippe II, Duc d'Orléans, and had become a favored person within the Orléans household and the royal court. His work and ideas, efficiently disseminated by numerous papers in the Académie's *Mémoires*, were being followed and discussed across Europe. He was living in the Palais Royal, in substantial apartments connected to the "most magnificent laboratory chymistry had ever had." The Duc d'Orléans, who lived in the adjoining apartments, financed and encouraged Homberg's research and worked regularly alongside him in the laboratory. Philippe helped collect chymical information and recipes for the two to pursue, even when he was off fighting his uncle's wars. The innumerable experiments carried out in the Palais Royal laboratory stretched the whole gamut of contemporaneous chymistry, ranging from refining techniques and chymical medicine to the search for the secrets of metallic transmutation and the philosophers' stone. At the same time, Homberg continued refining and extending his comprehensive new theory of chymistry, constantly writing and revising his *Élémens de chimie* in response to new experiments and observations.

This chapter covers the latter years of Homberg's life, from about 1708 until his death in 1715. Besides recounting significant events and discoveries that date from the period, this chapter also explores further what the Palais laboratory was like, not as a physical space, but as a locale of intellectual, social, and material exchange. Who visited Homberg there, and for what purposes?

The unique challenges of transmitting practical chymical knowledge often require personal visits to chymical workspaces in order fully to understand practical procedures. Such personal contacts prove especially important in chymistry. Homberg had demonstrated that fact early on when he amassed his store of chymical secrets and knowhow through years of travel and direct engagement with practitioners. Those early contacts covered the whole social and intellectual spectrum, from Leibniz, Kunckel, and Boyle to unnamed refiners, miners, and manufacturers. Now at the height of his career, Homberg continued to build and maintain contacts with a similar range of figures. Perhaps most unexpected are the contacts he, Philippe, and on occasion even the Académie itself had with characters sufficiently questionable that we know about them mostly because they ended up in the Bastille. These arrests point toward the deeper interest the French state began to show toward such chymical practitioners in the early eighteenth century, a previously largely unnoticed turn of events that, I argue, has particular relevance for the transmutations of chymistry in the period, especially the increased social opprobrium directed against chrysopoeia.

But first, in order to provide a full biography of Homberg, a few words have to be said about his personal life, namely, his late and slightly odd marriage. This episode throws further light on Homberg's personality and his relationships with colleagues both in the Académie and in the Orléans household.

Homberg's Marriage

In 1708, at the age of fifty-five, Homberg decided to marry. His bride was Marguerite-Angélique Dodart (1667–1751), who had herself reached the age of forty. In light of the advanced ages of both, the marriage is hardly likely to have been an attempt to start a family or to produce heirs. Indeed, given that Homberg was ready to take holy orders in 1687, marriage and procreation could scarcely have been of much importance to him. Thus, why he married at this late date is open to question. I suggest that the marriage was a gesture of generosity on Homberg's part to the Dodart family, whose members he had known for a long time. Marguerite-Angélique's father was Denis Dodart, Homberg's fellow academician. Thus, Homberg must have been more or less acquainted with Marguerite-Angélique for a long time, years before they actually married. The interactions between Homberg and Dodart began as soon as Homberg was admitted to the Académie in 1691; the Batavian chymist's first project, as described in chapter 2, was to assess the chymical analyses from the plants project that Dodart had been directing since the mid-1670s. It is difficult, however, to assess how close their relationship actually was.[1] Homberg may have been closer to Marguerite-Angélique's brother, Claude-Jean-Baptiste Dodart (1664–1730), who was Homberg's immediate predecessor as first physician to Philippe II, and so they presumably knew each other well and saw each other regularly at court.[2]

In early 1707, Denis Dodart fell seriously ill and, concerned about the welfare of his two children after his death, began putting his affairs in order.[3] When he died on 5 November 1707, he left behind a rather meager estate, a substantial debt, and an unmarried middle-aged daughter.[4] On the one hand, the financial and social position of his son, Claude-Jean-Baptiste, was secure. He had been appointed *premier médecin* to the Duc de Bretagne (Louis XIV's eldest great-grandson) in January 1707, and was granted the same position for the Duc de Bourgogne and the Duc de Berry in February 1708; he thus had a substantial income and a special place in the royal court.[5]

---

1. David Sturdy, *Science and Social Status* (Woodbridge: Boydell, 1995), 230–31, repeatedly suggests that Denis Dodart had significant influence upon Homberg, but without clear evidence. On p. 232, for example, he suggests that Homberg's 1696 investment in the Company for the Navigation of the Seine (see Homberg's Inventaire-après-décès, AN, Minutier Central LX-205, fol. 18r–v) was due to Dodart's influence, since he was also invested therein. But the Académie engineer Gilles Filleau des Billettes (1634–1720) and Pierre-Paul Gayot (see note 12, below) invested alongside Homberg in 1696, and there is clear documentation that Homberg and Gayot were extremely close at that time, and continued close for many years. These two represent at least equal if not more probable influences.

2. *L'État de la France*, 2 vols. (Paris, 1702), 2:121–22. The sources are unclear whether Philippe II initially hired Dodart, or if his father Philippe I did so and the son extended the appointment after his father's death in mid-1701 (see Gabriel Mareschal de Bièvre, *Georges Mareschal, Seigneur de Bièvre* [Paris: Librairie Plon, 1906], 389). Under what circumstances Homberg replaced Dodart in 1704 as *premier médecin* also remains unclear, as it was unusual for a *premier médecin* simply to give up his position.

3. Bernard de Fontenelle, "Éloge de M. Dodart," *HARS* (1707): 182–92; see Dodart's will, written 8 May 1707, and published in Étienne Charavay, *Revue des documents historiques* 4 (1877): 62–77; the documents drawn up around the same time pertaining to his children's inheritance are mentioned in Homberg's Inventaire-après-décès, AN, Minutier Central, LX-205, at fols. 18v–19v.

4. Dodart's will mentions recent losses of about 30,000 livres, twenty times his annual pension from the Académie; Charavay, *Documents historiques*, 63. As for other family members, Marguerite's paternal uncle, Toussaint Dodart, a lawyer, had killed himself with a pistol shot to the head in June 1699; see Elisabeth Charlotte to Sophie of Hannover, 2 July 1699, in *Aus den Briefen von Herzogin Elisabeth Charlotte von Orléans*, ed. Eduard Bodemann, 2 vols. (Hannover, 1891), 1:370–72, at 371; compare O/1/43, fol. 202v (5 July 1699).

5. AN, O/1/52, fol. 10v: "Retenue de Dodart premier médicin du duc de Bretagne" (9 January 1707); O/1/52, fol. 15r: "Dodart . . . premier médecin du duc de Bretagne, retenue de pre-

On the other hand, Marguerite-Angélique's position was far less secure. She had been living at the Hôtel de Conti with her father, who was physician to the Princesse de Conti.[6] As the elder Dodart's illness worsened, the Princesse de Conti promised to provide Marguerite-Angélique with a pension to supplement the low financial state [*modicité*] in which she would be left, but it is unclear if she could have remained living at the Hôtel. Her father left her a house on the rue Sainte-Croix-de-la-Bretonnerie near rue Bourg Tibourg, but her reliable income (from *rentes*) of only 1,000 livres per year would scarcely have sufficed to maintain and run it at a reasonable level.[7] Her poor financial situation was aggravated by a more serious issue that came to a head at just this time. Denis Dodart had long maintained close relations with the Abbey of Port-Royal-des-Champs, the notorious hotbed of Jansenism that so irritated Louis XIV and the Jesuits. He acted as physician to the community, received a stipend from them, and was known to be a follower of their insalubrious Calvinist-influenced theology. When Dodart's wife died in late 1669 or early 1670, the two-year-old Marguerite-Angélique became a pensionnaire of the abbey, meaning that she lived at and was raised and supported by the abbey. Not surprisingly, she thereafter followed her father's Jansenism. Her brother, in contrast, must have clearly repudiated such unorthodoxy, or he would never have obtained his position in the *maison du roi*. Indeed, Louis XIV wanted to expel Denis Dodart entirely from court because of his Jansenism, but protection by the Prince and Princesse de Conti, and Denis's very circumspect personal conduct, did not provide Louis with a suitable opportunity.[8] Concern over the orthodoxy of Port-Royal and of the Jansenists increased through the end of the seventeenth century and into the early eighteenth. In the same month that Denis died, a number of "defenders

---

mier médecin du duc de Bourgogne et du duc de Berry par le décès de Jean Poisson" (1 February 1708). He was reconfirmed in this position when the Duc de Bourgogne became Dauphin after the death of Monseigneur le Grand Dauphin in 1711 (see O/1/55, fols. 42r–43v, 4 May 1711), and given a pension after two of his royal charges died in 1712 (O/1/56, fol. 66r, 10 March 1712). He later became physician to Louis XV. See O/1/62, fol. 67v (6 April 1718); Louis de Rouvroy, Duc de Saint-Simon, *Mémoires*, ed. Yves Coirault, 8 vols. (Paris: Gallimard, 1983–88), 6:642–43.

6. Dodart was appointed physician to the Princesse de Conti in January 1680; see BNF, MS français 17051, fol. 181r, for a letter from Dodart dated 28 January 1680 mentioning his appointment "il y a 15 jours."

7. Fontenelle, "Éloge de Dodart," 190. The Princesse also reportedly promised to appoint Claude-Jean-Baptiste as her physician—although given his royal charges, an additional post would have been unnecessary financially, and perhaps impossible practically. Marguerite-Angélique's inheritance is listed in Charavay, *Documents historiques*, 64–65.

8. Saint-Simon, *Mémoires*, 6:643.

of Port-Royal" were rounded up and brought to the Bastille.[9] The tensions reached the breaking point on 27 March 1708, when Pope Clement XI issued a bull closing Port-Royal-des-Champs, suppressing its community, and transferring its goods and holdings.[10] Thus, in spring 1708 Marguerite-Angélique suddenly found herself without the protective father with whom she had been living, in a weak financial and unstable social position, and bereft of the religious community that had previously supported her and to which she may well have intended to return as a mature, unwed woman. It is, therefore, probably no coincidence that just three weeks after the papal bull closed the abbey, Homberg and Marguerite-Angélique were married on 21 April 1708. The marriage instantly gave Marguerite-Angélique financial and social stability, protection in the house of Orléans, and lodging at the Palais Royal.[11]

The marriage was witnessed by a remarkable constellation of court notables that included Elisabeth Charlotte, Philippe, his wife and children, the Prince and Princesse de Conti and their family, as well as Claude-Jean-Baptiste and his family, and Homberg's longtime friend Pierre-Paul Gayot.[12] Indeed, a document bearing so many notable signatures proved too great a temptation for a later autograph collector, and it was stolen from the archives. In 2005, I rediscovered the purloined document in the autograph collection of the Wellcome Library in London.[13] The contract stipulates that their goods

---

9. BNF, Arsenal MS 10582: "Jansenists défenseurs de Port-Royal" (November 1707).

10. *Lettres des religieuses de P. R. des Champs . . . touchant les Bulles de nôtre S. P. le Pape Clement XI du 27 mars 1708* (Paris, 1709). The nuns were forcibly removed in 1709, and the buildings razed in 1711 by order of Louis XIV.

11. Sturdy, *Science and Social Status*, 231–33, claims that Marguerite-Angélique was better off financially than Homberg, and thus that Homberg was the greater beneficiary from the marriage, but new evidence shows quite the opposite. Sturdy was not aware of the substantial salaries that Homberg received from Philippe (see chapter 4, note 14), had not seen their marriage contract, and did not note that Homberg himself paid off several of his wife's personal financial obligations, such as a staggering 1,947 livres he gave in 1710 to her *femme de chambre* Jacqueline Tiphanie (who had accompanied Marguerite-Angélique from childhood, and whom Denis Dodart had specifically entrusted to Marguerite-Angélique's care; see Charavay, *Documents historiques*, 66–67, 69) and whose annual salary, which was Marguerite-Angélique's responsibility to pay, was two years in arrears in 1715; see Homberg's Inventaire-après-décès, fols. 21v–22r.

12. Little is known about Homberg's personal friends, but Gayot was evidently one of Homberg's special and long-term friends. Des Billettes wrote that Gayot "finds himself happily fallen in with Mr. Homberg . . . thus both of them do everything they want together, and dream of nothing but amusing themselves"; Des Billettes to Leibniz, 23 February 1697, in Wilhelm Gottfried Leibniz, *Sämtliche Schriften und Briefe*, 1st ser. (Berlin: Akademie Verlag, 1987), 13:573–78, at 575.

13. Sturdy, *Science and Social Status*, 231n28, discovered that Homberg's marriage contract was missing from AN, Minutier Central L-240. There is, however, a registration version at AN, Y 281, fols. 142v–143r, and the recovered original is London, Wellcome Library, WMS/ALS,

and assets were not to be held communally. As would be expected from the fifteen-year difference in their ages, Homberg did not outlive Marguerite-Angélique. Upon Homberg's death in late September 1715, the widow wasted no time finding and marrying a second husband. Just seven months later, in April 1716, she married the infantry captain François de Burande, Sieur de Villeforge. Her new marriage contract contained the same terms as the former one with Homberg, and most of the document is devoted to a scrupulous enumeration of Marguerite-Angélique's possessions, most of them clearly inherited from Homberg—most notably a large portion of his very significant art collection.[14] After Homberg's death, the widow continued to live in the Palais Royal until her second marriage, although possibly not in Homberg's extensive apartments, at least part of which the Duchesse d'Orléans immediately reassigned upon Homberg's death as lodgings for members of her personal household.[15]

Chymical and Social Practices in the Laboratory: Visitors and Hands-On Experience

The laboratory at the Palais Royal was a busy place under Homberg's direction, not only in terms of chymical experimentation but also in terms of intellectual exchange and social interaction. Those at work interacted not only with their retorts, crucibles, and furnaces, but also with each other in a shared space. Homberg and the Duc d'Orléans worked together frequently on their chymical pursuits. They were assisted by Jean Grosse and perhaps other operators who performed various tasks and kept the facility operating. One of the honorary academicians, the Carmelite priest Jean Truchet (1657–1729), usually known by his clerical name Père Sébastien, also "worked assiduously on chymistry in the laboratory of Mr. Homberg," although we know nothing at present about what exactly he did there.[16] Geoffroy, of course, worked there

---

Box 21. It was purchased (Wellcome acquisition 68544) from the autograph dealer Charavay Frères in August 1935. Unfortunately, the inventories of Homberg's and Marguerite-Angélique's goods (*estat des biens*), needed for keeping them separate, which are mentioned (fol. 5v) as having been made, remain lost.

14. AN, Minutier Central LX-206; Mariage, 30 April 1716. My thanks to Justin Rivest for obtaining photos of this document for me.

15. AN, Minutier Central LX-206, fol. 1r (provides current residence of Marguerite-Angélique as the Palais Royal) and Homberg's Scellé après-décès, AN, Y//11647, fol. 3v: "la necessité il y a de vuider partie des lieux ou nous sommes dont Madame la duchesse d'orleans a disposé en faveur d'officiers de sa maison."

16. Bernard de Fontenelle, "Éloge du P. Sébastien Truchet," *HARS* (1729): 93–101, at 95.

extensively with Homberg, as did other academicians from time to time—for example, when Louis Lémery used the Tschirnhaus lens in connection with his own studies. The Duc d'Orléans invited friends and members of court to the laboratory, "to observe Homberg at work, and himself at work as well."[17] Their demonstration of transmutation in the laboratory's "secret place" before several of Philippe's associates at court provides one well-attested and remarkable example. Other members of the French state periodically also solicited work from Homberg's laboratory: the police asked him to analyze powders and liquids seized from suspicious persons, and finance ministers asked him to assay promising ores.[18] Visitors to court with scientific interests were put into Homberg's care, and surely were given tours of the laboratory, perhaps shown demonstrations, and possibly even invited to collaborate on experiments. Such was to be the case, for example, with Count Jörger, former first chamberlain to Emperor Joseph I and ambassadorial envoy from the imperial court, who had "extraordinary knowledge of all that part of natural philosophy that resolves substances by means of the fire," in other words, chymistry. When the Count, also a great admirer of the combinatorial art of Ramon Lull (and thus possibly also of the alchemical texts then attributed to Lull), planned to visit Paris, he was to be "put into the hands of our chymist Homberg."[19]

Several members of the court shared Homberg's and Philippe's interests in chymistry, including chrysopoeia, and they too may have visited the laboratory from time to time. Saint-Simon notes how Philippe "discussed it with everyone of the profession in the court and in the city," and reports how in the evenings Philippe and the Dauphin would often "set themselves in a corner to talk about the sciences, and no one could talk about them as precisely, intelligibly, and pleasantly as the Duc d'Orléans."[20] Specifically, in terms of chrysopoeia, one connection that has not been previously made is to Jean-Adrien Helvétius (1664–1727), one of Philippe's quarterly physicians

17. Saint-Simon, *Mémoires*, 4:456.

18. Pontchartrain to D'Argenson, 4 June 1710, in *Archives de la Bastille*, ed. François Ravaisson, 19 vols. (Paris, 1866–1904), 12:3 ("il faut les faire examiner par le sieur Homberg, qui semble à cela plus propre que tout autre"), and Homberg to D'Argenson, 21 July 1710, 12:4. A. M. Boislisle, ed., *Correspondance des contrôleurs généraux des finances*, 3 vols. (Paris, 1874–97), 2:294 (letter of Desmaretz, 27 December 1705).

19. Gottfried Wilhelm Leibniz to Nicolas-François Rémond, July 1714, in NLB, LBr 768, fols. 18r–19v, at 18v–19r: "il a une connoissance extraordinaire sur tout de cette partie de la physique, qui donne la resolution des corps par le feu"; and Rémond to Leibniz, 2 September 1714, fols. 23r–25v, at 24r: "je mettrai votre Comte aux mains avec nostre Chymiste Homberg."

20. Saint-Simon, *Mémoires*, 4:453, 456.

while Homberg was first physician.[21] Helvétius had the distinction of being
a son of Johann Friedrich Helvetius (1630–1709), physician to the Prince of
Orange (later William III of England) and author of one of the most celebrated eyewitness accounts of transmutation. Helvetius's 1667 *Vitulus aureus*
recounted how the physician, initially a skeptic in regard to transmutation,
was visited in his house in The Hague in late 1666 by an anonymous adept
who gave him a tiny fragment of the philosophers' stone. When the adept,
who had promised to return to demonstrate how to use the material, failed
to reappear, Helvetius and his wife melted some lead and threw the fragment of the stone upon the molten metal. In a few moments the base metal
was converted into gold. Helvetius's account provoked widespread interest,
notably from Robert Boyle and Benedict Spinoza.[22] It is hard to imagine that
Homberg and Philippe, having the younger Helvétius present as part of the
same household, did not interrogate him about his father's experience, and
perhaps may have been put in contact with the still-living elder Helvetius,
whom the younger visited regularly.

All these varied interactions represent only one part of the intellectual and
material exchanges and social contacts that took place in the laboratory of the
Palais Royal. Visitors not related either to the Académie or to the court also
visited the laboratory, and so it should not be thought of as a closed or especially privileged space—all the more reason for it to have a "secret room" for
special operations like transmutation. Given its position in the Palais Royal
and its patronage by the Duc d'Orléans, Homberg's laboratory was probably
an especially attractive destination for erudite visitors to Paris. Other chymical laboratories of the period were regular destinations for learned travelers,
and there is no reason to think that Homberg's was any different. Homberg
had himself visited many laboratories and workspaces during his formational
travels. Olaus Borrichius visited dozens of such sites during his own travels
and stay in Paris in the 1660s, and Martin Lister did the same while visiting
the city in 1698. By the end of the 1680s, Robert Boyle began receiving so
many visitors to his Pall Mall laboratory that he eventually had to post a sign
on the door restricting access to weekly "visiting hours."[23] A study of the visi-

21. On Helvétius, see Louis Lafond, *La dynastie des Helvétius: les remèdes du roi* (Paris,
1926); R. Krul, "Jean-Frédéric Helvétius et sa famille," *Janus* 1 (1896): 564–71; and Justin Rivest,
"Secret Remedies and the Medical Needs of the French State: The Career of Adrien Helvétius,
1662–1727," *Canadian Journal of History* 51 (2016): 473–99.

22. Johann Friedrich Helvetius, *Vitulus aureus* (Amsterdam, 1667); Principe, *Aspiring Adept*, 93–95; and M. Nierenstein, "Helvetius, Spinoza, and Transmutation," *Isis* 17 (1932): 408–11.

23. Michael Hunter, *Boyle: Between God and Science* (New Haven, CT: Yale University
Press, 2009), 237; and R. E. W. Maddison, "Studies in the Life of Robert Boyle, F.R.S.: Part I,

tors to the Palais Royal laboratory helps fill out a vivid picture of that busy place, as well as of the networks that Homberg and Philippe developed in terms of chymistry.

Because of the highly manual nature of chymistry, the written word often proved (and still proves) inadequate for the effective transfer of practical chymical knowhow, and it thus often needed to be supplemented with personal visits and hands-on experience.[24] Homberg's famous pyrophorus offers an illustrative example. In 1710, Homberg revisited his experiments of 1684 that first produced the spontaneously inflammable material, and streamlined the process for making it.[25] He first fried a mixture of alum and human excrement ("freshly made") in an iron skillet, stirring it constantly to keep it from sticking, until it became thoroughly dry. He then ground the material and heated the resultant powder in a long-necked flask for several hours. Still a showman and a lover of chymical exotica, Homberg began demonstrating the remarkable material to a variety of spectators. He would open a tightly sealed bottle, shake out a small quantity of powder onto a piece of folded paper, and within a few seconds the material would change color, begin to smoke, and then burst into flames. In late November 1710 he displayed it to the academicians at one of their meetings.[26] Soon thereafter, in December

---

Robert Boyle and His Foreign Visitors," *Notes and Records of the Royal Society* 9 (1951): 1–35. Olaus Borrichius, *Itinerarium 1660–1665*, ed. H. D. Schepelern, 4 vols. (Copenhagen: Danish Society of Language and Literature, 1983).and Martin Lister, *Journey to Paris in the Year 1698*, 3d ed. (London, 1699; reprint, ed. Raymond Phineas Stearnes, Urbana: University of Illinois, 1967).

24. This point touches on the well-worn issue of "tacit knowledge"; for a useful refinement of that concept as "unarticulated knowledge" and its several varieties, see Thijs Hagendijk, "Learning a Craft from Books: Historical Re-enactment of Functional Reading in Gold- and Silversmithing," *Nuncius* 33 (2018): 198–235. For more on the recovery of such knowledge through experimental reproduction of historical processes, see Hjalmar Fors, Lawrence M. Principe, and H. Otto Sibum, "From the Library to the Laboratory and Back Again: Experiment as a Tool for Historians of Science," *Ambix* 63 (2016): 85–97.

25. Homberg notes that he returned to the topic because he had recently seen another pyrophoric material (used as a cure for bladder problems!) that reminded him of his experiences in 1684. A letter published in the *Journal des sçavans* 36 (1715): 573–75 claims that the substance was invented by one Lyonnet and taught by him to Homberg and others, including Michel-Louys Reneaume (1675–1739), Académie botanist, who reportedly spoke about it at an Académie meeting. This letter might provide useful background information or it may be only a self-serving confection dependent upon information cribbed from Homberg's papers—it was written after the publication of Homberg's memoirs.

26. While this demonstration is not recorded in the procès-verbaux—yet another indication of how incomplete and selective a record of Académie meetings they really are—it was mentioned by Pierre Varignon in a letter to Leibniz dated 4 December 1710; see Georg Heinrich Pertz, ed., *Leibnizens gesammelte Werke*, 3d ser., 7 vols. (Halle, 1859), 4:170.

1710, he demonstrated the spectacle to Elisabeth Charlotte. One might question the propriety of pouring out a powder made from fried and roasted feces (presumably Homberg's own) onto a table in front of the sister-in-law of Louis XIV, but the dowager Duchesse d'Orléans was delighted by the effect (and she had an earthy sense of humor anyway). She wrote about it enthusiastically to her aunt Sophie, Electress of Hannover (1630–1714), and asked Homberg to write a memoir to send along with her letter with the request that Sophie share the information with Leibniz (who was part of the Hannover court at this time).[27] Leibniz swiftly wrote back to Homberg, asking for more information and a sample of the remarkable material. Homberg responded in February 1711, but, unable to send a sample due to its potential to self-ignite along the way, he instead fried up a fresh batch of alum and excrement for Leibniz and sent him the resultant dry powder with instructions describing how to roast it into the pyrophorus.[28]

All of Leibniz's attempts to produce the self-igniting material failed. So later that year he turned to C. R. Hasperg, secretary to the Duke of Wolfenbüttel, who was about to set off for France. In a lengthy laundry list of things for Hasperg to do, Leibniz wrote: "Mr. Homberg, physician to Monseigneur the Duc d'Orléans and a famous member of the Académie des Sciences, once sent me a prepared powder which had not yet had the final calcination after which it would ignite itself and other combustibles in the open air. I had this preparation carried out in Berlin as prescribed, but it did not work. I would therefore ask Mr. Hasperg please to visit him, greet him on my behalf, and have himself shown the operation to the extent possible, and also to bring along something appropriate for the trial [as a gift]."[29]

Accordingly, after Hasperg arrived in Paris, he went to the Palais Royal

27. Bodemann, *Aus den Briefen*, 2:261 (21 December 1710).

28. Homberg's first memoir, sent by Elisabeth Charlotte to Sophie, was lost before Leibniz could read it, and Leibniz was obliged to ask for a fresh copy, which Homberg sent along with his February letter and the sample: Homberg to Leibniz, 26 February 1711, NLB, LBr 420, fol. 1r–v; copy of Homberg's memoir for Sophie, 13 January 1711, fol. 2r–v. Homberg read his memoir for Sophie to the Académie on 17 January 1711 (PV 30, fols. 13r–14v). Before Homberg's letter and package arrived, Leibniz sent a second letter to thank Homberg in advance because he had been alerted (partly incorrectly) that Homberg would be sending him a sample of the completed pyrophorus; Leibniz to Homberg, 10 March 1711, LBr 420, fol. 3r–v. The package arrived shortly thereafter, as Leibniz informed Sophie on 20 March.

29. Leibniz to Hasperg, September 1711, NLB, LBr 374: "Monsieur Homberg Medecin de Mgr le Duc d'Orleans, und beruhmtes glied der Academie des Sciences hat mir einsmahls ein praeparirter pulver geschickt, welchem aber noch die letzte calcination gefehlet, nach welcher es sich und andere combustibilia von freyen lufft anzündet. Ich habe solche praeparation ex praescripto in Berlin thun laßen, ist mir aber nicht gerathen. Es wird demnach gebethen

to meet Homberg, and the two of them successfully prepared the pyropho-rus together.[30] A note among the Leibniz correspondence may be the result of Hasperg's visit to the Palais Royal laboratory, or perhaps of yet another person who visited Homberg on Leibniz's behalf. In either case, it provides details about the way Homberg received visitors to the laboratory. Surprised that Leibniz's attempts to follow Homberg's instructions did not give the ex-pected results, the writer recounted that "this morning in the Duc d'Orléans' laboratory, after we had taken some chocolate, Mr. Homberg shared with me a little part from the same mass some of which was sent to Mr. Leibniz. He let me see and read a copy of what he wrote to the said gentleman. I performed the operation, and when the said powder had cooled, I exposed it to the air on a folded piece of paper, and it instantly took fire and burned."[31]

This account underscores the importance of hands-on experience when trying to reproduce chymical processes. The same feature appeared in the case of white phosphorus. Homberg, who had performed the process along-side Kunckel in Berlin, succeeded admirably thereafter in preparing it, while the academicians in Paris—even though they had been sent the same rec-ipe more than once—constantly failed. The same was true of preparing the Bologna Stone. Those who (like Homberg) had worked in Italy alongside someone who knew how to make it succeeded, while virtually everyone else failed.[32] The note's charming reference to having first "taken some chocolate" indicates more about the social nature and conventions of such visits. Hom-berg received and entertained his guest rather than simply "getting down to business." They presumably also conversed, or examined other curiosities—surely Homberg liked to show off the Tschirnhaus lens—or perhaps strolled in the palace garden during the several hours the powder was roasting and while its vessel was cooling afterward. In the same vein, Leibniz specified that Hasperg had to bring Homberg a gift "appropriate" for the visit and

---

M. Hasperg wolle ihn besuchen, meinetwegen dienst. grußen, und sich die operation so viel erlaubt, zeigen laßen, auch etwas zugängliches zur probe mitbringen."

30. Hasperg was in Paris from roughly October 1711 until the end of January 1712, and then again a little later in the year. The visit to Homberg may have happened when Hasperg delivered a copy of Leibniz's *Essais de Theodicée* to Philippe and Elisabeth Charlotte in January 1712; Has-perg to Leibniz, 7 January 1712, LBr 374.

31. NLB, LBr 420, fol. 4r: "Ce jourdhuy matin dans le laboratoire de Mr. Le Duc Dor-lean apres la prise de Chocolat Mons. homberg ma fait part dune pettite partie de la Masse de laquelle on a envoyé a Mons. Lebenitz, il ma fait voire et lire une Copie de ce quil en a ecrit a Mondit Sieur, jay fait loperation et laditte poudre refroidie je lay exposé a lair sur un papier plié, et dans le meme moment elle y a mit le feu en flame."

32. See Lawrence M. Principe, "Chymical Exotica in the Seventeenth Century, or, How to Make the Bologna Stone," *Ambix* 63 (2016): 118–44, especially 119–20, 123–25.

demonstration. This protocol of exchange went further. Homberg's memoir to Sophie stated that "I would have sent the preparation [of the pyrophorus] with great pleasure to Mr. Leibniz, but I'm very sorry that I've been ordered not to communicate it to anyone." Rather than being ordered to conceal it—which seems very improbable since he described it openly to the Académie very soon afterward—it is more likely that Homberg kept the preparation secret in order to have something to trade, as he had done so frequently and regularly in the past. Leibniz, well aware of the usual game that needed to be played in order to obtain chymical secrets, took the hint and knew what to do. He quickly promised to send Homberg "something from here that will be worth your trouble . . . a new color, made by chymistry, which greatly approaches ultramarine"—one of the earliest references to the recently synthesized pigment later known as Prussian Blue.[33] Homberg thus became the first person in France to learn of this new German invention, and accordingly, the exchange having been offered, when Hasperg visited the Palais Royal, Homberg taught him the entire method of making the pyrophorus. Leibniz himself, again aware of the usual conditions for such exchanges, warned Hasperg thereafter to "conceal Mr. Homberg's fire." Hasperg reassured him that "I have not spoken of Mr. Homberg's fire to anyone; I am sending you here the method as I saw him prepare it."[34]

It is possible, however, that even sending Hasperg as a witness to the operation might not have been enough to guarantee Leibniz's own success with the process. As mentioned previously, shortly after Homberg's death, Leibniz wrote to his correspondent Nicolas-François Rémond, also a member of the household of the Duc d'Orléans, asking him to ensure that Homberg's *Nachlass* be carefully preserved. In addition to asking about any written materials, Leibniz very solicitously requested Rémond to "have the goodness to

33. Homberg's memoir to Sophie, 13 January 1711, fol. 2v: "Je suis bien faché qu'on m'ayt ordonné de ne la communiquer à personne, j'en aurois envoyé la preparation avec beaucoup de plasir"; Leibniz to Homberg, 10 March 1711, fol. 3r–v: "je souhaiterois de pouvoir vous mander quelque chose d'icy qui en valut la peine. On m'a promis des essais d'une nouvelle couleur, faite par la chymie qui approche extremement de l'Ultramarin." See Alexander Kraft, "On the Discovery and History of Prussian Blue," *Bulletin for the History of Chemistry* 33 (2008): 61–67. Hasperg visited Homberg again in 1714, at which time Homberg asked for more information about Prussian Blue; Hasperg to Leibniz, 17 August 1714, NLB, LBr 374, fols. 36r–37r.

34. Leibniz to Hasperg (notes), 14 October 1713, NLB, LBr 374: "Dissimuler le feu de Homberg"; and Hasperg to Leibniz, 11 December 1713: "Je n'ay pas encor parlé à quelqu'un touchant le Phosphore de Monsr. Homberg dont j'envoye par ce couvert la Methode, comme je l'ay vu preparer." Homberg described the method of making his pyrophorus to the Académie in February 1712, around the time he gave it to Hasperg, but that paper would not be made public until the *Mémoires* was published in 1714.

find out if he had any help from a collaborator informed a bit about his observations and views which he had not shared with the public; this is so that they are not lost in the least." After replying that Homberg's papers were safely in the hands of Geoffroy, Rémond added, "For the rest, you should address yourself to the Abbé Conti to know how a phosphorus is made from honey and from rye; I hope to inform him and send you shortly the explanations."[35] This is certainly a reference to revised methods of producing the pyrophorus from starting materials other than human excrement—both honey and rye flour were used successfully by Jean Lémery (1678–1716), with whom Homberg was possibly collaborating on this project during the last years of his life.[36] Of further interest is the mention of Abbé Conti, that is, Antonio Schinella Conti (1677–1749), who acted as an intermediary between Leibniz and Newton in the debate over the calculus. Conti was in Paris in 1714–15, and the way he is mentioned here in relation to questions about Homberg's collaborators implies that he too may have worked to some extent with Homberg on chymical preparations.[37] It is possible that Jean Lémery and Abbé Conti were helping Homberg in his last years when his ability to continue his research was seriously curtailed by his declining health and other factors.

35. Leibniz to Rémond, 6 December 1715, NLB, LBr 768, fol. 49r–v, at 49v: "Si Monsieur Homberg a laissé quelque chose par écrit, il sera bon de le conserver. Ayés la bonte de vous en informer, Monsieur, comme aussi s'il n'a pas eu quelque aide du collaborateur informé un peu de ses observations et veues, dont il n'a pas encore fait part au public; c'est a fin que elles ne se perdent point." Rémond to Leibniz, 23 December 1715, NLB, LBr 768, fols. 53r–54v, at 54r: "vous vous estes addressé a M. l'abbé Conti pour savoir comment on fait un phosphore du miel et du seigle j'espere que je le previendrai et que je vous enverrai bientost ces eclaircissemens."

36. Jean Lémery, "Experiences sur la diversité des matieres qui sont propres à faire un phosphore avec l'alun," *MARS* (1714): 402–8 (at 403, Lémery notes that he "believed with Mr. Homberg" that urine would work well and so began his researches with that substance, which may imply that they were working together at this point); and "Reflexions physiques sur un nouveau phosphore," *MARS* (1715): 23–41. This younger son of Nicolas Lémery is frequently referred to as Jacques and his date of death given as 1721; see *Index biographique de l'Académie des Sciences* (Paris: Gauthier-Villars, 1979). Jean-Dominique Bourzat has, however, determined that he actually died on 30 August 1716; see Patrice Bret, "Les chimistes à l'Académie royale des sciences à l'époque des Lémery (1699–1743)," *Revue d'histoire de la pharmacie* 64 (2016): 385–404, at 401n5. A contemporaneous record of his death (as Jean) is AN, Minutier Central, LXXIII-643; his father's Inventaire-après-décès also gives him the name Jean; AN, Minutier Central LXXIII-639 (19 June 1715), fol. 1r. My thanks to Patrice Bret and Jean-Dominique Bourzat for this information.

37. It seems, however, that if further information about the pyrophorus actually got to Leibniz, it came from Geoffroy instead (Rémond to Leibniz, 15 March 1716, NLB, LBr 768: "c'est [Geoffroy] qui m'a promis les differens phosphores que vous souhaitez"; by that date Conti had left Paris and was in England).

When Hasperg visited Homberg a second time, in April 1712, just before leaving Paris, the Batavian chymist asked him to carry a letter for him to Herman Boerhaave in Leiden. The letter told of Homberg's discovery of the pyrophorus and gave a description of how to prepare it. When Boerhaave later described the substance in his *Elementa chimiae*, he wrote: "[The pyrophorus] was first made known to me by a letter that the extraordinary Homberg wrote to me from Paris on 12 April 1712, which was hand-delivered to me by the noble Mr. Hasperg, who added in person at the same time some circumstances that must be noted [in its preparation]."[38]

Boerhaave's statement indicates again the importance of direct personal contact with an experienced practitioner for the transmission of important operational details of preparative processes. Boerhaave cites Hasperg's orally transmitted additions as necessary supplements to the written instructions of Homberg's letter. The fact that Homberg wrote to Boerhaave (the only evidence at present known of any direct contact between the two) suggests that Homberg was interested both in making contact with this then rising star in Leiden and in publicizing his new discovery—and perhaps also securing priority and notoriety for this remarkable substance.

The "Image Problem" for Chymistry and Its Practitioners

Hasperg's visit to Homberg's laboratory at the Palais Royal was surely only one example of a fairly regular occurrence, even though we currently lack the correspondence or notebooks to document such visits more fully. Most of the visitors to early modern laboratories about whom we normally have information were, unsurprisingly, other learned savants. Less visible in retrospect are visits by and connections to less-elevated persons, those either simply lower down the social ladder or of questionable reputation. Historians of science, particularly in the past generation, have been paying closer attention to the contributions of those outside the usual learned circles—artisans, workmen, miners, entrepreneurs, and others—and have thereby reclaimed an important part of the fabric of early modern science. Such contacts hold special importance for the history of chemistry, first because its practitioners ranged across every social and intellectual level, but also because the subject itself suffered from a less-than-exalted status in the early modern period. Part of chymistry's "image problem" resulted from the laborious manual operations

---

38. Herman Boerhaave, *Elementa chimiae*, 2 vols. (Paris, 1733), 1:205. Hasperg's presence in Paris in early April 1712 (and his departure just before 16 April) is documented in Jacques Lelong to Leibniz, 16 April 1712, NLB, LBr 549.

it required, often with dirty, malodorous, toxic, or otherwise unpleasant materials. Part came also from chymistry's lack of a classical heritage (unlike astronomy, for example) and its lack of an established position within the academic system (except when it was tied closely to medicine and pharmacy).[39] But a significant part also spilled over from the lower social standing of many of its practitioners and from its linkage—not entirely unwarranted—with immoral, illegal, or otherwise questionable activities. Boyle had faced this issue when "Learned Men" tried to dissuade him from wasting time on the "empty and deceitful study" of chymistry, based upon their opinion of "the Illiterateness, the Arrogance and the Impostures of too many of those that pretend skill in it."[40] Chymistry, for many outside observers at the time, was virtually inseparable from counterfeiting, poisoning, and cozenage.

Fears and accusations of counterfeiting were never far from chymistry, especially when its practitioners worked extensively with metals, as those inclined toward chrysopoeia tended to do. The explicit connection to counterfeiting and currency debasement dates at least to Pope John XXII's 1317 decretal *Spondent quae non exhibent*, and quite possibly as far back as late antiquity. Such worries continued unabated through the early modern period: Louis XIV's concerns about the French coinage, as mentioned in chapter 5, resulted in prohibiting the members of the Académie from studying or even discussing chrysopoeia.[41] As for poisoning, the France of Homberg's day remained deeply affected by the memory of the infamous "Affaire des Poisons" of the late 1670s, which uncovered an extensive network of poisoners and supposed enchanters that reached even into the royal court as far as Louis XIV's mistress Madame de Montespan. Although most historical treat-

39. Several early modern chymists endeavored to create an ancient lineage for their discipline, either by emphasizing the rediscovered *Corpus alchemicum graecum* dating from late antiquity, or by linking the practice with biblical figures or with Hermes Trismegestus, a supposed contemporary of Moses; see for example, Olaus Borrichius, *De ortu et progressu chemiae* (Copenhagen, 1668); and Daniel Georg Morhof, *Polyhistor* (Lübeck, 1688), 97–113. A few humanistic authors likewise endeavored to brush off the soot from chymistry by reclothing it in finer poetic or other literary dress; see Principe, *Secrets*, 173–81. On chymistry's place in universities, see Bruce T. Moran, *Chemical Pharmacy Enters the University* (Madison, WI: American Institute of the History of Pharmacy, 1991).

40. Robert Boyle, preface to the "Essay on Salt-Petre," in *Certain Physiological Essays*, in *Works of Boyle*, ed. Michael Hunter and Edward B. Davis, 14 vols. (London: Pickering & Chatto, 1999–2000), 2:85.

41. Principe, *Secrets of Alchemy*, 22–23, 50, 61–62, 179. For a detailed account of the legal status of transmutation and its relation to counterfeiting in late medieval England, see Jennifer M. Rampling, *The Making of English Alchemy*, chapter 1 (Chicago: University of Chicago Press, forthcoming 2020).

ments of the affair focus predominantly on the racier details of the magic, sorcery, murder, and sexual excesses the investigations uncovered, the ultimate fallout from the affair actually fell particularly heavily upon chymists through the resulting Edict of 1682.[42] This edict regulated the possession and trade of toxic materials and strictly forbade the possession of chymical furnaces and the making of chymical products and experiments without an official and registered license:

> We very expressly forbid everyone of whatever profession and condition they may be, excepting approved physicians in their residence, professors of chymistry, and master apothecaries, to have any laboratories or to work there upon any preparations of drugs or distillations under the pretext of producing chymical remedies, experiments, or particular secrets, searching for the philosophers' stone, the transmutation, multiplication, or refining of metals, the making of crystals or colored stones, or for other similar pretexts without previously having obtained permission from us in the form of Letters of the Great Seal to have such laboratories, and having presented such Letters and made a declaration to our local judges and officers of the police.[43]

The edict's wording is significant. Although the immediate cause for the edict was the availability of poisons and other materials criminally used in the "Affaire des Poisons," the law also explicitly forbade unlicensed "searching for the philosophers' stone, the transmutation, multiplication, and refining of metals." Since the affair did not involve any chrysopoetic activities, this inclusion suggests that the edict attempted to address simultaneously two worrisome problems of "unregulated chymistry": the immediate concern over

42. On the "Affaire des Poisons," see Lucien Nass, *Les empoisonnements sous Louis XIV* (Paris: Carre et Naud, 1898); Jean-Christian Petitfils, *L'Affaire des Poisons* (Paris: Albin Michel, 1977); Lynn Wood Mollenauer, *Strange Revelations: Magic, Poison and Sacrilege in Louis XIV's France* (University Park, PA: Pennsylvania State University Press, 2007); and Arlette Lebigre, *L'Affaire des Poisons* (Brussels: Editions Complexe, 1989).

43. *Edit du roy pour la punition de differents crimes qui sont devins, magiciens, sorciers, empoissoneurs* (Paris, 1682), 5: Faisons très-expresses defenses à toutes personnes de quelque profession et condition qu'elles soient, excepté aux Medecins approuvez, et dans le lieu de leur residence, aux Professeurs en Chimie, et aux Maistres Apothicaires d'avoir aucuns laboratoires, et d'y travailler à aucunes preparations de drogues ou distillations, sous pretexte de remedes chimiques, experiences, secrets particuliers, recherche de la pierre philosophale, conversion, multiplication ou rafinement des metaux, confection de cristaux ou pierres de couleur, et autres semblables pretextes, sans avoir auparavant obtenu de nous par Lettres du grand Sceau la permission d'avoir lesdites laboratoires, presenté lesdites Lettres et fait declaration en consequence à nos Juges et Officiers de Police des lieux." For a treatment of the Edict of 1682, see Roger Goulard, "À propos de l'affaire des poisons: le célèbre édit de 1682," *Bulletin de la Société Française de l'Histoire de la Médecine* 13 (1914): 260–68.

poisoning as well as ongoing concerns about counterfeiting and the currency. This reading is supported by the fact that licenses to practice chymistry were already being issued *before* the Edict of 1682, and the authority to grant such licenses was lodged specifically with the Cour des Monnaies. The mint continued issuing licenses for chymical furnaces and operations even after 1682.[44] The edict's wording implicitly, if inadvertently, cast further aspersions upon chymistry by juxtaposing it with the crimes of divination and sorcery. An analogous association appears in the contemporaneous comic play *La pierre philosophale*, written in 1681 by Thomas Corneille and Donneau de Visé during investigations of the "Affaire des Poisons," and following the success of their *La devineresse* (1679), which likewise drew upon popular sensationalism of the day relating to the affair. It bears mentioning that none other than Bernard de Fontenelle, who would become the most stridently anti-alchemical voice in the Académie, had a hand in these theatrical productions in the decade before he became perpetual secretary of the institution.[45]

Along with counterfeiting and poisoning, chymistry was also widely associated with cozenage. Stories of confidence tricksters and scam artists promising great riches through transmutation or health and longevity through chymical remedies were well known in the early modern period, and in fact date back at least to the Islamic Middle Ages. Multiple accounts and portrayals of the "false alchemist" were used in the sixteenth and seventeenth centuries by pro-chrysopoeia authors to establish their own distinctly higher status and as warnings against frauds, and by anti-chrysopoeia authors as arguments that transmutation was inherently fraudulent. In theater and literature, characters hoping to obtain the secret of transmutation were routinely ridiculed as fools duped by characters posing as adepts.[46] Tara Nummedal has usefully pointed out that some of the accused (and often executed) "false

44. For a useful study of authorizations of laboratories and their furnaces, and their connection to the Cour des Monnaies, see Maurice Bouvet, "Les laboratoires autorisés au XVIIe siècle," *Bulletin de la Société d'Histoire de la Pharmacie* 13 (1925): 10–16, 55–60; and Robert Scagliola, "Les apothicaires de Paris et les distillateurs" (Thèse de pharmacie, Université de Clermont-Ferrand, 1943), 36–42, and its compte-rendu in *Revue d'histoire de la pharmacie* 113 (1943): 38–40.

45. Didier Kahn, "L'Alchimie sur la scene française aux XVIe et XVIIe siècles," *Chrysopoeia* 2 (1988): 62–96; and Kahn, *Le Comte de Gabalis, ou Entretiens sur les sciences secrètes* (Paris: Honoré Champion, 2010), 128–29. On Fontenelle's role, Alain Niderst, *Fontenelle à la recherche de lui-même (1657–1702)* (Paris: Nizet, 1972), 24–25, 104–9.

46. Examples stretching over more than a century include Ben Jonson, *The Alchemist* (1610); Michel Chilliat, *Les Souffleurs, ou la pierre philosophale d'Arlequin* (1694); and John Gay, Alexander Pope, and John Arbuthnot, *Three Hours after Marriage* (1717). See Principe, *Secrets*, 186–88; and Kahn, "Scene française."

alchemists" simply made premature promises based on overly optimistic assessments of their abilities and the reliability of their ideas and recipes, rather than engaging in intentional fraud.[47] Nevertheless, there is also evidence (see below) of actual frauds criminally taking advantage of the allure of gold-making to extract money and favors from those willing to believe their promises and hoping to gain therefrom. Such characters represent an understudied and generally rather ill-documented demimonde of chymical hucksters and hopefuls that generally pass beneath the radar of historical investigation. The relevance of such characters for the present study lies first of all in their surprising intersections with Homberg, the Duc d'Orléans, the Palais Royal laboratory, and even the heady atmosphere of the Académie itself. Yet more strikingly, an array of government ministers, with the support of Louis XIV, also began dealing with such characters in the early eighteenth century, often under the cover of rooting out counterfeiters, poisoners, frauds, and diviners, but almost always with the intention of determining what chymical knowledge such figures actually had and how that knowledge—especially the transmutational—might benefit the French state. The increased attention paid to such figures in early eighteenth-century France, and the unsavory revelations and disappointing results that followed, played an important role in the increasing repudiation of chrysopoeia that came to pass in the 1720s. This "official" exiling of alchemy from the domain of chymistry represents one of the most dramatic transmutations of chymistry in the period, and one in need of further elucidation. The following sections of this chapter, therefore, deploy a range of archival sources to present remarkable accounts of these characters; their connections to Homberg, Philippe, and the French state; and the consequences for the transmutations of chymistry.

Interrogating Chymistry: *La Bastille Chymique*

In February 1704, one Étienne Vinache, described as "an empiric physician and chymist seeking the philosophers' stone," was arrested on suspicion of counterfeiting and *billonnage* (releasing defective or debased coinage).[48]

---

47. Tara Nummendal, *Alchemy and Authority in the Holy Roman Empire* (Chicago: University of Chicago Press, 2007), especially 40–72.

48. A document in the Bastille file on Vinache defines *billonage* as "a crime punished by death; it is the art of substituting defective coins for those of proper alloy [*un Crime puni de mort; c'est l'art de substituer des piéces défectueuses à celles qui sont d'aloy au titre*]," BNF, Arsenal MS 10548, fol. 111r. For an overview of the Vinache case see Roger Goulard, "Un mystère à la Bastille: Etienne Vinache, médecin empirique et alchimiste," *Bulletin de la Société Française de l'Histoire de la Médecine* 14 (1920): 360–72. Several of the documents are reworked into a

Vinache, a Neapolitan by birth (originally Vinaccio), lived in poverty until about 1698, when he took a trip lasting four or five months and returned to Paris a wealthy man. Over the next five years his wealth and prominence only increased. He bought a collection of diamonds from the estate of the deceased Philippe I, Duc d'Orléans, for 48,000 livres, lived in an extravagantly furnished house that included a gallery of paintings estimated at 25,000 ecus (75,000 livres), tried to purchase the Hôtel Mazarin just north of the Palais Royal, and offered 250,000 livres for a country estate. He boasted, moreover, that he knew how to make gold and silver, and that "if he believed that the King and his ministers had enough good faith not to demand his secret and would leave him in full liberty to come and go as he pleased with a guard such as one would want to provide him, he would give three hundred million louis d'or as easily as three . . . he would be delighted, if that happened, to see taxes decreased thanks to him, and he would like to be regarded like Joseph in Egypt."[49]

Following a denunciation of Vinache sent to Madame de Maintenon, a preliminary investigation turned up multiple causes for suspicion. Vinache had questionable contacts with a Dutch banker, several French government officials, and the very wealthy financier Samuel Bernard (1651–1739). A search of his house revealed several furnaces, fragments of melted gold and silver, and containers of mercury. Vinache was arrested and brought to the Bastille on 17 February 1704.[50] The interrogator asked Vinache who had given him permission to own and use these furnaces. He replied, rather shockingly, that his "three large ones . . . had been ordered by Monseigneur the Duc d'Orléans" and that he had received due authority from the mint to build and own them. Vinache stated that he had hired a young apothecary named Edmé Thuriac (or Thuriat) to help build these furnaces and act as "an op-

somewhat unreliable account in *Mémoires historiques et authentique sur la Bastille*, 3 vols. (London, 1789), 2:36–72; see also Ulrike Krampl, "Diplomaten, Kaufleute und ein Mann 'obskurer Herkunft,'" in *Nützliche Netzwerke und korrupte Seilschaften*, ed. Arne Karsten and Hillard von Thiessen (Göttingen: Vandenhoek & Ruprecht, 2006), 137–62.

49. Arsenal MS 10548, fols.145r–148v, quotation on 147r: "vinache . . . nous disant souvent que sil croyoit que le Roy et ses ministres fussent dassez bonne foy pour ne point exiger son secret et quon luy laissast une entiere liberté pour aller et venir ou il voudroit avec une garde tel quon voudroit luy donner quil donneroit trois cent millions aussy facilement que trois louis dor. Nous layant dit tres souvent audit sr et marconnet et a moy et me layant repeté aussy en mon particulier ajoutant quil seroit charmé si cela arivoit de voir diminuer par son moyen les Impots et quil voudroit estre regardé comme le Joseph dEgipte." On Vinache's huge wealth, see also *Archives of the Bastille*, ed. François Ravaisson, 19 vols. (Paris, 1866–1904), 11:138–40.

50. The letter of denunciation is Arsenal MS 10548, fols. 22r–25v. See also Goulard, "Mystère," 366–67.

erator in chymistry" for making medicinal essences, but Vinache denied that he had engaged in any transmutational endeavors.[51] Intriguingly, Jean Hellot owned a chrysopoetic manuscript entitled "True Remedy or Tincture for Silver to Make it One-Third Gold, by Père Mandreville and le Sieur Vinache" and dated 1687. If it is the same Vinache, as seems probable (even though he would have been only twenty-one at that date), this document would attest to an interest in transmutation earlier than the Bastille interrogators ever discovered.[52]

The apothecary Thuriac was then brought in for interrogation. He testified that he did help Vinache with the furnace construction, but that the three "portable and rolling furnaces" were designed and built primarily by none other than Martino Poli (1662–1714), the Italian chymist who would very shortly thereafter be made *associé étranger* of the Académie Royale des Sciences, in May 1704.[53] Poli later became involved in a dispute with Homberg over the presence of acids in the blood, and Fontenelle glancingly acknowledged the Italian's interest in metals and their transmutation.[54] Vinache's three interrogations are full of inconsistencies and equivocations, but the investigators were frustrated in their attempts to get the full truth because on 19 March 1704, a month after his arrest, Vinache died in the Bastille after reportedly cutting his own throat. His body was hastily buried under the false name of Étienne Durand and given the false age of sixty (Vinache was only

51. Arsenal MS 10548, fol. 40r: "neantmoins quil y en a dans sa maison trois considerables, qui luy ont esté commandez par Monseigr Le Duc Dorleans; mais Il nous prie d'observer quil ne sont pas achevez et que mesme quand ils le seroient Ils ne pourroient servir a la fonte des Mestaux mais seulement a la distillation des Essences et des Esprits"; fol 40v: "monseigr le Duc Dorleans l'ayant chargé de faire fabriquer les trois fourneaux dont il a esté parlé Il avoit pris ce jeune homme pour concourir a cet ouvrage"; fol. 45r: "Il a aussi dans sa maison de paris trois fourneaux imparfais qu'il a fait bastir pour Mr le Duc d'Orleans et dont il est demeuré depositaire de lautorite de la cour de monnoye"; and fols. 129r–137v, declarations of Edmé Thuriac (12 December 1703), on 135r: "artiste pour la Chimye."

52. *Catalogue des livres de feu M. Hellot* (Paris, 1766), 102, item 1821: "Véritable Remede ou Teinture pour la Lune au tiers d'or, par le P. Mandreville & le sieur Vinache, en 1687, *in-4 mss.*" I thank Didier Kahn for bringing this reference to my attention.

53. Arsenal MS 10548, fol. 132v: "Et le deposant qui travailloit en haut a des fourneaux de nouvelle fabrique avec le No[mm]é Martineau poly Italien"; fol. 137r: "Jestois pour lors avec un Italien nommé martineau polly avec lequel je faisois des fourneaux dune nouvelle fabrique; fol. 129r: "au 3e Estage de la maison dudit Vinache Il y avoit trois fourneaux portatifs et Roullans fabriques par martino poly Italien."

54. Bernard de Fontenelle, "Éloge de M. Poli," *HARS* (1714): 129–34, especially 130, 132; Martino Poli, *Il trionfo degli acidi vendicati dalle calunnie di molti moderni* (Rome, 1706); and Wilhelm Homberg, "Observations sur l'acide qui se trouve dans le sang," *MARS* (1712): 8–15; and "Suite des observations sur l'acide qui se trouve dans le sang," 267–75.

thirty-eight), while the king ordered that his wife be told he died of a sudden illness.[55]

Why did Philippe ask Vinache to build furnaces for him in late 1701 or early 1702? Did he plan to commission chymical work, possibly transmutational, from Vinache? How did their relationship begin? Vinache and Thuriac both noted that Poli's furnaces were left unfinished, which may have been because it was at this same time that Philippe engaged Homberg and began building the laboratory at the Palais Royal. Was Homberg's engagement perhaps linked to the same interests that had caused Philippe to commission furnaces from Vinache? Vinache and Philippe also had further contacts: Homberg reported that Vinache "had the insolence to propose to the Duc d'Orléans that he constellate some diamonds for him." The curious verb used here, *consteller*, means imbuing an object with astrological virtues, perhaps in this case to produce a "propitious" ring or other object—presumably to assist Philippe while engaged in his well-known penchant for gambling.[56] Vinache's wife, who never saw her husband's corpse because of its hasty and pseudonymous burial, believed for years afterward that he was still imprisoned secretly in the Bastille and so sent a petition for his release to Philippe in 1715 immediately after he became Regent. Her request refers to "the protection with which he was honored by your royal highness who knew his innocence," thus indicating further links between the two.[57] It was presumably links such as these, which surely were known to more people than just those questioned by the police, that in part spurred Saint-Simon to emphasize the innocent, public, and supposedly nontransmutational character of Philippe's interests in chymistry.

The investigation that began with Vinache quickly expanded to other characters around Paris, many of whom had further connections with the Palais Royal or directly with Homberg. The first was Isaac Thibaut, sieur de Marconnet. An array of chymical substances found in his possession were seized and sent for analysis, including the *caput mortuum* of colcothar—the residue left from roasting vitriol (iron and/or copper sulphate). Asked where he had obtained this material, Marconnet replied that he "got it from the Palais Royal, in the laboratory of Monseigneur the Duc d'Orléans . . . to make experiments with." Given this instance, one wonders to what extent Hom-

55. Chamillart to d'Argenson, 23 March 1704, Arsenal MS 10548, fol. 81r–v; summary documents, fol. 83r–v, 110r, and 118v; 119r reports that Vinache's wife was told her husband died of apoplexy. See *Archives de la Bastille*, 11:144n, for one interpretation of this strange affair.

56. *Archives de la Bastille*, 11:131; also Arsenal MS 10548, fols. 95v, 124r, 128v.

57. Arsenal MS 10548, fols 86r–87r, oat 87r: "La Protection dont il étoit honoré de Votre Altesse Royalle qui connoissoit son innocence."

berg and his laboratory were supplying materials, and presumably chymical advice, to various chymical practitioners around Paris. One of Marconnet's experiments aimed to produce a universal medicine. He told the interrogator that he had given a preliminary sample of it to the Académie, asking them to make trials of it, and "if they succeeded as he hoped, he would present it to the King as the fruit of work in which he has been engaged for more than twelve years." Marconnet was also familiar with Philippe's physician Helvétius and had visited him in his lodgings.[58] Marconnet worked on metallic transmutation as well, and, upset that his experiments in that regard had been ruined by the police just before they were complete, he asked to be freed from the Bastille "in order to redo them in the laboratory of the Académie or of the Jardin du Roi, where he would find the necessary vessels to make the preparations (as for the cost, he would do it at his expense), and [he asked] that he be given as an inspector Mr. Homberg, one of the most expert of the Académie. Thus he promises his Majesty that in less than a year he would furnish him with a salt that will fix mercury, and if he is given the time to have it brought to its perfect exaltation, he hopes to be in a position to furnish gold to his Majesty to help him overcome his enemies."[59]

The proposal to "overcome the enemies" of Louis XIV by means of metallic transmutation becomes a frequent refrain in dealings with chrysopoetic hopefuls. Despite earlier official reluctance to be thought dabbling in gold-making, in the early eighteenth century the French State apparently decided that chrysopoeia had now become a possibility not to be dismissed too quickly, undoubtedly due to the increasingly dire financial depletion in which France found itself as the War of the Spanish Succession (1701–14) continued. Thus, despite Marconnet's promise and a character reference from the

58. Arsenal MS 10548, interrogations of Marconnet, fols. 200r–210v (15 and 27 November 1704), at fol. 206v: "A dit qu'il a pris Ladite teste morte du palais Royal dans le Laboratoire de Monseigneur Le Duc d'Orleans, et que luy repondant n'a pris Ladite poudre que pour faire des Experiances Telles qu'il les auroit imaginés dans la suitte"; mention of Helvétius, third interrogation, fols. 215r–216v; fair copy, 217r–224v (22 December 1704), at fols. 217v–219r.

59. Arsenal MS 10548, Marconnet's petition to d'Argenson, fols. 225r–226v (read by Pontchartrain, 26 February 1705), on fol. 226r: "Elle auroit parfait en peu si on ne l'avoit pas interrompue, mais il est tellement persuadé de la veritté et bonté de son oeuvre que c'est ce qui luy donne la hardiesse de suplier tres humblement sa majesté de luy donner la liberté de la refaire dans son laboratoire de l'accademie ou du Jardin royal ou il trouvera les vesseaux nescessaires pour faire ses preparations et pour la despence il la fera a sa frais, et que la on luy donne pour inspecteur mons Humbert un des plus expert de l'accademie, et avec cela il promet a sa majesté quen moins d'un an de luy fournir un sel qui fixera le mercure, et si on luy donne le tems den faire faire la parfaitte exaltassion, il espere destre en estat de fournir sa Majesté de l'or pour luy ayder a surmonter ses enmies."

Académie's president, Jean-Paul Bignon, the chymist was not released.[60] Instead, he was allowed—or rather compelled—to conduct his promised experiments toward transmutation while remaining imprisoned in the Bastille, meaning that a laboratory must have been set up for him within its walls. He therefore carried on with his endeavors within the Bastille for the rest of 1705, producing at last a "salt of wisdom [*sel de sagesse*]." He was released only in April 1706, after he had decided, and declared under oath to the authorities in charge, that his ideas about how to achieve transmutation were flawed. Nevertheless, Pontchartrain still ordered the police to keep Marconnet's actions under surveillance after his release, "particularly in regard to chymistry."[61]

The investigation continued to spread, reaching two further associates of Vinache: Georg Conrad Schuster, a German from Leipzig, and Jean-Robert Tronchin, a lawyer of Swiss origin and Samuel Bernard's cashier. Schuster admitted that he was interested in acquiring the philosophers' stone and that he had awakened the same interest in Tronchin. At first they hoped for good results from a German count named Senenburg von Steinerbach, then living in Paris. The count promised them not only the philosophers' stone, but also perpetual motion and a vial of a luminous material that allowed one to read and write in the dark. After Tronchin advanced him considerable sums of money for five or six months, the count disappeared.[62] They then discovered that what little the count knew of value had actually come from an Armenian named Deodat, so they lodged him, fed him, and gave him the means to carry out his research, but soon enough Schuster figured out that he too was "ignorant of the knowledge of the philosophers' stone." Deodat then left Paris for Brussels, but promised to send a powder with which "you may make the projection with your own hands so that you will be convinced of the truth yourselves." Deodat sent the powder to Tronchin, but since Schuster could not afterward remember whether or not they tried it, its effects must not have been particularly memorable. Deodat then vanished with the 20,000 francs he had obtained from Tronchin.[63] This huge sum of money—more than three

60. Arsenal MS 10548, autograph letter of Bignon, fol. 231r–v (21 June 1705).

61. *Archives de la Bastille*, 11:146–48.

62. Arsenal MS 10548, Schuster's first interrogation, 14 March 1704, fols. 373r–376v; fair copy, 377r–382v, at 379v. Name of count given in second interrogation, 16 March 1704, fols. 383r–386v; fair copy, 387r–394v, at fol. 391r; Trochin to Schuster, undated, fol. 553r (luminous vial). The material on Schuster and Trochin is extensive; note that francophones had considerable difficulty spelling Schuster's name—it appears as Schute, Chuste, and Juste.

63. Arsenal MS 10548, Schuster's first interrogation, fols. 380r–v and 391v–392r; Schuster to Tronchin about Deodat, fol. 691r–v; Deodat's receipt for money from Schuster, 15 April 1700,

times the cost of the Tschirnhaus lens—on top of what had been advanced to Senenberg, is far too high to have come from Tronchin himself, and so it may well indicate that the fabulously wealthy Samuel Bernard (1651–1739)—who was partly financing the nearly bankrupt French state at this point—was himself exploring and bankrolling these transmutational schemes.

Undaunted, Schuster and Tronchin turned their attention to an Italian named Caraffe. Tronchin rented and furnished a house for Caraffe where "he worked on the powder of projection in the presence of the said Tronchin, who furnished him with all the apparatus, metals, substances, and generally everything necessary to work on the said powder of projection and the universal medicine."[64] When the interrogators asked who gave Caraffe permission to have furnaces, Schuster replied that he himself had "obtained permission through Mr. Homberg, director of the king's laboratory, by virtue of which Caraffe was allowed to work on chymistry for eight months or thereabouts." Once again, Homberg seems to have been facilitating or sponsoring transmutational work around Paris. Tronchin was very solicitous to ensure that Schuster obtain the necessary license from Homberg: "let me know please where it stands with his *lettres patentes*; please see Mr. Homberg without delay . . . neglect nothing in order to give me precise news . . . word for word what Mr. Homberg tells you about the state of things."[65] Schuster's interrogators then asked how well Homberg knew Caraffe, and if Homberg was present when Caraffe was at work. Schuster replied that he introduced Caraffe to Homberg, who judged the Italian to be a very able chymist. Homberg welcomed Caraffe and wrote him a complimentary and encouraging letter. Later,

---

fol. 720r; Deodat to Schuster, 2 June 1700, fols. 692r–693v; Deodat to Schuster[?] (from Brussels), 14 August 1700, fols. 677r–768v: "je luy envoieray de la poudre . . . pour que vous faciez la projection par vos propre mains afin que vous soyez convaincu de la verité par vous même."

64. Arsenal MS 10548, Schuster's first interrogation, fol. 381r–v: "travailla a la poudre de projection en presence dudit trouchin qui luy fournit tout les ustancilles metaux drogues et generallement touttes les choses necessaires pour travailler a ladite poudre de projection et a la medecine universelle."

65. Arsenal MS 10548, Schuster's second interrogation, fol. 388r: "ce fut luy repondant qui obtint par le moyen du Sieur Hombere directeur du laboratoire du roy le privilege en vertu duquel il fut permis audit Carafe de travailler a la chimie pendant huit mois ou environ." Caraffe began his work in June 1700 (fol. 388r), but did not obtain the permission through Homberg until mid-1701, as Caraffe was returning from a trip to Geneva (fol. 392v, and Tronchin to Schuster, fols. 511r and 689r: "faites moy savoir, je vous prie en quel estat les choses sont pour ses lettres patentes, voyés je vous prie monsieur humbert là dessus sans perte de temps affin que je le luy mande encore aujourd'huy s'il se peut, ne negligés rien pour me donner des nouvelles précises là dessus mais n'ajoustés rien de vôtre, & ne me mandés pas un billet, positivement mot à mot que ce que Monsieur Humb: vous aura dit de l'Estat de la Chose."

however, after "having seen [Caraffe] perform some chymical experiments," Homberg changed his opinion, "having recognized that he was not doing anything extraordinary."[66] Regardless of his subsequent assessment, Homberg nevertheless met with this supposed transmuter, successfully obtained a *privilege* for him, and was initially optimistic about his potential. This initial optimism also appears in a surprising place—the Académie's own *Histoire*. Surprisingly, the Académie itself became involved with Caraffe, described as "a chymist who is beginning to make a reputation for himself." For reasons that remain unclear, the academicians asked Homberg to visit his laboratory and provide a report on his work. The report, published in the *Histoire* for 1701, mentions only Caraffe's pharmaceutical preparations and the successful employment of one of them by Dodart and Morin. Homberg's report then concludes that "there is room to hope for many other useful things from this chymist," but without any hint of Caraffe's main purpose in Paris, namely, to work on transmutation and the "powder of projection."[67] Was anyone at the Académie besides Homberg aware of Caraffe's chrysopoetic endeavors? It is indeed striking that the same chymist who appears in Bastille interrogations for working on gold-making appears also in the Académie's *Histoire*.

At another time, Schuster and Tronchin engaged yet a further promising transmuter. This one, named Pussieu, presented them with two packets of a red powder that he claimed to be the powder of projection. These packages were seized from Schuster's lodgings and sent to the apothecary Antoine Lenoir to be tested.[68] Strikingly, the apothecary actually tried to transmute metals with the material: "I made many experiments upon mercury and upon different metals to see if this powder had a sympathetic or antipathic power to change or augment their natures." After failing to produce any precious metals with the red powder, Lenoir identified it as minium (red lead oxide) made to sparkle by the addition of tiny copper flakes.[69]

66. Arsenal MS 10548, Schuster's second interrogation, fol. 388r–v; Schuster to Caraffe, fols. 562r–563v: "Je ne vous marque que les sentimens de connoissance que l'on doit avoir pour des manieres aussi obligeantes que sont les vostres. Monsieur Homberg en est aussi tres persuadé, sa lettre cy-jointe le peut confirmer plus amplement." The letter seems to be written while Caraffe was traveling from Paris, presumably on his trip to Switzerland, which would date the letter to June/July 1701. Unfortunately, a copy of Homberg's letter is not preserved.

67. *HARS* (1701): 74.

68. Antoine Lenoir was frequently called in to analyze substances seized from accused prisoners; see also, for example, Arsenal MS 10590, fols. 65r–69v and 70r–72r (29 April 1709); *Archives de la Bastille*, 13:185 (8 May 1715).

69. Arsenal MS 10548, Schuster's second interrogation, fol. 391v; Lenoir's analysis, 20 March 1704, fol. 395r–v: "je fis plusieurs experiences sur le mercur, et sur differants metaux, pour voir,

The Vinache affair involved a wide cast of characters who fill out a remarkable portrait of the diversity of chymists and the various perceptions of them in early eighteenth-century Paris. Vinache was suspected of counterfeiting. Senenburg, Deodat, and Pussieu were probably swindlers who claimed knowledge that they did not possess and extracted money before suddenly disappearing. Schuster and Tronchin were their willing dupes. Thuriac was a (presumably) competent operator who made a living distilling pharmaceuticals. Poli was more advanced and eventually an academician, interested in both the theory and the practice of chymistry, and an active seeker after transmutation. Marconnet was thoughtful and optimistic, supported by Bignon, and confident that he saw the way toward achieving transmutation, but later realized that his methodology was mistaken. The obvious difficulty for an early modern observer was how to distinguish among these possibilities, and especially how to work out all the intermediate grades that could run from intentional criminals through well-meaning but mistaken practitioners to transmutational adepts. What about, for example, Caraffe? Where on the spectrum did he fall? From our distant perspective and with the limited information available, it is impossible to tell. Homberg apparently *could* tell, based on his personal expertise and direct contact—his initial support for Caraffe disappeared only after he watched him at work. But not everyone was a Homberg.

Similar affairs played out again and again with astonishing frequency throughout the opening decades of the eighteenth century. The police tended to assume that anyone with an interest in chymistry, especially its transmutational aspects, was an actual or potential criminal. Dozens of people were investigated, arrested, and interrogated during the period for "seeking the philosophers' stone" or simply for "being mixed up in chymistry." While the Vinache affair was playing out, another investigation pursued an even wider network of chymical practitioners. Most were released relatively quickly and, if they were from the provinces, told to leave Paris. But the attitudes of the police—and particularly of the government ministers like Pontchartrain who gave the orders—prove significant for understanding the contradictory ways in which chymistry was viewed at the time. This new investigation began with an Augustinian hermit named Pierre Meusnier who was arrested simply because a search of his pockets revealed powders and chymical writings. The investigator promptly concluded darkly that "he is a chymist," even though he claimed to be a surgeon. On the basis of what was found in his pockets he was

---

si cette poudre pouvoit, par une vertu simpatique, ou antipatique changer, ou augmenter leurs natures."

brought to the Bastille and "accused of working to make good or false coins, having secrets, having blown at the coals [*soufflé*] and worked, and moreover of seeking the philosophers' stone."[70] He was interrogated three times and imprisoned for eight months before being released and ordered out of Paris. All the interrogators learned was that he had inherited some medical recipes and chymical books from his father and, following the writings of Geber, Philalethes, Flamel, and others, tried to convert a certain purgative powder into a transmuting agent.[71]

Further investigations led to more arrests and interrogations. Meusnier's associate Desmures was also "accused of blowing at coals [*souffler*] for good and false coins, and seeking the philosophers' stone." Although he was freed after two and a half months of imprisonment, Pontchartrain ordered that care be taken thereafter to observe his conduct regarding furnaces and distillations, referring to the Edict of 1682 and encouraging the police to use it liberally for making more arrests. Another associate, Madame Montigny, was accused of "being in the company of *souffleurs* for the philosophers' stone and working on mercury and even on some metals."[72] Claude-Jean-Baptiste Vialet, Sieur de la Tournelle, was arrested once in 1702 and again in 1704, and interrogated about what he knew in regard to making the philosophers' stone and what successes he had achieved. He replied that after forty years of work and the expenditure of ten to twelve thousand ecus, he had given up. This answer did not win his release; a month later, still imprisoned, he was interrogated again about the philosophers' stone and transmutation. He said that

---

70. The verb *souffler* can be difficult to render correctly at this period. The *Dictionnaire de l'Académie Françoise* gives among its many meanings: "souffler, pris absolument, signifie quelquefois, Chercher le pierre philosophale, chercher à faire de l'or, de l'argent, par les operations de l'Alchimie." While the Bastille records do use the word in this specific sense, they also use it much more broadly, essentially meaning "to practice chymistry," which is clear in the present and following quotation where "seeking the philosophers' stone" is only a subset of *souffler*. It is used in the same broader and non-judgemental sense by Saint-Simon, *Mémoires*, 4:456, in reference to Philippe's chymistry. This broader contemporaneous meaning is borne out also in Guy Miege, *A New Dictionary, French and English* (London, 1677), which defines *soûfler* as "s'occuper à la Chymìe, to worke in Chymistry," and *soûfleur* as "Chymiste, a Chymist"; I thank Bill Newman for this reference.

71. The full dossier is in Arsenal MS 10555; some materials are published in *Archives of the Bastille*, 11:153–56. Meusnier's interrogations are 12 April 1704 (fols. 18r–20v), 29 April (70r–71v), 10 Sept 1704 (121r–122r). For a broader view of eighteenth-century police activities, with a focus on sorcery but including some material on chrysopoeia, see Ulrike Krampl, *Les secrets de faux sorciers: Police, magie, et escroquerie à Paris au XVIIIe siècle* (Paris: Éditions EHESS, 2011).

72. *Archives de la Bastille*, 11:157 (Montigny, 18 May 1704), 156–58 (Desmures charged, 9 May; released, 23 July).

the only valuable thing he had found was a potable gold that extended life and health, but he used it only on himself, except for a portion he had given to Philippe's physician Helvétius, who had used it to good effect. After extensive questioning about how to prepare this remedy, he was finally released after a month and a half of imprisonment in the Bastille. Perhaps Tournelle's potable gold did have some virtue—when he was arrested and imprisoned the second time, he was already ninety-three years old.[73]

The clear message that emerges from the Bastille records is that anyone found (or thought) to be a chymist was automatically an object of suspicion. One Tilly, a wine merchant, was accused simply of "being among those seeking the philosophers' stone and a good chymist." The same scrutiny fell upon one Boucheix, who is described as "a chymist . . . he has worked for a long time to find the philosophers' stone"; Pontchartrain ordered that he be interrogated and all his papers seized. The working assumption of the police seems to have been that chymistry either is a cover for criminal activity or leads eventually to bad ends. Thus, for the police, it seemed a logical sequence that the widow l'Huiller "consumed her goods in seeking out the secrets of chymistry, which led her to counterfeiting and poison, under the pretext of metals and the philosophers' stone."[74] The consumption of one's wealth is a common enough theme in early modern anecdotes about transmutational chymistry—such as the story of Bernard Penot, who was reputed to have said that the best way to ruin someone was to interest him in transmutation—but here in the records of the Bastille the literary topos takes on a more concrete existence. For example, in 1708 the family of the Abbé Chesnet complained to the police that their wealth was being used up in transmutational pursuits by the abbé who, according to them, was under the influence of one of his domestics, Antoine Bègue, a maker of crucibles and furnaces. Investigators eventually tracked down the abbé in the garret of Bègue's house on the rue Vieille-du-Temple surrounded by crucibles, furnaces, and "many books and manuscripts on the subject of chymistry." In this case, Pontchartrain's orders explicitly state the presumption that chymistry is a cover for illegality: "see if, under the pretext of chymistry, there might not be counterfeiting, as it often and almost always turns out."[75]

73. Arsenal MS 10555, fols. 95r–97v (19 July 1704), 105r–106r (11 August 1704).

74. *Archives de la Bastille*, 11:155 (Tilly, 24 April 1704), 161 (Boucheix, 24 October 1704), 169 (L'Huiller, 12 January 1706).

75. On Bernard Penot, see Eugène Olivier, "Bernard Penot (Du Port), médicin et alchimiste (1519–1617)," *Chrysopoeia* 5 (1992–96): 571–673. On Chesnet, *Archives de la Bastille*, 11:435–38 (21 September 1708 and following); Bègue died in the Bastille in 1709, and his wife was released only in 1711.

One might conclude that such investigations of real or supposed chymists were motivated entirely by fears of counterfeiting, or explain them away as the actions of a paranoid police state, but neither paranoia nor legitimate fears of counterfeiting were the sole motivations for government officials like Pontchartrain. Again and again, in close parallel with suspicions of counterfeiting, poisoning, and cozenage, there existed undeniable and persistent signs of hope that some of the accused and arrested might actually have knowledge and secrets useful to the state. This dual attitude is clear already in the case of Marconnet, who after his arrest was put to work in the Bastille on transmutational processes. Likewise, before his arrest, Meusnier was allowed to continue working with Montigny under covert surveillance because the investigators planned to surprise him, hoping that "several vessels of the powder with which it is said they make gold will surely be found." Montigny, acting as an informant, was told to probe his knowledge especially regarding his claim that "if he had someone to present him to the King, he would make the truth of his secret known to his Majesty, which would be very useful for the kingdom."[76] The same hopes appear yet again in the case of Bègue and Chesnet, who were interrogated "to get to the bottom of their chymistry." Similarly, when Pontchartrain ordered the arrest of one Marescot in May 1704, the officers of the Bastille were told to "make her explain [*raisonner*] her supposed secrets, and also put her to work if you think it fitting, in order to see how far the knowledge she claims to have will go."[77] While all these investigations were going on in Paris, an array of governmental officials, including Pontchartrain, were going to extraordinary lengths to indulge a metalworker from Provence named Jean Troin and to acquire his reputed gold-making skills for the state.

Hopes and Fears: Our Chymist in Provence

The case of Jean Troin (1672–1712), known as Delisle, provides the most stunning example of the involvement of the early eighteenth-century French state—and also of Homberg, Philippe, and even the Académie—with promising chrysopoeians. A dual attitude toward such chymists, waffling between skepticism and optimism, suspicion and enthusiasm, is on display throughout. Reports of Delisle's ability to turn base metals into gold and silver attracted intense interest, and many favors, from the French court for *more than seven years*. Although there are many accounts of transmuters in Ger-

---

76. *Archives de la Bastille*, 11:154 (13 April 1704).
77. *Archives de la Bastille*, 11:155–56 (1 May 1704).

man princely and imperial courts, there are very few known in the French context, particularly during the reign of Louis XIV.[78] Delisle's is the most important and well-documented case of a gold-maker related to the court of Louis XIV. The events played out while Homberg and Philippe were at work on their own transmutational endeavors at the Palais Royal, and they were fully aware of, and to some extent involved in, the affair. Delisle's story has been told in part before, but a richer account can be given here by adding new sources that have not previously been known or not previously been connected to the startling narrative of the gold-maker from Provence.[79]

In March 1705, the bishop of Glandèves (a town in Provence) wrote a letter to Michel Chamillart (1652–1721), the *contrôleur générale des finances* of the kingdom, informing him about a man who can "turn iron into silver with a powder he makes," and suggesting that such knowledge could serve "the interests of the King."[80] The letter made its way to the *intendant* of Provence, Cardin Lebret (1675–1734), who identified the man as Delisle, and found that he stood accused of counterfeiting. An arrest attempt was made, but failed. In late 1706, another Provençal ecclesiastic described Delisle's transmutations and reported that Lebret had given him protection and sent an ingot of his transmuted gold to the king. He thereafter specified that Delisle held "a safe-conduct [*sauf-conduit*] that the Court accorded him, but with the order to present himself there this coming spring," where he plans "to make a trial before the King worthy of his Majesty by changing in a moment a great quantity of lead into gold." The writer assured his correspondent that Delisle's

78. The most famous seventeenth-century example in France is that of Noël Picard, called Dubois, which occurred in the court of Louis XIII in 1636. See Principe, "Digby," 11–14; and Miguel López Pérez, "El alquimista Dubois y el Cardenal Richelieu," *Azogue* 7 (2010–13): 327–38. For alchemists in German courts, see Bruce Moran, *The Alchemical World of the German Court, Sudhoffs Archiv* 29 (1991); Pamela Smith, *The Business of Alchemy: Science and Culture in the Holy Roman Empire* (Princeton, NJ: Princeton University Press, 1994); and Tara Nummedal, *Alchemy and Authority in the Holy Roman Empire* (Chicago: University of Chicago Press, 2007).

79. For earlier accounts of Delisle, see Nicolas Lenglet de Fresnoy, *Histoire de la philosophie hermétique*, 3 vols. (Paris, 1742) 2:68–103; *La vie et les lettres de Messire Jean Soanen*, 2 vols. (Cologne, 1750), 1:21–22; Louis Figuier, *L'alchimie et les alchimistes* (Paris, 1856), 317–32; J. A. Durbec, "L'Alchimiste de Saint-Auban," *Annales de la Société Scientifique et Littéraire de Cannes et de l'Arrondissment de Grasse* 15 (1957–61):131–87; and Miguel López Pérez, "La trágica historia de Jean Delisle," *Azogue* 7 (2010–13): 402–48. Durbec, "L'Alchimiste," 182–85, provides a valuable list of archival materials relating to Delisle (although it does not include many newly discovered archival documents cited here).

80. AN, G/7/468, fol. 103r; César de Sabran, bishop of Glandéves, 19 March 1705. The addressee is not specified, although it is almost certainly Chamillart, and if not, the letter at least reached Chamillart in due course.

knowledge, once communicated to Louis XIV, would "humble the enemies of France."[81] It was in fact Chamillart, obviously interested by what he had been told, who ordered Lebret to issue Delisle the safe-conduct in November 1706.[82] It is significant that notice of this supposed gold-maker went primarily to the minister of finance. Chamillart's subsequent actions were undoubtedly motivated by his knowledge as finance minister that France was virtually bankrupt—a man able to make gold might be the only way to solve the problem. Unfortunately, Delisle did not appear at Versailles in 1707 to save the kingdom.

In early 1708, Delisle asked Lebret to renew his safe-conduct. He explained that his preparations required exposure to bright sunlight and since he could not guarantee the weather, he would like a safe-conduct without an expiration date. The transmuter assured the intendant that as soon as his preparations were complete he would "go to give His Majesty all the proofs that are asked of me." Lebret forwarded the request to Chamillart, who granted a new safe-conduct.[83] In May 1708, a visitor claiming to be Delisle's servant appeared at Versailles and met with Pontchartrain to inform him of Delisle's abilities. Pontchartrain briefed Louis XIV, and sent a letter to François de Grignan (1632–1714), the king's lieutenant-general in Provence. The letter stated that while the king "has little belief in this sort of secret," Delisle should nevertheless be investigated. The minister noted that Delisle might be a counterfeiter—the same assumption that Pontchartrain expressed to investigators at the Bastille—but also that Grignan should be careful not to frighten Delisle into flight since he lived close "to the borders of the states of the Duke of Savoy."[84] Pontchartrain left this concern unexplained, but it was more than concern about the escape of a possible criminal; it was probably the first intimation of the worry (later expressed explicitly) that Delisle's transmutational abilities might fall into the hands of a foreign power. Pont-

81. De Cerisy to the Vicar of St. Jacques du Haut-Pas, 18 November 1706 and 27 January 1707, published in Lenglet de Fresnoy, *Histoire*, 69–74. The *intendant* mentioned is certainly LeBret.

82. The issuance of the *sauf-conduit* in November 1706 can be inferred from the statements made in AN, G/7/472, fols. 42r–v, 48r–v, and enclosure.

83. AN, G/7/472, fols. 42r–v (summary of Delisle's request), 48r–v (Lebret to Chamillart, 28 February 1708), and enclosure (Delisle to Lebret, 21 February 1708); and BNF, MS NAF 4302, fol. 4r (date of *sauf-conduit*; Jean Soanen to Nicolas Desmaretz, 15 November 1709).

84. Pontchartrain to de Grignan, 30 May 1708; published in *Archives de la Bastille*, a 12:52–53. Pontchartrain's visitor was named Troin, suggesting that he was either relative of the transmuter or perhaps even Delisle himself.

chartrain's concern was timely: Victor Amadeus II of Savoy had invaded Provence only a few months earlier. The uneasily juxtaposed parallel concerns that Delisle was merely a fraud or a counterfeiter *and* that his chrysopoetic knowledge must be secured for France recur throughout the subsequent governmental correspondence.

Delisle was now an object of interest to two arms of the state, which seem initially not to have known of each other's involvement. On the one hand, there was the finance minister Chamillart working through the intendant Lebret, and on the other, the minister Pontchartrain working through the lieutenant-general Grignan—not to mention the king's involvement. Following Pontchartrain's orders, Grignan investigated Delisle, and replied that he considered him a fraud and so tried to arrest him, but discovered that Delisle held a "safe-conduct to let him work without being bothered." Pontchartrain replied rather sharply to Grignan that Louis XIV found Grignan's negative assessment at odds with the positive accounts he had received from Jean Soanen (1647–1740), Bishop of Senez, "and from others who regard [Delisle] as a marvelous man."[85] At Pontchartrain's request, Soanen had invited Delisle to Senez to perform a transmutation under close observation, and reported that he was "perfectly convinced of the excellence of his art and of the honesty of his intentions for his Majesty." Soanen also described a nail that Delisle had produced a year earlier: half of its shank had been transmuted into silver and half of its head into gold.[86] This nail was one of many such items that Delisle produced as evidence of his transmutations. Such "proofs" were not original to Delisle; the primary example, and undoubtedly Delisle's inspiration, was an iron nail half-transmuted into gold by Leonhard Thurneysser (1531–96) in the presence of Cardinal Ferdinando de' Medici in Rome in 1586. Thurneysser accomplished this feat by dipping the nail halfway into a supposedly transmutatory oil he had prepared.[87] Pontchartrain told

85. De Grignan to Pontchartrain, 12 September 1708, and Pontchartrain to de Grignan, 26 September 1708, in *Archives de la Bastille*, 12:53–54.

86. Jean Soanen to Pontchartrain, 18 October 1708, BNF, MS NAF 31, fol. 157r: "je suis parfaitement convaincu de l'excellence de son art, et de la droiture de ses intentions pour sa majesté."

87. On Thurneysser's nail, see Didier Kahn, "The Significance of Transmutation in Alchemy: The Case of Thurneysser's Half-Gold Nail," in *Fakes!? Hoaxes, Counterfeits and Deception in Early Modern Science*, ed. Marco Beretta and Maria Conforti (Sagamore Beach, MA: Science History Publications, 2014), 35–68, and references therein. Soanen mentions Thurneysser's nail in his letter, saying that Delisle's is a greater wonder because it combines *three* metals rather than just two.

Soanen that the king would like to see "the nail that [Delisle] has made" and asked for it to be sent and for the bishop to find out whether Delisle "would be in a state to work to render his discovery (too excellent to be believed without seeing it) useful to the king." Here again Pontchartrain combines skepticism about transmutation with cautious optimism that gold-making could benefit Louis XIV.

Both 1708 and the summer of 1709 came and went without Delisle making an appearance at Versailles. In November 1709, Soanen wrote a long letter to the new *contrôleur général des finances*, Nicolas Desmaretz (1648–1721), who had replaced Chamillart in early 1708, and was skeptical of Delisle's abilities.[88] The bishop recounted how Delisle turned iron nails into silver, and transmuted lead into gold and silver in front of a dozen witnesses. Soanen had sent these transmuted metals to Pontchartrain, who had them assayed successfully "by the best goldsmiths in Paris." As for Desmaretz's concern that Delisle was accused of counterfeiting, Soanen explained that Delisle merely had the misfortune of lodging with a man convicted of restamping louis d'or, and so the authorities had simply assumed that Delisle was his accomplice. Soanen explained that Delisle had not yet appeared at Versailles because he did not have enough materials to make a demonstration suitable for Louis XIV. His transmuting agents could be prepared only with exposure to the summer sun, and Delisle had been prevented from working during the two previous summers. In 1707 his work was disrupted by the Duke of Savoy's invasion; in summer 1708, Grignan's surprise attempt to arrest him caused him to break a bottle of his preparations, and he was thereafter unable to settle down to work out of fear for his safety. Now he was in need of a third safe-conduct.[89]

---

88. There is some confusion about the addressee and date of this letter. Lenglet du Fresnoy (*Histoire*, 2:76–83) states it is to Desmaretz and dated simply 1709 without a month; Figuier (*L'Alchimie*, 359) copies Lenglet's version but adds the month of April without explanation. *Archives de la Bastille* (12:54–56) prints a significantly truncated version with Daniel François Voysin (1654–1717) as addressee and 15 November 1709 as the date. A contemporaneous copy (Arsenal MS 2064, fols. 174r–175r) gives the same truncated text but names no addressee. BNF, MS NAF 4302, fols. 3r–4v, however, provides the most complete version; it is dated 15 November 1709 and addressed to Desmaretz. Internal evidence—that the addressee had been elevated to a new position over a year before—points clearly to Desmaretz (who became *contrôleur général des finances* on 20 February 1708) rather than to Voysin, and remarks made by the Marquis de Dangeau (see note 92) support the November date.

89. Jean Soanen to Nicolas Desmaretz, 15 November 1709, MS NAF 4302, fols. 3r–4v; the breakage of the bottle is noted in Lebret to Desmaretz, 5 September 1708, AN G/7/472, piece 231; and the delay noted by Delisle to Lebret, 2 September 1708, AN G/7/472, piece 232: "Jaurois este en estat daller a paris a la noel pour faire voir au roy et a mgr le contourleur general la verite de mon travail si lon me navoit auté [that is, oté] le moien par linsulte quon ma fait faire."

The finance minister Desmaretz agreed and granted a new safe-conduct, and just in the nick of time since Grignan was setting up another arrest attempt.[90]

The bishop's nephew, du Bourget, had recently visited, and he too witnessed Delisle's transmutations. Accordingly, du Bourget supplied the finance minister with an affidavit that included precise details of how he performed the transmutation while Delisle gave instructions, and how he was rendered speechless when he poured gold out of the crucible that he had himself initially filled with lead.[91] A contemporaneous source—the journal of the Marquis de Dangeau—records that du Bourget brought this transmuted gold to Paris in November 1709 and that it was found to be "the best gold in the world and of the quality of our louis [d'or]." Desmaretz ordered du Bourget back to Provence at once to tell his uncle to bring Delisle to Paris personally.[92]

Significantly, at the same time, letters about Delisle and the philosophers' stone were read at two meetings of the Académie Royale de Sciences. Not surprisingly, there is no record of these letters whatsoever in the *procès-verbaux*, even though they record that letters from Hans Sloane and others were read on the same day. Only Bourdelin's diary records that "a letter from Mr. Courtois about the philosophers' stone" and "a letter from the bishop of Senez to Mr. Desmaretz about the philosophers' stone of Delisle" were read to the assembly.[93] Since Soanen wrote his letter just eight days before it was read to the Académie, it was obviously transmitted to the Académie immediately upon its arrival. Desmaretz must have asked the academicians to weigh in about Delisle's transmutations. Unfortunately, Bourdelin does not record how the academicians responded. Soanen's next letter, however, indicates that the response was not entirely positive, since the bishop strove to reassure Desmaretz's brother-in-law Louis Béchameil, Marquis de Nointel (1649–1718), of the reality of such transmutations based on his own eyewitness and that of other reliable people, "even though several great philosophers of the Académie des Sciences judge it otherwise at Paris."[94] Of course, some

90. Grignan to Desmaretz, 3 December 1709, AN G/7/474, piece 198; and Lebret to Desmaretz, 26 December 1709, AN G/7/474, piece 234: "Ainsy vostre intention est apparemment qu'on ne fasse plus aucune désmarche pour larrester."

91. Arsenal MS 10599, fols. 195r–202r , especially 201v.

92. *Journal du marquis de Dangeau*, 35 vols. (Clérmont-Ferrand: Éditions Paleo, 2002), 22:276–77 (25 November 1709).

93. Bourdelin's Diary, fol. 90r (16 and 23 November 1709); the letter read on 23 November is surely Soanen's letter of 15 November. It is not yet clear who Mr. Courtois is, but given the proximity to the letter about Delisle, that letter presumably relates to the same affair.

94. Soanen to Nointel, 30 April 1710, Arsenal MS 10599, fols. 162r–167v; at fol. 165v: "puisque plusieurs grands Philosophes de l'Academie des Sciences en jugent autrement a Paris."

academicians—Nicolas Lémery and Fontenelle in particular—famously rejected transmutation even while Homberg and others supported it. Homberg
was present when the bishop's letter was read, and so he certainly heard about
Delisle at this time, if he had not been informed about him previously; he
would have more direct contact with Delisle in the following year.

Another eyewitness soon seconded Soanen's support of Delisle—no
less a person than Nicolas Foy de Saint-Maurice (b. 1672), president of the
mint of Lyons. Saint-Maurice had been charged in December 1709 to "seek
out false productions of coins in Provence," an order he carried out with a
vengeance—the archives record many executions resulting from his investigations. Such duties presumably explain why he contacted Delisle.[95] In May
1710, Delisle invited Saint-Maurice to the village of Saint-Auban (where he
was lodging with the local abbé). Accompanied by his *prevôt* and a represen-
tative of the Provençal intendant Lebret, Saint-Maurice helped carry out the
final operations to prepare Delisle's transmuting agent. He first exposed a cin-
dery material (which had been buried in the garden) to the sun and distilled a
drop of liquid—the "philosophical mercury"—from it. He mixed this drop
with 3 ounces of quicksilver, added 2 drops of "oil of the sun" from a bottle
Delisle handed him, and put the whole into a furnace. Upon pouring out the
contents, he obtained a 3 ounce ingot of gold. Delisle then melted 3 ounces of
lead bullets taken from Saint-Maurice's valet and gave the president a powder and 2 drops of the "oil of the sun" to throw in. Saltpeter was added, and
the whole heated for a quarter of an hour, then poured out, providing more
gold. Delisle then produced silver from lead. An astonished Saint-Maurice
immediately sent a letter "from amid the mountains" to Desmaretz by special
courier, assuring him "that nothing is more certain, Sir, than his secret for
making gold and silver." He wrote a fuller letter a few days later promising
many more things he thought should be relayed only verbally, and reiterated
the concern that Delisle must not "pass into foreign lands."[96] Saint-Maurice
provided detailed documents describing his witness of the transmutation

---

A summary on fols. 155r–163r states (at fol. 158r) that Desmaretz had asked Nointel to undertake
a closer investigation of Delisle.

95. François Abot de Bazinghen, *Traité des monnoies et de la jurisdiction de la cour des
monnoies*, 2 vols. (Paris, 1764), 1:290; Jean-François Jolibois, *Histoire de la ville et du canton de
Trévoux* (Lyons, 1853), 30. Soanen to Nointel, 1 August 1710, Arsenal MS 10599, fols.169r–171v,
published in *Archives de la Bastille*, 12:56.

96. Saint-Maurice to Desmaretz, 21 May 1710 (from Cannes, about to leave for Saint-
Auban); ditto (from Saint-Auban, also dated 21 May, so the date on this or the previous letter
must be a mistake): "Rien n'est plus seur Monseigneur, que son secret pour faire de l'or et de
l'argent"; ditto (from Cannes, 26 May 1710); all in G/7/1463.

and Delisle's procedures, and presented a formal affidavit to Louis XIV at Versailles later that year.[97] He sent the gold and silver to the Lyons mint. The mint master there coined the silver into "2 écus, 2 demi-écus, 5 quarter-écus, and 3 pieces of 10 [sol]" which he gave to Saint-Maurice, fearing to release coins made of "philosophical materials" into the general currency. The gold, however, proved too brittle to coin. The mintmaster wanted to refine it, but Saint-Maurice sent it instead to Desmaretz, who gave part of it to Nicolas de Launay (1647–1727), director of the mint for medallions, who had it successfully coined into three medallions bearing a bust of Louis XIV, the date 1710, and the inscription "aurum arte factum" ("gold made by art"). These medallions were given to the king.[98]

In August 1710, five years into the affair, Delisle still claimed not to have enough transmuting agents for the king. Nevertheless, he kept demonstrating transmutation for others, which convinced the bishop that he had more transmuting agents than he would admit. Therefore, Soanen urged him to go to Versailles in September and gave him written orders from the court forbidding him to carry out any more transmutations in order "to reserve everything for the king." At the same time, a troubling rumor reached Versailles: Delisle had been seen in Turin at the court of the Duke of Savoy. Soanen protested that, as far he knew, the transmuter had gone only to Monaco and Menton to see his wife, "in the lands of an entirely French prince."[99] Soanen at once conferred with Saint-Maurice, who had his own news both good and bad. On the good side, Delisle now suggested that he was ready to accompany the president of the mint to Paris. On the bad side, the rumor was true: the *procureur* of Turin verified that Delisle had been there. "You will agree that this trip does not signify anything good, and makes one think that Delisle is thinking about leaving the kingdom. . . . I would think that the court ought not wait for the expiration of his safe-conduct but ought to assure itself of this man in a respectable [*honnête*] way, as well as of all the prepared materi-

97. "Maniere dont le Sr. de l'Isle a dit a M. de St. Maurice qu'il faisoit son huile de soleil" (undated, two copies), "Les Epreuves et experiences qui ont été faittes par le sieur President de St. Maurice au chateau de St. Auban dans le mois de May de l'année 1710 au sujet de la mutation des meteaux . . ." (two copies, dated 14 December 1710 at Versailles), G/7/1463. A copy of the latter exists in Arsenal MS 10599, fols. 185r–189r, and the text is published in Lenglet de Fresnoy, *Histoire*, 2:86–90.

98. Lenglet de Fresnoy, *Histoire*, 2:94–95; and *Vie de Soanen*, 1:22.

99. Soanen to Nointel, 1 August 1710, Arsenal MS 10599, fols. 169r–172r; published in *Archives de la Bastille*, 12:56–58. Indeed, that "entirely French Prince," the Prince of Monaco, kept Delisle under surveillance when he was in Menton, and sent information about him to Desmaretz on 17 December 1709 (G/7/1435) and 11 April 1710 (G/7/1438).

als he might have." In reality, things were worse than Saint-Maurice or the court knew; it emerged later that Delisle had developed relations also with the king of Portugal, who gave him a gold medallion and a title.[100] Saint-Maurice requested explicit orders to bring Delisle to Paris.[101]

Saint-Maurice brought Delisle to Paris—whether willingly or unwillingly—in November 1710. This trip is not mentioned anywhere in the archival sources, but is reported by prominent observers at court. The Marquis de Dangeau and the Duc de Saint-Simon both report that upon his arrival in Paris, Delisle was put under the supervision of Jean Boudin, the first physician to Monseigneur, Louis XIV's eldest son. Boudin was himself a chymical practitioner and avid seeker after the philosophers' stone. As such, he represents yet another chrysopoeian in court circles of the time.[102] Delisle was neither the first nor the last transmuter Boudin supervised for the French state. Two years earlier, in 1708, a transmuter convinced Boudin that he had the secret of the stone. Boudin informed Louis XIV, and the transmuter was placed under guard at Boudin's house on the rue de Montreuil, close to the gates of the palace of Versailles. He worked there for two months, and was visited by many members of the court, including Chamillart, then *contrôleur général des finances*, and princes of the blood (perhaps including Philippe). As time passed without the production of any gold, "it was necessary to save the honor of the king and his ministers, and make him work in the château de Noisy without anyone knowing about it; for, due to the poverty of the finances and the sad state of the armies, enemies would believe that the kingdom was in the worst state of depletion [*épuisement*] because it had sought to procure money for itself by such extraordinary and uncertain means."[103]

This report echoes the political concerns behind Colbert's and Louis XIV's earlier prohibitions against chrysopoeia at the Académie, and clearly expresses once again the uncomfortable dilemma presented by metallic trans-

100. Précis of seizure of Delisle's effects at Saint-Auban by Calvy, judge in Grasse, AN, G/7/477, piece 85. This précis is undated, but a date of 29 March is provided in Grignan's letter of 13 April 1711, AN, G/7/477, piece 106.

101. Saint-Maurice to Nointel, 17 August 1710, Arsenal MS 10599, fols. 179r–180v; published in *Archives de la Bastille*, 12:59–60. The *Archives* implies that the letter is to Voysin, but it is certainly to Nointel, who is given as the addressee of a letter from Soanen written on the same day from Saint-Maurice's chamber, and the two letters clearly share the same addressee.

102. *Journal de Dangeau*, 23:262; and Saint-Simon, *Mémoires*, 3:1042–43.

103. Pierre Narbonne, *Journal des règnes de Louis XIV et Louis XV*, ed. J. A. Le Roi (Paris, 1866), 4–5. This account from Narbonne, the commissioner of police at Versailles, has sometimes been thought to refer to Delisle, but this is impossible. Narbonne clearly gives the date 1708 for these events, and notes that Chamillart was then *contrôleur général des finances*; Chamillart was dismissed from that post in 1708, well before Delisle arrived at the end of 1710.

mutation. Chrysopoeia's promises of wealth, especially for a state in such dire financial straits as early eighteenth-century France, were too alluring to ignore, but at the same time their pursuit could lead to dangerous political and economic consequences.

This earlier experience probably explains why Delisle was entrusted to Boudin in November 1710.[104] In Boudin's house Delisle was supplied with materials, furnaces, and crucibles, and expected to demonstrate his transmutations. Hitherto unnoticed letters from Elisabeth Charlotte provide further information and connect Delisle more closely to Philippe—and hence to Homberg—than has previously been known. In mid-November 1710, Elisabeth Charlotte wrote that "a man has arrived who says that he can make gold. I think he's a rogue and a fool, for he says that he is 'regenerated' [*rejeneres*], that he has abandoned the old Adam, and for this reason, on account of the pure life he now leads, that God has given him the grace to find the philosophers' stone. I think that if he had it, he wouldn't talk about it and wouldn't present himself to the court."[105] She sent her correspondent one of Delisle's half-iron and half-gold nails. He returned the nail, and Madame gave it to someone else, but promised to send "another piece of his making."[106] Apparently, Delisle was doing little more at this time in Paris than providing his standard demonstrations with iron nails, rather than appearing before the king to perform a substantial transmutation—something he kept protesting he did not have sufficient materials to do. While Madame was skeptical, others were convinced. "I've heard a lot said about Mr. Delisle this morning; a president of the mint named Saint-Maurice came to find my son [the Duc d'Orléans] in order to assure him that Delisle surely has the powder of projection, and that he made silver once in his presence and very good gold twice. He [Delisle] will not go to the galleys, having only lodged with a counterfeiter; he [Saint-Maurice] has given all this in writing to the King."[107] But just after Christmas, Elisabeth Charlotte's negative opinion was over-

---

104. Nor was Delisle the last transmuter entrusted to Boudin. In the summer of 1715, Diesback, who had been imprisoned in the Bastille because of rumors of his transmutational prowess, was released in order to work under the observation of Boudin; see *Archives de la Bastille*, 13:185–87.

105. Elisabeth Charlotte to Étienne Polier de Bottens, 16 November 1710, in *Aus den Briefen der Herzogin Elizabeth Charlotte von Orléans an Étienne Polier de Bottens*, ed. S. Hellman, *Bibliothek des litterarischen Vereins in Stuttgart* 231 (1903): 104. This is the only mention of any religious sentiment connected with Delisle.

106. Elisabeth Charlotte to Étienne Polier de Bottens, 6 December 1710; *Aus den Briefen*, 105.

107. Elisabeth Charlotte to Étienne Polier de Bottens, 16 December 1710; *Aus den Briefen*, 107. Saint-Maurice presented his certificate on 14 December; see note 97.

turned by report of success. "Mr. Boudin's gold-maker has finally made a sufficient quantity of gold which has survived all the assays, except that of aquafort, but the man says that the crucibles and the coals are no good. The crucibles cracked and some gold was found in the broken pieces; this is why he is going to begin to work again. My son is not persuaded that he can succeed; we shall soon see what comes of it, and I'll let you know."[108] These letters clearly indicate the Orléans household's interest in and involvement with Delisle's endeavors—Saint-Maurice even sought out Philippe specifically to relay information to him—and so Homberg would certainly have been kept well informed if not yet more involved.

Saint-Maurice's testimony also appeared in the popular press in January 1711, indicating how widespread the information became. The editor of the *Mercure galant*, Charles Dufresny (1654–1724), in the context of his lengthy review of a recently published anti-chrysopoetic tract, reported that "I have had several conversations with Monsieur le President de Saint-Maurice about this subject. He was in that state of uncertainty that one ought to have in regard to obscure knowledge when he was charged to make an experiment in the valley of Barcelonette. He took every possible precaution in order not to be duped, and he really saw two ounces of lead changed into gold by means of a little pinch of powder. One should not fail to pay attention to this. It is still a curious secret; it remains to be known if it will be useful, and if the cost of the operation will exceed the profit."[109]

Delisle left Paris shortly after Christmas 1710—although it is unclear whether he was allowed to leave freely or somehow escaped his "hosts." He returned to Provence and met with the bishop of Senez in early January 1711. Saint-Maurice wrote to express his disappointment with Delisle, but Soanen pleaded for leniency and another safe-conduct. Fear of losing him to France's enemies was now more intense because "many foreign princes are offering him their protection."[110] A new safe-conduct was not issued; instead, the order was given to arrest Delisle. He was captured near Nice, apparently on his way out of France. He was brought to Marseille before Grignan, and his effects at Saint-Auban were seized. He wrote to Soanen asking him to send

108. Elisabeth Charlotte to Étienne Polier de Bottens, 26 December 1710; *Aus den Briefen*, 108.

109. *Mercure galant* (January 1711), 3–28, at 18–19. The book under review was François Pousse, *Examen des principes des alchymistes sur la pierre philosophale* (Paris, 1711). See Alain Mothu, "L'alchimie en examen: *L'examen des principes des alchymistes sur la pierre philosophale* (1711) d'après les journaux de l'époque," *Chrysopoeia* 5 (1992–96): 739–50.

110. Soanen to de Saint-Maurice, 9 January 1711, Arsenal MS 10599, fols. 191r–194r; printed in *Archives de la Bastille*, 12:61–64.

on to Paris the materials that Delisle had entrusted to him; Nointel had already advised the bishop that if Delisle was brought to Paris, the king expected him to come as well with the powders. After begging Nointel to treat Delisle kindly, Soanen set out for Paris. "Our chymist"—as he had come to be called—was then transported to Paris "bound hand and foot." He arrived on 3 April and was put into the Bastille the next day.[111]

Rather than being imprisoned in a cell and subjected to interrogation as a fraud or a counterfeiter, Delisle was lodged in the "grand appartement" within the Bastille where he was made comfortable. He slept in a large "bed with four columns and curtains of green serge," and was attended by servants.[112] Once he had recovered from injuries suffered during his attempt to escape on the way to Paris (he was shot in the thigh), he set to work again on his transmutational agents. Supplied with a laboratory, apparatus, charcoal, and materials, he worked in the Bastille through the rest of 1711. Desmaretz sent the materials confiscated from Saint-Auban. Soanen, his nephew du Bourget, and de Launay acted as witnesses and assistants. Charles le Fournière Bernaville (governor of the Bastille) recorded the processes as they were carried out. Soanen and du Bourget sent regular reports to Nointel, who remained hopeful that Delisle would soon "bear out the truth of his secret."[113] On 31 October, de Nointel himself, de Launay, Soanen, du Bourget, and three officers of the mint assembled at the Bastille for a trial of Delisle's preparations. Nointel brought crucibles and lead, de Launay brought retorts for distilling the "mercure philosophique," and Soanen brought the *poudre métallique* he helped Delisle prepare. The first step involved heating the *poudre métallique* in order to isolate the needed *mercure philosophique*, but after two hours of heating, only a watery distillate appeared. Delisle resumed his work and tried again on 11 December, but the process failed again. Delisle blamed these failures on the fact that the autumn sun of Paris was no match for the summer sun of Provence—too weak to make the materials combine as they should to generate the necessary *mercure philosophique*. Delisle's only recourse was to try putting the flask containing the material near the fire as

111. Soanen to de Nointel, 14 March 1711, Arsenal MS 10599, fols. 204r–205v; printed in *Archives de la Bastille*, 12:64–65. Delisle made an escape attempt on the way, but it ended badly; he was recaptured after having been struck by a bullet in the thigh; see Grignan to ?, 31 March 1711, G/7/477, piece 86; and Durbec, "L'Alchimiste," 163–65, for a fuller account of this misadventure.

112. Arsenal MS 10599, fols. 231v and 233r.

113. Nointel to Soanen or Bernaville, 10 July, 12 and 23 August 1711, MS NAF 4302, fols. 8r–10r; Bernaville's notes, fair copy, Arsenal MS 10599, fols. 219r–224r (originals, fols. 210r–213v).

a substitute, but this method also failed because the heat of the fire did not "have the virtue of that of the sun."[114]

After this third failure, Desmaretz ordered the lieutenant-general of police, Marc-René de Voyer de Paulmy d'Argenson (1652–1721), to interrogate Delisle. But the interrogation was delayed because Delisle wanted one more chance—he wrote to Abbé de Saint-Auban, who, he believed, still possessed some of his preparations, in the hopes that those materials would be more effective. The abbé delivered everything to the intendant Lebret, who forwarded them to Desmaretz at Versailles with "all the precautions that I am able, so that the treasures that this box may contain will not be spoiled on the road." Clearly, despite the failures, the government officials still held out a gleam of hope for Delisle's success. Such hope did not last long. The box arrived, was delivered to Delisle at the Bastille, but was found to contain little of use. The materials were tried on 20 January 1712, but to no effect.[115]

His last recourse gone, Delisle was interrogated a few days later. He revealed that his secret came from an Italian named Denis, whom he had met in Nice and then lived with in Avignon for eight months in 1701. He admitted that while he had followed Denis's instructions several times, the process had succeeded for him only once, and that his successful demonstrations in Provence were performed using materials prepared by the Italian. His answers often seemed contradictory to the interrogator (although Delisle explained how they were not), and in the end he concluded that Delisle was lying.[116] Three days later, before dawn on 31 January 1712, Delisle, exhausted and deeply disheartened by his interrogation, began to vomit, and at ten o'clock that evening he died.

After an autopsy, Delisle's death was attributed to natural causes. Yet his manner of death strongly suggests poisoning, whether administered by the transmuter himself or by others. Immediately after his interrogation he said that he wanted to die and wished he were already dead, and several times before his arrest he had declared that if he were treated poorly, he would rather

114. Nointel to Bernaville, 28 and 30 October 1711, MS NAF 4302, fols. 11r–12v; Soanen to Bernaville[?], 30 October 1711, MS NAF 4302, fols. 19r–20r; Bernaville's notes (the date here is erroneously given as 29 rather than 31 October), Arsenal MS 10599, fol. 221r–v and 156v–157r; and d'Argenson to Pontchartrain, 1 November 1711, *Archives de la Bastille*, 12:65–66.

115. Desmaretz to d'Argenson, 22 December 1711, Arsenal MS 10599, fol. 216r; Lebret to Desmaretz, 2 January 1712, Arsenal MS 10599, fol. 217r–v, and printed in *Archives de la Bastille*, 12:66; and Bernaville's notes, Arsenal MS 10599, fol. 222r–v.

116. Interrogation transcript, 27 January 1712, Arsenal MS 10599, fols. 226r–229v, 156r.

kill himself than reveal what he knew.[117] Delisle had access during his experiments to arsenic, antimony, and *nux vomica*, any of which, if ingested, would have caused vomiting and proven fatal. When submitting the final report, d'Argenson remarked "that I still suspect that his death was hastened along." Despite these qualms, the police chief abruptly concluded the investigation by labeling Delisle "a notable scoundrel [*fripon*] who preferred to die rather than reveal the secret of his deceits," and declared coldly that "at last he will not fool anyone anymore."[118]

This affair illustrates the contradictory ways the French court viewed chrysopoeia at the start of the eighteenth century. While the king and his ministers feared that transmuters were frauds or counterfeiters, they remained unwilling not to indulge, even patronize, them in case their abilities were authentic and could solve the kingdom's financial crisis, and lest their knowledge wind up with a foreign power. Interestingly, all the known examples of serious interest by the ministers in transmutation during Louis XIV's reign date from the early eighteenth century (none during the seventeenth), suggesting that France's steadily deepening financial crisis eventually overcame qualms and skepticism enough for the possibility to be entertained that transmutation might resupply the treasury. Madame's letters show that the Orléans household was closely involved with Delisle while he was in Paris. If Philippe and his mother knew of events as they played out, then certainly Homberg did as well, besides having heard of Delisle's claims at the Académie itself. Likewise, if one of Delisle's half-transmuted nails was passed around by Madame, it is inconceivable that Homberg did not handle and examine it as well. This likelihood suggests a further connection. When Geoffroy gave his famous paper about false transmuters in 1722 (discussed in chapter 7), he mentioned Thurneysser's half-transmuted nail and noted that "such are the nails I display today to the company, half silver and half iron." Geoffroy then immediately referred to the objects that "a famous charlatan spread about in Provence a few years ago," making it clear that the nails Geoffroy displayed to the Académie were Delisle's.[119] Remarkably, these same nails were offered at public sale more than fifty years later, in 1777, by Bernard Azéma, Geoffroy's successor. Azéma described them as "three horse-cart nails [*clous de charette*;

117. D'Argenson to Nointel, 1 February 1712, Arsenal MS 10599, fol. 230r–v, and printed in *Archives de la Bastille*, 12:67.

118. D'Argenson to Nointel, 3 February 1712, Arsenal MS 10599, fol. 235r; to Desmaretz, 14 February 1712, fol. 241r; both letters printed in *Archives de la Bastille*, 12:67–68. Delisle's death certificate, 24 June 1712, Arsenal MS 10599, fol. 242r.

119. Geoffroy, "Supercheries," 63–64; see also *Mercure galant* (May 1722): 122.

that is, large-headed nails for securing iron rim-bands to cartwheels] converted by transmutation into silver—one at its head, another at its point, and the third in its entirety."[120] This account agrees very well with a description of Delisle's nails dating from 1710.[121]

Azéma also provided intriguing information about where the three nails were produced: "This operation took place in Mr. Geoffroy's laboratory in the presence of this great man and of a number of chymists . . . these savants really saw this transmutation of the nails into silver take place when they were dipped into a liquor, the knowledge of which, of course, the author of the phenomenon concealed." If Azéma's information is correct, then Geoffroy was present at a transmutation performed by Delisle, presumably during the latter's visit to Paris in late 1710. If Geoffroy was present, it is inconceivable that his mentor Homberg—even more interested in transmutation than his younger associate—was not also present. What remains open to question is where this demonstration took place. At that early date, it is doubtful that Geoffroy had a laboratory of his own, as Azéma suggested, and he certainly did not yet have sufficient standing to preside over a group of chymists acting as witnesses. The transmutation might have taken place at Boudin's house, but would he have thought to invite Geoffroy to the event? It seems more likely that Geoffroy was present thanks to the elder Homberg, and that the demonstration might therefore actually have taken place in the laboratory at the Palais Royal. This locale would neatly explain how Elisabeth Charlotte owned one of the half-transmuted nails and was so confident of being able to obtain another, and how Geoffroy obtained his, namely, through Homberg.

What happened to these nails after 1777 is unknown; however, another of Delisle's alchemical products can be traced further. In 1886, one of the three medallions coined from the gold produced by Saint-Maurice resurfaced and was examined by Albert de Rochas (1837–1914). He provided a drawing of the coin (fig. 6.1) and reported that it seemed too light to be pure gold. The coin also showed some "spots of verdigris," which one would expect from the corrosion of a copper-containing alloy. Upon treatment with nitric acid, the spots of verdigris dissolved promptly, but the coin itself remained untouched.[122] This use of nitric acid represents the proverbial "acid test" for

120. *Journal de Paris*, no. 229 (17 August 1777): 2–3. See also Marco Beretta, "Transmutations and Frauds in Enlightened Paris: Lavoisier and Alchemy," in *Fakes!?*, ed. Beretta and Conforti, 69–107, especially 69–73.

121. Letter to Mr. Ricard, 19 July 1710, published in Lenglet de Fresnoy, *Histoire*, 2:84–85: "six large nails, one of the six . . . was transmuted into silver from its head to the middle, below that it remains iron. The other six were entirely converted into silver."

122. Albert de Rochas, "L'or alchimique," *La Nature* (1886): 339–43, at 341.

FIGURE 6.1. A nineteenth-century drawing of one of the medallions struck in Paris from the gold produced by Saint-Maurice and the transmuter Delisle in May 1710. From Albert de Rochas, "L'or alchimique," *La Nature* (1886): 339–43, at 341. This coin was last seen in Paris in 1928—*where is it now?*

gold—all the base metals and silver are quickly corroded by nitric acid, while gold is not, hence the composition of Delisle's metal remains a mystery. The coins can be traced still further. One of them, perhaps the one examined by de Rochas, was in the famous nineteenth-century Feuardent Collection of French *jetons*. This collection—including the Delisle coin, described as being of "a very beautiful metal"—was auctioned off in Paris in 1928; the coin's purchaser and its present whereabouts remain unknown.[123]

Homberg's Narrow Escape: The Danger of Being a Chymist

Delisle, Vinache, Marconnet, Meusnier, and many others ended up in the Bastille in the early eighteenth century because of their real or purported chymical abilities—either as potential gold-makers or as criminal counterfeiters. Delisle and Marconnet were put to work on metallic transmutation within its walls, and Vinache and Delisle died under questionable circumstances while confined there. Less than three weeks after Delisle's death in the Bastille, Homberg himself narrowly escaped being imprisoned and interrogated there, and once again primarily because he was a chymist, although not for transmutational reasons. This barely averted disaster resulted both

123. Félix Feuardent, *Jetons et méreaux*, 4 vols. (Paris: Rollin et Feuardent, 1904–15), 1:386, cat. no. 4909; *Collection Feuardent: Jetons et méreaux, vente aux enchères publiques* (Paris, 1928), 51 (Delisle's coin was part of lot 879). If any reader possesses a clue about the current location of this coin, the present author would be eager and appreciative to hear about it.

from intrigues in the unstable world of the royal court and from the usual suspicions that dogged chymists of the period. Although Homberg was saved by Louis XIV at the last moment, the events that played out in early 1712 had disastrous consequences for the rest of his life.

In 1711 and 1712, a sequence of tragedies struck the royal family. In April 1711, Louis, Monseigneur le Grand Dauphin, Louis XIV's only legitimate son and the heir apparent to the throne of France, died—probably of smallpox. The mantle of heir apparent and the title of Dauphin passed to his son Louis, Duc de Bourgogne. In mid-January 1712, Jean Boudin, who had been physician to the late Grand Dauphin and was now physician to the Dauphine, announced that a "credible threat" had been conveyed to him that someone was trying to poison the Dauphin and Dauphine. Boudin's warning seemed justified when, in early February, the Dauphine suddenly fell ill, and after a short and excruciating sickness, died on 12 February. As she was dying, the Dauphin himself began to fall ill in a similar way, and on 18 February he died as well. The heir apparent was now their orphaned five-year-old son, the Duc de Bretagne, Louis XIV's great-grandson.

The bodies of the deceased Dauphin and Dauphine were autopsied by order of the king. Considerable damage to their internal organs was found— the Dauphin's body was found internally gangrened, and his heart collapsed into a stinking pulp as it was removed. Boudin declared before the king and Madame de Maintenon that the deaths resulted from poisoning. The king's chief physician, Guy-Crescent Fagon, concurred in this assessment, and Gilles-François Boulduc did not contradict them. Only the king's surgeon, Georges Mareschal (1658–1736), objected, saying that he had seen similar internal damage where there was no suspicion of poisoning.[124] Despite Mareschal's objections, Boudin and Fagon persisted in their diagnosis of poisoning—probably in no small measure to cover up their own obvious failure to save the lives of the Dauphin and Dauphine. The claim was embraced and repeated by one of Louis XIV's bastards, the Duc de Maine (1670–1736), as well as by Madame de Maintenon, who habitually advocated on his behalf. Given that both were antagonistic to Philippe and his mother, and the obvious fact that the deaths of the previous ten months moved the Duc d'Orléans two steps closer to the crown, Maintenon did not hesitate to name Philippe explicitly as the guilty party. This accusation spread quickly through the court, the city of Paris, and beyond. Philippe was jeered by angry

124. Saint-Simon, *Mémoires*, 4:386–87, 411–12, 444–48. On Mareschal, see also Alexandre Lunel, *La maison médicale du roi* (Seyssel: Champ Vallon, 2008); Mareschal de Bièvre, *Georges Mareschal*.

mobs when he appeared during funerals for the deceased and was shunned at court.[125] Elisabeth Charlotte wrote that "it is not enough that I am heartbroken over the deaths of the Dauphine and Dauphin . . . something more must be thrust upon me that is yet more painful and that pierces my soul: malicious persons have spread it about throughout Paris that my son poisoned the Dauphin and Dauphine."[126]

Fingers quickly pointed at Homberg as the source of the poison. Elisabeth Charlotte laid the blame for this accusation upon Philippe's servants: "many of my son's domestics were jealous of Homberg because my son esteemed him so much . . . they wanted to hurt Homberg when they spread it around that he knew all about poisons [*mitt gifft umbgeht*]."[127] Even without such intentional malice in the palace, the simple fact that Homberg was a chymist and busy in the laboratory beneath the apartments of the Duc d'Orléans was enough to call down accusations against him. At this point, one of Philippe's companions—Antoine Coiffier, Marquis d'Effiat—urged the Duc d'Orléans to go to the king to try to clear his name by volunteering to be interrogated at the Bastille and having Homberg arrested and interrogated. Louis XIV initially refused, but upon being pressed, he agreed to give an order that if Homberg went to the Bastille voluntarily, he would be taken in.[128] Elisabeth Charlotte confirms that "my son wanted to send Homberg into the Bastille because it was being said that he prepared the poison upon my son's orders, on account of which my son wanted him to be rigorously examined."[129] Thus, Philippe informed Homberg that he should turn himself in at the Bastille. Madame, given her great fondness for Homberg, was "struck to the heart" and beside herself with grief. She recounted the next morning that "I could not close my eyes for the whole night" at the thought that Homberg was in the Bastille.[130] As for Philippe, he was "found stretched out on the floor, weep-

125. Saint-Simon, *Mémoires*, 4:448–52, 457; Elisabeth Charlotte to Sophie von Hannover, 8 April 1712; *Aus den Briefen*, 2:309–11.

126. Elisabeth Charlotte to Sophie, 20 February 1712; *Aus den Briefen*, 2:302; Saint-Simon, *Mémoires*, 4:457, 464–65.

127. Elisabeth Charlotte to Sophie, 19 March 1712; *Aus den Briefen*, 2:307.

128. Saint-Simon, *Mémoires*, 4:459–60.

129. Elisabeth Charlotte to Sophie, 17 March 1712; *Aus den Briefen*, 2:307. On the day of the event, her account was slightly different, namely, that when Philippe appeared before the king, Louis XIV reassured him that he did not believe the rumors but "nevertheless advised my son to send his chymist, the poor and learned Homberg to the Bastille"; Elisabeth Charlotte to Sophie, 20 February; *Aus den Briefen*, 2:302. Philippe may initially have been afraid or embarrassed to tell his mother that sending Homberg to the Bastille was his idea rather than the king's order.

130. Elisabeth Charlotte to Sophie, 20 February 1712: "ich bin recht auß mir selber"; 21 February 1712: "habe ich die gantze nacht kein aug zugethan"; *Aus den Briefen* 2:302–3.

ing and out of his mind with despair; his chymist Homberg was on the way to turn himself over to the Bastille and to give himself up as a prisoner."[131]

Homberg dutifully went to the Bastille either late on 20 February or early the next morning. But thanks to a last-minute change of heart on the part of Louis XIV, he was not admitted. Mareschal, seemingly the sanest and most judicious medical adviser around the king, was responsible for this reversal. Shortly after meeting with Philippe, the king returned to his chamber, where he found Fagon and Mareschal, and told them what had happened. Mareschal exclaimed against sending Homberg to the Bastille, and pointed out emphatically that such an action was to no purpose; since both Homberg and Philippe were innocent, it could only bring shame upon Louis XIV himself. The king was convinced and immediately sent orders to the Bastille that Homberg was not to be admitted when he appeared there. When Louis saw Philippe some hours later, he ordered him to tell Homberg "not to think any more about the Bastille."[132]

The records of the Académie reveal no hint of this terrible crisis for one of its most prominent members. The academicians did cancel their meeting on 13 February 1712 due to the death of the Dauphine the day before, but on 20 February, while the decision to send Homberg to the Bastille was being made at Marly, the Batavian chymist was sitting in a regular Saturday meeting of the Académie listening to a paper about marine plants. Perhaps the order to go to the Bastille was waiting for him when he returned to the Palais Royal that evening. Despite this drama, which must have affected Homberg deeply, just four days later he soldiered on, apparently unfazed, and presented a new memoir about his pyrophorus.[133] Fontenelle, of course, makes no mention of so scandalous a matter in Homberg's éloge. The invisibility in these sources of so important an event underscores how limited they really are on their own for documenting the lives of academicians. Homberg himself was caught in the same web of suspicions that so regularly entrapped chymists of the period, and that the police were using at the same time to arrest and interrogate both real and merely suspected chymists around Paris.

Although the immediate crisis for Homberg was averted, subsequent events perpetuated the rumors. Just two weeks later the five-year-old Duc

131. This is the testimony of the Duc d'Orléans's friend Philippe de Montboissier-Beaufort, Marquis de Canillac (1669–1725), as recounted to Voltaire and recorded in the latter's *Siècle de Louis XIV*, ed. Émile Bourgeois (Paris: Hachette, 1910), 530–31.

132. Saint-Simon, *Mémoires*, 4:465–66.

133. PV 31, fol. 35r (cancellation of meeting on 13 February); 53r (20 February); 71r–75r (Homberg, "Nouveau phosphore, ou Suite des observations sur la matière fecale," 24 February 1712).

de Bretagne, the Petit Dauphin, the newest heir apparent to the throne of France, fell ill with symptoms much like those of his parents, and on 8 March 1712, he too died. It is generally thought today that the Dauphine, Dauphin, and Petit Dauphin all succumbed to measles, a diagnosis that was made by some at the time, but that failed to quell the rumors of poisoning. With the death of yet a *third* heir apparent, the suspicions of poisoning by Philippe redoubled: "the fury against my son is more violent than ever before," wrote Elisabeth Charlotte, "he is supposed to have poisoned him too."[134] Posters were hung on the walls of the Palais Royal (where Homberg must have seen them) reading: "Here's where the best poisons are found."[135] The line of succession now passed to the Petit Dauphin's younger brother, the two-year-old Duc d'Anjou, but soon he too was showing similar symptoms. The physicians rushed to bleed him—the medical treatment most favored by Fagon—as they had done to his mother, father, and elder brother. Fortunately, his governess, Madame de Ventadour (1654–1744), seized the child to keep him away from the doctors and locked herself in her chamber with him and three nursemaids to feed him. It was rumored at the time that she gave the child an antidote to poison.[136] Thanks to her care and her contempt for the court physicians, the Duc d'Anjou recovered, and eventually became Louis XV.

All these deaths meant that within a year Philippe had moved from a distant sixth in line to the throne to third, and that gap narrowed further in 1714 when Charles, Duc de Berry, the last of Louis XIV's eligible grandsons, died following a hunting accident. The implication of Philippe in these deaths never entirely vanished. For the next decade Elisabeth Charlotte continued to refer to the rumors and their consequences (as well as the role of Madame de Maintenon in perpetuating them). The satirical title of a supposedly forthcoming volume (dedicated to Philippe) referred to them unambiguously in 1720: *The Art of Using Poisons according to their Type and Particular Qualities, with a Physical Consideration of their Efficacy, enriched by Remarkable Histories that have come to pass in Europe since 1711.*[137] The drama of these accusations, the rumors of Philippe's attempts to gain the crown by means of

---

134. Elisabeth Charlotte to Sophie, 17 March 1712, *Aus den Briefen* 2:306–7; and Saint-Simon, *Mémoires*, 4:457–58.

135. Elisabeth Charlotte to Sophie, 27 March 1712; *Aus den Briefen* 2:308–9, at 309. The same posters also made the scurrilous and loathsome claim of incest between Philippe and his daughter.

136. Saint-Simon, *Mémoires*, 4:436–37.

137. BNF, Clairambault, F. Fr. MS 12697, 483–84: *"L'art d'user des poisons suivant leur espèce et qualités particulières, avec un raisonnement physique sur leur efficace, enrichi d'histoires remarquables arrivées en Europe depuis 1711,* par le Duc de Noailles, dédié au Régent."

poisoning, his supposed dabbling in the black arts as well as in chymistry, and Homberg's role in the affair even spawned a nineteenth-century "historical novel" entitled *La chambre des poisons* in which Homberg appears as a major character![138]

Philippe was therefore obliged to comport himself more carefully to avoid any appearance of corroborating the malicious gossip. One of his actions in this regard had the worst possible consequences for Homberg: Philippe shut down the chymical laboratory at the Palais Royal. This closure has not previously been known. But evidence both for the closure and for the connection of the closure to the royal deaths appears in Jean Hellot's notes. Hellot reports of the laboratory work that "everything ceased by [Philippe's] orders in 1712 or 1713 following the death of the Duc and Duchesse de Bourgogne."[139] The fact that Homberg's laboratory operator Jean Grosse, who would have lived next to the laboratory, found new lodgings at this very same time—around 1713—is probably a consequence of Philippe's action.[140] Hellot's note also provides background to an otherwise slightly obscure comment by Saint-Simon regarding how Philippe "threw himself into painting after his great love of chymistry was over, or rather, deadened by everything that was so cruelly publicized about it."[141] How completely the Palais Royal laboratory was shut down remains unclear from Hellot's notes alone—was it actually dismantled at this time, or was it simply put behind lock and key, or was Homberg allowed to continue to work there quietly and alone?

One piece of indirect evidence strongly suggests that the Palais Royal laboratory was closed entirely, perhaps even dismantled. In early 1713, there was a sudden and dramatic increase in the costs of running the Académie's old laboratory at the Bibliothèque du Roi on the rue Vivienne. Homberg had remained the official head of that laboratory even though he moved his own experimental operations to the far better workspace at the Palais Royal soon after it was completed. Some low level of activity must have continued at the Bibliothèque laboratories (by whom remains uncertain), and Homberg, as laboratory director, remained obliged to fund its operating costs from his own pocket and then request reimbursement from the government. In 1709 he submitted a request for a mere 268 livres, and subsequent requests indi-

138. Paul L. Jacob [Paul Lacroix], *La chambre des poisons: Histoire du temps de Louis XIV (1712)*, 2 vols. (Paris: Victor Magen, 1839).

139. Hellot Caen Notebooks, 5:144r: "tout avoit cessé par ses ordres en 1712 ou 1713 apres la mort du Duc et de la Duchesse de Bourgogne."

140. Paul Dorveaux, "Jean Grosse, médecin Allemand, et l'invention de l'éther sulfurique," *Bulletin de la Société d'Histoire de la Pharmacie* 17 (1929): 182–87.

141. Saint-Simon, *Mémoires*, 5:245.

cate that the laboratory costs averaged about 800 livres per year in 1710, 1711, and 1712—about half of this amount would have gone just to pay a laboratory operator, if one were still employed there. But at the start of 1713 these costs abruptly skyrocketed. In the first five months of 1713 alone, Homberg submitted three reimbursement requests totaling 2,450 livres; in other words, the costs of running the Académie's laboratory suddenly increased more than sixfold.[142] This sudden increase suggests that Homberg was attempting to relocate his operations back to the site on the rue Vivienne that he had vacated a decade earlier.[143] This substantial figure also indicates the high level of financial support, now brought to an end, that the Duc d'Orléans had been giving to Homberg's work in previous years.

The year 1712 that proved so dismal for the royal family thus proved tragic for Homberg as well. The busy workshop where he and the Duc d'Orléans collaborated; where Geoffroy, Lémery *fils*, Père Sebastien, and others worked; where Grosse assisted and other operators tended the furnaces; and where visitors from around Paris and farther afield came to be instructed, entertained, and supplied with materials was forced to extinguish its fires and shut its doors for good. The exalted position of the Duc d'Orléans had allowed him to build the "most magnificent laboratory chymistry had ever had," but that very same position also forced him to end its days prematurely, and to Homberg's enormous loss. Indeed, it probably curtailed chymical work for others at the Académie as well, since so much of their own work had been carried out at the Palais Royal.

Final Years: 1713–15

The attempted return to the Bibliothèque du Roi did not, however, remedy the terrible situation in which Homberg found himself. His productivity—as gauged by his Académie presentations and publications—plunged dramatically in 1713. A simple count of the number of pages Homberg published in the *Mémoires* after 1712 clearly indicates this decline (see figure 4.5), but this numerical assessment falls far short of expressing the gravity of the situation. Homberg's presentations and publications after 1712 contain virtually no recent results. For example, his "Observations on the Sublimation

142. "Memoir des ordonnance de Mr homberg pour les depence du laboratoire de lacademie des Sience," AdS, pochette de séances 1716.

143. A further piece of suggestive evidence is that Homberg converted one of the private rooms in his apartment into a makeshift laboratory and storage room for chymical materials, but it is unlikely he could have have accomplished much in such a space; Homberg's Inventaire-après-décès, fol. 15v: "un petit cabinet servant de Laboratoire."

FIGURE 6.2. The conclusion of Homberg's memoir on the sublimation of mercury, showing the endorsement that it was presented twice—first in 1697, and again in 1713. Courtesy of the Archives of the Académie des Sciences, Paris.

of Mercury," read on 6 September 1713 and published in the *Mémoires*, had already been presented to the academicians in 1697. The 1713 version even preserves unchanged the remark from 1697 that the experiment was done "a little while ago," even though sixteen years had elapsed. In fact, the manuscript memoir survives, bearing the original endorsement "read by Mr. Homberg 13 August [November] [16]97" which has been struck through and replaced, in Fontenelle's handwriting, with "read by Mr. Homberg 6 September 1713" (see figure 6.2).[144] His paper of the same year on the refining of gold and silver is clearly related to work he presented in 1697 and 1700, and may have been excerpted from his lost treatise on assaying precious metals.[145] A third paper presented in 1713, on substances that can penetrate metals without corroding them, also relies on work done in the 1690s. Homberg incorporated a 1695 memoir into it largely unchanged, but did rewrite the section explaining the phenomenon—updating his explanations to incorporate the findings from his 1703 analysis of sulphur and his new chymical theory. This paper was Homberg's last presentation at an *assemblée publique*.[146] Because Hom-

144. Wilhelm Homberg, "Observation sur une sublimation de mercure," *MARS* (1713): 265–67; PV 17, fols. 1r–2v (13 November 1697); AdS, pochette de séances 1713.

145. Wilhelm Homberg, "Observation sur une separation de l'Or d'avec l'Argent par la Fonte," *MARS* (1713): 67–70; compare Homberg, "Observations sur le rafinage de l'argent," *MARS* (1700): 42–46; PV 15, fols. 272r–274v (16 January 1697), 295r–296r (30 January); PV 16, fols. 8r–10v (20 February 1697). On his lost manuscript on refining, see chapter 2.

146. Wilhelm Homberg, "Observations sur des matières qui pénetrent & qui traversent les métaux sans les fondre," *MARS* (1713): 306–13. Compare Homberg, "Sel qui travers le fer sans le fondre," PV 14, fols. 212r–213r (14 December 1695); this material appears on 308–9 of the published version. It also refers (212r) to a demonstration of the phenomenon that Homberg performed in 1692. Homberg did a similar demonstration on 4 January 1696 (PV 14, fol. 224r) but using mercury of antimony and a silver plate; it is therefore likely that this work was related

berg mentioned that one of his preparations turned silver the color of lead, it provoked a remark from Thomas Gouye (1650–1725), the only Jesuit in the Académie, who wittily "reproached chymists in the person of Mr. Homberg that in place of the riches that they promise us, he has not yet taught us more than how to turn gold into glass in previous years' experiments, and silver into lead in the present one."[147]

Homberg remained obliged to fulfill his regular turns (*tour de rolle*) to present papers, but his 1714 papers continued to depend on recycled results. When his turn came around again on 12 May, he reread a memoir he had presented originally in 1698 that stemmed from a demonstration performed in 1694 with his air pump—a device he had not mentioned in over a decade and that possibly no longer even existed by 1714.[148] Seven academicians were in attendance for the paper both in 1698 and in 1714—Jean Marchant, Jacques Cassini, Giacomo Maraldi, the de la Hires *père et fils*, Pierre Varignon, and Fontenelle—and so would have heard it twice. Unfortunately we cannot know whether they had forgotten the first reading, had a sense of déjà vu at the second, or felt regret that the great Homberg, who had demonstrated so much productivity and innovation for twenty years as an academician, was now reduced to rereading a sixteen-year-old memoir on a topic he had left long ago simply to fulfill his duty to present regular communications. Despite Homberg's inability to regale his colleagues with fresh results, he continued to be active in the Académie's affairs. He was named to the committee overseeing publications in January, and assigned the examination of a paper on the growth of plants in June.[149]

The last paper that Homberg presented to the Académie merits special mention. Although it too has roots in much earlier work, unlike his other post-1712 papers, it showcases many of the features that characterize Homberg's scientific work: painstaking observation, measurement, and experimentation, obvious relish in telling detailed anecdotes, wonder at unexpected results, and the tireless pursuit of choicest secrets in chymistry. The

---

to Homberg's chrysopoetic endeavors, even though Fontenelle links it solely to demonstrating the corpuscular structure of matter, *HARS* (1713): 37–39.

147. *Mercure galant* (November 1713): 117–18.

148. PV 33, fols. 155r–158v (12 May 1714): "De l'effet du Siphon dans un lieu vuide d'air," originally PV 17, fols. 358v–362r (20 August 1698): "De l'effet du Siphon dans un lieu vuide d'air." The initial demonstration is described in PV 14, fol. 23v (14 August 1694).

149. PV 33, fol. 15v (20 January 1714), fols. 159r–166r (16 May 1714). A document dated 29 April 1713 making Claude Rigaud the Académie's printer and signed by the committee (including Homberg) indicates that he served on the committee in 1713 as well (AdS, pochette de séances 1713).

paper deals with the volatilization of fixed alkalies, a topic he had touched upon in the 1690s (see chapter 2), but here he explicitly connects this work to one of the traditional *arcana maiora* of seventeenth-century chymistry. Van Helmont had consoled his followers who were unsuccessful in preparing the alkahest by telling them that if they could volatilize a fixed alkali salt, they could prepare a solvent with nearly the properties of the alkahest. George Starkey devoted himself to this endeavor throughout the mid-1650s.[150] Now Homberg reveals his own similar experiments. He recounts how, dissatisfied with the smell of the soap he used for shaving, he mixed it with fragrant oils and set it aside to reharden. To his surprise, after two months, the soap had sprouted crystals. Homberg put some of these crystals into the fire, and they vanished into smoke. Since the only salt used in making soap is a fixed alkali (soda, today sodium carbonate), he concluded that the addition of oil must have caused this salt to become volatile, so he embarked on closer investigations. Rather than working on soda, he chose the similar salt of tartar (potassium carbonate) because, he wrote candidly, "I thought that by this same operation I could have a volatile salt from the fixed salt of tartar which Paracelsus and Van Helmont so greatly extolled."[151] After several attempts to use expressed oils, such as olive oil, to volatilize the salt, he turned to distilled oils. This was the same pathway taken by Starkey in the 1650s. But whereas Starkey used oil of turpentine, Homberg chose a more interesting route based on his more modern theory. Using his theory of salts—namely, that salts become fixed because the fire has expelled all their volatile components—he tried to restore exactly the materials lost when salt of tartar is made. So he began with tartar (that is, the crust from wine barrels, potassium bitartrate) and destructively distilled it. He calcined the residue to obtain salt of tartar. The distillate, initially a stinking black oil, he then purified by repeated distillation from slaked lime, a process he had described in 1692.[152] Then he recombined

150. On Starkey's work and Van Helmont's promises, see William R. Newman and Lawrence M. Principe, *Alchemy Tried in the Fire: Starkey, Boyle, and the Fate of Helmontian Chymistry* (Chicago: University of Chicago Press, 2002), 136–52. Starkey's work on the process in 1655–56 is recorded in George Starkey, *Alchemical Laboratory Notebooks and Correspondence*, ed. William R. Newman and Lawrence M. Principe (Chicago: University of Chicago Press, 2004). Intriguingly, Starkey recorded this work in the same notebook owned by Boyle that also contained the unpublished method of making philosophical mercury that Homberg appears to have received from Boyle; perhaps Boyle gave Homberg access to the whole document?

151. Wilhelm Homberg, "Memoire touchant la volatilisation des sels fixes des plantes," *MARS* (1714): 186–95, quotation at 190. Part of the original autograph memoir survives in AdS, pochette de séances 1714.

152. See chapter 2.

the salt and oil in a dish left open to the air, following very closely the procedure Starkey used in the 1650s with oil of turpentine. By gentle distillation of this mixture, Homberg finally obtained the desired volatile salt, a point Starkey does not appear to have reached. Readers "in the know" who recognized what Homberg was really doing—like the readers of his 1700 paper on making the philosophical mercury, or listeners in the public audience when he slyly mentioned "mercury purified with antimony and iron" in 1705—would have been on the edge of their seats. In the former cases, they were awaiting word of the mercury's elaboration into the philosophers' stone; in this case, they expected the use of this volatile salt as a substitute for the alkahest. Such readers were disappointed. Homberg ended the paper with the promise of a follow-up memoir, but never provided one, perhaps due to his now failing health.[153]

Near the end of 1714 Homberg fell ill with an intestinal malady, one that had apparently plagued him before and was described as a kind of dysentery. He was absent from most of the Académie's meetings in December. He did attend on 5 December 1714, when he had the pleasure of hearing the results of a long study of his pyrophorus by Jean Lémery, who succeeded in finding less disagreeable starting materials than fresh human feces.[154] Homberg managed to attend again on 22 December, but that proved to be his last appearance at the Académie. His illness worsened through 1715, and there is no evidence of any contact with the Académie throughout that year nor any other indication of his activities, which suggests that he was largely confined to his residence. At last, in the early hours of 24 September 1715, during the Académie's annual recess, Wilhelm Homberg died in his apartments at the Palais Royal at the age of sixty-two.[155]

Homberg's body was autopsied a few days later. The physicians found that one kidney and its urinary tract were nearly blocked with stones, part of his intestines were gangrened, and one lung was seriously damaged. His body was emaciated from hard work, lengthy illness, and strict regimen.[156] Hom-

153. Fontenelle noted that Homberg said nothing about the use of the volatilized salt; *HARS* (1714): 32.

154. PV 33, fols. 393r–397r (5 December 1714): "Experiences sur la diversité des matières qui sont propres à faire un Phosphore avec l'alum." He gave a more developed version of the paper to the same public assembly in November 1715 at which Homberg's éloge was read; see below.

155. Homberg's death in the early morning is verified in the Scellé-après-décès, fol. 1r, which records that the notaries who arrived at seven in the morning saw the "corps mort dudit deffunt estendu sur son lit."

156. Pierre Huard, "L'autopsie de Guillaume Homberg (1652–1715)," in *89e Congrès des sociétés savantes* (Lyons: Imprimerie Nationale, 1964), 155–56. The original document "Mémoire

berg's mortal remains were interred in the church of Saint-Eustache, the parish of which the Palais Royal was part. Eighteenth-century guides to Parisian monuments regularly mentioned Homberg, along with his first French patron, Jean-Baptiste Colbert, among the notables buried in Saint-Eustache.[157] Today, the location of his burial within the church is not known. Whatever indication there was of his final resting place—a grave slab or plate, a wall plaque, or a monument—has since disappeared, perhaps during the desecrations of the church by the revolutionary *canaille* at the end of the eighteenth century.

Homberg's death went entirely unremarked in the periodical literature, undoubtedly because it was overshadowed by the death of Louis XIV just three weeks earlier. Elisabeth Charlotte's letters, however, show that she was heartbroken by the loss of Homberg. Writing just two days after his death, she told Leibniz how not only she and her son were mourning his loss, but "that everyone who knew him mourns him." Leibniz replied that he too was deeply saddened by Homberg's death, noting that "the Académie has clever people, but they are not all like Homberg was."[158] The academician Pierre Varignon also lamented the loss of "Homberg, one of the most capable chymists in Europe" and expressed concern over the difficult job of replacing him in the Académie with anyone of equal caliber.[159]

When the Académie returned from its annual break, it held the customary *assemblée publique* on 13 November 1715, at which Fontenelle read his éloge for Homberg.[160] Homberg did live just long enough to see his patron, collaborator, and friend Philippe II confirmed by Parlement as Regent of France. The Académie trooped dutifully to the Palais Royal four days before Homberg's death to greet the newly proclaimed Regent. Homberg was presumably unable to attend this event—he is not listed in the procès-verbaux among those present. He did however, perhaps in lieu of attending the for-

abrégé de ce qu'il est trouvé à l'ouverture du cadavre de M. Humbert" is preserved in Carton 2 of the Archives de l'Académie Royale de Chirurgie, held at the Académie Nationale de la Médecine, Paris.

157. For example, Jean-Aymar Piganiol de la Force, *Description historique de la ville de Paris*, 10 vols. (Paris, 1765), 3:189; there follows a biographical sketch of Homberg cribbed from Fontenelle.

158. Elisabeth Charlotte to Leibniz, 26 September 1715, and Leibniz to Elisabeth Charlotte, undated, in Eduard Bodemann, "Briefwechsel zwischen Leibniz und der Herzogin Elisabeth Charlotte von Orleans 1715/16," *Zeitschrift des historischen Vereins für Niedersachsen* 49 (1884): 1–66, at 21 and 24.

159. Pierre Varignon to Leibniz, 9 November 1715, LBr 951, fols. 87r–88v: "chimiste des plus habiles de l'Europe, et aussi tres difficile à remplacer en son genre."

160. PV 34, fol. 223r (13 November 1715); the éloge for Nicolas Lémery was read the same day.

mal visit, write Philippe a long and familiar letter upon his becoming Regent; that letter, as informative as it would be about their relationship, has not survived. But Elisabeth Charlotte noted how Homberg had shared her joy at Philippe becoming Regent.[161] The following week, the academicians proposed three candidates to the new Regent to fill Homberg's position, and Homberg's longtime associate Étienne-François Geoffroy was chosen—technically by the king, but effectively by Philippe, since Louis XV was then but five years old.[162]

161. PV 34, fol 255v (20 November 1715): "L'Académie est allée en corps au Palais Royal pour avoir l'honneur de salüer Monseigneur Le Duc d'Orléans"; Fontenelle, "Éloge de Homberg," *MARS* (1715): 91; and Elisabeth Charlotte to Leibniz, 26 September 1715.

162. PV 34, fol. 259r (27 November 1715); PV 35, fol. 1r (8 January 1716; letter from the Regent dated 7 December 1715). See also Philippe to Chevalier Renau, 7 December 1715, AdS, pochette de séances 1716: "C'est un grand Eloge pour les 3 Sujets qui ont esté Elus que davoir esté trouvez dignes de succeder a un personage d'un merite si distingué."

7

Homberg's Legacy

As detailed in the last chapter, Homberg maintained a broad range of contacts and collaborators. Some of these contacts sought his help and advice in chymical matters, others offered him potentially valuable reports and recipes, and yet others Homberg simply observed for indications that they might possess special knowledge. The reciprocal question to be addressed in this chapter is how others responded to Homberg and his ideas. Many savants engaged with Homberg's publications, claims, theories, and research programs in various ways both during his lifetime and after his death, even many years afterward. His most dramatic experiments were repeated, both refuted and confirmed, and reworked into new directions. His comprehensive theory of chymistry centering on the identity of the sulphur principle as light and its ability to combine with and transform other matter was adopted and adapted in several ways by various followers. In regard to transmutational endeavors, the subsequent actions of his two closest collaborators—Philippe and Geoffroy—prove especially significant. Besides indicating more about Homberg's ongoing impact and legacy, their actions into the 1720s played an important role in one of the most significant transmutations of chymistry during the period, namely, the apparent disappearance of chrysopoeia from the normal activities of chymists.

Homberg's ideas reached multiple audiences by multiple routes. Although little of his personal correspondence survives, it is nevertheless clear that he corresponded at various times with Kunckel in Hamburg, Leibniz in Hannover, Boerhaave in Leiden, Friedrich Benedikt Carpzov in Leipzig, Tschirnhaus in Saxony, Johann Bernoulli in Groningen, and even Maria Sibylla Merian in the South American Dutch colony of Surinam.[1] The annual *Histoire et mémoires* played a major role in disseminating Homberg's work. Homberg is unusual for the period in that *all* of his publications were articles (about fifty of them) rather than books. While we lack precise information about the print run and distribution of the *Mémoires*, frequent references to it across Europe argue that it circulated quickly and widely. Lengthy reviews of these volumes in other learned periodicals—the *Journal des sçavans*, the *Nouvelles de la république des lettres*, the *Journal de Trévoux*, the *Acta eruditorum*, and elsewhere—enhanced this diffusion. Popular publications, such as the *Mercure galant*, also regularly devoted several pages to the Académie's public assemblies, thus broadcasting news of the events and their content to yet wider audiences. These paths of dissemination were complemented by newsletters sent out by the Académie to sister institutions, most particularly to the Royal Society of London. Geoffroy, thanks to the facility with English he had acquired during a stay in Britain and his status as a Fellow of the Royal Society, served as the Académie's official correspondent with the British institution. Many of his missives survive in the Royal Society archive, as well as dozens of letters to the society's secretary Sir Hans Sloane (1660–1753), now in the British Library. Such newsletters served a crucial function by compensating for the usual delay of two years or more in publishing the *Mémoires*. For example, Geoffroy sent the recipe for Homberg's pyrophorus to the Royal Society in August 1712, at which time (as Geoffroy notes) the 1710 *Mémoires* had not yet been printed, and the 1711 issue (containing the pyrophorus recipe) remained over a year in the future.[2] Royal Society Fellows were therefore kept especially well apprised of the goings-on at the Académie Royale. Indeed, one of the more significant examples of Homberg's broad impact appears in the English context, and illustrates how one did not necessarily have to wait for the *Mémoires* in order to know—and respond to—their contents.

1. Kunckel and Carpzov (1649–99) are mentioned in PV 18, fol. 149v (4 March 1699); Carpzov is an unexpected contact for Homberg—perhaps they met while Homberg was a student in Leipzig, or the correspondence was related to Carpzov's work with Otto Mencke on the *Acta eruditorum*. The *Index biographique de l'Académie des Sciences* (Paris: Gauthier, 1979), 174, claims that Homberg's correspondent was *Christian* Benedikt Carpzov (1684–1749) but it seems rather unlikely that Homberg would have scientific correspondence with a fifteen-year-old. Mention of Homberg's correspondence with Tschirnhaus appears in Pierre Varignon's letters to Leibniz; see Georg Heinrich Pertz, ed., *Leibnizens gesammelte Werke*, 3d ser., 7 vols. (Halle, 1859), 4:129, 147. A letter obviously from Merian is excerpted in PV 20, fols. 181v–182v (25 May 1701).

2. Geoffroy to Sloane, 31 August 1712; British Library, Sloane MS 4043, fols. 85r–86v.

Responses to the Lens Experiments: Newton and Hartsoeker

Homberg's results from his experiments with the great Tschirnhaus lens were closely followed all around Europe, providing a key locus for engagement with the Batavian chymist's ideas. Contemporaneous experiments and analogous ideas across the Channel in England offer one of the most striking instances of such engagement. In early 1704, Sir Isaac Newton (1643–1727) built and used a compound burning mirror at the same time Homberg was pursuing his experiments with the Tschirnhaus lens in the garden of the Palais Royal.[3] Newton's device, in contrast to the massive Tschirnhaus lens, used an array of seven small parabolic mirrors to reflect sunlight into a single focus. Strikingly, Newton designed and built this device *in direct response* to reports of Homberg's results. As the astronomer David Gregory (1661–1708) recorded in May 1704, "M. Geoffrey writes from France, that by a Large Lens the Sun's beams reduce Gold to a transparent Sky coloured Stone or Glass. This was the occasion of Mr. Newton's 7 Miroirs."[4] Thus, even though the 1702 *Mémoires* containing Homberg's results with the lens had not yet been published in early 1704, Newton heard about Homberg's work in advance through the medium of Geoffroy's newsletters and was sufficiently inspired by it to carry out his own similar experiments. Geoffroy had indeed reported Homberg's results with the lens in a letter he wrote in August 1703, providing the exciting news that "metals calcined in the focus of this lens evaporate for the most part in smoke, and this smoke is mercury. It has been drawn out abundantly from lead, tin, and silver. . . . Fine gold reduced to a calx by means of spirit of salt changes into a very dark violet-colored glass after having let the greater part of its mercury escape in smoke."[5]

Newton, who had extensively pursued the decomposition of metals into their metallic principles in the context of his own chymical researches—

3. D. L. Simms and P. L. Hinckley, "Brighter Than How Many Suns? Sir Isaac Newton's Burning Mirror," *Notes and Records of the Royal Society of London* 43 (1989): 31–51; John Harris, *Lexicon technicum*, 2 vols. (London, 1704 and 1710), 2: s.v. "Burning mirror."

4. W. G. Hiscock, *David Gregory, Isaac Newton and Their Circle: Extracts from David Gregory's Memoranda 1677–1708* (Oxford: Printed for the editor, 1937), 17; cited in D. L. Simms and P. L. Hinckley, "David Gregory on Newton's Burning Mirror," *Notes and Records of the Royal Society of London* 55 (2001): 185–90.

5. Geoffroy to Sloane, 10 August 1703, Sloane MS 4039, fol. 171r–v: "Les métaux calcinés dans le foyer de ce verre s'en vont pour la plus grande partie en fumée, et cette fumée est du mercure. on en a tiré abondamment du plomb, de l'etain, et de largent. . . . Lor fin reduit en chaux par lesprit de sel, se change en verre de couleur violette tres foncée apres avoir laissé echaper en fumée la plus grande partie de son mercure."

particularly when interpreting and following the work of various chrysopoetic authors—would certainly have been riveted by this report of apparently
successful metallic decompositions carried out in France.[6] Indeed, as soon
as Newton had assembled his burning mirror, he used it to repeat Homberg's
experiments on the metals, and obtained similar results, particularly in the
apparent decomposition of gold and other metals into their principles as revealed by their "smoking" during exposure to the focus. The results were
apparently significant enough that Newton continued to improve his apparatus; in mid-1705 he hoped to add twelve more mirrors to it, as well as a
reflector to cast the focus downward in order to make it easier to work with
and more like Homberg's refracting lens.[7] John Flamsteed (1646–1719) also
connected Newton's mirror with Homberg's Parisian experiments, although
he compared Newton's apparatus unfavorably (if wittily) with the "extraordinary" Tschirnhaus lens.

> Our R Society produces nothing worth imparteing to you nor I fear is like
> to doe until its present constitution be altered: Mr Newton at this time is
> president. Mr. Thirnhaus has contrived an extraordinary burning glass. The
> French have bought it at a great price & sent over an account of its perfor-
> mances. Mr. N[ewton] has contrived one that consists of 6 concave glass spec-
> ulums placed about a 7th each near a foot in diameter. . . . I am apt to thinke
> a burning glass made to performe by Refractions would performe better, but
> wee are fond here of Theorys & reflections, tho our success in either is not
> much to be boasted of.[8]

Homberg's ongoing experiments, and particularly his newly announced
chymical theory of light, may also explain the sudden appearance of what
are Newton's best known remarks in regard to chymistry: the two so-called
chymical queries of his *Opticks*. These queries have long been seen as foundational for later Newtonian notions of chymistry. The English first edition of
the *Opticks*, published in 1704, did not contain the chymical queries; Newton
added them for the first time only to the 1706 Latin translation of the work
prepared by Samuel Clarke. The first of these two queries, the twenty-second
in the series present in the 1706 *Optice*, begins "are not dense bodies and light
able to be converted and transmuted into one another? And is it not pos-

6. On Newton's experiments and his chymistry more broadly, see William R. Newman,
Newton the Alchemist: Science, Enigma, and the Quest for Nature's "Secret Fire" (Princeton, NJ:
Princeton University Press, 2019).

7. Hiscock, *David Gregory*, 18, 26.

8. Flamsteed to James Pound, 15 November 1704, in *The Correspondence of Isaac Newton*,
7 vols. (Cambridge: Cambridge University Press, 1959–77), 4:424–28, at 425–26.

sible that bodies receive their active power from particles of light that exist within their compositions?"[9] These ideas sound very similar to, if not identical with, Homberg's own ideas about the incorporation of light as the sulphur principle in compound bodies where it functions as the sole source of activity in matter. To what extent are Newton's ideas expressed in 1706 related to or dependent upon Homberg's contemporaneous work?[10]

On the one hand, the timing works out extremely well and so is highly suggestive. As described at the start of this book, Homberg announced his sulphur as light theory at a public assembly in April 1705, just about a year before Newton's *Optice* appeared. Homberg's announcement thus falls squarely in the period between the 1704 and 1706 versions of Newton's book when the chymical queries took shape.[11] Although the *Mémoires* carrying Homberg's paper would not appear until later in 1706, news of Homberg's announcement traveled quickly by means of learned journals, the popular press, and, more specifically for Newton's circle, Geoffroy's newsletters. As is clear from his immediate undertaking to build his own burning mirror to replicate Homberg's experiments, Newton had already become interested in Homberg's work by early 1704. But there is further clear evidence that Newton kept up with results from the Académie. In the second of the "chymical queries," Query 23, Newton cites the division of common sulphur into a "fatty bitumen, an acid salt, a fixed earth, and a little metal." This information comes directly from Homberg's analysis of common sulphur as described in the 1703 *Mémoires*. That volume was published in 1705, again neatly fitting between the 1704 and 1706 versions of the queries, and furthermore indicating that the new chymical queries were actually written during this interval. In the same query, Newton also describes the demonstration performed by Nicolas Lémery of producing an "artificial volcano" from iron filings mixed with sulphur and buried in the earth, as well as recapitulating Lémery's related ideas about the

9. Isaac Newton, *Optice* (London, 1706), 319.

10. The possible influence of Homberg upon Newton's chymical queries was raised originally in my "Wilhelm Homberg et la chimie de la lumière," *Methodos* 8 (2008), https://journals .openedition.org/methodos/1223. The *Journal de Trévoux* (1709): 200 recognized the correspondence of Newton's ideas with Homberg's in its review of *Optice*: "[Newton] auroit bien profité des belles decouvertes de Monsieur Homberg sur les rayons du Soleil, s'il en avoit eu connoissance." Newton did undoubtedly have *connoissance* of Homberg's work, as indicated in what follows here.

11. Newton's *Optice* was announced in the Term Catalogues for the Easter term of 1706; *Term Catalogues, 1668–1709*, ed. Edward Arber, 3 vols. (London, 1903–6), 3:504. The chymical queries were apparently written by 21 December 1705, when Newton mentioned them to Gregory; see Hiscock, *David Gregory*, 29–30.

cause of meteorological phenomena.[12] Even though Newton does not name either of his French sources, his inclusion of this material shows that he was mining the Académie's work at this time for chymical materials and incorporating them into his writing.

Newton also endowed light with attractive forces, which of course have no direct counterpart in the Batavian chymist's system. The second paragraph of Query 22 of 1706 states that light is composed (as Homberg also asserted) of "the very smallest of all bodies." Drawing an analogy with lodestones, where it had long been noticed that smaller lodestones are more powerful relative to their size than large ones, Newton concludes that light particles, as the smallest particles of matter, must have the greatest attractive power. Thus, light incorporated in bodies essentially holds compound substances together by attraction, and the same force governs their activity and mutual interactions. Similarly for Homberg, his light/sulphur also holds compound substances together, but by means of its "natural glue [*gluten naturel*]." In both cases, however, it is incorporated light that is the unique source of activity in matter—in Newton's case through the action of immaterial forces, and in Homberg's either by a more mechanical means or by some inherent (but unspecified) motion or property of its particles. Newton's remarks on the attractive power of light particles serves as his introduction to the following Query 23, which deals with various chymical reactions and the driving forces behind them, but he deleted this section from later editions, and so it appears only in the Latin 1706 edition. Perhaps he later changed his mind about light's predominant role in driving chymical interactions, but in 1706 he certainly saw incorporated light particles as the chief cause of a broad array of well-known chymical phenomena.[13] It is also the case that Newton had been thinking about the interaction of light with matter before 1706—as Boyle had done also in the 1670s—and even its possible relationship to the sulphur principle specifically. In the 1704 *Opticks* Newton related the refractive index of various substances to their degree of sulphureous character, suggesting that the sulphur principle interacted more powerfully with light than the other chymical principles do.[14]

12. Newton, *Optice*, 325–26, 330–31. Newton's material is drawn from Wilhelm Homberg, "Essay de l'analyse du souffre commun," *MARS* (1703): 31–40; and Nicolas Lémery, "Explication physique et chimique des feux sousterrains, des Tremblement de terre, des ouragans, des eclairs et du tonnerre," *MARS* (1700): 101–10; Principe, "La chimie de la lumière," para. 27.

13. Newton, *Optice*, 320–21.

14. Isaac Newton, *Opticks* (London, 1704), 75–76; and Robert Boyle, "New Experiments to make the parts of Fire and Flame Stable and Ponderable," in *Essays of Effluviums*, in *Works of Robert Boyle*, ed. Michael Hunter and Edward B. Davis, 14 vols. (Pickering & Chatto: London,

Given this background, it would overstate the case to claim that Newton simply coopted ideas wholesale from Homberg. Newton had his own ideas, and even modified Homberg's interpretation of his analysis of common sulphur (in terms of the cause of combustibility) to fit them.[15] Nevertheless, it does indeed appear that accounts of Homberg's ideas and experiments impelled Newton to pursue his own thinking further, to construct a new instrument and conduct new experiments, and to bring his own ideas into print by inserting the two additional queries into the 1706 *Optice*. It is significant that Newton's experimentation and his famous chymical queries can be so directly connected with the Académie Royale des Sciences, and specifically with Homberg's work. It may be that more of the many chymical illustrations Newton expressed in the lengthy Query 23 were partly drawn from or inspired by the work going on in Paris, but they are sufficiently nonspecific or reworked that they are not as easily recognizable as the material Newton obviously adopted from Homberg and Lémery.

The lens experiments provoked other responses from around Europe. In Italy, the *Giornale de' letterati*, reporting on experiments conducted with a smaller burning lens in Florence, briefly mentioned the "one acquired by the Duc d'Orléans, of which curious things [*strane cose*] have been written."[16] But the most extended engagement with Homberg's lens experiments came from Nicolas Hartsoeker (1656–1725), who not only devoted a significant part of one of his books to refuting Homberg, but also claimed to have tried to replicate his experiments with one of the two other large Tschirnhaus lenses in existence. Hartsoeker had associated with the academicians in Paris for many years, predominantly as an optical instrument-maker, had become an *associé étranger* of the Académie in 1699, and as Alice Stroup has argued, probably also acted as a spy in the Netherlands for Louis XIV.[17] He also became, es-

1999–2000), 7:299–333. See Newman, *Newton the Alchemist*, 443–59, for an analysis of Newton's changing ideas about the role of light in chymistry and on differences between Newton's ideas of combustion, as expressed in the chymical queries, and Homberg's.

15. Newman, *Newton the Alchemist*, 458–59.

16. *Giornale de' letterati d'Italia* 8 (1711): 222–23. I thank Michael Bycroft for this reference.

17. On Hartsoeker, see Alice Stroup, "Nicolas Hartsoeker, savant hollandais associé de l'Académie et espion de Louis XIV," in *De la diffusion des sciences à l'espionnage industriel, XVe–XXe siècle* ed. André Guillerme (Lyon: ENS Editions, 1999), 205–28; Stroup, "Science, politique et conscience aux débuts de l'Académie royale des sciences," *Revue de synthèse* 114 (1993): 423–53; Catherine Abou-Nemeh, "The Natural Philosopher and the Microscope: Nicolas Hartsoeker Unravels Nature's 'Admirable Oeconomy,'" *History of Science* 51 (2013): 1–28; and Abou-Nemeh, "Réaumur's Crayfish Experiments in Hartsoeker's *Système*," in *The Life Sciences in Early Modern Philosophy*, ed. Ohad Nachtomy and Justin E. H. Smith (Oxford: Oxford University Press, 2014), 157–80.

pecially after the start of the eighteenth century, one of natural philosophy's *bêtes noires*, criticizing and generally picking fights with virtually everyone dead or alive—Leibniz, Huygens, Dortous de Mairan, Mariotte, Leeuwenhoek, Newton, to name just a few. He also weighed in on the debate between Johann Bernoulli and Homberg about the cause of the luminosity exhibited by mercury in some barometers, and his criticisms in that instance provoked the vituperative wrath of the reliably grumpy and vitriolic Bernoulli, who called him "a man most arrogant, most impudent, most vain, and above all, most unskilled in geometry."[18] The Académie itself was displeased with Hartsoeker's *impolitesse*. Fontenelle warned him that he had violated the rule set down in the 1699 Règlement that disagreements were to be handled "without terms of contempt or bitterness," and the Académie stopped sending him the copies of the annual *Histoire et mémoires* he was due as an associate member—presumably to avoid offering his biting critiques more to chew upon.[19] Hence, it is not surprising that Homberg's highly celebrated lens experiments became one of Hartsoeker's targets.

The exact relationship between Hartsoeker and Homberg, and how it changed over time, would benefit from closer investigation—they were undoubtedly acquainted personally. Both were in Paris, and both moved in the Académie's penumbra in 1684. Homberg popularized one of Hartsoeker's new microscope designs in Italy in 1686. Both were interested in microscopy at that time, and both wrote on animal generation. Both made lenses and were experienced in optics. When Homberg returned to Paris in the late 1680s, and certainly after he became an academician in 1691, he and Hartsoeker once again moved in the same circles until the latter's departure from Paris in 1696. Hartsoeker's *Principes de physique* (1696) contains many pieces of informa-

18. Bernoulli to Jacob Hermann, 1 March 1712, Universitätsbibliothek Basel, L Ia 659, Nr. 12: "hominum arrogantissimum, impudentissimum, vanissimum et imprimis Geometriae imperitissimum." The insults only get worse in his letter to Dortous de Mairan of 17 June 1723 (Universitätsbibliothek Basel, L Ia 661, Nr. 4), where Bernoulli refers to Hartsoeker as "n'ayant ny merite ny naissance ni education, rempli qu'il est de vanité et rusticité."

19. Hartsoeker's lengthy response to Fontenelle—defending his actions, bemoaning wrongs done to him, and accusing a long list of academicians of stealing his ideas and of being more guilty of bad manners than he—is printed in the preface to his posthumously published *Cours de physique* (The Hague, 1730), sig. ***2v–****2r. Fontenelle remarks that Hartsoeker "just before his death wanted to justify himself with a sort of an *apologia*" in response to reproaches from the Académie but did not complete it; "Éloge de M. Hartsoeker," *HARS* (1725): 137–53, at 153. I propose that the "Lettre de M. Hartsoeker à M. de Fontenelle" published in the preface to the *Cours de physique* represents the core of that document. The preface, written by Hartsoeker's sons Theodore and Christopher (see *Journal des sçavans* [1730]: 384 and [1731]: 125), is almost entirely devoted to a spirited defense of their father.

tion and conclusions gleaned from contact with Homberg and other academicians, some of which the academicians published on their own only later.
Hartsoeker mentions, for example, how the Danube flows more slowly in
the morning and attributes this phenomenon to the impact of the sunlight
on the flowing water—as Homberg would explain in 1707 from his own observations made during his travels. Hartsoeker notes also how a watch spring
vibrates when placed near the focus of a burning mirror (due to the collisions
from light particles)—as Homberg reported in greater detail in 1708—and
Hartsoeker even describes what is certainly Lémery's "artificial volcano" accompanied with the same conclusions, which Lémery himself would publish
fully only in 1700.[20]

The disagreement between Hartsoeker and Homberg began with a letter sent to Homberg in late 1706. The letter, sent by "a Dutch philosopher
who had witnessed part of my experiments," requested more information
about the vitrification of gold brought about by the Tschirnhaus lens. The
writer claimed that when the experiment was being done in his presence,
he occasionally saw a bit of ash fly up from the charcoal used to support the
gold and land on the molten gold. He therefore suggested that it was this
ash that vitrified to form the observed glass, not the gold itself. Homberg
responded that if that were the case, then the same vitrification would be observed when silver is treated in the same way, but in that case only a whitish
powder is produced rather than a glass. The Dutchman, apparently satisfied
with Homberg's reply, wrote back to ask for an explanation of the production of the glassy material and why the results with gold and silver differed.
Homberg replied that the minute, fast-moving particles of the matter of light
penetrate the gold and disjoin its principles of mercury and sulphur. (Here
again one can see the matter of light working for Homberg like a solvent, and
specifically like the alkahest whose unparalleled solvent ability depends upon
the extremely small size of its particles.) The volatile mercury then departs,
along with a portion of the sulphur, as the smoke seen rising from the fused
metal. Gold, however, contains a small quantity of fixed earth, and this earthy
residue combines with some of the remaining sulphur to produce a drop of
glass. Since silver contains less sulphur than gold, there is not a sufficient
quantity remaining to combine with the residual earth to produce a glass. According to Homberg's account, another Dutchman wrote to him two weeks
later, saying that he was not satisfied with the explanation and claiming once
again that the whitish powder seen on silver and the glass seen on gold were
both produced from the charcoal ash. Since gold is less porous that silver, he

20. Nicolas Hartsoeker, *Principes de physique* (Paris, 1696), 83, 137, 209.

wrote, it reflects more of the light striking it and so the combined heat of the impingent and reflected light then suffices to melt the ash into a glass on the gold, but on the silver, which reflects less light, the temperature is too low to fuse the ash into a glass, and so it remains as a white powder. Homberg answered these objections with several experimental observations, including by noting that the white ash produced on charcoal itself vanishes instantly when placed at the focus of the lens, so if it does so on porous, nonreflective charcoal, it would do so on silver as well.[21]

As in other instances, the text in the *Mémoires* diverges significantly from that in the procès-verbaux. The procès-verbaux provide an entirely third-person account of the correspondence, while the *Mémoires* transcribes the text of Homberg's letter to the first Dutchman verbatim, replete with the "Je suis, Monsieur, &c" at the end. But oddly, while the published version tells of *two* letter-writing Dutchmen, the procès-verbaux report that all the objections came from just *one*. It is not obvious how to resolve this discrepancy. Given the more private nature of the procès-verbaux, it seems reasonable to accept its account of a *single* Dutchman. Perhaps it seemed better in public to appear to have satisfied the objections of one Dutchman out of two rather than none? Whatever the resolution, it is clear that Hartsoeker was either the only Dutchman or the unconvinced second one; Fontenelle later revealed his identity explicitly, and Hartsoeker continued to object to Homberg's vitrification of gold not only in his 1710 *Eclaircissemens sur les Conjectures physiques* but until his dying days.[22]

Hartsoeker's *Eclaircissemens* is composed entirely of his replies to objections that had been made to his earlier *Conjectures physiques*; these replies generally tumble out in no obvious order.[23] Many academicians serve as targets for Hartsoeker's criticisms, but he returns most frequently to Homberg, finding fault not only with his vitrification of gold, but also with his explanations of dissolution, his sulphur principle, his study of salts, and other topics. Some of this criticism is based upon using Fontenelle's sometimes imprecise

21. Wilhelm Homberg, "Eclaircissemens touchant la vitrification de l'or au verre ardant," *MARS* (1707): 40–48; PV 20, fols. 53r–58r (16 February 1707).

22. Fontenelle, "Éloge de Hartsoeker," 147–48. Leibniz asked Hartsoeker on 10 March 1707 whether his correspondence with Homberg on this subject had been published; Carl Immanuel Gerhardt, ed., *Die philosophische Schriften von Gottfried Wilhelm Leibniz*, 7 vols. (Berlin, 1887), 3:493.

23. Nicolas Hartsoeker, *Eclaircissemens sur les Conjectures physiques* (Amsterdam, 1710) and *Conjectures physiques* (Amsterdam, 1706). The latter work is a rearranged version, largely verbatim, of his 1696 *Principes de physique* addressed to his new patron, Johann Wilhelm II, Elector Palatine.

summaries in the *Histoire* as if they were Homberg's own text, or from mis-representing Homberg's claims, or from finding "contradictions" between Homberg's ideas expressed in different papers without acknowledging that new experimental results caused Homberg constantly to revise his chymical theory (as shown in chapters 3 and 4).[24] But the vitrification of gold provides the most interesting case because only in this instance does Hartsoeker claim to have repeated Homberg's experiments, rather than simply advancing alter-native interpretations of experiments based on his own "conjectures."

In his 1710 *Eclaircissemens*, Hartsoeker reports that he traveled to Kassel in order to use the Tschirnhaus lens bought by Karl I, Landgrave of Hessen-Kassel (1654–1730), which was roughly the same size as the one owned by the Duc d'Orléans (see figure 7.1). Using this lens, and in front of witnesses that included the Landgrave, Hartsoeker melted gold on a cupel—rather than on charcoal as Homberg had done—placed at the focus and left it there in fusion for over an hour. He saw no glass produced, no significant smoke re-leased, and he found that after the gold had cooled, it had not lost any weight. He did the same with lead, and again found no vitrification or other change.[25] Hartsoeker therefore asserted that the glass Homberg observed resulted not from the gold itself but from a combination of ash from the charcoal, which he considered "a type of sand," with salts drawn out of the air.[26]

Quite a different account of these experiments, however, is provided by the travel diary of Zacharias Conrad von Uffenbach (1683–1734), who visited Kassel in November 1709 and met there with the mathematics professor Lo-thar Zumbach von Koesfeld (1661–1727). After von Uffenbach surveyed the Landgrave's considerable collection of scientific instruments, which included the Tschirnhaus lens, Zumbach showed him boxes containing the samples of materials that had been exposed to the lens. He explained that he himself had performed experiments with the lens "primarily to please Mr. Hartsoeker in Düsseldorf because he was in a dispute with Mr. Homberg in Paris." Zum-bach informed von Uffenbach that he had made experiments only upon lead. Exposed to the lens in an earthenware crucible, the lead did produce a glass, although it did not do so when placed in an iron crucible or in a hollowed-out piece of chalk. Zumbach thus concluded that the vitrification of lead re-

---

24. The major criticisms of Homberg appear on 64–70, 79–82, 121, 130, 137–43, 166–67, 175–77; see 57–59 for criticism of Tschirnhaus's claim to have vitrified gold; examples of mistak-ing Fontenelle's statements for Homberg's are at 140 and 154.

25. Hartsoeker, *Eclaircissemens*, 166–67.

26. Hartsoeker, *Eclaircissemens*, 58.

quired "the sand which is in the earth of an ordinary crucible," a sentiment seemingly echoed by Hartsoeker's reference to the "sand" of the charcoal ash. But Zumbach did *not* experiment on gold or the other metals, nor did he report that Hartsoeker had ever come himself to Kassel to conduct experiments with the Tschirnhaus lens, much less in the presence of the Landgrave, which surely would have been an event significant enough to merit mention to Uffenbach.

In later publications, Hartsoeker continued his attacks on Homberg, and went yet further with his claims. He reported that his patron, Johann Wilhelm II, Elector Palatine (1658–1716), ordered him to make his own burning lenses "in order to verify what is said in Paris about the transmutation and vitrification of metals." (Johann Wilhelm is the same Elector Palatine who attempted to attract Homberg away from Paris in 1704; see chapter 4.) Hartsoeker claimed that he then produced three huge lenses, five Rheinland inches greater in diameter than any Tschirnhaus and Fremel had ever produced (thus roughly 42 inches, or 107 cm, in diameter). The Elector let Hartsoeker keep one of them. These lenses would therefore have been produced sometime between the 1710 *Eclaircissements* (which do not mention any such lens) and Johann Wilhelm's death in 1716. Hartsoeker then reported that he exposed gold and all the other metals to the focus of this new lens, with the result that none of them was altered in the least. He wrote, just before his death, that he had carried out "all the experiments that are spoken of in the *Mémoires* of the Académie Royale des Sciences," presumably meaning Geoffroy's as well as Homberg's, "and found almost all of them false."[27]

Hartsoeker's claims are problematic. In the first case, he does not mention that it was Zumbach who actually carried out the experiments with the Tschirnhaus lens on his behalf and that Zumbach experimented only on lead. If Hartsoeker traveled to Kassel to do the experiments himself on other metals, and in the presence of the Landgrave, as he claims, that must have occurred in the very short period between von Uffenbach's visit near the end of

27. Nicolas Hartsoeker, "Lettre à Fontenelle," in *Cours de physique*, sig. ****r–****v. See above, note 19, on this text. Fontenelle, "Éloge de Hartsoeker," 148, provides slightly more information about this lens. Homberg's and Tschirnhaus's vitrification of gold is actually problematic, since it would be difficult to explain how elemental gold gave rise to a glassy material on its own. The information in Geoffroy's letter (above, note 5) that the gold used had first been reduced to a powder through dissolution in aqua regia and precipitation may offer an explanation: residual salts from this treatment may have provided a vitreous material in combination with some of the gold. Unfortunately, without a Tschirnhaus lens available for reproducing the experiment, any suggested explanations must remain conjectural.

FIGURE 7.1. The Tschirnhaus lens bought by Karl I, Landgrave of Hessen-Kassel, now in the Naturwissenschaftlich-Technische Sammlung kept at the Orangerie in Kassel, and reportedly used by Hartsoeker to reproduce Homberg's experiments in Paris. The lens is 81 cm (32 inches) in diameter. The smaller focusing lens (not pictured) is 21 cm (8 inches) in diameter. Photo by the author.

November 1709 and the publication of the *Eclaircissemens* in early 1710.[28] That timing would situate the supposed experiments in the winter or early spring months, which are unsuitable for using the lens. Additionally, when placed at the focus, as Hartsoeker states, gold should have boiled or splattered away, as it did reliably for Homberg, Newton, and other investigators. Homberg's first memoir about the Tschirnhaus lens describes in detail how at the focus gold boiled away unchanged and could be collected as a powder on a piece of paper held nearby. For this reason Homberg chose to do his experiments away from the focus itself, in what he called the "vitrifying region" where the effect of the lens was reportedly different.[29] It is, moreover, inconceivable that

28. A lengthy review of Hartsoeker's *Eclaircissemens* appeared in the June 1710 issue of *Nouvelles de la république des lettres* (603–28), implying that it was printed no later than April 1710. The review in the *Journal des sçavans* appeared in July (475–79).

29. Wilhelm Homberg, "Suite des observations faites au Miroir ardent," PV 21, fols. 341r–343r (12 August 1702), at 343r; and "Observations faites par le moyen du Verre ardant," *MARS* (1702): 141–49, at 143–44.

none of the metals was changed in the least by exposure to the focus either of the Tschirnhaus lens or the yet larger one Hartsoeker claims to have manu-factured. The base metals would at least have oxidized, if not evaporated at least in part, as they did for Geoffroy and Lémery at the Palais Royal.

Hartsoeker's claim to have produced even larger lenses than Tschirnhaus and Fremel raises yet more questions. Fremel's success in casting high-quality glass blanks of enormous size was a significant technological achievement, and Tschirnhaus's massive, water-powered machines for grinding and pol-ishing the blanks into biconvex lenses were no less a marvel. It is therefore hard to imagine how Hartsoeker managed successfully to cast and cool a blank heavier than Fremel's, and then grind and polish it not with a spe-cially designed, water-powered machine, but, as he claims, in an 18-foot-wide (5.5-meter-wide) copper basin, using nothing but paper covered with polish-ing powder.[30] Hartsoeker was indeed an experienced lens-maker, but those he produced were generally small, hand-sized objects that he ground and polished by hand in a basin, and even work on that scale was time-consuming and wearisome.[31] Given the notoriety that followed the success of Fremel and Tschirnhaus in casting, cooling, shaping, and polishing large burning lenses, it is extremely odd that there are no reports of Hartsoeker's surpassing their achievement. Both the feat itself and the lack of contemporaneous reports about it are enough to make one question Hartsoeker's account, especially given how his related claim to have traveled to Kassel to perform lens experi-ments is contradicted by Uffenbach's report. Nevertheless, a lens of exactly the dimensions he quotes, and weighing 200 pounds, is reported to have ex-isted among his possessions at the time of his death. The lens is listed in a supplement to the auction catalogue of his library, where it is described as something "that has never been heard spoken of."[32] Why Hartsoeker would have kept this instrument unseen and unmentioned for over a decade only adds to the questions.

Why was Hartsoeker so keen to criticize Homberg's work? He himself provides the answer. In his letter to Fontenelle, he writes that if Homberg's results are accurate, then "everything that I have written about natural phi-losophy would fall into ruin, as Mr. Leibniz and others have very correctly

30. Hartsoeker, "Lettre à Fontenelle," in *Cours de physique*, sig. ****r and 176–77; also *Recueil de plusieurs pieces de physique* (Utrecht, 1722), 136–37.

31. Abou-Nemeh, "Philosopher and Microscope," treats Hartsoeker's work as an instrument-maker (particularly of lenses) at length, see especially 7–9.

32. "Catalogue des verres ardens," 1, in Jan Swart, *Bibliotheque de feu Mr. N. Hartsoeker* (The Hague, 1727). My thanks to Christine Lehman and Catherine Abou-Nemeh for directing me to this source.

judged."[33] Note how Hartsoeker remarked that he made his supposed lens to refute the claims not only about the vitrification of metals, but also about their *transmutation*. This is the crux of the matter. Hartsoeker's matter theory—built upon a fluid "first matter" and permanent, indivisible atoms—denied the mutability of substances, and not just of the metals that we today consider elemental: "Every metal is nothing other than an assemblage of an infinity of homogeneous, eternal, and immutable particles, just like air, water, salts, and an infinity of bodies in this visible world . . . thus the transmutation as well as the vitrification of metals are only imaginary [*chimeres*]."[34] Homberg's and Geoffroy's claims to have decomposed the metals with the lens would indicate that these substances are compound bodies, rather than being made up of Hartsoeker's "homogeneous, eternal, and immutable particles." For the same reason, Hartsoeker also criticized Homberg's interconversions of salts, because again, Hartsoeker denied that such transformations of one salt into another were possible. Hartsoeker's confidence in the immutability of substances led him—six years before any of Homberg's lens experiments—to declare that "the chymists who labor to make [gold] lose their efforts in vain, and that those who boast of having the secret are charlatans and liars who seek only to entrap credulous persons."[35]

Hartsoeker's admission to Fontenelle that Leibniz recognized the threat that Homberg's experiments posed to his system is highly relevant. The Leibniz correspondence reveals that many of the initial objections around which the *Eclaircissemens* is built came in fact from Leibniz. Indeed, it was Leibniz, in objecting to Hartsoeker's claim that all substances are wholly immutable, who brought up the vitrification of gold performed by Tschirnhaus and Homberg as specific evidence to the contrary.[36] Hence, it was particularly important for Hartsoeker to oppose that result.[37] Leibniz was convinced of the reality of the vitrification because he had been shown a piece of the glassy ma-

33. Hartsoeker, "Lettre à Fontenelle," in *Cours de physique*, sig. ****r.

34. Hartsoeker, *Eclaircissemens*, 168–69, see also 175–78; compare Hartsoeker to Leibniz, 8 July 1710, in Gerhardt, *Leibniz*, 3:499.

35. Hartsoeker, *Principes de physique*, 96.

36. The Leibniz-Hartsoeker correspondence is published in Gerhardt, *Leibniz*, 3:483–535; see particularly the letters of 4 October 1706 (490–92) and 10 March 1707 (492–94). For a fuller analysis of Leibniz's ideas and his objections to Hartsoeker, see Peter Anstey, "The Coherence of Cohesion in the Later Leibniz," *British Journal for the History of Philosophy* 24 (2016): 594–613.

37. Hartsoeker's son reported from Paris in August 1710 that "Mr. Homberg and the other chymists of the Académie Royale des Sciences begin to see that the vitrification of metals about which they have made so much noise is a fantasy"; Gerhardt, *Leibniz*, 3:504 (22 August 1710). This claim is obviously false, since Homberg included the vitrification in the final version of his *Élémens de chimie*, written after this date.

terial by Tschirnhaus himself, but he allowed that there were other possible interpretations for its formation than Homberg's. Nevertheless, Leibniz was adamant in his opposition to the idea of immutable, indivisible atoms, writing to Hartsoeker that "it is easy to make up fictions but difficult to render them reasonable . . . [your] atoms are just such a fiction."[38] Although most of Leibniz's own arguments against Hartsoeker's brand of atomism were based on mathematics and logical reasoning, he also endeavored to obtain supporting experimental evidence from none other than Homberg himself. In the midst of his continuing epistolary arguments with Hartsoeker, he wrote to Homberg to say that "I hope that you may continue your beautiful discoveries for a long time, and if you have anything about the transmutation of metals and of salts that proves its truth, even without any profit, I hope that one may learn about it. For Mr. Guglielmini, Mr. Hartsoeker, and other capable people, prepossessed [*prevenus*] by their atoms, consider these transmutations impossible."[39]

The other "prepossessed" person whom Leibniz mentions is Domenico Guglielmini (1655–1710). Guglielmini, like Homberg, wrote about the chymical principles of salt and sulphur. His 1705 *De salibus dissertatio* appeared so soon after the publication of Homberg's 1702 essay on the salt principle that the Italian chymist could do little more than add an appendix to acknowledge (and dissent from) it. Guglielmini argued, for example (as Leibniz implies), that salts were fixed in their identity and could not be transformed into other salts, clearly in contradiction to the work Homberg and others were doing at the Académie on the mutual interconversions of salts. In a hastily composed appendix, Guglielmini offered a Latin translation of part of Homberg's 1702 essay on salt, and recognized that Homberg's work "would seem to destroy my doctrine of salts."[40] His work on the sulphur principle appeared

38. Gerhardt, *Leibniz*, 3:506 (30 October 1710).

39. Gottfried Wilhelm Leibniz to Homberg, 10 March 1711; NLB, LBr 420, fol. 3r–v: "Je souhaitte que vous puissés continuer long temps vos belles decouvertes, et si vous aves quelque chose sur la transmutation des metaux et des sels, qui en prouve la verité, quoyque sans aucun profit, je souhaitte qu'on le puisse apprendre. Car M. Guglielmini, Mons. Hartsoeker et quelques autres habiles gens, prevenus par leur atomes tiennent ces transmutations pour impossibles." For more on Leibniz and chymistry, see Anne-Lise Rey, "Alchemy and Chemistry," in *The Oxford Handbook of Leibniz*, ed. Maria Rosa Antognazza (published online 2016): http://www.oxfordhandbooks.com. For further remarks by Leibniz on transmutation, in particular regarding reports of the success of Johann Friedrich Böttger (named only a *jeune garçon apothecaire*) in Berlin in 1701, see Gerhardt, *Leibniz*, 3: 489–90 (12 December 1706) and 500–501 (summer 1710).

40. Pierre Varignon to Leibniz, 10 October 1706, in *Leibnizens gesammelte Werke*, ed. Pertz, 3d ser., 4:150–51: "On aura vû à Paris le livre de M. Guglielmini des Sels, où il soutient contre

only posthumously in 1710, and does have some affinities with Homberg's ideas. Guglielmini does not cite Homberg, which impelled his editor to insist vigorously that the Italian reached his conclusions independently.[41]

Continuations of Homberg's Studies

Given the multiple clashes between Homberg and Nicolas Lémery, and the multiyear debate between Homberg's protégé Geoffroy and Nicolas's son Louis over the artificial generation of iron, it is perhaps surprising that the clearest embrace, defense, and elaboration of Homberg's chymical ideas within the Académie came from Louis Lémery. In 1709, the younger Lémery spoke to a public assembly about the nature of the "matter of fire or of light" and its incorporation with compound substances.[42] He criticized those who deny that such incorporation is possible, and presented evidence in favor of it.[43] His evidence repeats or parallels Homberg's: the increase in weight of metals upon calcination, whether done over a fire or in the light of a burning lens, and the liberation of heat when quicklime slakes and alkalies dissolve in water. As the paper proceeds, Lémery's initial references to the *matière du feu* are completely replaced with references to the *matière de la lumière*, pulling his ideas ever closer to the central theme of Homberg's chymical theory. Lémery cites experiments with the Tschirnhaus lens—he is the only academician known to have used the instrument besides Homberg and Geoffroy— and treats at length the composition and activity of the Sun (which he calls "a great reservoir of the matter of light") and its light. In accord with Homberg, Lémery states that light incorporates itself variously with various substances, thereby producing a variety of oily and fatty products—that is, the class of

---

M. Homberg et autres Chymistes de l'Académie, qu'on ne peut point changer les sels." Domenico Guglielmini, *De salibus dissertatio* (Venice, 1705), 269–72. On Guglielmini and salts, see Reijer Hooykaas, "Domenico Guglielmini et le développement de la cristallographie," *Atti della Fondazione Ronchi* 8 (1953): 5–20; and Alberto Vanzo, "Corpuscularism and Experimental Philosophy in Domenico Guglielmini's 'Reflections' on Salts," in *The Idea of Principles in Early Modern Thought*, ed. Peter R. Anstey (New York: Routledge, 2017), 147–71.

41. Domenico Guglielmini, *De principio sulphureo dissertationes* (Venice, 1710).

42. Louis Lémery, "Conjectures & reflections sur la matiere du feu ou de la lumiere," *MARS* (1709): 400–418.

43. It is not entirely clear who was critical of Homberg's light/sulphur theory. It is possible there were critics within the Académie whose views have not been recorded (perhaps Louis Lémery's own father, who had argued with Homberg repeatedly?). The only contemporaneous critique I have located appears in the anonymous "Reflexions physiques sur le souffre principe de Mr. Homberg," sent to the *Journal de Trévoux* (1706): 1621–27; its criticisms are firmly and explicitly based in Cartesian physics.

sulphurs—that are inflammable. Indeed, for Lémery as for Homberg, the chymical principles of earth, water, and salt act primarily as vehicles for the active matter of light, which in turn "gives [compound bodies] all their properties."[44] Even Homberg's "economy of sulphur" reappears clearly when Lémery concludes his paper by noting that combustible materials on Earth are in effect "little Suns," that is, they are terrestrial repositories of sunlight in fixed form. These materials can then release the sunlight they have captured when they are burned, and thus provide heat and light for human life during the season when the Sun does not provide a sufficient amount on its own.[45]

Lémery does not merely repeat Homberg's idea about the chymical nature of light, but also strengthens it with his own contributions. He provides a more developed mechanism for the incorporation and liberation of light in solid bodies to answer those who doubt how something so active as light can become fixed in bodies. He suggests that particles of light become trapped inside *cellules* or *vesicules* within the texture of solid bodies. Light's characteristic rapid motion is preserved within these *cellules* because its particles continue spinning rapidly about their center, and such motion is perpetuated by the constant flow of a *matière subtile*, a substance a hundred times finer than light itself and whose function is to provide motion to other bodies. The light particles can thereafter be liberated when the solid structure is disrupted by burning, fusion, or dissolution. Phosphorescent substances, which he calls "sponges of light," have a weak structure that allows the light to seep out slowly from the *cellules*.[46] The ideas of a *matière subtile* and of particles of light spinning about their center do not, of course, have counterparts in Homberg's system. But Lémery, in defending and extending Homberg's system, joins it with Cartesian ideas—reintroducing Descartes's "subtle matter" and providing a "particular shape" to the particles of light, thereby allowing for the inclusion of some of Descartes's ideas.[47]

44. Lémery, "Conjectures," 414–16.

45. Lémery, "Conjectures," 418. The idea of candles and fireplaces making up for dark and cold weather may have been especially on the audience's mind in November 1709 as they recalled the deadly and unprecedented cold that gripped Europe during the previous winter, known as "le Grand Hiver." The Académie suspended its usual meeting on 26 January 1709 because of the "grand dégel"; PV 28, fol. 21r. One is also reminded of the "sooty" member of the Academy of Lagado endeavoring to extract sunbeams from cucumbers, as described in *Gulliver's Travels*. A connection with Homberg's ideas is possible, given the origin of other Swiftean absurdities in contemporaneous scientific publications; see Marjorie Nicholson, "The Scientific Background to Swift's *Voyage to Laputa*," *Annals of Science* 2 (1937): 299–334 (where the cucumber experiment is, not entirely convincingly, connected to Stephen Hales).

46. Lémery, "Conjectures," 407–8.

47. See Fontenelle's summary of Lémery's paper, *HARS* (1709): 6–8.

The younger Lémery also revisited the old analysis of plants project, eventually presenting four lengthy memoirs on the topic between 1719 and 1721.[48] Homberg's final conclusion about the project (as noted in chapter 2) was that fire analysis—as Van Helmont had stated and Duclos had reiterated—so greatly altered the components of plant materials that the results offered little or no insight into either the original composition of the plant or its medicinal properties. Lémery, twenty years after Homberg's conclusions, endeavored to pursue Homberg's findings further by exploring exactly how fire changes materials. Like Homberg, he chose to focus on the salts naturally present in plant materials. Observing that these were *sels moyens*, composed of an acid combined with an alkali, he noted that the fire ordinarily broke apart such salts into their acidic and alkaline constituents. If one were able, he suggested, to isolate each *sel moyen* in plants, then subject each salt separately to thermal decomposition and identify the products, then one might be able to work backward from Bourdelin's voluminous results to determine what *sels moyens* existed originally in the plants. But this, he lamented, would be a project "of vast extent, and require a very scrupulous level of detail," even though it promised to reveal "how the disguising and alteration [by fire] takes place."[49] Yet because many different *sels moyens* are present simultaneously in plants, fire analysis releases a mixture of several acids and alkalies that can subsequently react with one another to create new *sels moyens*, thereby complicating the interpretation. Despite Lémery's initial enthusiasm for learning something chymically interesting from the plants project, by the time of his fourth memoir, he had circled back to Homberg's earlier conclusions. He demonstrated, for example, that sorrel leaves treated in different ways gave different analytical results—just as Homberg had done with grape juice twenty years earlier. Lémery clearly reveals his dependence on Homberg when he ends his project by noting "that deadly nightshade and cabbage, one a poison, the other a foodstuff" give virtually the same analytical results—an example taken verbatim from Homberg's 1701 conclusions.[50] What is additionally worth noting is that the younger Lémery once again—like Homberg

48. Louis Lémery, "Reflexions physiques sur le défaut & le peu d'utilité des analises ordinaires des plantes & des animaux," *MARS* (1719): 173–88; "Second memoire sur les analises ordinaires de chimie," *MARS* (1720): 98–107; "Troisième memoire sur les analises de chimie," *MARS* (1720): 166–78; "Quatrième memoire sur les analises ordinaires des plantes et des animaux," *MARS* (1721): 22–44.

49. Lémery, "Second memoire," 106.

50. Lémery, "Quatrième memoire," 24–25, 44; compare Wilhelm Homberg, "Observations sur les analyses des plantes," *MARS* (1701): 113–17, at 113.

and Duclos—was interested primarily in the more general chymical principles that might be gleaned from the plants project—namely, what *sels moyens*
exist in plants and exactly how they decompose thermally—rather than in
the pursuit of medicinal utility. While Lémery was a physician who had written on the medicinal value of foodstuffs, his orientation in the Académie was
not—unlike the majority of the Académie's *chimistes* a generation earlier—
that of an apothecary.[51]

While Louis Lémery endeavored to strengthen Homberg's theory by inserting Cartesian ideas, Jean-Jacques Dortous de Mairan (1678–1771) added
Newtonian ones. Just before his admission to the Académie in 1718, Mairan
wrote an essay entitled *Dissertation on the Cause of the Light of Phosphori*,
which won first prize in a competition held by the Académie Royale des
Sciences, Belles-lettres, et Arts de Bordeaux. In speaking of the composition
of matter, Mairan echoes Homberg in preferring principles that are "more
material" than those of the physicists, namely, "those that are called chymical," and among these more material chymical principles he includes "the
matter of light." He then adopts Homberg's light as sulphur theory and borrows nearly verbatim from his "Essais." "I believe then that the matter of light
consists of a very subtle and very agitated sulphur, and is none other than the
active principle of the chymists, so called because it alone acts and makes the
others [of the principles] act."[52]

Despite his obvious source in Homberg (whom he does not cite at this
point), Mairan nevertheless refers to "*my* conjecture that sulphur is the matter of light." He also includes the further notion that the particles of light
"turn about their center," which Louis Lémery had added to Homberg's
theory in order to preserve light's motion even when combined with grosser
matter.[53] Mairan then discusses luminous materials, particularly the Bologna
Stone so beloved by Homberg. He tries to explain why the Bologna Stone, after its exposure to white light, emits light of only one particular color (usually
orange). Citing Newton's demonstration of the existence of heterogeneous
colors of light within white light, Mairan suggests that these various colors
depend upon the size of the light particles. Preparation of the Bologna Stone
produces pores in the mineral of a uniform size that can capture only those

51. Louis Lémery, *Traité des aliments* (Paris, 1702).

52. Jean-Jacques Dortous de Mairan, *Dissertation sur la cause de la lumiere des phosphores*
(Bordeaux, 1717), 21–22. This dissertation won him the Bordeaux Academy's annual essay prize
for the third year in a row, forcing a change in the rules so that no one could win more than
three times.

53. Mairan, *Dissertation sur la cause de la lumiere des phosphores* 24 (emphasis added), 34.

particles of light that fit into them snugly; thus, the calcined stone can absorb and hold light of only a single color, and thus it subsequently emits light only of that color.[54]

In addition to these younger members of the Académie who carried on Homberg's intellectual program in various ways—of course, with modifications, new experiments, and the addition of other influences—Homberg's name, ideas, and experiments were also cited as authoritative by a wide range of chymical authors all across Europe throughout the 1710s, 1720s, 1730s, and later. In addition to the authors already named, the list includes Johann Heinrich Pott (1692–1777) and Caspar Neumann (1683–1737) in Germany, Peter Shaw (1694–1763) in England, Herman Boerhaave in the Netherlands, and many others. Some other members of the Académie also continued lines of research that Homberg initiated, for example, Charles François de Cisternay Dufay's (1698–1739) further examination of luminous substances, and Jean Lémery's (1678–1721) improvements to Homberg's pyrophorus.[55]

One of the most remarkable deployments of Homberg's chymical theory, however, appears in an anonymous treatise written probably in the 1720s and never published, although the existence of multiple manuscript copies suggests that it was quite widely distributed and read. The "Essay to Elaborate upon the Science and Practice of the Work of the Chymical Philosophers" is a book-length work about metallic transmutation that examines various routes toward preparing the philosophers' stone.[56] What distinguishes this particular chrysopoetic treatise is its wholesale adoption of Homberg's theory of chymistry regarding the sulphur principle as light, including frequent citations to his papers in the *Mémoires*. The author not only sees Homberg as fully part of the tradition of Mercurialist chrysopoeians that includes Philalethes and others, but also attributes to Homberg significant theoretical and practical advances in the ongoing search for the philosophers' stone. Indeed, when explaining the history of attempts to prepare the stone (much as Homberg had done in his early textbook), the au-

---

54. Mairan, *Dissertation sur la cause de la lumiere des phosphores* 46–53.

55. Charles François de Cisternay Dufay, "Mémoire sur un grand nombre de phosphores nouveaux," *MARS* (1730): 524–35; and Jean Lémery, "Experiences sur la diversité des matieres qui sont propres à faire un phosphore avec l'alun," *MARS* (1714): 402–8; and "Reflexions physiques sur un nouveau phosphore," *MARS* (1715): 23–41.

56. London, Wellcome Institute, MS 2298, "Essai pour développer la science et la pratique de l'Oeuvre des Philosophes chimiques." Another copy existed at Bordeaux, Bibliothèque Universitaire des Sciences et Techniques, MS 23, but is among the many important manuscripts carelessly lost by the science library; see chapter 5, note 16. The dating is suggested by the frequent reference to "le feu M. Homberg."

thor smoothly includes the Batavian chymist and his central idea about the sulphur principle as light in the sequence of "chymical philosophers" who had contributed toward finding the secret. Alongside an explicit reference to Homberg's 1705 paper, the anonymous author writes that "endeavoring then to reduce sulphur into its first matter in order to know its principle, and not having ever been able to arrive at this end, this [sulphur] principle always escaping from them, they concluded that the sulphur principle must be light." Apparently forgetting that Homberg's identification of the sulphur principle with light was a recent innovation—or perhaps implying that it was known to earlier adepts and revealed openly only by Homberg—the author explains that "for this reason also they [the adepts] call their Elixir the son of the Sun or the son of fire."[57]

The text then asserts that this sulphur/light can bind itself in the pores of "homogeneous mercury," whereupon that mercury becomes a true metal. This statement is simply a summary of Homberg's explanation of how his "specially prepared" mercury produced a small quantity of gold upon extended heating. Indeed, the manuscript's author adopts Homberg's primarily mechanical explanation of this transmutation: "when the aforesaid homogeneous mercury receives the particles of the sulphur principle in the pores of its globules . . . it follows that the metal produced should be the most perfect that nature can produce in its mines, which is what is called perfect gold of 24 karats."[58] The second part of the treatise deals with the practical issue of how to prepare this "homogeneous mercury." While the author cites the expected chrysopoetic authors, such as Philalethes and his interpreter Pantaleon, he also references Homberg's 1700 paper, which he recognizes as containing a straightforward recipe for the same philosophical mercury.[59] He discusses Homberg's use of copper in that process (which Homberg eventually rejected) and proposes—again basing himself on principles drawn from Homberg's own chymical theory—an improved method of attaining

57. Wellcome MS 2298, 5: "Mais voulant ensuite reduire le soufre en sa premiere matiere, a fin d'en connoitre le principe, et n'ayant jamais pu en venir à bout, ce principe leur échapant toujours, ils ont conclu que le soufre principe devoit être la lumiere. . . . Et c'est pour cette raison aussi qu'ils apellent leur Elixir le fils du Soleil ou le fils du feu."

58. Wellcome MS 2298, 7: "Quand le susdit Mercure homogène reçoit dans les pores de ses globules les particules du souphre principe autant que la nature peut lui en donner suivant son degre de chaleur interne, il s'ensuit que le métal qui en provient, doit être le plus parfait que la Nature puisse produire dans ses mines, qui est ce qu'on appelle or parfait à 24 Karats."

59. On Pantaleon and Philalethes, see William R. Newman, *Gehennical Fire: The Lives of George Starkey, An American Alchemist in the Scientific Revolution* (Cambridge, MA: Harvard University Press, 1994), 224–26.

the necessary mercury.[60] Elsewhere he cites Homberg's experiments with the Tschirnhaus burning lens as evidence for the composition of metals, and Homberg's remarks on the composition of gold and silver specifically, all in order to propose a surer method for successfully preparing the philosophers' stone.[61]

The "Essai pour développer" is significant on several levels. First, it indicates clearly that some readers recognized the chrysopoetic import of Homberg's publications and their close connection with the foregoing transmutational tradition, even though Homberg did not acknowledge such connections explicitly in print. Second, it shows how early eighteenth-century pursuers of chrysopoeia ("alchemy") continued to avail themselves of the newest results and theories drawn from the broader field of chymistry in order to assist in the ancient quest of achieving the philosophers' stone and producing gold. The author of the "Essai" obviously read the Académie's *Mémoires*, and mined it for potentially helpful information. Finally, the document shows that there was not a clear division, even around 1720, between transmutational endeavors and the work being done at esteemed institutions like the Académie. The "Essai pour développer" provides evidence of the continuation of an integrated "chymistry," at least outside of the Acádemie. What about within the Académie?

Chymistry Becomes Alchemy and Chemistry

Many general histories of chemistry written during the twentieth century speak of the "demise" or "end" of alchemy, or of a division taking place between "alchemy" and "chemistry," or of the replacement of alchemy by chemistry. Generally, such events are attributed to the seventeenth or eighteenth century, although various authors chose various dates, persons, or incidents as marking this transition. For many anglophones, the choice fell on Robert Boyle or, more narrowly, his 1661 *Sceptical Chymist*.[62] For many francophones, the choice was often Boyle's French contemporary Nicolas Lémery and his *Cours de chymie*, which supposedly made chemistry "rational" (a term so beloved by the French) by making it Cartesian.[63] Both of these claims have

60. Wellcome MS 2298, especially 46–49, 84–85.

61. Wellcome MS 2298, 7, 194.

62. For example, E. J. Holmyard, *Alchemy* (Harmondsworth, UK: Penguin, 1957), 273, calls the book alchemy's "death warrant."

63. For example, Hélène Metzger traces a route for chemistry from obscure Paracelsianism through rational Cartesianism (exemplified by Lémery) to law-like Newtonianism in her *Les doctrines chimiques en France du début du XVIIe à la fin du XVIIIe siècle* (Paris: Les Presses Uni-

been exploded by more recent studies. In fact, the very terms of the question have become problematic. It has been demonstrated repeatedly that the criteria such authors used for what counted as "alchemy," or alternatively as "chemistry," were anachronistic, or ambiguous, or self-serving, or at the very least not informed by actors' categories.[64] Nevertheless, setting aside the claims rooted primarily in triumphalist approaches, many of these earlier formulations were attempting to describe or explain *something significant* going on in the history of chemistry, even if these honest attempts were hampered by both the lack of a rigorous language to describe it and a limited understanding of "alchemy" in its historical context. This "something" is not difficult to detect, even if the exposition of its causes and determination of its exact contours remain problematic: namely, the apparent disappearance of chrysopoeia—a more precise word than "alchemy"—or, more generally speaking, metallic transmutation, from the usual activities and interests of chymists, and the sequestering of such endeavors as something *other than* mainstream chemistry instead of as part of it. The linguistic emergence of something close to the modern and distinct connotations of the words "alchemy" and "chemistry" is certainly a related development.

While these twin developments occurred at different times and rates in different parts of Europe (presumably linked to differing constellations of causes), in regard to France it seems correct to localize them in the 1720s. This is not to say that interest in or pursuit of transmutation died out—far from it, as several authors have noted, and as we shall see in greater detail shortly. But significant changes do indeed seem to have been taking place around this time, shortly after Homberg's death. One widely cited indication is when the Académie Royale des Sciences seems to have taken an official position against transmutation in 1722 through Geoffroy's paper on cheating practices relating to the philosophers' stone. Likewise, the words *chimie* and *alchimie*, and the practices classed under those titles definitively parted company by the 1730s.

---

versitaires, 1923) and *Newton, Stahl, Boerhaave et la doctrine chimique* (Paris: Albert Blanchard, 1930). This historiography is treated in Lawrence M. Principe, "A Revolution Nobody Noticed? Changes in Early Eighteenth-Century Chymistry," in *New Narratives in Eighteenth-Century Chemistry*, ed. Principe (Dordrecht: Springer, 2007), 1–22, at 3–7. Metzger's formulation has precedents in Fontenelle's rhetoric, which persisted at the Académie; see Dortous de Mairan, "Éloge de Boulduc," *HARS* (1742): 167–71, at 167–68, where he claims that "the principles of Descartes" are the best protection "against the mysterious and sublime pretensions of the former chymistry," and credits Nicolas Lémery with the "fatal attack" on transmutation; compare his "Éloge de M. [Louis] Lémery," *HARS* (1743): 195–208, at 195.

64. William R. Newman and Lawrence M. Principe, "Alchemy vs. Chemistry: The Etymological Origins of a Historiographic Mistake," *Early Science and Medicine* 3 (1998): 32–65.

Given its long-standing prominence within chymistry, the disappearance of chrysopoeia from the normal and visible activities of chymists represents one of the major transmutations of chymistry in this period. Given what is now known about the sophisticated practices and rational bases present within chrysopoeia, and its continuing pursuit into the early eighteenth century by notables such as Homberg, it can no longer be viewed as a weakly supported idea easily defeated by scientific progress, and therefore not in need of historical explanation. On the contrary, we must now examine more closely how, why, and to what real extent this transformative disappearance actually took place. While I have no illusions of being able to present here a comprehensive view or explanation of this transformation, a closer look at the activities of Homberg's two closest collaborators—the Duc d'Orléans, now Regent of France, and Étienne François Geoffroy—offers important new insights as well as new information that needs to be taken into consideration.

The Regent, France's Economic Woes, and Transmutation

Very shortly after becoming Regent, Philippe II chose to reserve the administrative oversight of the Académie to himself rather than designating a government minister to do the job. Philippe quickly instituted several reforms to the institution, increasing the number of honoraires and replacing the class of three élèves with that of two adjoints.[65] More significantly, toward the end of 1715, just a few months after having become Regent, he set the academicians the task of participating in a huge project to catalogue the natural resources of France, in particular its mineral wealth. This project, known as the "Enquête du Régent," lasted from 1716 until 1718 with various continuations by specific academicians, particularly Réaumur, through the 1720s. Following Philippe's orders, the Académie sent questionnaires and instructions to provincial intendants all around France. The intendants subsequently returned not only the reports that now fill many boxes of papers in the Académie's archives, but also a variety of mineral samples that Réaumur and his assistant assayed for metal content or assessed for other uses.[66]

The enquête was part of Philippe's program for addressing the dire finan-

---

65. See the "Règlement of 1716" in *MARS* (1716): 2–5. Philippe's administration of the Académie is mentioned in several letters between his mother and Leibniz; see Eduard Bodemann, "Briefwechsel zwischen Leibniz und der Herzogin Elisabeth Charlotte von Orleans 1715/16," *Zeitschrift des historischen Vereins für Niedersachsen* 49 (1884): 1–66.

66. Christiane Demeulenaere-Douyère and David J. Sturdy, *L'Enquête du Régent 1716–1718: sciences, techniques et politique dans la France pré-industrielle* (Turnhout: Brepols, 2008). On Réaumur and his work on chemistry and the improvement of trades and manufacture,

cial condition of France which he had inherited from Louis XIV. Its purpose was to assess what goods and resources the French state had at its disposal and that it might exploit. Philippe's interest in overseeing the Académie himself and in launching this project rested to a large extent not only upon his own personal interest in the sciences, but also upon his years of collaboration and association with Homberg, and all he had learned from him. The speed and specificity of Philippe's actions indicate that they were well planned out before he became Regent, making it probable that he had formulated them during his close association with Homberg. Homberg may even have suggested specific courses of action to Philippe in the (lost) letter he wrote to him just before his death; Fontenelle implies a link between the Regent's administration of the Académie and that letter.[67] The enquête may in fact have begun unofficially under Homberg's own direction in the last months of his life, perhaps as a replacement for the experimental activity that the closure of the Palais Royal laboratory and his own declining health had denied him. In December 1715, Geoffroy read the academicians a letter written to the recently deceased Homberg from Fonsjean, the director general of silver mines in Alsace, that gave an account of the mines there. The letter thus closely resembles, and may be a prototype of, the hundreds of similar reports requested during the official enquête. Geoffroy wrote back with more questions, and the same Fonsjean came to Paris in early 1716 and thereafter played a major role in the enquête, working with Réaumur and carrying out assays of the ores sent to the Académie from all over France.[68] Thus Homberg, through the actions of Philippe, continued to exert an indirect influence over the Académie for years after his death.[69] It is indeed remarkable that the Regent, given all the pressing needs of the kingdom both at home and abroad—"more thorny and difficult than they had been for several centuries"—still considered the Académie and its services to the state so important that he made special efforts to attend to it and to direct its activities himself.[70]

The enquête did not, however, solve France's financial crisis—nor would it have been expected to do so on its own. Thus, as the Académie was moving forward with the enquête, Philippe was moving forward with another innova-

---

see Paola Bertucci's excellent *Artisanal Enlightenment: Science and the Mechanical Arts in Old Regime France* (New Haven: Yale University Press, 2017).

67. Fontenelle, "Éloge de Homberg," 91.

68. PV 34, fol. 271r (14 December 1715); PV 35, fol. 9v (15 January 1716). On Fonsjean, listed there as Fousjean, see Demeulenaere-Douyère and Sturdy, *L'Enquête du Régent*, 37–39, 628–30.

69. Demeulenaere-Douyère and Sturdy, *L'Enquête du Régent*, 17.

70. *HARS* (1716): 2.

tive solution—the banking schemes engineered by John Law (1671–1729).[71] In 1716, Law was allowed to open the Banque Générale, which commenced to issue paper currency in the form of banknotes redeemable in gold and silver coinage. In the following year he organized the Compagnie d'Occident, which would control the trade with the French holdings in North America, from Canada down the Mississippi basin to the Gulf of Mexico, and over the following months he added to this monopoly virtually all of France's extra-European trade.[72] Revenues from this company, generally known as the Mississippi Company, were intended to pay down the enormous French national debt. Law brought in further revenue by selling shares in the company, initially at the rate of 500 livres per share in early 1719. Interest in the shares was so great, however, that by the end of the year share prices had skyrocketed to nearly 10,000 livres. Unwisely, but in order to facilitate the purchase of shares, Law printed more banknotes, which could be used to buy shares, hence the number of banknotes issued far exceeded the available amount of gold that was supposed to back them up. At the same time, precious metals that were supposed to flow into the bank from the Mississippi territory failed to materialize. By this time, the Banque Générale had become the state-operated Banque Royale, meaning that its notes were now guaranteed by the French state. In early 1720, investors began cashing in on their gains, causing stock prices to fall and requiring that redemptions in gold be severely restricted. Panic set in, and the situation quickly went from bad to worse; toward the end of 1720 the bank stopped payment on its notes. Philippe dismissed Law from the position he had acquired as *contrôleur général des finances* (in addition to the rank of honoraire in the Académie Royale des Sciences), and the disgraced Law finally fled from France. By 1721, the price of the shares had returned to their starting figure of 500 livres, but they were nearly as worthless as the paper notes of the Banque Royale, given that there was not enough precious metal in the treasury for exchanging either into hard currency.

The collapse of the "Mississippi Bubble" is a well-known event in French economic history. It may, however, have a hitherto unnoticed connection to the topics of this book. The direct evidence is scanty, but what does exist is highly suggestive when placed in the context of Philippe's experiences with

71. On Law and his programs, see Antoin E. Murphy, *John Law: Economic Theorist and Policy-Maker* (Oxford: Clarendon Press, 1997); and Arnaud Orain, *La Politique du merveilleux: une autre histoire du système de Law (1695–1795)* (Fayard: Paris, 2018).

72. The Regent's corresponding interest in promoting France's foreign trade and shipping is manifested in his establishment of a prize of 100,000 livres for a workable method of determining longitude; he sent various proposals to the Académie for evaluation; PV 35, fol. 99r–v (15 March 1716).

Homberg, his interest in the Académie, and the French crown's endeavors just a few years earlier (as described in chapter 6) to address the financial crisis by means of metallic transmutation. During lectures given in Giessen in 1722, shortly after the banking collapse in France, the chymist Johann Thomas Hensing (1683–1726) reported that "the current Regent has himself requested the members of the Académie Royale who apply themselves to chemistry to seek out the philosophers' stone."[73] The records of the Académie, of course, provide no verification of this statement—but such is hardly to be expected of an institution that so carefully policed the contents of its procès-verbaux. But if Hensing heard this claim in Germany, the account (whether true or merely a rumor) must have been widespread. It might also explain why an agent of the English ambassador to France sent home a special account of the Regent's abilities in chymistry and his work with Homberg.[74]

Placed in context, the report of Philippe's request to the Académie's chymists is completely plausible. Following the collapse of the Mississippi Bubble, Philippe was faced with a stark but relatively straightforward problem— there was simply not enough gold and silver in France to support the shares and notes that had been issued. His enquête had failed to identify any significant deposits of gold or silver within the kingdom, and it had become clear by 1721 that there was none to be found in the Mississippi territory either, despite previous reports to the contrary. As documented in chapter 6, Philippe knew of the earlier governmental attempts to obtain supplies of gold from supposed transmuters like Delisle. He had himself carried out with Homberg a transmutation of mercury into gold within the secret room of his own Palais Royal laboratory. He had acquired and sent more recipes for transmutation to Homberg, and Homberg worked through not only the recipes that Philippe provided, but also those he acquired himself, in addition to laboring persistently over the methods of Philalethes and other cryptic writers. Homberg had published his success in producing a small quantity of gold from

73. Hensing's "Discurs von dem Stein der Weisen," prefixed to his November 1722 chemistry lectures at Giessen, was published in Georg von Welling, *Opus mago-cabbalisticum et theosophicum* (Homburg, 1735), 518: "hat der jetzige *Regent*, diejenigen Glieder der Königlichen *Societ*ät, so die *Chemie excoli*ren, selbst angehalten, den beruffenen *Lap. Phil.* zu untersuchen." A French version was published later as Johann Thomas Hensing, "Dissertation sur la pierre philosophale," in *Mémoires littéraires*, ed. Marc Antoine Eidous (Paris, 1750), 121–54, at 122–23: "M. le Régent a voulu que les Membres de l'Académie qui s'appliquent à la Chymie, travaillassent de tout leur pouvoir à découvrir la Pierre Philosophale." I first suggested this connection in my 2007 article "A Revolution Nobody Noticed?"

74. "Mr. Pulteney's Private Letter No. 1 of 12 September 1720 N.S.," Kew, National Archives, State Papers 78/166, fol. 340r–v; the text is excerpted from Fontenelle's éloge for Homberg.

mercury in the Académie's *Mémoires* as the centerpiece of his new theory of chymistry. Elisabeth Charlotte wrote in a similar way of how Philippe could make gold, but not in a sufficient quantity to be profitable.[75] Surely Philippe and Homberg discussed the problem of transmutation frequently over the course of their long collaboration. Since Philippe was certain that transmutation worked on a small scale, he could reasonably enough hope that more dedicated research could render it profitable on a larger scale—as the adepts had always promised. Philippe's early connection to Vinache—when he asked him to build furnaces for some undisclosed purpose—is very likely to have been transmutational in nature, suggesting therefore that transmutation was a driving interest in Philippe's chymical work and may even stand behind his immediately subsequent engagement of Homberg (as the Abbé de Chalucet had done previously). Given Philippe's position as administrator of the Académie and his experience with transmutation, it is not at all implausible that he would have asked the chymists of the Académie to apply themselves to the problem. He would have been especially driven to do so around 1721 when Law's actions to improve the financial condition of France had only made matters worse and underscored the insufficiency of France's supply of precious metals.

If the report about the Regent's request is accurate, it marks a further and final step in the transformation of the relationship between the state and the Académie in regard to transmutation. In the 1660s and 1680s, Colbert and Louvois had strictly forbidden any research into or even discussion of the matter. In the 1690s, Louis XIV, worried about devaluation of the French coinage by rumors that it might contain artificial gold, reiterated the prohibition. By the early eighteenth century, however, the economic situation had become sufficiently dire that the previously forbidden became a potential way out, and as a result, promising or merely suspected transmuters were investigated, encouraged, or interrogated, and several of them set to work on their processes in the Bastille or under guard at Boudin's residence in Versailles. In the case of Delisle, ministers of state spent more than seven years indulging and pursuing his promises in the hope that they would yield golden fruit, and they called upon the Académie to discuss and offer advice on the matter. While Louis XIV and his ministers were hopeful but agnostic about the possibility of metallic transmutation, Philippe was confident of it because he had actually performed it himself with Homberg, although in small quan-

75. Elisabeth Charlotte to Sophie, 27 October 1709, in Eduard Bodemann, *Aus der Briefe der Herzogin Elisabeth Charlotte von Orléans an die Kurfürstin Sophie von Hannover*, 2 vols. (Hannover, 1891), 2:231.

tity. Thus, around 1721, if the rumors are true, the Regent took the final step in overturning the previous bans on chrysopoetic investigations at the Académie by actively requesting its chymists to direct their research toward discovering how to prepare the philosophers' stone for the good of the kingdom.

Geoffroy's "Supercheries" and Chrysopoeia at the Académie in the 1720s

It is probably not mere coincidence, then, that the event long seen as signaling the decline or demise of transmutational alchemy in France took place right on the heels of this report about the Regent. Connecting the two events remains equally plausible whether the report about the Regent's request is accurate or only a rumor. On 15 April 1722, at the Académie's public assembly, Étienne-François Geoffroy read his famous paper "Des supercheries concernant la pierre philosophale" ("Some Cheats concerning the Philosophers' Stone").[76] While virtually any historical treatment of transmutational alchemy in the eighteenth century cites this paper as a signal event, the paper, its contents, and its author deserve to be examined more closely in context. The greater understanding of transmutational alchemy that has been developed over the past generation means that Geoffroy's paper can no longer be casually cited as an unsurprising or unproblematic indication of metallic transmutation's "overdue" demise at the hands of scientific progress.

Geoffroy's 1722 paper describes methods used by fraudulent would-be transmuters to fool people into believing that they have witnessed a true transmutation. Such methods, according to his account, include using crucibles that contain gold hidden under a false bottom, or using a hollow stirring rod or pieces of charcoal with gold hidden inside. In each case, the hidden gold is made to appear at the right moment in order to convince onlookers. It would indeed be interesting to locate any accounts of a supposed transmuter who was actually caught using one of these methods; I am currently unaware of any such. The paper also recounts ways of making coins and medallions that appear to have been half-transmuted into gold. During the presentation Geoffroy displayed half-transmuted iron nails produced by Delisle as an example of such practices (as mentioned in chapter 6). Geoffroy also told of a recipe for a supposed "destruction" of gold offered to the Académie which he, Réaumur, and Lémery examined and found false, although it is unclear if the claims made about the process to Bignon were actively fraudulent or merely mistaken. Geoffroy even explains (correctly) the results that Boyle obtained

76. Étienne-François Geoffroy, "Des supercheries concernant la pierre philosophale," *MARS* (1722): 61–70.

in the 1660s with his *menstruum peracutum* that made him believe that he had turned gold into silver—Geoffroy does not mention Boyle by name, and clearly fraud was not involved in that case.[77]

In 1992, Wolfgang Beck showed that much of Geoffroy's paper was cribbed from the 1617 *Examen fucorum pseudochymicorum*, a well-known work by Michael Maier. Maier had intended this catalogue of sleights-of-hand to help his fellow chrysopoeians distinguish true from false transmutations, and thus to *defend* true transmutation. Much of Maier's work was in turn borrowed from Heinrich Khunrath's 1597 *Trewhertzige Warnungs-Vermahnung* of 1597.[78] Indeed, Geoffroy's paper—like Khunrath's and Maier's publications—can be read as falling into a developed tradition within transmutational alchemy of identifying and "calling out" the false and the fraudulent, primarily as a means of setting apart and protecting the authentic.[79] Homberg himself did not hesitate—also at a public assembly—to debunk an erroneous claim for transmutation without ever dismissing the reality of authentic transmutations, and he did the same in his early textbook.[80] Geoffroy does not state that either transmutation or the philosophers' stone are impossibilities, merely that the characters who boast of them are likely to be fraudsters well stocked with a range of tricks, and that their actions must be observed with great care. As noted in chapter 5, Geoffroy had himself been involved in transmutational

77. Geoffroy, "Supercheries," 67–68. On this transformation (and brief mention of its modern reproduction), see Lawrence M. Principe, "The Gold Process: Directions in the Study of Robert Boyle's Alchemy," in *Alchemy Revisited*, ed. Z. R. W. M. van Martels (Leiden: E. J. Brill, 1990), 200–205.

78. Wolfgang Beck, *Michael Maiers Examen Fucorum Pseudo-chymicorum: eine Schrift wider die falschen Alchemisten*, Ph.D. diss., Technische Universität München, 1992; Michael Maier, *Examen fucorum pseudochymicorum detectorum et in gratiam veritatis amantium succincte refutatorum* (Frankfurt, 1617); Heinrich Khunrath, *Trewhertzige Warnungs-Vermahnung* (Magdeburg, 1597).

79. Tara Nummedal, *Alchemy and Authority in the Holy Roman Empire* (Chicago: University of Chicago Press, 2007), especially 62–71; Nummedal, "The Problem of Fraud in Early Modern Alchemy," in *Shell Games: Studies in Scams, Frauds, and Deceits (1300–1650)*, ed. Mark Crane, Richard Raiswell, and Margaret Reeves (Toronto: Centre for Reformation and Renaissance Studies, 2004), 37–58; Robert Halleux, "L'alchimiste et l'essayeur," in *Die Alchemie in der europäischen Kultur- und Wissenschaftsgeschichte*, ed. Christoph Meinel (Wiesbaden: Otto Harrassowitz, 1986), 277–91, at 291.

80. For example, Homberg provides an analysis of a process that "seduced one of the greatest chymists of Europe" into believing he had found "a true transmutation of silver into gold"; Wilhelm Homberg, "Observations sur une dissolution de l'argent," *MARS* (1706): 102–7, at 107. The procès-verbaux version of the memoir (PV 25, fol. 117v) identifies the chymist left unnnamed in public as Otto Tachenius, as does Bignon's report on the public assembly sent to Pontchartrain; BNF, Clairambault MS 566, fols. 229r–32v, at 230v–231r.

endeavors, had translated and collected chrysopoetic texts, and continued such interests even after 1722. Geoffroy's posthumously published treatise on medicinal substances includes a cautious treatment similar to that of his "Supercheries." There he warns readers against "the talkative who promise much with golden words" but really "try to sell smoke and steal gold," and concludes that "the prudent man will diligently guard himself against their frauds and sleights-of-hand." Yet when speaking of the universal medicine and the philosophers' stone itself, he does not deny the possibility of either, but merely says he does not know how to prepare them: "since this universal medicine remains unknown to us, we shall fall silent regarding it. In regard to the philosophers' stone, the material from which it ought to be prepared still remains uncertain, as does the method of its preparation."[81]

Given that Geoffroy's paper is largely a reprise of material already more than a century old, it invites the questions of why he presented it at all and why he did so at this particular time. It is, moreover, highly anomalous for an academician to present a paper not based on his scientific research—I am not aware of any other such paper published in the *Mémoires* of the period. The presentation of this paper specifically at a *public* assembly suggests that it was intended to be a public response to some contemporaneous public event or set of events. Geoffroy explicitly cites the public utility of his presentation when he remarks that he provides the paper "to prevent the public from letting themselves be abused by these pretended chymical philosophers."[82] But was such abuse especially a problem in the early 1720s? It is conceivable that the widespread financial ruin following the collapse of Law's programs might have set the stage for an increased number of "projectors" promising quick riches and financial recovery through transmutation. Clear evidence for such a surge in chrysopoetic cozenage is lacking at present, but it would be extremely interesting and informative to look more closely for it. Chrysopoetic books continued to be published and reviewed in learned journals in the early 1720s, but that seems an unlikely trigger for the Académie's intervention, since the number of such publications did not notably increase at the time, and anti-transmutational treatises were also being published alongside them, as had been the case for nearly all of transmutational alchemy's history.[83] The

81. Étienne-François Geoffroy, *Tractatus de materia medica*, 3 vols. (Paris, 1741), 2:318.

82. Geoffroy, "Supercheries," 62.

83. Allen G. Debus suggested concern over pro-transmutational books as background to Geoffroy's paper in "French Alchemy in the Early Enlightenment," in *Ésotérisme, gnoses, et imaginaire symbolique*, ed. Richard Caron, Joscelyn Godwin, Wouter J. Hanegraaff, and Jean-Louis Veillard-Baron (Leuven: Peeters, 2001), 47–59, at 48. The anti-transmutational treatises of the period include François Pousse, *L'examen des principes des alchymistes sur la pierre phi-*

contemporaneous rumor of the Regent's request to the Académie's chymists could offer a better background to the paper. Geoffroy's paper, presented publicly by the Académie's chief chymist of the day, would serve to counteract any rumors that the academicians were spending their time working on transmutation at the request of the state. Whether the account of Philippe's request is true or false, the rumor itself would have threatened to spark all the worries that Philippe's uncle had previously had about the gold in French currency, potentially igniting yet another round of bad consequences for France's economy. Hence, there would have been a very practical and urgent reason for stamping out the idea. Likewise, reports (again, whether true or false) that the Académie's chymists were at work on making gold for the state would cast the esteemed members of the Académie into the same lot in the public mind with various unsavory claimants to transmutation, like those who had been confined to the Bastille just a few years earlier—and who remained the subject of police investigations into the 1720s. Such connections would present a public relations nightmare for the Académie, and so would again require a public disavowal.

This issue of "being in bad company" returns us to the problem of chymistry's poor public image. The popular linkages to counterfeiting, poisoning, and cozenage described in chapter 6 persisted, as did, of course, the reality of the laborious and often dirty character of chymical work. The connections to illegal and immoral behavior were reinforced during the first decades of the eighteenth century by the actions of the state and the Paris police in seeking out and arresting anyone suspected of having chymical expertise or interests. The chrysopoeian Francesco Maria Pompeo Colonna, for example, well aware of these police actions (as most of the Paris public must have been), makes bitter references to them in his 1722 *Les secrets les plus cachés*.[84] Public knowledge of Philippe's chymical expertise, together with persistent rumors that he had poisoned the Dauphin, Dauphine, and Petit Dauphin, further reinforced such connections. Seen in the context of chymistry's reputation, Colbert's decision to include chymists in the foundation of the Académie Royale des Sciences emerges as a bolder and more innovative move than we today might otherwise recognize. The move provided

---

losophale (Paris, 1711)—for contemporaneous reviews of which, see Alain Mothu, "L'alchimie en examen: *L'examen des principes des alchymistes sur la pierre philosophale* (1711) d'après les journaux de l'époque," *Chrysopoeia* 5 (1992–96): 739–50.

84. Francesco Maria Pompeo Colonna, *Les secrets les plus caches* (Paris, 1722), 4–5. Colonna's reference to the lieutenant general of the police, Marc-René de Voyer de Paulmy d'Argenson, as having "rare mérite" is undoubtedly ironic, since he had been disgraced in the Law affair and his 1721 funeral had been disrupted by an angry mob.

the discipline with a visible, respectable, and state-sponsored presence. His prohibition of the controversial topic of transmutation was no doubt in part a means of safeguarding that newly engineered respectability. The Académie's 1699 Règlement further elevated chymistry's status and profile when it guaranteed *les chimistes* the same number of positions and ranks in the Académie as the practitioners of well-established and better respected disciplines, such as astronomy, geometry, anatomy, and botany. The significance of this institutionalization of the subject and its linkage to the crown should not be underestimated. Yet these successive institutionalizations on their own could not make the ambiguous public reputation of chymical practitioners go away.

Several authors of the period pointed specifically toward transmutation (and its fraudulent hawkers) as the immediate source for chymistry's continuing bad reputation. For example, Jean Mongin's 1704 *Le chimiste physicien* lays the blame for chymistry's low status squarely at the feet of failed chrysopoeians: "the extreme poverty to which some of the former chymists reduced themselves by laboring out of greed on operations whose success is no less imaginary than uncertain, is the reason that so many people care little for chymistry."[85] Antoine Furetière (1618–88) is even more direct in the context of his definition of *le chimiste*: "People have great scorn for chymists because they do not judge them save by the measure of some ignorant rogues and tricksters [*ignorants gueux et affronteurs*] who say that they have the philosophers' stone. It just the same as if one should judge astronomers by almanacmakers, or poets and musicians by the singers on the Pont-Neuf. Rather, one owes to chymistry the discovery of the most necessary things for life, such as the preparation of metals and of most medicines."[86]

The questionable repute of chymistry as a result of fraudulent transmuters persisted in some quarters even into mid-century when Gabriel François Venel (1723–75) wrote his entry for *chymie* in the *Encyclopédie*: "they do not distinguish the chemist from the puffer [*souffleur*], both of these names are equally ill-sounding to their ears." At least by this time such ill repute was limited, according to Venel, to "the least educated persons," but it perpetuated views that were more widely held earlier in the century.[87]

As chymistry was increasingly institutionalized at the Académie, its ambiguous public reputation had to be addressed. Geoffroy's 1722 paper can be

---

85. Jean Mongin, *Le chimiste physicien* (Paris, 1704), 2.

86. Antoine Furetière, *Dictionnaire universel*, 3 vols. (The Hague, 1690), s.v.

87. For more on Venel and this subject, see Christine Lehman and François Pépin, eds., *La chimie et l'*Encyclopédie, *Corpus, revue de philosophie* 56 (2009). Note that by this time *souffleur* had also become a term of opprobrium; compare chapter 6, note 70.

seen as part of that endeavor. But the crafting and presentation of a public face for the Académie was primarily the responsibility of its perpetual secretary, Bernard de Fontenelle. Fontenelle's views on chymistry are complex and have already received substantial scholarly attention, so it is unnecessary to reexamine them here in extenso.[88] It suffices simply to cite some of his specific actions relevant to the topics of this chapter, and contrast his views with those of Homberg and other practicing chymists at the Académie. In the first place, it is clear that Fontenelle had an extremely dim view of transmutation. Accordingly, he paid special attention to Geoffroy's 1722 paper, taking full advantage of it as a platform from which to launch some of his most sarcastic condemnations of transmutation.[89] His summary in the *Histoire* asserts that the "alchimistes" have never made a single grain even of an imperfect metal, much less of the perfect metals silver and gold—in fact, they could not make even an ordinary pebble. Nature, he writes, has reserved such productions to herself. He then claims, rather oddly, that the "extreme difficulty" of doing something should be considered proof of its practical impossibility. Transmutation has proven extremely difficult to carry out, *ergo* it is an impossibility. Two years later, remarking on the inherent difficulty of chymistry itself, he rules transmutation specifically and unambiguously out of the realm of chymistry, writing definitively that "one should not despair of anything in chymistry, except the philosophers' stone."[90] When summarizing Geoffroy's paper, Fontenelle also ridicules claims to have decomposed the metals into their components, saying that no one has ever succeeded in separating metals into their principles. Apparently realizing that he was now contradicting the work of prominent academicians—or, perhaps more seriously, contradicting what he himself had written about those contributions in previous *Histoires*—he then concedes that Homberg *did* decompose metals into their principles using the burning lens. But Fontenelle quickly adds that since the

88. For treatments of Fontenelle and chymistry, see especially Luc Peterschmitt, "Fontenelle et la chimie: la recherche d'une 'loi fondamentale' pour la chimie," *Methodos* 12 (2012): https://journals.openedition.org/methodos/2873; Peterschmitt, "Fontenelle, the Idea of Science and the Spirit of Chemistry," in *Chymia: Science and Nature in Early Modern Europe*, ed. Miguel López-Pérez, Didier Kahn, and Mar Rey Bueno (Cambridge: Cambridge Scholars Publishing, 2010), 367–85; François Pépin, "Fontenelle, l'Académie et le devenir scientifique de la chimie," *Methodos* 12 (2012): https://journals.openedition.org/methodos/2898. For more on Fontenelle and his views of the sciences, see the classic work by Leonard M. Marsak, *Bernard de Fontenelle: The Idea of Science in the French Enlightenment*, *Transactions of the American Philosophical Society* 49 (1959); as well as Simone Mazauric, *Fontenelle et l'invention de l'histoire de sciences à l'aube des Lumières* (Paris: Fayard, 2007).

89. *HARS* (1722): 37–39.

90. *HARS* (1724): 39.

mercury and sulphur were volatilized and lost leaving behind only a vitrified and useless residue, this work would not be of any help to *un alchimiste*,
and so, apparently, it is irrelevant. He does not, however, acknowledge that
his claim that no perfect metals have ever been produced chymically contradicts Homberg's 1706 "Essai," which includes the production of gold from
mercury—an article that, interestingly enough, Fontenelle chose not to summarize *at all* in his *Histoire*.[91] The *Mercure galant* also gave Geoffroy's paper a
high profile. While that popular periodical routinely mentioned the Académie's public meetings in one or two paragraphs (as it did in April 1722), in this
particular case, anomalously, the following issue of May 1722 carried a detailed, five-page reprise of Geoffroy's paper—a unique treatment for a paper
given at a public assembly in this period.[92] Fontenelle, who had long-standing
personal ties to the *Mercure*, may have orchestrated this enhanced coverage,
and clearly he felt that it was necessary to spread Geoffroy's comments to the
broadest possible audience.

   For Fontenelle, metallic transmutation and the stone belonged to a discredited chymical past, and their persistence only perpetuated the (at best)
ambiguous reputation of chymistry. Fontenelle's rhetorically crafted version
of chymistry's history endeavored to make its contemporaneous developments—particularly the chymistry of the Académie—appear as a fresh start
with little connection to the dark, secretive, unsavory notions of the past. In
1701, he wrote that "chymistry has at last emerged from the mysterious shadows in which false philosophers intentionally enveloped it." The following
year he declared that "most of the former chymists were at least a bit extravagant in their imaginings [*visionnaires*] . . . the time has come that chymists who
are more sensible and of better faith have dissipated its artificial shadows."[93]
The same rhetoric is at work in his "revisions" to Homberg's biographical details, as described in chapter 1, aimed at reshaping him into an idealized figure
conforming to what the perpetual secretary thought appropriate and respectable for an Académie chymist. This revision meant denying that Homberg
ever had anything to do with transmutation or the philosophers' stone, to
the point of making him flee to Italy to put distance between himself and
a Parisian chrysopoeian (when in reality Homberg was then attempting to

91. *HARS* (1706): 41.
92. *Mercure galant* (April 1722): 97, and (May 1722): 121–25. The latter notice adds that the
Cardinal de Polignac, who presided over the public assembly, remarked that if the time and talent spent on astrological prognostication and seeking for the stone had been directed toward
improving astronomy and chymistry, then we would have better, or rather more of the good,
chymists and astronomers.
93. *HARS* (1701): 66, and (1702): 45.

transmute mercury into silver). Fontenelle treated Nicolas Lémery similarly in the same year, although to a lesser extent, since Lémery had himself publicly rejected transmutation. Even much later, in 1739, another of Fontenelle's éloges declared emphatically that the chymist Dufay "never aspired to the Great Work," when in fact Dufay was deeply involved in transmutation, and discussed with other academicians his witnessing of a conversion of mercury into gold using the philosophers' stone.[94] Also in the 1730s, Fontenelle did not miss the opportunity to launch a final vitriolic critique when summarizing a paper by Boerhaave.[95] In his *Histoire*, the perpetual secretary asserted that the "former hope [of transmutation] . . . abused by so many imposters to engage gullible and greedy persons in infinite labors and ruinous expenses . . . has up to the present day deceived everyone who has given himself up to it." He concluded that "the true chemists shall leave the alchemists with nothing but the refuge of invincible stubbornness, a refuge always open to those who can profit from it, where in fact an infinity of people boldly take their stand."[96]

It is worthy of note that in addition to the language of darkness and obscurity, Fontenelle repeatedly accuses the *anciens chimistes* and seekers after the philosophers' stone of specifically *moral and social* failings, rather than merely of scientific errors. Earlier chymists jealously chose to be obscure and unhelpful. Christophle Glaser was a "true chymist, full of obscure notions, greedy with such ideas, and unsociable." Homberg, in contrast, was "too honest" to put the "vain idea" of the philosophers' stone into anybody's head.[97] The alchemists arrogantly hide themselves in a "refuge of invincible stubbornness." Fontenelle was not alone in this moral disapprobation of chrysopoeians. Bignon used a similar moralizing language when praising Homberg for revealing that he had made a certain discovery by accident rather than by design: "conduct so modest is indeed worth all the charlatanry customary for

94. Bernard de Fontenelle, "Éloge de M. Dufay," *HARS* (1739): 73–83, at 74; and Lawrence M. Principe, "The End of Alchemy? The Repudiation and Persistence of Chrysopoeia at the Académie Royale des Sciences in the Eighteenth Century," *Osiris* 29 (2014): 96–116, at 109–11.

95. On this Boerhaave paper, its content, and background, see John Powers, *Inventing Chemistry: Herman Boerhaave and the Reform of the Chemical Arts* (Chicago: University of Chicago Press, 2012), especially 170–91; and Powers, "Scrutinizing the Alchemists: Herman Boerhaave and the Testing of Chymistry," in *Chymists and Chymistry: Studies in the History of Alchemy and Early Modern Chemistry*, ed. Lawrence M. Principe (Canton, MA.: Science History Publications/Chemical Heritage Foundation, 2007), 227–38.

96. *HARS* (1734): 55–57, quotations on 55 and 57.

97. Fontenelle, *HARS* (1701): 66; "Éloge de Lémery," 73; and "Éloge de Homberg," 88.

the art [of chymistry]."[98] Note how even Bignon declares charlatanry to be "customary" for chymistry! Other early eighteenth-century critics of transmutation deploy similar moralizing language rather than logical or scientific argument, like Mongin and Furetière, mentioned above. Geoffroy's 1722 paper, with its catalogue of cheating practices, continued this depiction of the pursuit of transmutation as a morally suspect and socially subversive activity. From the perspective of government ministers, even *successful* chrysopoeia would be subversive politically and economically—unless, of course, they were able to hide it in the Bastille and control it for themselves. I would argue that it is this repeated characterization of chrysopoeia and its practitioners as morally and socially unacceptable that best explains the disappearance of transmutational endeavors from the normal activities of "respectable" chymists in the 1720s.

The Fate of Chrysopoeia at the Académie

Does Geoffroy's "Supercheries" actually mark a transformative rupture within chymistry between its transmutational and nontransmutational aspects? It would, of course, be overly reductionist to point at Geoffroy's communication alone without recognizing Fontenelle's anti-transmutational rhetoric and revisionist history, or the growing frustration with chrysopoeia's abuses expressed by a wider circle of writers and thinkers, and even the broader public, in early eighteenth-century France. That said, can historians legitimately point to the 1720s as the time when transmutation, after a millennium and half of existence at the core of chymical activities, disappeared from the agenda of serious and learned chymists (at least in France)? Examination of the records of the Académie seems to give an affirmative reply. No further positive mention of metallic transmutation or of the possibility of separating metals into their principles appears in the *Mémoires* after Geoffroy's paper. So, to all appearances, *chymistry* appears to have been purged of its longstanding transmutational aspirations, and thereby in this period approached more "modern" notions of *chemistry*.

Changes in vocabulary reinforce this conclusion. The words *chimie* and *alchimie* definitively parted company in French at this same time, taking on their separate and more modern meanings—the former designating something useful and scientific, and the latter something archaic, applied pejo-

---

98. BNF, Clairambault MS 566, fol. 231r; Bignon to Pontchartrain: "une conduitte si modeste vaut bien toutte la Charlatenerie ordinaire au mestier."

ratively and specifically to seekers after metallic transmutation. Fontenelle's evolving usage illustrates this linguistic realignment. His earliest sustained deployment of the term *alchimiste* in the *Histoire* appears in his summary of Geoffroy's 1722 "Supercheries," where he uses it four times in short order—even though Geoffroy himself *never* uses the word. By 1734, Fontenelle explicitly sets "les vrais chimistes" in unbridgeable opposition to "les alchimistes."[99] By the end of the 1730s this stark division had become ubiquitous in French. Noël Antoine Pluche's very popular *Histoire du ciel* warns readers not to "confuse the alchemists, or seekers after the philosophers' stone, with chemists." The chemists and their discoveries, he continues, are highly beneficial to society, while the alchemists are immoral frauds and misguided failures. In the lengthy examination of the "principles of the alchemists" that follows, Pluche even cites Geoffroy's claim to have synthesized iron and praises him for seeing the error of his ways. (But did he?) Homberg does not get off so easily. Pluche ridicules Homberg's discovery of the pyrophorus while attempting to turn mercury into silver, and, basing himself on Hartsoeker's critique, completely rejects Homberg's vitrification of gold. Pluche even goes so far as to call Homberg's idea that chemical substances can be transformed one into another a "dangerous opinion."[100] Clearly, the meanings and connotations of the two words had changed significantly and definitively in the years since Homberg defined *alchimie* explicitly as a "particular part of *chimie*" in the early 1690s. Pro-chrysopoeia authors were also sensitive to this linguistic shift. By the 1720s, they tend to refer to themselves as *chimistes philosophes*, both to distinguish themselves from the nontransmutational and less exalted *chimistes*, and to avoid the increasingly pejorative connotation of the word *alchimiste*. Even Geoffroy uses the term *chimiste philosophe* ironically in 1722 to refer to fraudulent transmuters. These "chimistes philosophes" themselves did not hesitate to condemn charlatans, and often in terms strikingly reminiscent of Geoffroy's "Supercheries." One of them even coins the novel term *chymiastres*, for use where Fontenelle would have written *alchimistes*, for "those who have no knowledge save how to cheat."[101]

99. *HARS* (1734): 55–57, at 57. This text is Fontenelle's presentation of Herman Boerhaave, "Sur le mercure," *MARS* 36 (1734): 539–52, reporting that long-term heating of common mercury does not yield gold. Fontenelle's summary contains his final and most acrid anti-alchemical remarks. On Boerhaave's paper, see Powers, "Scrutinizing the Alchemists."

100. Noël Antoine Pluche, *Histoire du ciel*, 2 vols. (Paris, 1739), 2:85–118.

101. Geoffroy, "Supercheries," 62 and 66: "pretendus Philosophes Chimistes." For uses of *philosophe chimiste* by chrysopoeians, see, for example, "Essai pour développer," Wellcome MS 2298, passim; Colonna, *Les secrets*, 4, 25, 95; *Abrégé de la doctrine de Paracelse* (Paris, 1724), i, x, xiv, xxiv, xxxv, xl, xlv, xlix, l. The term is a slight modification of the traditional use of "phi-

The charged rhetoric of moral opprobrium and inherent dishonesty meant that it became injudicious for respectable chemists—such as the academicians—to be seen publicly as studying, pursuing, or defending transmutation. The earlier, explicit, and largely ineffective prohibitions against chrysopoeia from administrators like Colbert and Louvois now reemerged in an implicit and more effective form policed by moral opprobrium and social convention. The claims and concepts that Homberg published in the first decade of the century regarding metals, their composition, and their transformation one into another no longer appeared in the pages of the *Mémoires* or of the procès-verbaux. The indictment of transmutation as a means of "cleaning up" chymistry's image thus left chrysopoeia, for the first time in its long history, without serious advocates willing to speak up publicly to counteract the rhetoric, opprobrium, skepticism, and denunciations aimed against it. Negative assessments of chrysopoeia, it is worth underlining, were not something new at this time; they dated back to the Middle Ages. But the loss of its serious, distinguished, and visible advocates is a new development. Thus, the 1720s *do* mark a critical time of transformation for the history of chemistry in regard to metallic transmutation.

Yet appearances can be deceiving. A closer inspection reveals this transformation to be only skin-deep—it was in this sense a "false transmutation" of the discipline. A deeper look reveals that, rather than the emergence of a radical division between Fontenelle's supposed *vrais chimistes* who firmly rejected transmutation and the immoral *alchimistes* who ignorantly and fraudulently continued to pursue it, the split that really developed in the 1720s was instead between *public image* and *private practice*. While the public face of the discipline was markedly and irreversibly changed, the private work the academicians were actually doing was a different matter. Just as the explicit administrative prohibitions against chrysopoeia were perpetuated in a newly implicit way, so also the previous disconnect between administrators and chymists in regard to transmutation changed in character. Transmutation was not expunged from the Académie, it merely went underground *within* the institution. Duclos and Homberg had few qualms about their chrysopoetic endeavors being known publicly, even while the Académie's administrators fumed. In the 1720s, while such activities as theirs did retreat entirely from public view and public expression, they persisted in private laboratories and private conversations. Of course, transmutational alchemy had always been secretive to various degrees, but now at the Académie it became completely

---

losophers" in chrysopoetic texts to refer to the adepts. "Les miserables Chymiastres" appears in Colonna, *Secrets*, xiii; see also 12–15.

hidden from public view. This situation allowed for the subsequent historical conclusion that transmutational alchemy died out in serious and institutionalized circles like the Académie in the 1720s. That conclusion, naturally
enough, remained largely unchallenged in the context of triumphalist and
positivistic narratives of the history of science. Contrary evidence was scarce,
due to the lack of private notes and papers documenting what the Académie's chemists were actually *doing* rather than only what they were publishing
and presenting officially. But thanks to the rediscovery of Jean Hellot's voluminous transcripts of his colleagues' private papers and of the oral reports
he received from them, a glimpse is now available of this otherwise hidden
world of the academicians and their continuing pursuit of ancient chymical
arcana.[102] Hellot's notes have already proven useful in revealing otherwise
lost information about Duclos, Homberg, Philippe, and Geoffroy, but they
say even more surprising things about the work of the *next* generation of academicians, with whom Hellot was in direct personal contact. Hellot's notebooks, supplemented with other contemporaneous sources, lead inevitably
to a truly surprising result: *from the 1720s until the 1760s, more academicians
seriously pursued or defended transmutational endeavors and the philosophers'
stone than had done so before.*

The newly discovered Hellot manuscripts at the Bibliothèque de l'Arsenal
are highly revealing in this regard. They constitute a coherent three-volume
compilation, written in the 1730s, that deals almost exclusively with metallic
transmutation and the separation of metallic mercuries and sulphurs, generally en route to some transmutational goal.[103] Hellot organized these three
notebooks into sections, each dealing with a particular starting material for
preparing either transmutational *particularia* or the philosophers' stone itself. He devotes the first notebook exclusively to the use of gold and silver,
and to various approaches to the stone. In the second he records processes
beginning with salts, sulphurs, and *nostoch.* The third notebook contains
routes using base metals, semimetals (bismuth, zinc, antimony, and so on),
minerals, and organic materials.[104] Significantly, he cites about half the en-

102. Much of the documentation that follows here appeared first in Principe, "End of Alchemy." For the continuation of interest in transmutation at the Académie into the nineteenth
century, see Lawrence M. Principe, *Alchemy and Chemistry: Breaking Up and Making Up (Again
and Again)* (Washington, DC: Dibner Library Publications, Smithsonian Institution, 2017).

103. In regard to dating these three manuscripts, Étienne-François Geoffroy is referred to as
"le feu Mr Geoffroy" (MS 3008, 53, 185−87), indicating a date after 1731. MS 3006 contains the
dates 1732 and 1734, at 20 and 24, respectively.

104. Hellot labeled these sections throughout by writing the symbol for the featured material
in the upper right corner of the first page and then of every other folio of each section.

tries as originating from sources *within the Académie*. While he copied some material from the surviving papers of earlier academicians, he also references a yet larger quantity as coming from his Académie contemporaries.

A large amount of material originates from Antoine-Tristan Danty d'Isnard (1663–1743). Danty d'Isnard was admitted to the Académie in 1716 as a botanist, and then promoted to associé chimiste in 1721, and associé botaniste in 1722. Hellot's notebooks record Danty d'Isnard's processes for such things as the extraction of the sulphur of gold and the mercury of anti-mony, for the transmutation of tin into gold and mercury into silver, for the alkahest, and for the preparation of several different transmuting powders from gold. Some of this material came from manuscripts that Danty d'Isnard owned and that he shared with Hellot—such as a method for the alkahest that Hellot cites as "drawn from the manuscript that Mr. Danty d'Isnard lent me."[105] Many other entries, however, are in the first person and give circum-stantial details of working through the process—such as a case where the digesting vessels broke three times in a row such that "we did not obtain any profit"—indicating that Danty d'Isnard was actively experimenting with these transmutational processes himself.[106] There are also many accounts and processes that Danty d'Isnard collected from experimenters outside the Aca-démie, indicating his participation—like Duclos, Homberg, and Philippe before him—in a network of exchanges with other hopeful chrysopoeians. One process, for example, describing how to "exalt" gold into a transmuting agent able to convert silver and mercury into gold, he obtained from "Mr. de la Motte"; Hellot highlighted this process by drawing a hand pointing to it in the margin.[107] Other processes include one for potable gold from "Mr. Clecy," another for a powder to turn silver into gold from "Mr. de Rasé," a trans-mutation of mercury into silver from "Mr. Lambert," a "magnet of the phi-losophers for the celestial mercury" from "Mr. Jacques," and two for making mercury of antimony, the first from a "Sieur Denis" and the second from "Benato."[108] Thus it is clear that Danty d'Isnard maintained an active interest in the topic of transmutation—collecting processes and experimenting with them.

Hellot's notebooks likewise preserve dozens of chrysopoetic processes from Charles François de Cisternay Dufay (1698–1739), pensionnaire chimiste

---

105. Hellot Arsenal MS 3006, 95–96, 101–7; MS 3007, 33–37 ("tiré du manuscrit que ma presté Mr Danty d'Isnard"), and 51; MS 3008, 154, 157, 188–89.

106. Hellot Arsenal MS 3007, 111–12: "n'y trouvâmes aucun profit."

107. Hellot Arsenal MS 3006, 96–98.

108. Hellot Arsenal MS 3006, 99, 110, 111–15; MS 3007, 74; and MS 3008, 189, 190–91.

of the Académie during the 1730s, intendant of the Jardin du Roi, and a close associate of Hellot.[109] Materials from Dufay record, for example, the transmutation of silver into gold and the preparation of a red oil of vitriol able to convert silver into gold.[110] Dufay's own labor on such processes is clear in an account where Hellot records that "this operation is very certain, for Mr. Dufay used one marc of silver to do it, from which he himself separated the gold by quartation." The same entry also contains Dufay's explanation of the operation in theoretical terms, which he then deployed to improve the procedure so that from 1 ounce of fine silver "he found after quartation 4 grains of very beautiful gold."[111] A process for preparing a salt capable of extracting the "soul of silver" contains first-person narrative, suggesting that Dufay carried out this operation himself as well.[112] Another first-person narrative from Dufay tells of a partly miscarried experiment that nevertheless seemed to have produced some gold from silver calx, and is given the rubric "proof of transmutation."[113] Hellot also records that while Dufay was in Strasbourg in 1732, a skilled chymist "who appeared to know the [Great] Work" gave him the first steps of a process for making the philosophers' stone, which involved sealing up amalgams in flasks and allowing them to digest in a box without external heating. Dufay and Hellot discussed the progress of this experiment, underway in May 1739, while Dufay was awaiting further instructions from the Alsatian chymist. Unfortunately, such instructions did not arrive by the time of Dufay's premature death from smallpox two months later, in July, but Hellot subsequently found the four flasks still containing the slowly reacting materials when he sorted Dufay's effects. Dufay had even labeled the box containing the flasks "Peru," presumably in hopeful reference to the riches of precious metals they would produce, just as South America had provided the

109. Dufay's sharing of secrets with Hellot is mentioned in Jean-Paul Grandjean de Fouchy, "Éloge de M. Hellot," *HARS* 68 (1766): 167–79, at 174, in regard to a process for making gold bas-reliefs, not to be communicated until after Dufay's death. Hellot was also Dufay's executor, further testifying to their close relationship; see Fontenelle, "Éloge de Dufay," 82.

110. Hellot Arsenal MS 3006, 12–14, 128–31; MS 3007, 80.

111. Hellot Arsenal MS 3006, 128–30, at 129: "cette operation est tres certain, car Mr Dufay s'est servi pour la faire d'un marc d'argent dont il a separé luy meme l'or par le depart"; and 130: "par le Depart il a trouvé 4 graines de tres bel or." Since previously he had obtained 4 grains from 1 marc of silver, that is, 8 ounces, the improved process gave eight times as much transmuted gold.

112. Hellot Arsenal MS 3006, 132–34, at 133–34: "Jai pris dit Mr Dufay, la matiere blanche resteé dans la matras."

113. Hellot Arsenal MS 3008, 89–91.

Spanish Empire.[114] Here and elsewhere, Hellot's expressions make it clear that he was not just copying out information from written remains; rather, he and Dufay actively discussed these procedures and their results.

Both the Arsenal and the Caen notebooks also record Dufay's close association with Francesco Maria Pompeo Colonna (1646–1726), one of the most famous chrysopoeians in early eighteenth-century France. Colonna authored several books about chymical medicine and metallic transmutation in the decade prior to his sudden death in a fire in the early morning hours of 6 March 1726 (ironically enough, an Ash Wednesday).[115] Colonna was on familiar terms with several other persons connected to the Académie. He had collaborated with Mathieu-François Geoffroy (1644–1708; Étienne-François's father) and was a personal friend of the astronomer Gian Domenico Cassini (1625–1712) and of the brother of Jean-Paul Bignon.[116] Hellot's voluminous notebooks now in Caen, dating from the 1740s and 1750s, continue the testimony of those at the Arsenal. There Hellot records Dufay's report that he witnessed two transmutations performed by Colonna. One case involved the conversion of some silver into gold, described simply as "seen by Mr. Dufay."[117] The other is more significant: Dufay's witness of the transmutation of lead into gold using the philosophers' stone itself. Hellot underscored this account's importance with a marginal note reading "proof of the Great Work." "It is claimed that by means of this philosophical mercury [described in the previous entry] Mr. Colonna succeeded in the Great Work. Mr. Dufay has assured me that he was present at the Hôtel de Richelieu for the conversion of 1 ounce

114. Hellot Caen MSS, 1:85v: "Feu Mr. dufai etant a Strasbourg en 1732 y fit connoissance avec un tres galand homme amis de Mr D'angevilliers, ce galand homme qui s'amusoit depuis 30 ans a la chimie parroissoit scavoir l'oeuvre. . . . Mr. Dufay a fait ces amalgames dans de petits bouteilles. . . . A sa mort en juillet 1739 j'ai trouvé une boete ronde qu'il avoit etiqueté *le Perou* et dans cette boete 4 petits bouteilles bien fermées avec liege et cire d'espagne. . . . Il m'avoit dit au mois de Mai precedent qu'il avoit vu un des ces amalgames passer du noir au gris blanchatre de cendres le tout sans feu: qu'il attendoit de Strasbourg la reponse du Galand homme qui lui avoit promis la suite du procedé mais au 16 juillet 1749 [*sic,* for 1739] jour de sa mort il n'en avoit pas encore recu de reponse." My thanks to Michael Bycroft who alerted me to this entry.

115. *Mercure de France* (March 1726): 603–7.

116. Gustavo Costa, "Un Collaboratore italiano del Conte di Boulainviller: Francesco Maria Pompeo Colonna (1644–1726)," *Atti e memorie dell'Accademia Toscano di Scienze e Lettere* 29 (1964): 207–95, at 218–19.

117. Hellot Caen MSS, 9:105v: "Transmutation de Mr. Colonne, vuë par Mr duFay de l'arg[en]t en Or." This citation appears in Hellot's index to his notes with a reference to "Chym. p. 281," one of his notebooks that has not been located (see note on sources) and which would have contained a fuller account of the event.

of molten lead, from which 4 gros [half an ounce] and a few grains of gold were obtained by throwing into it a little ball of wax into which had been introduced, in his presence, in Mr. Colonna's chamber, a very small quantity of a powder of a very beautiful red."[118]

Another process gathered from Dufay, for a transmutation of silver into gold using "fixed sulphur," suddenly switches in mid-sentence from French to Italian, suggesting that it might also have come ultimately from Colonna. It is preceded by an account of "silver tinged into gold with orpiment and fusible salt" that Hellot attributes explicitly to "Colonne," and an Italian recipe exists elsewhere in the manuscript for a "solfo fisso" drawn from Dufay's papers.[119]

Fontenelle's 1739 éloge for Dufay recalls that his grandfather "got chemistry into his head, with the design, truth be told, to achieve the Great Work." The perpetual secretary did not miss the chance at this point to emphasize that Dufay's grandfather "worked a great deal and spent a great deal, with the usual degree of success." He then notes with a tone of concern that "perhaps the blood of that grandfather of whom we have just spoken acted in" the grandson, but Fontenelle immediately reassures his (undoubtedly very worried) readers that "it was corrected in that grandson, who never aspired to the Great Work."[120] Even in 1739—while Hellot was examining Dufay's sealed flasks of materials for making the philosophers' stone—Fontenelle continued to believe it important to emphasize, for public consumption, that Dufay had no interest whatsoever in chrysopoeia. It is hard to believe that Fontenelle was not aware of the truth, and many who sat around the table listening to him read Dufay's éloge knew better and may have smirked in response. Hellot's notebooks unambiguously bear witness to Dufay's real interests, as a careful reader might well have suspected when Fontenelle yet again "doth protest too much."

Hellot's notes do not explicitly indicate his own opinion of transmutational chymistry. But although there is currently no direct evidence that Hellot himself made practical attempts at gold-making, his notes on chrysopoeia are so extensive and so well organized that I think it unlikely that his interest

118. Hellot Caen MSS, 5:189r: "On a pretendu que Mr. Colonna avoit réussi par ce mercure philosophique à la grande operation. Mr du Fay m'a assuré qu'il avoit esté présent à l'hotel de richelieu à une conversion d'une once de plomb fondu, dont on tira 4 gros quelques grains d'or en y jettant une petite boule de cire, où on avoit introduit, devant lui, dans la chambre de Mr Colonna une tres petite quantité d'une poudre d'un tres beau rouge."

119. Hellot Arsenal MS 3006, 130–31; MS 3007, 91. There is another transmutative process, using liver of antimony and mercury precipitate, attributed to Colonna at MS 3008, 18–19.

120. Fontenelle, "Éloge de Dufay," 73–74.

was no more than historical. He records reports and conversations about transmutation without negative remark, even though he offers judgments about other debated subjects.[121] He was well read in the earlier transmutational literature, chides those who would ignore the writings and work of "les alchimistes," and declares that such writings directed him toward specific lines of experimentation. His papers on the properties of zinc, for example, grew out of an interest in testing claims about that metal he found in Respour's 1668 *Rares experiences sur l'esprit minerale*, a book promising to teach the preparation of the "necessary agents" for making the philosophers' stone and alkahest from zinc.[122] The experiments described in the first of these papers include Hellot's repeat of the process Homberg reported in 1710 about isolating the flammable sulphur principle from zinc in the form of a combustible oil. Using vinegar as the extracting solvent, the process worked just as Homberg described, although using oil of vitriol instead (as Homberg reports he did) did not produce the oil. Hellot stated that he needed to repeat the process "before concluding anything against a fact advanced by so celebrated a chymist."[123]

Hellot's grandfather, another Jean Hellot, also had an interest in chymistry and translated William Davisson's 1640 *Philosophia pyrotechnica* into French.[124] In accord with this information, the younger Hellot's notebooks occasionally cite entries taken from "an old manuscript in the hand of my grandfather." In one case, for example, he copies out a process for making the mercuries of antimony and of lead.[125] Indeed, the official éloge for the younger Hellot mentions that he was diverted from his initial interest in pursuing theology when he discovered his grandfather's chymical papers.[126] It would be significant (and somewhat amusing) if one of the most productive and prominent chemists of mid-eighteenth-century France actually took up the study of chemistry by becoming fascinated with transmutational processes.

---

121. For example, he criticizes Duclos as being "bien crédule" after transcribing some material from his notes about the divining rod; Hellot Caen MSS, 3:275r.

122. Jean Hellot, "Analise du zinc. Premier mémoire," *MARS* (1735): 12–31, at 12; "Analise chimique du zinc. Second mémoire," *MARS* (1735): 221–43, at 221; and Respour, *Rares experiences sur l'esprit minerale pour la preparation et transmutation des corps metaliques* (Paris, 1668). Hellot also cites *Alchymia denudata* (a work that had been translated by Geoffroy) at 14, and lists the various names given by "les Alchymistes" to the flowers of zinc (sublimed zinc oxide) at 16.

123. Hellot, "Premier mémoire," 20–21. On Homberg's experiment, see chapter 5, note 108.

124. *Les elemens de la philosophie de l'art du feu ou chemie . . . traduit du Latin du Sieur Davissone . . . par Jean Hellot* (Paris, 1651). A second edition appeared in Paris in 1657, and a third under the title *Le cours de chymie* at Amiens in 1675.

125. Hellot Arsenal MS 3008, 186–87; "ancien MSC de la main de mon ayeul."

126. Jean-Paul Grandjean de Fouchy, "Éloge de M. Hellot," *HARS* 68 (1766): 167–79, at 167.

Might one even speculate that Hellot—forced to work as editor of the *Gazette de France* throughout the 1720s in order to maintain a livelihood due to the loss of his fortune during the collapse of John Law's banking schemes—saw in transmutational alchemy a way to restore his former financial security?[127] It would also be extremely interesting to learn if his interest in transmutation had anything to do with his position as *contrôleur des essais* at the royal mint. There is no clear evidence to answer these questions. Yet however they might be answered, Hellot provides yet another example of a highly active and important academician who maintained a serious interest in chrysopoeia—albeit privately—and now into the 1760s.

Yet even in the 1760s Hellot was not alone in these interests. Guillaume-François Rouelle (1703–70), both by his own admission and by the testimony of others, worked diligently on transmutational endeavors. Writing in 1771, Denis Diderot (1713–84) recalled that Rouelle "believed in alchemy; he used the last two lectures of a course that lasted eight months to demonstrate its reality by facts and principles." Although he urged his students not to pursue chrysopoeia, "nevertheless," continued Diderot, "he often confided to me that it would be the object of his work in his later years."[128] Further testimony comes from the long-lived Académie chemist Michel-Eugène Chevreul (1786–1889). He recalled that the brother of the chemist Joseph-Louis Proust (1754–1826) "many times affirmed" to him that Rouelle worked on transmutation secretly in a special laboratory he kept for this purpose in Paris. In this laboratory on the rue Copeau (now rue Lacépède), Rouelle "delivered himself over, in the greatest secrecy, to his alchemical labors."[129] Located not far from the Jardin du Roi, this special workspace was clearly a different locale than Rouelle's well-known private laboratory on the rue Jacob—over a mile away in St. Germain—where he customarily worked and where he taught

127. Grandjean de Fouchy, "Éloge de Hellot," 168.

128. Denis Diderot, *Oeuvres complètes*, ed. Jane Marsh Dieckmann et al., 25 vols. (Paris: Hermann, 1975), 20:630; on Diderot and Rouelle's course, see Jean Jacques, "Le Cours de chimie de G.-F. Rouelle recuelli par Diderot," *Révue d'histoire des sciences* 38 (1985): 43–53; Marco Beretta, "Rinman, Diderot, and Lavoisier: New Evidence Regarding Guillaume-François Rouelle's Private Laboratory and Chemistry Course," *Nuncius* 26 (2011): 355–79, especially 370–79; and Christine Lehman, "Innovation in Chemistry Courses in France in the Mid-Eighteenth Century," *Ambix* 57 (2010): 3–26. On Rouelle more generally see Rhoda Rappaport, "G.-F. Rouelle: An Eighteenth Century Teacher and Chemist," *Chymia* 6 (1960): 68–101.

129. Chevreul's comments appear in the context of his four-part essay on alchemy in the guise of a review of L. P. François Cambriel's *Cours de philosophie hermétique* (Paris, 1843): Michel-François Chevreul, *Journal des savants* (May 1851): 284–98, (June 1851): 337–52, (August 1851): 492–506, and (December 1851): 752–68. The quotation cited above is at 293.

his private chemistry course (as opposed to the well-attended public one he taught at the Jardin du Roi).[130] These claims are confirmed by surviving student notes from Rouelle's lectures given in Paris through the 1750s and 1760s.[131] In regard to transmutational alchemy, Rouelle asserts that "the generality of natural philosophers [*commun des phisiciens*] doubt the truth of this science, but they are not able to be judges of a subject that is entirely unknown to them; the most knowledgeable chemists, even those who do not possess its principles, do not call it into doubt." When discussing the power of the philosophers' stone to transmute base metals into gold, Rouelle declares emphatically that "the most sensible and knowledgeable chemists have believed it; the ignorant ones and uneducated people have denied it."[132] Significantly, rather than merely perpetuating the ideas and explanatory systems of earlier alchemical authors, Rouelle "updates" chrysopoeia into the theoretical context of contemporaneous mid-eighteenth-century chemistry. He applies Stahlian principles to the constitution of metals and remarks that "the philosophers' stone is nothing other than the result of a fermentation of gold with mercury—not common mercury but a special mercury oversaturated with phlogiston."[133] This special overphlogisticated mercury is identical to the "animated" philosophical

130. For more on Rouelle's laboratory on rue Jacob, see Beretta, "Rinman, Diderot, and Lavoisier," 362–70.

131. There are numerous copies of notes from Rouelle's course preserving his comments on transmutational alchemy. I have used the Clifton College Science Library (Bristol, UK) manuscript "Cours de chimie, ou Leçons de Monsieur Rouelle, recueilles pendant les années 1754, 1755, et rédigées en 1756, revües, et corrigées, en 1757 et 1758" (alchemy section on 894–903), with comparisons to Bibliothèque Interuniversitaire de Santé, Paris, Pharmacie MS 19 (fols. 394v–401r), and Bibliothèque Interuniversitaire de Santé, Paris, Médecine MS 5022 (126–40). The Clifton College MS is attributed to "Messrs. Roux et Darcet," who were preparing Rouelle's lectures for publication. I was able to use photocopies of this MS made for the late Larry Holmes, and given to me from his *Nachlass* by Alan Rocke. For more on these and other MSS of Rouelle's course, see Rappaport, "G.-F. Rouelle," 86–92; and Christine Lehman, "Mid-Eighteenth-Century Chemistry in France as Seen through Student Notes from the Courses of Gabriel-François Venel and Guillaume-François Rouelle," *Ambix* 56 (2009): 163–89.

132. Guillaume-François Rouelle, Clifton College MS, 894: "le commun des phisiciens doute de la verité de cette science; mais ils ne peuvent pas être juges dans une matiere qui leur est entierement inconnüe; les plus sçavans d'entre les Chimistes, ceux même qui n'ont pas possedé ces principes ne le revoquent pas en doute." And on 898: "Les plus sensés et les plus sçavants chimistes l'ont crû, les ignorans et les gens peu instruits l'ont nié."

133. Rouelle, Clifton College MS, 898: "M. Rouelle pense que la pierre philosophale, n'est autre chose que le résultat d'une fermentation de l'or, avec le mercure; non pas le mercure ordinaire, mais un mercure particulier surchargé de phlogistique."

mercury described by chrysopoetic writers—the same material with which Homberg had worked so assiduously. Unfortunately, Rouelle continues, the written accounts left by the adepts, such as Philalethes, are enigmatic and ambiguous, and so it is possible that no one to date has interpreted them correctly—a conclusion suspected also by Homberg during his experiments at the Palais Royal. Just as Diderot recounts, Rouelle discouraged his students from going off in pursuit of the stone, but only because success in so expensive an undertaking was very uncertain "without a sure guide to lead one through an operation that is preserved only in oral transmission."[134]

Unlike his immediate predecessors at the Académie, Rouelle seems to have been willing to speak out in favor of transmutation in a somewhat more public setting—his classroom. His colleague Pierre-Joseph Macquer (1718–84) also prepared a public intervention on behalf of transmutation at about the same time, but seems to have thought better of publishing it. In response to a dismissive essay written by Pierre-Louis Moreau de Maupertuis (1698–1759) on the subject of transmutation, Macquer penned a short essay entitled "Sur la pierre philosophale" sometime between 1753 and 1766. Although he did not publish this text, its manuscript—displaying a degree of rewriting that suggests he intended to publish it—survives in the Bibliothèque Nationale, and has recently been studied and published by Christine Lehman.[135] Maupertuis had written that "among the most capable chemists, although some few have spent their lives in this search, others laugh at it and believe it an impossible thing," and characterized those who seek it as "insane [foux]."[136] Macquer, for his part, responds by asking, "Can one consider insane a person who labors for the philosophers' stone only for the pleasure of resolving one of the most beautiful and most difficult problems in natural philosophy?"[137]

134. Rouelle, Clifton College MS, 903: "faute d'un guide seur pour se conduire dans une operation qui n'est conservée que par tradition."

135. BNF, MS français 9132, fols. 98r–106v. For an initial description of this manuscript, see William Smeaton, "Macquer on the Composition of Metals and the Artificial Production of Gold and Silver," *Chymia* 11 (1966): 81–88; and for more complete analysis and contextualization, see Christine Lehman, "Alchemy Revisited by the Mid-Eighteenth Century Chemists in France: An Unpublished Manuscript by Pierre-Joseph Macquer," *Nuncius* 28 (2013): 165–216, which includes a full transcription of the manuscript at 202–16. Much of Macquer's rewriting tones down his criticism of Maupertuis and deletes references to him by name—presumably in accord with the Académie's regulations on managing disagreements. For more on Macquer, see Christine Lehman, "Pierre-Joseph Macquer: Chemistry in the French Enlightenment," *Osiris* 29 (2014): 245–61.

136. Pierre-Louis Moreau de Maupertuis, "Sur la pierre philosophale," in *Lettres de M. Maupertuis* (Dresden, 1752), 81–85.

137. Lehman, "Alchemy Revisited," 204.

Macquer recounts the frauds with which chrysopoeia had become associated, along lines similar to Geoffroy's 1722 "Supercheries," yet pairs these with famous accounts of successful transmutations—in regard to the veracity of the latter he "suspends judgment."

While there is at present no evidence that Macquer himself actively dabbled in the aurific art, the fact that he wrote such an essay indicates once again that transmutation was by no means a dead issue in the mid-eighteenth century among the chemists of the Académie Royale. Significantly, Lehman has convincingly connected this manuscript with Macquer's interest in further studies of the metallic composition.[138] She notes how a few years after writing this essay, Macquer would take full advantage of an unexpected opportunity to explore the composition of metals experimentally and at considerable length. What Macquer did in 1772 is of special importance for this book because it involved repeating some of the fundamental experiments that Homberg had carried out seventy years earlier. Moreover, this episode connects Homberg and his work more directly than previously recognized with the most celebrated chemical experiments and discoveries of the late eighteenth century.

Homberg Redivivus

During the early 1770s, the academicians expended great efforts—and undoubtedly a considerable sum of money—studying the nature and composition of diamonds, particularly in exploring the strange phenomenon that they disappeared entirely when exposed to air and high temperatures.[139] Could something as hard and as seemingly immutable as a diamond actually evaporate? In June 1772, after multiple experiments carried out in various furnaces, the academicians Louis-Claude Cadet de Gassicourt (1731–99) and Mathurin-Jacques Brisson (1723–1806) requested the Académie's permission to take the great Tschirnhaus lens out of storage in order to use its intense heat for their study of diamonds.

The lens had not been used for more than fifty years. In 1772 it lay crated up in storage at the Louvre. At some point after the closure of the Palais Royal laboratory, Philippe had given the Tschirnhaus apparatus to Louis-Léon Pajot, Comte d'Onsenbray (1678–1754). D'Onsenbray, made an honoraire of the Académie in 1716, had a serious interest in natural philosophy, and

---

138. Lehman, "Alchemy Revisited," 190–200.

139. On these experiments, see Christine Lehman, "What Is the 'True' Nature of Diamond," *Nuncius* 31 (2016): 361–407.

unlike many honoraires, he actually attended Académie meetings regularly and published several original papers in the *Mémoires*. During a royal visit to d'Onsenbray's chateau, cabinet of instruments, and laboratory at Bercy in July 1717, the Count demonstrated the power of the lens to a seven-year-old Louis XV.[140] D'Onsenbray's éloge mentions his long association with Père Sébastien (Truchet), and notes that Geoffroy worked for four years in the chymical laboratory that d'Onsenbray had installed at Bercy. When Caspar Neumann visited Paris in 1717 and associated with the Geoffroys, he and Étienne-François used the burning lens together in the gardens of the Château de Bercy.[141] One wonders, given Geoffroy's extended work at Bercy and the new home of the lens there, whether the new laboratory served in some sense as a successor to the one at the Palais Royal.[142] However that question might be resolved, the Tschirnhaus lens remained part of d'Onsenbray's celebrated cabinet of scientific instruments. He arranged for the entire collection to be transferred to the Académie upon his death, where it was to be reinstalled and opened to the public. Thus, in 1756, the lens apparatus, along with hundreds of other scientific instruments, arrived at the Palais du Louvre and was stored near the Académie's meeting rooms. But the promised installation never took place, and the collection remained in its crates; most of the objects were dispersed at the Revolution and consequently perished.[143]

Cadet and Brisson's request to take the lens out of storage was approved by the academicians, who also assigned Macquer and Antoine-Laurent Lavoisier (1743–94) to participate in the experiments. The lens, mounted in its carriage—presumably the original one, which both Macquer and

140. *Le mercure nouveau* (August 1717), 110–12.

141. "Lebens-Beschreibung" in Caspar Neumann, *Chymiae medicae dogmatico-experimentalis*, 6 vols. (Züllichau, 1749–55), vol. 1, sig. ****[4]r–****[4]v.

142. At present, very little is known about the laboratory at Bercy. Further evidence of a connection with the Palais Royal laboratory is that d'Onsenbray also had access to at least some of Homberg's manuscripts—perhaps shared with him by Geoffroy. A sequence of four recipes from Homberg's unpublished 1690s textbook (VMA MS 130, fols. 216v–218v) reappears nearly verbatim in the Hellot Caen notebooks (5:11r–v) where the material is referenced as coming from the "papers of d'Onsenbray (Homberg)" indicating clearly that d'Onsenbray owned Homberg's manuscript, either the original or a copy parallel to the one now in St. Petersburg.

143. Jean-Paul Grandjean de Fouchy, "Éloge de M. d'Onsenbray," *HARS* (1754): 143–54; Jean-Dominique Augarde, "The Scientific Cabinet of Comte d'Ons-en-Bray and a Clock by Domenico Cucci," *Cleveland Studies in the History of Art* 8 (2003): 80–95 (with photos of some of the surviving objects); and David Sturdy, *Science and Social Status: The Members of the Académie des Sciences, 1666–1750* (Woodbridge, UK: Boydell Press, 1995), 296n. The promise to bequeath the collection to the Académie is already mentioned in 1717; see *Mercure nouveau* (August 1717), 111.

Lavoisier describe as "very large and cumbersome"—was placed in the Jardin de l'Infante, a convenient location on the south side of the east end of the Louvre, between the palace and the Seine, and thus just downstairs from the Académie's meeting rooms. A wooden shed was built in the garden for its storage in order to protect it from the weather, and a room adjoining the garden was made available to the experimenters.[144] The necessary preparations were completed in short order, and the experiments began in mid-August 1772 and continued on every sunny day until the end of good weather in mid-October. Notes from these experiments survive among Macquer's papers, and what is immediately striking is that they record only a single experiment with a diamond—the original purpose that justified rehabilitating the Tschirnhaus apparatus. Instead, Macquer's notes record dozens of trials done in rapid succession with a wide range of mineral and other substances, but it is gold and silver that end up most frequently in the focus of the lens. Gold was exposed to the lens on August 15, 29, and 30, again on September 5, 16, and 28, and yet again on October 10 and 13; Macquer reports that they experimented with gold "at least twenty times." Silver was subjected to the lens on August 27, September 13, and October 6, 7, and 17.[145] No other substances reappeared so frequently. Thus, it is clear that Macquer, the most senior of the experimenters, took advantage of the availability of the lens to divert the intended experimental program away from diamonds in order to repeat Homberg's experiments on gold and silver and explore the composition of metals.[146] Years earlier, Macquer had already expressed his keen desire to repeat Homberg's experiments. In 1766, he wrote of their importance but noted that the reported results were "contested" and therefore "they need to be repeated with greater exactness."[147]

Macquer's subsequent reports on the experiments make clear that his focus was on gold, silver, and the other metals. These reports themselves are rather complex and occasionally seem contradictory; this character is prob-

144. See Lavoisier's unfinished memoir about setting up the Tschirnhaus apparatus as transcribed in Henry Guerlac, *Lavoisier: The Crucial Year* (Ithaca, NY: Cornell University Press, 1961), 204–7. For a more comprehensive view, see the lucid account of the setting up of the lens and the first experiments with it in Lehman, "Diamond," 375–82.

145. BNF, français MS 9132, fols. 261r-89v and "Nouvelle suite d'experiences a foyers des grands verres ardens de Tschirnhaus par Mrs. Cadet, Brisson, Lavoisier, et Macquer," fols. 192r–196r (at 194v). See also Antoine-Laurent Lavoisier, *Oeuvres*, 6 vols. (Paris: Imprimerie Impériale, 1862–93), 3:284–348.

146. The promised experiments with diamonds did eventually take place, but not until 1773; see Lehman, "Diamond," 382 and following.

147. Pierre-Joseph Macquer, *Dictionnaire de chimie*, 2 vols. (Paris, 1766), 2:167.

ably due in part to the communal nature of the work, which meant that the claims and content had to be agreed upon by all four academicians whose names appeared on it. What complicated matters further was that the gold behaved differently on different days—some days it smoked, some days it did not, some days the gold produced a glass, some days it remained unchanged, meaning that one either had to choose which results were the "authentic" ones or remain uncertain about them.[148] Macquer's autograph draft, which forms a basis for the memoir he read to a public assembly on 14 November 1772, provides the best depiction of Macquer's own views in the immediate aftermath of the experiments, and his final thoughts appear in new material he inserted under the headings of "Or" and "Verre Ardant" in the 1778 edition of his *Dictionnaire*.[149] Macquer's draft report ignores all the other experiments the group performed and focuses immediately upon Homberg's 1702 vitrification of gold and Geoffroy's 1709 work with the imperfect metals. Macquer was especially interested in testing Homberg's claims that the lens caused gold to smoke and to turn into a purplish glass. Using Homberg's own instrument, Macquer verified Homberg's observations: gold and silver both smoked when in the focus, and gold produced a violet-colored glassy substance. Macquer had previously accepted as valid the objections that Hartsoeker made against Homberg's results, but after repeating the experiments himself, he now dismisses those objections and vindicates Homberg: "There are none [of Homberg's observations] that I have not observed with pure gold held in a support of hard porcelain in the focus of the same lens that Homberg used."[150]

Although Macquer concluded that the experiments of 1772 confirmed Homberg's *observations*, he did not embrace Homberg's *interpretations* of them. In order to understand Macquer's departure from Homberg's interpretation of the same experimental results, it is necessary to recognize how chem-

148. Homberg's original report also describes a variation of results—for example, gold and silver purified in different ways gave different results; see Wilhelm Homberg, "Observations faites par le moyen du Verre ardant," *MARS* (1702): 141–49.

149. Macquer, "Nouvelle suite d'experiences" (BNF, français MS 9132, fols. 192r–196r). The November 1772 paper was never published in the *Mémoires*, but did appear five years later as "Mémoire . . . sur des expériences faites en commun, au foyer des grands lentilles de Tchirnhausen" in the 1772 issue of *Introduction aux observations sur la physique*, 2 vols. (Paris, 1777), 2:612–16. The "many interesting memoirs" based on the 1772 experiments that it promised never appeared.

150. Pierre-Joseph Macquer, *Dictionnaire de chimie*, 4 vols. (Paris, 1778), 4:152; compare with his *Élémens de chymie-théorique* (Paris, 1756), 86–88. Note that Macquer's specification of a porcelain support responds to the objections made by Hartsoeker, whom Macquer never mentions by name, although he does refer to the critical "Dutchman" of Homberg's 1707 paper at 156, and remarks that Homberg's responses were *assez satisfaisantes*.

ical theories had changed over the course of the eighteenth century. Homberg began working from the five chymical principles—mercury, sulphur, salt, earth, and water—current in the late seventeenth century. Although he would constantly refine this vision of chymical composition in response to his experimental results, this basic framework was his starting position. Within this context, he saw the smoke as gold's volatile mercury and sulphur being expelled, and the glass as a result of the combination of gold's earth with some of its residual sulphur. Gold—generally assumed at the time to be a compound body—was being decompounded by the lens, thus revealing its components of mercury, sulphur, and earth.[151] As described earlier in this chapter, Hartsoeker attacked Homberg's results because they conflicted with his own idea that gold was composed of uniform and indestructible atoms. He then claimed that he made his own experiments with a comparable lens, and that these trials gave no smoke, no glass, no change to the gold—thus reinforcing his own theory of composition. By the time of Macquer, Stahlian ideas of composition had come to the fore, according to which metals were composed of an earth combined with the inflammable principle phlogiston. Macquer's essay "On the Philosophers' Stone" embraces this viewpoint and therefore proposes seeking out every sort of naturally occurring earth and treating each one with a phlogiston-containing substance (such as charcoal) to see what metals would thereby be produced. He holds out the hope that some such earths, once joined to phlogiston, might thereby produce gold or silver.[152] Thus, in 1772, in the context of Stahlian theory, Macquer did not see the smoke from gold and silver as their mercury—such a component of metals did not exist in his theoretical framework. Instead, the smoke was simply the volatilized but compositionally unchanged metal, which he demonstrated by holding a cold sheet of silver in the smoke, which became covered with a fine deposit of gold.[153]

Macquer's interpretation of the glass he observed forming on gold is more interesting. Exposed to heat of a fire, the base metals (lead, tin, copper, iron, mercury) all convert into a powdery calx, which, according to Stahlian principles, is the constituent earth of the metal left behind after its phlogiston has been expelled by the fire. The noble metals (gold and silver) are not altered by fire, but reasoning by analogy from the behavior of base metals, Macquer sus-

151. As noted in chapter 6, Homberg would later remove earth from the composition of gold and silver, or more precisely, theorize that mercury could be transformed into earth when its particles were destructively altered.

152. Lehman, "Alchemy Revisited," 212–13.

153. Macquer, "Mémoire," 615.

pected that the purple glassy material he and Homberg observed was gold's earth vitrified by the heat of the lens after its phlogiston had been expelled. In other words, the lens could do what no ordinary fire could do; it could expel the phlogiston from the noble metals. Because of the frustrating variability of the results, however, Macquer could not be entirely confident in this conclusion, and so he hoped that one day he would be able to convert an entire sample of gold into the violet glass "in sufficient quantity that so one could be certain if it will be turned back [*revivifiera*] into gold through the addition of some phlogistic substance as other metallic glasses do."[154] Significantly, although the *identities* of gold's components changed between Homberg and Macquer, Homberg's main conclusion—that the light of the lens causes the decomposition of gold and silver into their components—persists unchanged in Macquer's thinking.

Other elements of Homberg's chymical theory reappear even more strikingly in two of Macquer's other conclusions. Macquer observed a phenomenon that Homberg had not mentioned: that when gold melts at the focus, the molten drop adopts a spherical shape, and this sphere immediately begins to spin rapidly. The direction of its spin, Macquer noted, is determined by where on the sphere the light impinges most strongly. Similarly, he noticed that lightweight materials are driven away from the focus as if "by a puff of air"—an observation made by Homberg as well. Macquer concluded that these motions were caused by "the impulsion of the solar rays."[155] In other words, as in Homberg's system, light is composed of minute solid particles in rapid motion capable of imparting motion to other, more massive bodies. Another of Homberg's ideas reappears in Macquer's only mention in his draft report and public address of a lens experiment that did *not* involve gold or silver. When "ferruginous earths" (such as clays) were exposed to the focus of the Tschirnhaus lens, they were transformed into a substance with the color of iron that was "equally attracted by a magnet as the best iron." In order for these earths to revert to the state of metallic iron, they must have somehow acquired the phlogiston necessary for that transformation. Having ensured that the support upon which the earths were held when exposed to the lens could not provide any phlogiston, Macquer concluded that "the fire of the Sun now performs the role of phlogiston and becomes capable of entering

154. Macquer, *Dictionnaire* (1778), 3:54. The phrase "other metallic glasses" refers to the well-known vitrifications of base metal calxes, especially those easily formed from lead and tin. Macquer rather despondently feared that he might not live long enough to complete this endeavor.

155. Macquer, "Nouvelle suite d'experiences," fol. 195r ("la impulsion des rayons solaires").

into the composition of the metals as a principle."[156] Macquer's 1772 interpretation is thus *virtually identical* to Homberg's 1705 announcement that "the matter of light itself is our sulphur principle" and that light acts a component principle of compound bodies. The names have changed—Macquer's phlogiston vs. Homberg's sulphur principle—but the idea, the conclusion from the lens experiments, remains the same.

Crucially, one thing that Macquer (like most phlogistonists) did *not* preserve from Homberg's system—or rather from Homberg's entire approach to experimentation—was the Batavian chymist's routine comparison of weights of materials before and after an experiment as a means of probing their transformations. As is well known, the phlogiston theory did not place great importance upon the observation that metal calxes weigh *more* than the metals from which they are produced—potentially a problem for a system that claims that calcination involves the *expulsion* of a component from the metals. As mentioned previously, Duclos observed this weight increase when he calcined antimony regulus using a burning lens in 1667, and concluded that the metal must absorb sulphurs from the air during its calcination that add to its weight. In the 1690s, Homberg recognized the increase of weight, but attributed it to the mechanical incorporation of foreign material from the vessels in which the metal was calcined. But in the opening years of the eighteenth century, convinced by accumulated results from his experiments with quicklime, the report of Duclos's result, and especially by his work with the Tschirnhaus lens and parallel trials carried out using fire, Homberg concluded that the increased weight was due to the incorporation of particles of light. Hence, light's ability to augment the weight of substances with which it combines became a primary feature of his comprehensive chymical theory.[157] All throughout his career, Homberg never lost sight of weight results, their importance, or the need to explain them thoroughly within his theory of chymistry. For him, during calcination, the dense, fine-grained sulphur (as it exists in light) drives out the less-dense, coarser sulphur found in base metals and packs more tightly into the vacuities left behind, the weight naturally increases. When that dense sulphur is forced out again and replaced by a less-dense sulphur (for example from oil or charcoal), the weight decreases and returns to its original value. The eventual Achilles' heel of the phlogiston theory proved to be its failure to acknowledge the importance of comparative

---

156. Macquer, "Nouvelle suite d'experiences," fol. 195v: "ce qui indique que le feu du Soleil fait alors fonction de phlogistique et devient capable d'entrer comme principe dans la composition des métaux."

157. Homberg, "Suite des essays de chimie," *MARS* (1705): 88–96, at 89.

weight determinations and its consequent inability to account for observed changes of weight.

But there was one member of the team using the Tschirnhaus lens in 1772 who *did* recognize the importance of weight determinations in very much the same way as Homberg had done—namely, Lavoisier. As soon as the first sunny weather returned in March 1773, he began working again with the burning lens, carrying out a series of experiments that Macquer had mentioned previously as desirable, that is, "a study of the nature of the vapors that the focus causes to arise from the most fixed bodies and that we will endeavor to collect in closed vessels."[158] In early August 1772, while the Tschirnhaus lens was still being readied, Lavoisier himself had proposed that "independently of Homberg's and Geoffroy's experiments on the metals that it will be necessary to repeat, it would be good to try if the metals calcine in sealed vessels. Since they all give off a vapor or smoke by means of the burning lens, it would be very interesting to find an apparatus able to retain it."[159]

Thus, the following March, in a chamber of the queen's apartments adjoining the garden that had been made available to the experimenters, Lavoisier focused the light of the burning lens onto a range of materials not in the open air, but held in a cup enclosed under a bell jar. Although his initial results proved inconclusive, Lavoisier kept experimenting along these lines, hoping to capture and study whatever might be expelled upon heating. The story of his subsequent discoveries, which led to the oxygen theory of combustion and the eventual replacement of the phlogiston theory, is well known, and so need not be repeated here.[160] For the present purpose it suffices to underscore the close connections between Homberg and Lavoisier's foundational work in 1772 and 1773. Lavoisier's March 1773 experiments with the calcination of metals form an integral part of the replication of Homberg's experiments with the Tschirnhaus lens that had begun the previous August, and for which he used at least a part of the same apparatus Homberg had used.[161] Within

158. Macquer, "Nouvelle suite d'experiences," fol. 195v: "recherche sur la nature des vapeurs que le foyer fait émaner des corps les plus fixes et que nous tâcherons de recueillir dans les vaisseaux clos."

159. Antoine-Laurent Lavoisier, "Reflexions sur les experiences qu'on peut tenter a l'aide du miroir ardent," transcribed in Guerlac, *Crucial Year*, 208–14, at 212.

160. For details of these experiments, see Guerlac, *Crucial Year*, and Frederic Lawrence Holmes, *Antoine Lavoisier: The Next Crucial Year* (Princeton, NJ: Princeton University Press, 1998), 22–29.

161. I think it is unlikely that Lavoisier used the entire Tschirnhaus apparatus for these experiments. First, it would have been difficult, to say the least, to move the entire, "very large and cumbersome" apparatus indoors and such that it could still collect sunlight. Second, the heat from the

the broadening scope of his investigations of combustion and the role of gaseous substances in chemical transformations, Lavoisier would later conduct a series of experiments with Homberg's pyrophorus, and present a memoir on this topic in the same year that he presented his more widely cited papers on respiration, the burning of phosphorus, and combustion in general.[162] Notably, Lavoisier's so-called balance-sheet method for comparing starting and ending weights of reacting substances, which emerged as crucial for such combustion experiments, actually represents a *return* to the evidentiary power of weight determinations that was fundamental to Homberg's own theory of combustion and calcination contained within his comprehensive "light as sulphur" theory.[163]

This chapter has followed the fate of several of Homberg's ideas and interests through much of the eighteenth century and down to the time of Lavoisier. Throughout this period, Homberg's experiments with the Tschirnhaus lens apparatus continued to fascinate. Newton built his own apparatus to repeat them, Hartsoeker endeavored to refute them, Geoffroy and Neumann explored them, and finally, after a hiatus of nearly sixty years, a group of academicians reerected Homberg's apparatus along the banks of the Seine and reinvestigated them, thereby guiding Lavoisier to the crucial experiments that led to his oxygen theory of combustion. Homberg's comprehensive theory of chymistry, emulated and extended by several younger chymists during and shortly after Homberg's life, had several of its aspects revived within a new framework by Macquer in the 1770s. As for the Batavian chymist's dogged pursuit of metallic transmutation, it too persisted in various guises in the work of a broad roster of later academicians down at least to the 1760s, despite carefully constructed images to the contrary.

---

lens was so great that it would have shattered the glass bell jar. It is more plausible that Lavoisier dismounted and used only the small secondary lens, which on its own would have provided sufficient heat for the experiments he carried out.

162. Antoine-Laurent Lavoisier, "Expériences sur la combinaison de l'alun avec les matières charbonneuses," *MARS* (1777):363–72.

163. This statement is *not* to imply that Homberg was the originator of the method; William Newman and I showed that a very similar method was employed in chrysopoetic processes pursued by George Starkey in the 1650s, and that weight measurements were likewise crucial for Van Helmont. Homberg read both of these carefully; see *Alchemy Tried in the Fire* (Chicago: University of Chicago Press, 2004), especially 71–91, 120–25, 296–314. A similar use of the problem of "missing weight" as a probe of chemical reactions appears even in the work of the fourteenth-century John of Rupescissa; see Principe, *Secrets of Alchemy*, 65–66.

Epilogue

Homberg and the Transmutations of Chymistry at the Académie

In 1653, when Wilhelm Homberg was born in far-off Java, the discipline of chymistry looked very different than it would at the time of his death in 1715. This book has traced Homberg's life and the life of chymistry itself in France across much of that period. Integral to both those stories is that of the Académie Royale des Sciences, which served as an institutional home for Homberg and for chymistry. After having followed the interconnected journeys of Homberg, chymistry, and the Académie through the period, it is now time to take stock of some of the notable features presented by these three lives—one human, one disciplinary, and one institutional—and review the transformations that chymistry underwent in that period and shortly thereafter.

Wilhelm Homberg has emerged as a remarkable and central figure for both chymistry and the early Académie Royale des Sciences, even though he has previously received little coverage in general narratives of the history of chemistry. His unwavering devotion to chymistry as an independent discipline and as a uniquely explanatory and exploratory part of natural philosophy marks him as unusual for the period. As one of his associates jocularly recalled, "he didn't esteem any but the chymists," in terms of how to explore and understand the natural world.[1] Homberg envisioned chymistry as the discipline of "infinite extent" and the source of the most secure and plausible understanding of the material world. Historians have rightly seen Robert Boyle as a key voice in advocating for chymistry's place within natural philosophy; Homberg was another such voice. But Homberg went even further than Boyle by boldly coopting new territories into chymistry's domain: light

---

1. Nicolas-François Rémond to Gottfried Wilhelm Leibniz, 23 December 1715; NLB, LBr 768, fols. 53r–54v, at 54r.

became a chymical reagent and principle, the Sun and stars became chymical engines that power the universe, and everything that could be sensed became inherently chymical. In the long (and apparently unending) territorial battle between chemistry and physics, Homberg took no prisoners. Talented as he was in other fields, it was *la chimie* that for him provided the most certain knowledge of nature, thanks to its direct and sensual contact with matter, while the speculations of *les physiciens* about invisible entities beyond the range of sense perception remained "at best suppositions," and "too problematic" to offer solid truth.

This is not to say that Homberg had a much more generous evaluation of many of his contemporaries who self-identified as chymists, for "among the chymists he didn't esteem any but himself."[2] The predominantly French didactic tradition of chymistry that had developed since the start of the seventeenth century did not suit Homberg (as it had not satisfied Boyle). It tended to see material production, mostly of pharmaceuticals, as the primary purpose of chymical operations. This predominantly practical agenda remained too narrow and unsatisfying for Homberg; he wanted chymistry to pursue a natural philosophical agenda that offered coherent explanation and deeper understanding of the way things are and behave. Within the didactic tradition, theoretical structures and practical production were inadequately linked, and so theoretical systems were accepted too uncritically. The usually cited principles of mercury, sulphur, and salt were not the *true* chymical principles that should rightly bear those names. Concepts of composition lacked sure foundations, and explanations of experimental results were too ad hoc or based more on supposition than on measurement, like Tachenius's and Lémery's pointy and porous particles. Likewise, a failure to understand chymical transformation and composition at a deeper level hampered the practical and productive operations that chymistry could effect. In his first attempt to write a textbook, therefore, Homberg departed significantly from the didactic tradition by including a broader range of chymical phenomena, and used preparative processes primarily to explore and illustrate general principles of chymical behavior and reactivity. He reworked the tradition still further in his later "Essais" and final *Élémens de chimie*. He included fewer and fewer preparative processes, and eventually dispensed with them entirely, in favor of building up a comprehensive explanatory theory of chymistry. Of course, "making stuff" remained extremely important for him and for his vision of chymistry. He collected and traded chymical recipes constantly. The recipe was a crucial means for transferring chymical knowledge and dis-

2. Rémond to Leibniz, 23 December 1715; NLB, LBr 768, fols. 53r–54v, at 54r.

coveries in the early modern period, even if success in obtaining the promised results often required hands-on training with the original practitioner. Homberg showed a consistent interest in processes and production, whether of chymical exotica and transmuting agents or of manufactured goods like glass, metals, and medicines. But material production was not enough on its own; explanation had to have an equal profile. For Homberg, productive processes yielded not only a material result but also new information and insight about chymistry and its possibilities, new data that had to be fitted into more general concepts. This point is useful to bear in mind during the current (and generally useful) "material turn" in the history of science. Homberg loved his materials, especially the unusual and exotic ones, but they had an intellectual purpose and importance for him as well.

Homberg's way forward for establishing a broader, more certain, and more explanatory understanding of chymistry lay squarely in practical experience and experiment—and lots of it. "All philosophy according to him lay in the use of the fire-tongs."[3] Homberg was especially insistent that explanatory and conceptual principles had to be as closely tied as possible to practical experimentation. This is not to say that some of his chymical predecessors did not think the same; among its most distinguished exponents, chymistry had always been a joint exercise of head and hand, theory and practice.[4] But within the context of the early Académie and among a large proportion of late seventeenth-century chymical practitioners, Homberg stood out in this regard. The chymical theories he inherited from his predecessors, that his contemporaries proposed, or that he devised himself had to withstand persistent experimentation and testing. He therefore spent much of the 1690s testing the Helmontian theories of matter and composition that held a high profile in the early Académie. Despite early successes, parts of the system eventually failed to correspond adequately to his laboratory results, and so Homberg looked elsewhere. He explored the claims of the acid-alkali theory, and the contours of the *tria prima* and the pentad. He wrestled with mechanical explanations, always wary of wandering too far with them from directly observable evidence, but was eventually unable to do without their explanatory power. He repeatedly modified his own ideas in the aftermath of new experiments and phenomena. The tight dynamic between laboratory

3. Rémond to Leibniz, 23 December 1715; NLB, LBr 768, fols. 53r–54v, at 54r.

4. Examples can be cited from at least the medieval Geber to the mid-seventeenth-century George Starkey (whom Homberg greatly respected); see William R. Newman and Lawrence M. Principe, *Alchemy Tried in the Fire: Starkey, Boyle, and the Fate of Helmontian Chymistry* (Chicago: University of Chicago Press, 2004), especially 92–155.

practice and theory creation appears strikingly in the successive versions of the chymical system he labored to craft throughout the twenty-five years of his mature career. New experiments, new results, and new instrumentation forced him to modify his explanatory framework, and then to pursue new experiments and lines of inquiry. It is rare that the cycle of laboratory results and the reformulation of an explanatory theory can be charted so clearly in this period. Newly discovered manuscripts and the Académie's multiple levels of records made such analysis possible for this book, and the matter might be even better illuminated if more of Homberg's and Philippe's records from the Palais Royal laboratory had survived.

Homberg's chymical approach to the material world relied emphatically upon the sensible, the measurable, and the observable. Suspicious of speculations and rationalizations too distant from direct human experience, Homberg emphasized the testimony of human senses and sensory experience. This preference expressed itself most significantly in his constant monitoring of comparative weights. Invisible components that eluded the direct sense of sight could be tracked by their weight as they changed partners during various chymical transformations. Weight discrepancies pointed toward overlooked features and losses that had to be explored and explained. Changes in weight signaled otherwise undetectable chymical changes. The measurement of weights, densities, and concentrations was as routine and important a part of his exploration of the chymical world as were the colors, smells, sounds, and tastes he witnessed with his other senses. While histories of chemistry most readily connect such numerical monitoring of chymical processes with the later "balance sheet" method of Lavoisier, reliance on such methods was a matter of course already for Homberg in the late seventeenth century, and rested upon the gravimetric traditions established as a useful tool in chymistry by such esteemed earlier figures as Van Helmont and Starkey.[5]

Of course, in order to engage in so much practical experimentation, Homberg needed physical space and financial support. Homberg worked in at least four laboratories, each of them supported by patronage. At first, in the 1680s, he had a private space in Paris that he was able to maintain probably thanks to support from Colbert. Thereafter he worked briefly in a rented space paid for by the Abbé de Chalucet. After his admission to the Académie in 1691, Homberg gained control of the institution's well-furnished laboratories at the Bibliothèque du Roi, whose maintenance, outfitting, and operating expenses were reimbursed by the crown. Finally, from 1702, Homberg had a magnificent new laboratory at the Palais Royal. He surely had a major role

5. See Newman and Principe, *Alchemy Tried in the Fire* 35–91.

in designing it to suit his research goals, while Philippe II provided the space
and financing for its construction and continuing support. There Homberg
enjoyed not only an enviable workspace but also access to the most costly
and spectacular scientific instrument in Europe, the Tschirnhaus double-lens
apparatus, as well as the assistance of at least one operator, the cooperation of
other savants and academicians, collaboration by the Duc d'Orléans, and visits from numerous others. While royal and other patronage of chymistry was
not something new, the collaboration and support Homberg enjoyed with
the Duc d'Orléans were at a level rarely if ever equaled elsewhere. Alongside
this steady elevation of his ability to do chymistry came a parallel rise in his
notoriety and his social and financial status. Intelligence, talent, showman-
ship, and a winning personality carried Homberg from his humble origins in
Java as the child of a war refugee to his eventual status as an esteemed pensionnaire chimiste of the Académie Royale, a resident of the Palais Royal, and
a favored member of the royal household of France. But the lofty status he
achieved eventually exacted a tragic cost when, denounced as a poisoner in
1712, he narrowly escaped going to the Bastille and his beloved laboratory was
forced to close. The nearly total collapse of his scientific productivity thereafter signals how crucial the Palais Royal laboratory was to his endeavors.

Yet before that disaster struck, Homberg's curiosity led him to many discoveries and innovations. Some could be fully explained, or had to be rediscovered or made practicable, only centuries later, such as luminescence
poisoning and doping, and the now routine practice of reduced-pressure
distillation. His reputation for "exactitude" led to inventive and painstaking attempts to standardize reagents by weights and volumes, and to strive
(although not always successfully) to assure the purity, homogeneity, and
consistency of starting materials—even of human excrement.[6] Ironically
enough, his insistence on exactitude and purity may have only increased his
confidence that he had actually produced a small quantity of gold from purified mercury by extended heating, since it would have been nearly impossible
for him, practically speaking, to detect the presence of trace amounts of gold
in the antimony ore he had employed many steps earlier in the sequence,
even if he had thought to look for it there. Other innovations included new or
streamlined techniques—for determining the density of liquids, for assaying
gold and silver, for separating metal alloys, for carrying out sublimations, and
so on—as well as the discovery of new substances that bore his name long

---

6. Note his remarks on the importance of such standardization in his "Observations sur
la quantité exacte des sels volatiles acides," MARS (1699): 44–51, at 44.

afterward: Homberg's sedative salt (boric acid), Homberg's pyrophorus, and Homberg's phosphorus (triboluminescent calcium chloride).

While Homberg made many innovations and desired to strike out in new directions, he never attempted to make a "clean break" with chymistry's past. Some contemporaries, such as Fontenelle, were keen for such a break to happen in order to purge chymistry of its less reputable connections, but Homberg would have none of it. He stands instead as a transitional (or integrating) figure during an important period of change for the natural sciences socially, institutionally, and intellectually. His intellectual formation was very much one of the seventeenth century, despite Fontenelle's endeavor to recast it in more "respectable" and "enlightened" terms. Whether consciously or not, Homberg followed the famous Paracelsian advice to learn by travel, direct experience, and conversation with all sorts of people high and low who had direct knowledge of things. Rather than acquiring any formal education beyond his law degree, Homberg spent his formational years crisscrossing Europe, accumulating and trading secrets and rare knowledge everywhere he went. He learned from the likes of Cellio, Kunckel, Boyle, and Leibniz, as well as from Swedish refiners, Austrian gilders, German miners, and French vinegar-makers. His wide travels, diverse contacts, and facility in multiple languages provided him with a special degree of internationalism in the republic of letters. The fact that the major theoretical influences on his younger collaborator Geoffroy came from Becher, Kunckel, and Stahl—all Germans—may be due to Homberg's own native relationship with Germany. Homberg exemplifies the importance of transnational exchanges in early modern science—the knowledge he gained in travels across Europe all ended up with him in France.

Probably emulating his early mentor von Guericke, Homberg initially acquired his knowledge and chymical secrets through trade, and he knew how to display his knowledge without giving it away, even in a period often celebrated for moving toward greater openness in scientific matters. He never gave up on the method of trading secrets. He was thus among the last of the "chymical secretists" and the last "empiric" to be employed as physician in the royal household of France. When he arrived in Paris in the early 1680s, he captivated members of the Académie with his accumulated treasury of marvels and knowledge. Having attracted their attention, he was thereafter able to collaborate with some of them as an "unofficial member" within the institution's diffuse penumbra. Homberg continued to be a showman throughout his life—not in the sense of creating a false façade, but of knowing how to showcase his knowledge in ways that would evoke the sense of wonder that

powers intellectual curiosity and that could captivate onlookers able to provide him entry into the circles he wished to join. The eventual (but delayed)
result of the gavottes he danced in Paris was his admission to the Académie
as soon as Louvois's death cleared the way for the institution's revitalization
at the hands of Pontchartrain and Bignon. A sense of wonder and delight at
the natural world was something that Homberg never lost. Glowing Bologna Stones, self-igniting powders, spontaneously exploding oils, luminous
phosphorus, surprising experiments with the air pump—all were marvels
to capture attention, provoke wonder, and increase knowledge, as was the
showy Tschirnhaus lens, the Palais Royal laboratory, and certainly the "Indian Virtuoso" himself.

Homberg engaged seriously with the intellectual heritage of chymistry not only by exploring and testing ideas received from the likes of Van
Helmont, Tachenius, Becher, Kunckel, and many others but, most strikingly,
by intently pursuing chymistry's ancient quest for metallic transmutation
and the philosophers' stone. From his early days he labored over the enigmatic writings of Philalethes, striving to put them into practice, first through
the "interpretations" of them by George Starkey and then, when these failed
to reach the promised goal, by creating his own interpretations and putting
them to the test in the laboratory. He acquired recipes from many others all
over Europe, often with the help of his patron, Philippe II, Duc d'Orléans,
and helped facilitate parallel work being carried out around Paris, either by
supplying materials and advice or by negotiating the acquisition of the licenses needed for building furnaces and practicing chymistry. While both
Homberg's and Philippe's keen interest in chrysopoeia was known at the
time to both the academicians and the wider world, little of such research
appears in the Académie's records, presumably due to the multiple prohibitions against academicians engaging in such matters. That said, the core illustrative experiment of his "Essais," as published in the Académie's *Mémoires*
and upon which his novel chymical theory of light as sulphur rested most
emphatically, was nothing other than the key step in the preparation of the
philosophers' stone according to the principles of Mercurialist chrysopoeians. Simply stated, Homberg built a completely new theory of chymistry
upon results from a "traditionally alchemical" process. Significantly, his new
theory of light as the true sulphur principle not only explained and justified
the transmutation of metals, but also accounted for the changing weights
of metals during their calcination and reduction that would so vex the later
eighteenth century. In fact, it did so far better than Stahlian phlogiston was
ever able to do, given Stahl's dismissal of the relevance of weight determinations—an outlook very much at odds with Homberg's practice. Homberg's

application of his principles uniformly to both problems underscores once again how metallic transmutation remained an integral and important part of chymistry into the eighteenth century.

That unity would, however, scarcely outlive Homberg. The links to counterfeiting, poisoning, and cozenage that had dogged chymistry for centuries intensified in France during Homberg's lifetime. Legal strictures responding to the Affaire des Poisons had codified the link between chymical practice and criminality. Vigorous police investigations and arrests in Paris of those suspected of "being chymists" reinforced such connections. Government attempts to recruit chrysopoeians into service for the bankrupt state, and the unsatisfactory outcomes from such endeavors, further depressed chymistry's reputation, particularly in regard to transmutation. Finally, knowledge of the Regent's transmutational interests while he was administering the Académie may well have triggered the Académie's apparent denunciation of transmutational claims in 1722. This intervention, and the anxieties leading up to it, may explain why Homberg's completed *Élémens de chimie*—built around a reportedly successful transmutation of mercury into gold using an established chrysopoetic pathway—never saw the light of day despite two separate promises that its publication was imminent.[7] Thereafter, abetted by Fontenelle's increasingly shrill rhetorical performances, the practices of chymistry divided into a respectable chemistry and a repudiated alchemy . . . at least on the surface. In reality, many of the *vrais chimistes* (in Fontenelle's terms) of the Académie continued seeking for the ancient secrets of transmutation and the philosophers' stone, as Homberg had done, but they now did so more quietly. Remarkably, *more* academicians continued such transmutational investigations *after* 1722 than had done so before. Chrysopoeia most emphatically did not die out, but it did retreat from view, making it *seem* to disappear from learned circles. Moral opprobrium and public reputation nevertheless transformed the discipline of chymistry in ways that sober scientific study, experiment, and administrative regulation did not.[8]

Despite Homberg's talent and robust vision for chymistry, his immediate impact in reshaping the discipline was mixed. His degree of success or failure in this regard, however, had at least as much to do with institutional policies, available personnel and their interests, changing views about chy-

7. Perhaps now that the manuscript has been recovered after nearly three hundred years of remaining lost, it is time for it to be published. But given the previous two unfulfilled promises for its publication, I dare not risk a third.

8. The Académie would flirt openly with alchemy again a hundred years later; see Lawrence M. Principe, *Alchemy and Chemistry: Breaking Up and Making Up (Again and Again)* (Washington, DC: Dibner Library Publications, Smithsonian Institution, 2017).

mistry, and a host of uncontrollable factors as it did with Homberg's ideas themselves. Developments in chymistry at the Académie after Homberg's lifetime showcase the inseparability of the lives of individuals, the discipline, and the institution, as well as the impact of highly contingent factors that affected the course that the transmutations of chymistry could take. Although a more detailed study of chymistry at the Académie in the decades immediately after Homberg's death is desirable, a quick glance back to the 1660s and forward to the 1760s will help clarify what became of Homberg's vision and contributions.

The Académie's significant role in the history of chemistry should not be underestimated. Colbert's inclusion of *la chimie* on an equal footing with other, better established scientific disciplines at the institution's founding in 1666, and the subsequent establishment of eight positions for chymists in the 1699 Règlement were key moments for a subject that did not have a clear place in the academic system or a generally positive reputation among the wider public. Most chymists recruited to the Académie during the seventeenth century—Bourdelin, both Boulducs, Charas, and Nicolas Lémery— were apothecaries interested primarily in pharmaceuticals and the application of chymistry to medicine. Even Geoffroy, Homberg's closest associate, was initially trained as an apothecary, though his interests and work quickly moved away from the pharmaceutical, presumably due to Homberg's influence. The clear outlier during the Académie's early years was Duclos who, although a licensed doctor of medicine and *médecin du roi*, did not like to practice medicine, preferring to spend his time instead on chymical investigations, including the quest for *arcana majora* such as metallic transmutation and the alkahest.[9] From the institution's very first meetings, Duclos wanted the Académie's chymical work, soon to be exemplified in the chymical analysis of plants, to address fundamental questions about the nature and identity of the chymical principles and the origins and composition of material substances. Although a majority of the academicians agreed initially, his advocacy of a more natural philosophical and exploratory agenda was eventually overwhelmed by personalities and institutional forces pushing a more practical and result-driven agenda with direct applications to medicine. When Homberg was admitted in 1691, six years after Duclos's death, he was also an outlier. Unlike Duclos, he held no medical degree, having picked up his medical knowledge by collecting recipes and by gaining experience in

---

9. I have excluded Borelly from this account because there is so little documentation about him; he was not an apothecary, and, as I suggested in chapter 2, he may have shared Duclos's broader interests, but this remains merely conjectural.

Provençal hospitals. Like Duclos, he did not much care to put what medical knowledge he had into practice; even though he ultimately became Philippe's *premier médecin*, the appointment was probably not due to medical prowess. Homberg shared Duclos's broader vision of chymistry as a key part of natural philosophy. Therefore, he soon redirected the plants project back toward Duclos's aims of identifying the chymical principles and understanding chymical composition, while the more practically inclined physician Dodart lost interest and the apothecary Bourdelin continued accumulating results from a flawed analytical methodology without much concern over what they might or might not mean.

Through the opening decades of the eighteenth century, the apothecaries occupying the Académie's lines for *la chimie* were gradually replaced by licensed physicians, but this change of primary profession did nothing for the discipline. Quite a few of the physicians scarcely contributed at all to the institution because of the greater attention they paid to their medical practices.[10] Some others did make contributions, but sooner or later their attention turned preferentially toward their more lucrative and socially advancing medical careers. In fact, the apothecaries, generally speaking, rendered better service to the discipline and to the Académie. Significantly, the three occupants of the Académie's chymistry positions whose work and interests align best with Homberg's had also collaborated with him: Louis and Jean Lémery, and Étienne-Francois Geoffroy. Although both Lémerys were élèves to their father Nicolas, their work is strikingly—and surprisingly—much more along Homberg's lines of interest, indicating the kind of influence he must have exerted within the institution while he lived.[11] All of Jean Lémery's publications extend lines of work Homberg had initiated: on the pyrophorus and on phosphorescent materials. Louis Lémery and Geoffroy, both licensed physicians, adopted and adapted Homberg's comprehensive theory of chymistry and chymical composition, and pursued research programs similar to Homberg's. In the years immediately following Homberg's death, Lémery and Geoffroy continued the study of salts that Homberg had championed from his earliest days at the Académie.[12] From 1716 to the mid-1720s, Lémery wrote

10. For a recent prosopography of the Académie's *chimistes* and other academicians involved in chymical topics, see Patrice Bret, "Les chimistes à l'Académie royale des sciences à l'époque des Lémery (1699–1743)," *Revue d'histoire de la pharmacie* 64 (2016): 385–404.

11. For an earlier divergence of Louis from his father Nicolas, see Bernard Joly, "À propos d'une querelle concernant la production artificielle du fer: les divergences entre Nicolas Lémery et son fils Louis," *Revue d'histoire de la pharmacie* 64 (2016): 375–84.

12. Frederic Lawrence Holmes, *Eighteenth-Century Chemistry as an Investigative Enterprise* (Berkeley, CA: Office for History of Science and Technology, 1989), 35–41.

on saltpeter as well as on the volatilization of fixed salts, and offered mechanical explanations for observations made when dissolving salts. From 1719 until
1721, he revisited the old issue of plant analyses from a fully chymical perspective, just as Homberg had done, and at the end of these investigations came to
the same conclusion that the Batavian chymist had reached twenty years earlier.[13] As for Geoffroy, he also pursued the study of salts, actually completing
Homberg's unfinished business by accomplishing the sole transformation of
one kind of salt into another—turning an acid salt into a urinous salt—that
Homberg had not achieved during his decade-long work on such interconversions. To do so, Geoffroy followed Homberg's methodology for determining the "exact quantity of acid salt" in a sample of aqua fort, used arguments
from "missing mass," combined observation of naturally occurring syntheses
with laboratory analysis and synthesis, and deployed experimental combina-
torics, all as Homberg himself would have done, which clearly indicates how
Geoffroy had made Homberg's experimental methodology his own.[14] In 1718
and 1721, Geoffroy published his two highly influential papers on chemical
affinity. As noted in chapter 3, Geoffroy's notions of rapport are extremely
similar to Homberg's notion of semblance, which undergirds the fundamental explanatory principle of his first textbook. The subsequent popularity of
"affinity theory" through the eighteenth century may thus mark one example
of Homberg's influence, if at one remove, on the direction that chemistry
would take. In any event, Geoffroy's endeavor to put forth a comprehensive
organizing principle for chymistry represents part of the same goal to which
Homberg had devoted himself. Subsequent scholars have remarked on how
Geoffroy based his "Table" directly upon observed laboratory results and
avoided speculation about the unseen causation behind such phenomena,
notably avoiding arguments based on particle shapes and mechanical factors.
Geoffroy's choice in this regard may reflect something of Homberg's own
resistance to proposing explanations of observables that rely too heavily on
factors that cannot be directly sensed, particularly invisible particle shapes
not directly deducible from macroscopic properties. Thus, judging from the
research directions and publications by Geoffroy and the two Lémerys in the

13. Louis Lémery, "Explication mechanique . . . de la dissolution des differents sels," *MARS*
(1716): 154–72; "Premiere memoire sur le nitre," *MARS* (1717): 31–51; "Seconde memoire,"
122–46; "Sur la volatilisation vraye ou apparente des sels fixes," 246–56; "Reflexions physiques
sur . . . analyses ordinaire des plantes et des animaux," *MARS* (1719): 173–88, (1720): 98–107,
166–78, and (1721): 22–44.

14. Étienne-François Geoffroy, "Du changement des sels acides en sels alkalis volatils urineux," *MARS* (1717): 226–38; "Table des differens rapports observés en chymie entre differentes
substances," *MARS* (1718): 202–12; and "Eclaircissement sur la Table," *MARS* (1720): 20–34.

decade after Homberg's death, it would appear that the Batavian had successfully left his mark on the direction that chymistry—or at least chymical
investigations at the Académie—would take.

But this situation did not last. Jean Lémery, despite a very promising start
and his regular participation and contributions to the Académie, died suddenly in mid-1716 at the age of thirty-eight, less than a year after Homberg
and after only three years as an academician.[15] His brother Louis purchased
the charge of *médecin du roi* in 1721, and his productivity accordingly began
to decline after the mid-1720s, as he started giving greater attention to his
medical career. Geoffroy followed a similar trajectory, at first continuing the
kind of chymical work Homberg had done, but after becoming dean of the
Parisian medical faculty in 1726, he quickly became embroiled in controversies there, and did not publish again. He died only a few years later, at the
start of 1731.

What about the other five positions "guaranteed" for *la chimie* within the
Académie? The third pensionnaire besides Geoffroy and Lémery was Simon
Boulduc, twenty years older than they. He had held the position since its
creation in 1699 and would keep it until 1723. As noted in chapter 2, Boulduc
had been admitted in 1694 to be Homberg's assistant, and he made use of that
experience in the first years of the eighteenth century by applying Homberg's
ideas of semblance to the practical deployment of alcohol and water to better analyze plant materials through successive solvent extractions. But as an
apothecary, Boulduc was interested in preparing and improving pharmaceuticals. He devoted virtually his entire independent career at the Académie
to studying purgatives and emetics—a class of the materia medica with an
especially (and distressingly) high profile in early modern pharmacopoeia.
Boulduc thus continued the lines of work characteristic of the seventeenth-
century Académie.

Following the two associés lines and the entry-level élèves (transformed
in 1716 into two adjoints) points to a major, and hitherto largely unnoticed,
factor in the development of the discipline at the Académie, and one that
helps explain why it did not continue the course advocated by Homberg and
followed initially by Geoffroy and the Lémerys. In short, the institution failed
to recruit appropriate talent. Shortly after Homberg's death, both Varignon
and Leibniz expressed concern about being able to replace Homberg with
a chymist of comparable talent, and subsequent events proved worse than
their fears. In the immediately succeeding years, the Académie chose almost

15. On his correct date of death, see chapter 6, note 36. The attendance lists in the PV
indicate that he last appeared at the Académie on 19 August 1716 (PV 35, fol. 283r).

no new members with any discernible chymical interests or talent. After Jean Lémery's premature death in 1716, the Académie left the position of associé chimiste vacant for *five years.* Strikingly, when in 1719 the academicians chose candidates to fill three recently vacated associé positions in astronomy, geometry, and mechanics, they did not even mention the still empty associé position in *chimie.*[16] It remained unfilled until 1721, when Danty d'Isnard, a botany élève, was promoted into that position. He held it for less than a year before being moved back to botany as pensionnaire. Ironically, while Danty d'Isnard's seven publications in the *Mémoires* are all strictly botanical, he was in fact very active in chymical matters, but chrysopoetic ones—as witnessed by Hellot's notebooks—topics that by the 1720s were not to be mentioned openly at the Académie. In 1722, the position was given to François Pourfour du Petit (1664–1741), a licensed physician on the faculty at Montpellier, who was shifted from the anatomy section and then moved back into it three years later as pensionnaire anatomiste. The position then sat vacant again for eighteen months until Gilles-François Boulduc (1675–1742) was promoted into it in 1727 after having remained élève/adjoint since 1699. He could have been proposed for the position eleven years earlier in 1716, thereby providing room for fresh talent as a new adjoint chimiste—but that did not happen, for reasons that remain unknown. As royal apothecary, Gilles-François followed in his father Simon's footsteps, devoting himself almost exclusively to describing purgatives—not botanical simples in his case, but mineral salts and waters—thereby further perpetuating the view of chymistry as primarily an adjunct to pharmacy.

The other associé position passed to Claude-Joseph Geoffroy (*le cadet*) when his brother Étienne-François moved into the pensionnaire position that had been Homberg's. As a *maître apothecaire*, he, like the Boulducs, was more inclined to applied pharmacy, and expended much of his efforts on medicinal substances, such as essential oils and therapeutic mineral waters. Although extremely productive, he continued this inclination when he became pensionnaire chimiste in 1723.[17] In 1724, the associé position passed to Dufay, a very active academician who shared many of Homberg's diverse interests (such as phosphorescence, optics, glass, and artificial gems), and who came from a noble and military background rather than from pharmacy or medi-

16. PV 38, fol. 292r–v (2 December 1719).

17. He was also involved in a rather acrimonious debate with Louis Lémery over the production of sal ammoniac; he eventually won the debate, and his paper, presented in 1716 but suppressed due to the controversy with Lémery, was eventually published in the 1720 *Mémoires.* Geoffroy reexamined Homberg's *sel sedatif,* itself a pharmaceutical, in 1732; "Nouvelles expériences sur le borax," *MARS* (1732): 398–418.

cine. Dufay became pensionnaire chimiste in 1731, but like Danty d'Isnard he devoted much of his chymical activity to metallic transmutation, a pursuit that he could not share openly or publish.

Dufay had moved up from adjoint chimiste, but that line had suffered the same neglect as the associé line. From 1712 until 1722 it was held by Henri Imbert, another practicing physician on the Paris faculty of medicine, who published but a single paper in the *Mémoires* (describing a medical case), rarely attended meetings, and did not merit an éloge.[18] After Dufay held it for less than a year, it went to François-Joseph Hunauld (1701–42), another physician, this time from the medical faculty of Montpellier. He held it for less than two years before becoming associé in anatomy. His éloge states plainly that he was given the position of adjoint chimiste simply because it was "the only one open, even though it was well known that the class of *chimie* is not where he aspired to be nor where it suited him to be placed."[19] He published exactly two papers, both narrowly anatomical, in his seventeen years as an academician. In 1728, the position went to Henri-Louis Duhamel du Monceau (1700–1782), whose interests were mainly agricultural—accordingly, most of his published papers treat fruits and vegetables. Of his remaining papers, the three that do have chemical content are co-authored with Jean Grosse, Homberg's former laboratory assistant.[20] Duhamel du Monceau remained adjoint chimiste for two years before being shifted over to the more suitable botany branch. He was replaced by Charles-Marie de la Condamine (1701–74) who, although renowned for his later expedition to South America regarding the shape of the Earth and his role as a mathematician, displayed little interest in anything chemical. He occupied the position for five years until a place opened up for him in geometry.[21]

Thus, throughout the 1720s, the Académie used the entry-level positions for *la chimie* merely as convenient holding positions for anybody, regardless of whether or not they had any interest or talent in chemistry. It did the same with one of the two associé lines, or simply left it vacant. From 1715 until the

18. Henri Imbert, "Histoire d'un assoupissement extraordinaire," *MARS* (1713): 313–17.

19. Jean-Jacques Dortous de Mairan, "Éloge de M. Hunauld," *HARS* (1742): 206–12, at 207. See also David Sturdy, *Science and Social Status: The Members of the Académie des Sciences, 1666–1750* (Woodbridge, UK: Boydell Press, 1995), 408.

20. See chapter 5, note 62. He also wrote on sal ammoniac, as had several academicians, in 1735.

21. He was shifted back as pensionnaire chimiste in 1739 as a replacement for Dufay, again a strange move given his lack of interest in chemistry and significant work in other areas. His sole chemical communication is "Sur une nouvelle espèce de végétation métallique," *MARS* (1731): 466–82.

1730s, therefore, these positions were little more than revolving doors, as a host of characters, most of them unqualified or unsuitable, filled them only until other positions opened up.[22] The institution did not treat the adjoint positions in the other disciplines in this fashion. Likewise, no adjoint position in another branch was ever used to hold someone waiting for a permanent position to open up within *la chimie*. The rank of adjoint chimiste appeared open to everyone *except* those with a serious interest in the discipline. The only two new academicians of the period who were actually active in practicing chymistry, Dufay and Danty d'Isnard, did so within the realm of chrysopoeia and therefore quietly. Thus, there was actually no new blood in the Académie through the 1720s to pursue Homberg's ideas and methods.

Did the institution simply not care enough about *la chimie* to recruit appropriate new members? Did the academicians actually intend to let that branch languish? Did they consider it more useful to use up the spaces for *la chimie* as "swing space" for new academicians in other disciplines rather for their intended recipients? Given recent events regarding the public impression of chymistry, did they consider it better simply not to admit any new *chimistes* in case they turned to the pursuit of transmutation? Were there perhaps simply no qualified candidates available? If the last was the case, then had the Académie shown the initiative (and had the financial resources) to recruit talent wherever it resided—for example, in Germany with its abundance of talented chymists—as it had done during Colbert's leadership, rather than taking the easy route of resorting so frequently to the local medical faculties, or to the nearby sons and brothers of existing academicians, the discipline would have had greater success at the Académie in the 1720s.[23]

These questions are difficult to answer, given the paucity of information available about what exactly went on during private discussions at the Académie and what, if anything, the academicians actually planned for their institution's future in regard to *la chimie*.[24] Yet while the causes must now remain open to question, one thing is clear: the situation changed abruptly at the start of the 1730s. De la Condamine was the last nonchemist the Académie put into the chemistry section. The list of the succeeding adjoints chimistes then starts to resemble a roster of the most significant contributors to

---

22. Modern academics should be familiar with the unhealthy consequences of such a situation, given the analogous pattern on display with deanships and other administrative positions in so many of today's universities.

23. On the chemists in Germany at this time, see Karl Hufbauer, *The Formation of the German Chemical Community* (Berkeley: University of California Press, 1982).

24. The plumatifs, which might have given details of discussions at the meetings deemed unfit for the procès-verbaux, are missing from 1720 until the late 1750s; see note on sources.

French chemistry of the mid-eighteenth century: Grosse (1731), Hellot (1735), Malouin (1742), Rouelle (1744), and Macquer (1745). Whether this was the result of an active decision on the part of the Académie or the felicitous appearance of a more qualified pool of candidates remains unclear. This new list reinforces the observations made previously about the weaker appointees of the 1720s. Out of the five new adjoints admitted between 1731 and 1745, only one—Malouin—was a devoted medical practitioner. Accordingly, he is also the least productive of the five in terms of chemical contributions and publications. Condorcet's rather lefthanded éloge for him reports that his publications provided "few new truths," that he "soon renounced chemistry to give himself entirely over to medicine," and that "he did not contribute to the rapid progress chemistry has recently made in France."[25] In contrast, Rouelle, although himself an *apothecaire-privilégé*, refused a position as *premier apothecaire du roi* on the grounds that it would have obliged him to curtail or give up his studies of chemistry.[26] Macquer, who held a medical degree, also actively chose the pursuit of chemical studies over medical practice, as Duclos had done long before.

A comparison of two textbooks from these new academicians proves revealing, Malouin wrote a *Traité de chimie* prior to his admission to the Académie. A glance at its contents reveals it to be a throwback to the seventeenth-century didactic tradition. It provides little organizational, explanatory, or theoretical content. Instead, it is primarily devoted to the recitation of pharmaceutical recipes, a content and orientation signaled clearly by its subtitle: *The Manner to Prepare the Remedies Most in Use in the Practice of Medicine.* It is scarcely different in style or intent from the texts of Glaser and Lémery dating from the 1660s and 1670s.[27] In stark contrast, Macquer's 1749 *Élémens de chymie-théoretique* explicitly excludes preparative recipes, deferring them to another publication, just as Homberg's own "Essais" had done. Instead, it presents a coherent natural philosophical system of chemistry based largely

---

25. Condorcet, "Éloge de M. Malouin," *HARS* (1778): 57–65, at 61. Only during his first few years at the Académie was Malouin active in chemical matters, contributing three papers on metals, particularly zinc, and one on salts, before shifting (as Condorcet remarks) to entirely medical matters around 1745.

26. Jean-Paul Grandjean de Fouchy, "Éloge de M. Rouelle," *HARS* (1770): 137–49, at 144; and Jean Mayer, "Portrait d'un chimiste: Guillaume-François Rouelle," *Revue d'histoire des scienes et leurs applications* 23 (1970): 305–32.

27. Paul-Jacques Malouin, *Traité de chimie* (Paris, 1734). Malouin contributed a communication to the Académie about metals before he was admitted; *HARS* (1740): 61–62. This communication, together with his *Traité* and his experiments in the early 1740s with metals, would have justified his admission to the chemistry class.

on Geoffroy's affinities and Stahl's phlogiston, and even resuscitates some of Homberg's own work and ideas.[28] While Homberg ultimately decided to exclude preparative recipes entirely from his final *Élémens*, Macquer did produce a pair of follow-up volumes devoted to preparative operations. Nevertheless, Macquer's choice of processes to include contrasts dramatically with Malouin's *Traité* and Lémery's still-in-print *Cours*, even if it adopts the latter's format of following the processes with explanatory "remarques." Only a fraction of Macquer's processes deal with pharmaceuticals, while the majority describe the production of a range of chemical materials and operations useful for trades, assaying, refining, and further chemical research. It reflects Macquer's own broad interest across all the various practical applications of chemistry, a breadth very similar to Homberg's own.[29] Macquer's choice of what to include actually resembles Homberg's own selection of processes for his early 1690s textbook, both based on a broader scope and understanding of the discipline.

A more detailed investigation of the changing character of chemistry at the Académie from the 1730s to the 1750s would prove revealing but lies beyond the timeframe of this already lengthy book. The main purpose of the cursory overview presented here is only to help assess the fate and influence of Homberg's work and vision after his lifetime. When combined with material presented earlier in this book, this quick survey suffices to point out some significant features. First of all, the predominance of apothecaries among the Académie's *chimistes* in the seventeenth century gradually gave way to licensed physicians in the eighteenth. Second, the less distinguished of these latter, generally admitted in the 1720s, showed less interest in chemistry than did the apothecaries, yet were nevertheless allowed to occupy the places supposedly reserved for *la chimie*—whether by design, neglect, or a lack of more suitable candidates. They were neither interested in nor capable of following up on the directions for the discipline that Homberg had advocated and pursued. In contrast, the academicians who had actually worked in one way or another with Homberg—Geoffroy, the two Lémerys, and (by 1731) Grosse—all pursued the directions, and even some of the same topics, that Homberg had done. They thus represent the first and most direct

28. The topic of the changing character and content of the *cours de chimie* in eighteenth-century France is a large and important topic that cannot be covered in detail here; see Bernadette Bensaude-Vincent and Christine Lehman, "Public Lectures of Chemistry in Mid-Eighteenth-Century France," in *New Narratives in Eighteenth-Century Chemistry*, ed. Lawrence M. Principe (Dordrecht: Springer, 2007), 77–96.

29. On this point, see Christine Lehman, "Pierre-Joseph Macquer: Chemistry in the French Enlightenment," *Osiris* 29 (2014): 245–61.

influence that Homberg continued to exert. Philippe, as administrator of the Académie, although no longer able to be a practitioner himself, should also be considered part of this group. His changes to the structure of the Académie and his directives to the academicians (such as the enquête and its advocacy of the greater application of chemistry to France's economic health and the trades, pursued diligently by Réaumur and later by Hellot) probably stem ultimately from his close collaboration with Homberg. Third, it appears that external forces, including the uncontrollable ones that affect human life, curtailed the impact that these five associates of Homberg could have had on the Académie and on chemistry's future directions. The younger Lémery died prematurely. Philippe, who had planned to continue guiding the Académie after Louis XV's majority, died suddenly in late 1723, ten months after the end of the Regency.[30] Louis Lémery and Geoffroy turned increasingly to their professional medical duties after the mid-1720s, while the reclusive Grosse was admitted to the Académie only in the 1730s. Hence Homberg's vision and directions evaporated from the Académie in the 1720s, in large part due to a lack of personnel capable or willing to perpetuate them. Ironically, the one place where studies akin to Homberg's did live on vigorously was precisely the one place where they were not "allowed" to do so—namely, in the chrysopoetic endeavors carried on privately.

Finally, the tide began to turn at the start of the 1730s. The Académie stopped using the entry-level adjoint chimiste positions as a convenient places for physicians and others to be lodged temporarily. The newly chosen adjoints tended now to come from outside the professions of either apothecary or physician, and those who did hold such training willingly chose chemistry over practicing those professions, as Duclos had done. Thus, the liberation of the discipline of chemistry from service to medicine and pharmacy—one of its key transformations in the eighteenth century—advocated by Homberg (and Boyle before him) also required that the Académie's chemistry positions themselves be liberated from practicing physicians and apothecaries. When that change finally came to pass, it is striking how the new cadre of académiciens chimistes, most notably Hellot, Rouelle, and Macquer, did return not only to Homberg's vision for the discipline but also to research lines similar to his and even to several of his specific experiments and pursuits. Starting in the 1740s, they revived his keen interest in uncovering the composition of the metals and salts, identifying the true chymical principles, and devising a more comprehensive, experimentally grounded, explanatory system for the discipline—not to mention perpetuating his interest in chrysopoeia.

30. On Philippe's continuation, see HARS (1716): 2.

Perhaps most dramatically, the redoing of Homberg's famous Tschirnhaus lens experiments on the metals by Macquer and Lavoisier in the early 1770s, which led directly to Lavoisier's landmark experiments with metals and their calcination, shows clearly how the Batavian's ideas, results, and practices held renewed importance and relevance, and could spark further fundamental questions about chemistry and chemical composition. In short, Homberg had to wait for successors who shared his priorities, interests, and programs to come along, and perhaps for the Académie to catch up with his vision of a chemistry transmuted.

Note on Sources

The documentary evidence for the early Académie Royale des Sciences and its members is both complex and fragmentary, and some explanation will be helpful for understanding what the main primary sources are, their origins, their interrelationships, and how I have used them. This note first explains how the contents of Académie meetings were transformed through various reworkings into the annually published *Histoire et mémoires*. This overview is essential for understanding what was lost or altered along the way, and why the *Histoire et mémoires* is therefore insufficient on its own for understanding what really took place at the Académie and how academicians' ideas developed—a point not always fully appreciated. This overview will also explain the references given throughout this book, making them easier for readers to understand and to follow. I hope this note will also prove useful as an orientation for those wishing to work on the early Académie. I thereafter introduce several other archival materials relevant to the Académie that I have used in this study—some of them newly discovered—that shed additional light on the work of particular academicians or the workings of the institution.

This note is not intended to be comprehensive; its purpose is merely to clarify the nature, origin, value, and limitations of the available sources used in this study. For a broader treatment of archival and other source materials relating to the Académie, see the extremely useful *Histoire et Mémoire de l'Académie des Sciences: Guide des recherches* by Éric Brian and Christiane Demeulenaere-Douyère (Paris: Tec & Doc, 1996).

Académie Manuscripts: Plumatifs, Procès-verbaux, and Pochettes de Séances

The members of the Académie Royale des Sciences met twice weekly, on Wednesdays and Saturdays, except during their recess in September and October, first at the Bibliothèque du Roi on the rue Vivienne, and then, starting in 1699, at the Palais du Louvre. The content of these meetings was taken down by a secretary in running notes called the *plumatif*. These notes, together with the text of any memoirs presented by the members and submitted to be entered into the permanent record, were transferred into the official register known as the *procès-verbaux*, generally by a professional copyist and sometimes months or even years after the meetings took place. No plumatifs dating from before 1713 survive; they were probably discarded when the corresponding procès-verbaux entries were compiled from them. A few plumatifs survive from 1713, more from 1714, and for the following few years, but only until early 1719. There follows a long period for which no plumatifs survive; the next extant plumatifs date from the late 1750s. The surviving plumatifs from the 1710s are written in the hand of René-Antoine Ferchault de Réaumur, who took down these notes first on scraps of paper, and then rewrote them (with modifications) into a fairer copy on full sheets. Upon comparing the original scraps with the rewritten versions, and the rewritten versions with the procès-verbaux, it becomes immediately apparent that significant amounts of material were omitted, transformed, or garbled at each transfer. For example, during the meeting on 14 July 1714, Réaumur recorded observations by Homberg about the rusting of iron, but then chose, for no apparent reason, to exclude this material from his rewritten second plumatif. Some information in this second plumatif—for example, about experiments Claude-Joseph Geoffroy and Louis Lémery carried out to refute claims made by Martino Poli—were not carried forward by the copyist into the procès-verbaux, presumably on the instruction of the perpetual secretary Fontenelle whose duty it was to see that the procès-verbaux were compiled. Other details were modified as well between versions: the first plumatif attributes the critique of Poli to Nicolas Lémery, the second plumatif to an unspecified Lémery and Geoffroy, while the procès-verbaux specifies "le fils" and "le cadet," respectively.[1] At this distance it is not always possible to tell whether such modifications made during transmission are corrections or errors.

The fact that Geoffroy and Lémery began critiquing Poli before he had even finished reading his paper on the subject implies that the biweekly meet-

---

1. AdS, pochette de séances 1714 and PV 33, fol. 251r–v (14 July 1714).

ings were more lively and controversial than is suggested by the ever placid procès-verbaux. The latter do not explicitly connect the critique with the ongoing paper, unlike both plumatifs. As Brian and Demeulenaere-Douyère elegantly express it, the procès-verbaux are only "an echo" of the original meetings and remain "very discreet" about internal dissensions.[2] The steady erosion of content from the meeting itself, to the plumatif, to the procès-verbaux explains why so many entries in the procès-verbaux occupy one folio, one page, or even less—as if the meeting had lasted only a few minutes. A comparison of the plumatifs with the procès-verbaux might have clarified more about the selection criteria employed, but such a project is hamstrung by the paucity of extant plumatifs.

For the years from 1699 to 1707, the diary of Académie meetings kept privately by Claude Bourdelin the younger (BNF, NAF MS 5148, see below in this note), provides many important and otherwise lost details, some of which might have been present in the lost plumatifs but do not appear anywhere in the procès-verbaux. His diary once again underscores the highly selective nature of the content of the procès-verbaux.

The surviving plumatifs are now filed in the *pochettes de séances*—files containing the documents relating to specific meetings—along with any extant manuscript memoirs submitted by the academicians, letters received, and other materials. For the period covered in this study each pochette covers an entire year; in later periods, each meeting has its own pochette. Some of the memoirs preserved in the pochettes were later published, but many were not, nor were they transcribed in the procès-verbaux, especially those submitted in the 1690s.[3] As for the procès-verbaux, they survive back to the first meetings of the Académie in 1666, with the exception of the lost (or perhaps never compiled) volume for the years 1670–74. There are also shorter gaps, such as the several weeks missing in January and November–December 1694; these gaps may be the result of plumatifs that were lost before the procès-verbaux were compiled. For these lacunae, one can often consult Jean-Baptiste Duhamel's *Historia* (or the later *HMARS 1666–99 Histoire*, see below), which almost always contains some account of the missing months or years. Since Duhamel was able to summarize material not found in the procès-verbaux, he must have worked from the plumatifs, indicating that these plumatifs must in fact have been taken—probably by Duhamel himself, acting as secretary—during those "missing" periods. The full run of the

---

2. Brian and Demeulenaere-Douyère, *Guide des recherches*, 62.

3. For the period 1667–1723, one pochette contains all the materials for an entire year. Thereafter, there is one per month, and starting in 1755 there is a pochette for each séance.

procès-verbaux is consultable as bound photocopies in the Archive of the Académie des Sciences, Paris. Ninety-three of the volumes are available on Gallica, but fourteen are randomly (and annoyingly) missing at present.[4]

Académie Publications: Monthly *Mémoires,* Annual *Histoire et mémoires,* and *Histoire et mémoires 1666–99*

The various publications of the *ancien régime* Académie have been excellently treated in Brian and Demeulenaere-Douyère's *Guide des recherches.*[5] Here, I provide supplementary information relevant to the present study and clarifications regarding how I have cited these publications.

As explained in chapter 2, the Académie endeavored to produce a monthly publication—the *Mémoires de mathématique et de physique tirez des registres de l'Académie Royale des Sciences*—but managed to do so for only two years, 1692 and 1693. A sustained serial publication succeeded only after the Académie's *renouvellement* in 1699. This new serial differed from the earlier attempt in two ways. First, it was annual rather than monthly; second, it was divided into two separately paginated sections, the *Histoire* (*HARS*) and the *Mémoires* (*MARS*). The latter of these two sections contains academicians' scientific papers, as the 1692–93 issues had done, and under the same title. The former section, however, entitled *Histoire de l'Académie Royale des Sciences,* contains summaries (more or less) written by the perpetual secretary—Bernard de Fontenelle for the period covered in this study—of many, but not all, of the papers in the *Mémoires.* The *Histoire* also contains various other materials, such as short notices of miscellaneous topics mentioned at the weekly meetings, royal *règlements,* and éloges for deceased academicians. Fontenelle's serial *Histoire* thus continued but significantly redesigned the foregoing tradition of the *Regiae scientiarum academiae historia* written by Fontenelle's predecessor as Académie secretary, Jean-Baptiste Duhamel, which initially covered the years 1666 to 1696, and was then extended to 1698 in a second edition.[6]

The key point to note is that the *Histoire* should not be used as a "shortcut" to the content of the academicians' papers (even though that was part of

---

4. The volumes currently missing from Gallica are 1 and 2 (1666–67), 15 (3/1696–2/1697), 25 (1706), 32 (1713), 39 (1720), 41 (1722), 50 (1730), 57 (1738), 70 (1751), 72 (1753), 89 (1770), 93 (1774), and 94 (1775).

5. Brian and Demeulenaere-Douyère, *Guide des recherches,* 107–27. One may also consult Robert Halleux, James McClellan, Daniela Berariu, and Geneviève Xhayet, *Les publications de l'Académie Royale des Sciences (1666–1793),* 2 vols. (Turnhout: Brepols, 2001).

6. Jean-Baptiste Duhamel, *Regiae scientiarum academiae historia* (Paris, 1698; 2d ed., Paris, 1701).

its original intent). The *Histoire* is very much the composition of Fontenelle, who applied his own particular criteria for deciding what to highlight, what to ignore, and how to present the materials—and he occasionally made judgments that conflict with the ideas of the papers' authors. Consequently, the *Histoire* sometimes reveals more about Fontenelle's prejudices and preoccupations than about the content and intent of the academicians' papers. In some cases, he made his summary from a version of the paper that differed from the one published in the *Mémoires*, such that the *Histoire* sometimes contradicts the *Mémoires*. The *Histoire* and the *Mémoires* are different sources from different authors with different audiences and intents, and need to be deployed as such by historians.

The *Mémoires* also presents potential problems for the historian that need to be made explicit. It might seem reasonable to assume that its contents were "drawn from the registers of the Académie Royale des Sciences" as their titles announce—that is, copied from the memoirs transcribed into the procès-verbaux—but this is often *not* the case. A gap of one to three years intervened between the presentation of a paper as found in the procès-verbaux and its publication in the *Mémoires*, during which time authors often made considerable changes to their texts and their conclusions. Thus, the procès-verbaux and the *Mémoires* sometimes provide quite different versions of the "same" paper. Crucial parts of chapters 3 and 4 are based upon the substantial differences between versions of the "same" paper recorded in the procès-verbaux and published in the *Mémoires*. I am not aware of other historians who have recognized and made use of such differences to chart the changing ideas of an academician as revealed by two "snapshots" of his work and ideas captured at different times. While I have used such comparisons to shed light specifically on Homberg's intellectual evolution, the study of other academicians would probably benefit from similar comparisons. In short, the papers transcribed into the procès-verbaux and those published in the *Mémoires* should be considered *complementary* sources until proven otherwise by a close comparison of the texts in the two series.

The need for caution does not end there. The *Mémoires* appeared both in multiple editions separated by many years, and in sequential printings within the same year, and in both cases differences can appear. For example, early printings of the 1700 *Mémoires* (published in 1703) omit one of Homberg's important papers. This early printing contains the "Continuation" [*Suite*] of a paper but not the main paper itself. Although the main paper is not listed in the table of contents, both it and its continuation are nevertheless summarized in the accompanying *Histoire*. The omission was presumably caught while the sheets were still being printed off, and so later printings of the first edition (with the same publication date) insert the missing paper in the proper place, but without repaginating the balance of the volume (hence creating two sets of pages 191–96) or correcting the table of contents. The second edition (published 1719) regularizes the pagination and provides a new table of contents, but thereby causes problems for anyone providing or following page references because all subsequent papers in the volume occupy different page ranges in the two editions. Some published papers were even altered significantly between successive editions (again provoking pagination discrepancies). For example, a paper by Pierre Varignon in the 1702 *Mémoires* was expanded by *six pages* between the first edition in 1704 and the second in 1720.[7] Hence, it can be important for scholars to check multiple editions.

Further to complicate matters, the annual volumes appeared in several pirated editions both in Amsterdam, a capital of printing piracy, and in Paris. Some of these pirated editions omit the "mathematical" papers, publishing only the "natural philosophical" [*de physique*] ones. All the pirated editions are printed in duodecimo format, while the authorized Paris editions are in quarto; hence there are again differences in pagination depending on which edition is used, although some of the pirated editions usefully provide running marginal notes giving the corresponding page numbers in the quartos.

For this study I have always used and referenced the first editions of the authorized Parisian quartos, unless otherwise noted.

The start of the regular annual publications only in 1699 meant that roughly thirty years of the Académie's previous work remained either unpublished or difficult to access. Hence, in 1727 the Académie decided to publish or republish a large portion of that material. The result was an eleven-volume set (bound in fourteen parts) in quarto format, published between 1728 and 1734. This work I cite collectively as the *Histoire et mémoires 1666–99* (*HMARS 1666–99*) in order to distinguish it from the annual series. The first two volumes of the set provide a *Histoire* for the years 1666 to 1699, written partly by Fontenelle and partly by Louis Godin; it diverges in various ways from the two editions of the *Historia* of Duhamel that cover the same period, and so it is wise to consult both sources and, of course, always in conjunction with the corresponding procès-verbaux. The tenth volume reprints the papers published in the monthly *Mémoires* of 1692–93 as well as the articles about the Académie's work published from 1665 to 1698 in the *Journal des sçavans*, which served during those years as a regular outlet for news and find-

---

7. Pierre Varignon, "De la résistance des solides," *MARS* (1702): 66–94 (1st ed., 1704) and 66–100 (2d ed., 1720).

ings from the academicians.[8] The other volumes contain numerous works by various academicians (such as Cassini, de la Hire, and Picard), most of them previously published but with some newly published materials, as well as re-editions of the "communal" works on plants and animals from the 1670s. The first volume contains a useful, and mostly reliable, table of contents for the entire set. The various volumes, especially the first two, exist also in pirated editions, in both quarto and duodecimo formats.

Manuscripts of Claude Bourdelin, *père* and *fils*

Extremely valuable records of the activities of the early Académie Royale des Sciences were kept by the two Claude Bourdelins, father and son, especially the elder (1621–99) who performed many of the institution's chymical experiments. The most extensive of these documents are his laboratory notebooks, which span virtually the whole of the Académie's seventeenth-century existence. These records exist in multiple copies located in various repositories; there are (at present) *five* known versions of Bourdelin's laboratory notebooks. These versions differ from one another in various ways that reveal how the Académie chose to record and revisit the results from its chymical laboratory. All five versions have not hitherto been listed in one place and compared; the present comparison should be considered preliminary rather than definitive.[9] The relationships between the versions, as well as their voluminous contents, deserve further study. The five versions of the laboratory notebooks, as well as the additional material originating from the two Bourdelins, are listed below.

1. PARIS, ADS, FONDS BOURDELIN, CARTONS 1 AND 2, 11 VOLS.: "REGISTRES DES ANALYSES FAITES"

These eleven quarto volumes, bound in limp vellum with leather ties, are Bourdelin's original laboratory registers. The earliest entry is dated 23 June 1672, and the entries, primarily of plant analyses, continue until 2 September 1699, just a few weeks before Bourdelin's death. Bourdelin probably took

8. Some of the earliest articles republished from the *Journal des sçavans* date from some months before the official founding of the Académie, but relate to the soon-to-be academician Adrien Auzout.

9. A previous treatment is Christiane Demeulenaere-Douyère and David J. Sturdy, "Image versus Reality: The Archives of the French Académie des Sciences," *Archives of the Scientific Revolution: The Formation and Exchange of Ideas in Seventeenth-Century Europe*, ed. Michael Hunter (Woodbridge, UK: Boydell, 1998), 185–208, at 198–99. See also Brian and Demeulenaere-Douyère, *Guide des recherches*, 65, 103, 198–99.

loose notes while performing the analyses, and then recopied them into these notebooks as the official record of his work.[10] The entries were then read, verified, and signed or initialed, generally by Duhamel, Perrault, Duclos, Mariotte, or Dodart. All the volumes are almost exclusively in Bourdelin's own hand, except for volume 9, which is in the hand of a copyist. Scattered through most volumes are brief notes or reflections in the hands of Duhamel, Duclos, and other members of the Académie. Several volumes also contain entries written back to front and upside-down relative to the majority of the text; these appear to record experiments done by laboratory workers other than Bourdelin, particularly Borelly and Duclos.

Carton 1

Vol. 1: June 1672–January 1674
Vol. 2: January–July 1674
Vol. 3: July–December 1674
Vol. 4: December 1674–October 1675
Vol. 5: November 1675–June 1677
Vol. 6: October 1676–January 1678

Carton 2

Vol. 7: February 1678–July 1679
Vol. 8: July 1679–September 1682
Vol. 9: October 1682–December 1686
Vol. 10: December 1686–August 1692
Vol. 11: September 1692–September 1699

2. PARIS, BNF, NAF MS 5133–5149, 17 VOLS.

Although often catalogued and referenced all together simply as Bourdelin's laboratory notebooks, this collection of seventeen manuscripts actually contains diverse materials. The entire set came from the Bourdelin family itself, having been sold as a unit at auction upon the death of Louis-Claude Bourdelin in 1777.[11]

NAF MS 5133 is misleadingly catalogued as "Procès-verbaux des séances et du laboratoire, 1667–68." In reality, it consists of two separately paginated

---

10. What is probably one of these originals, in Bourdelin's autograph and dated 25 November 1672, is preserved tipped into item 3 (Paris, AdS, Fonds Bourdelin, Carton 3, 10 vols., vol. 6, between 4 and 5).

11. Lot no. 29 in *Notice des principaux articles de la bibliothèque de feu M. L. Cl. Bourdelin . . . la vente se fera en la maniere accoutumée, au plus offrant & dernier enchérisseur, le jeudi 23 octobre, & jours suivants, à trois heures de relevée, en sa maison, rue Mazarine* (Paris: Didot, 1777), 27.

documents, both in Bourdelin's own hand, bound together. The first does contain accounts of experiments performed by Bourdelin between March 1667 and November 1668 (pp. 1–42), but also has recipes for a universal medicine and potable gold recorded in 1676 (pp. 45–70). The second document contains more experiments, dating from March 1667 to April 1669, sometimes performed before the assembly and often ordered by Duclos (pp. 1*–49*).

NAF MSS 5134–5146 are Bourdelin's autograph laboratory notebooks running from 1672 to 1699. They are copies of the official volumes now in Cartons 1 and 2 of the Fonds Bourdelin at the Académie des Sciences Archives. Their secondary status is signaled by the fact that the verification signatures are written in the same hand as the entries (rather than being original autographs) and the notes from Duhamel, Duclos, and others are rendered here in Bourdelin's own handwriting. Bourdelin made these copies probably as his own personal records, separate from the official records; accordingly, they remained in his family after his death. Many of these volumes are now in a poor state of conservation that prevents their being made available to readers without special request; however, the same material in the same order is available in the originals of the Fonds Bourdelin at the Académie des Sciences.

NAF MS 5147 is a unique and extremely valuable record, again in Bourdelin's own hand, of the expenses for the chymical laboratories. It bears the title "Memoire de la despence faitte par Monsieur Bourdelin pour le Laboratoire de l'Academie Royalle des sciences." The first dated entry is 6 March 1667; it is preceded by two folios of undated entries enumerating "startup" materials going back to the origins of the Académie's chymical experimentation.

NAF MS 5148 has been the most useful document of the series for this study. It is a running account kept by Claude Bourdelin the younger (1667–1711) of the meetings of the Académie from the time of his admission in 1699 until 1709, and is cited here as the "Bourdelin Diary." It preserves rich details of the meetings, which were not recorded in the more formal procès-verbaux, such as arguments between academicians and candid evaluations of the quality of various presentations. It sheds new light on the way the meetings were actually run and the variety of tasks carried out at the biweekly meetings. It shows, for example, that some meetings were given over entirely to reading correspondence that had been received. It occasionally mentions guests who were admitted to the otherwise closed meetings. It indicates how voting was done, how academicians were assigned specific functions, and how questions sent to the Académie by the crown (for instance, about gunpowder) or by important persons (for instance, about amber) were farmed out to individual academicians and how their reports were gathered, collated, and synthesized into a unified response. It mentions the materials and objects brought in by academicians and others for display to the assembly, and reveals that much more time was devoted to the examination of proposals and claims submitted to the Académie for evaluation than the procès-verbaux would suggest. In short, as an eye-opening supplement to the procès-verbaux, it deserves to be much more closely studied by those writing about the Académie.

NAF MS 5149 presents a miscellany of materials and is adequately described in the BNF catalogue. Among the most useful of these items is the inventory, in Bourdelin's hand, of the Académie's chymical laboratories (21r–34r) carried out upon the death of Jacques Borelly: "Memoire de tout ce qui a esté trouvé aux laboratoires apres la mort de Monsieur Borelly desquels Monsieur de la Chapelle en a donné les clefs a Bourdelin le mecredy[sic] 23ᵉ jour de novembre 1689."[12]

3. PARIS, ADS, FONDS BOURDELIN, CARTON 3, 10 VOLS.

This carton holds ten volumes, in cardboard covers, containing the material from the original notebooks in Cartons 1 and 2, but reworked and rearranged. The first four volumes contain Bourdelin's plant analyses reshuffled into alphabetical order (vol. 1, C–L; vol. 2, A–C; vol. 3, S–V [plus analyses of animal substances, waters, and other substances]; vol. 4, M–R). Volumes 5 and 7–10 contain copies of Bourdelin's laboratory reports but often not in any strict chronological or other obvious order. Volume 6 is a miscellany that would bear closer scrutiny. It begins (pp. 1–49) with an attempt by an unidentified person to work through the results of the plant analyses. Thereafter, some original laboratory notes in Bourdelin's hand and parts of letters to Dodart are bound in, as well as other items from 1673–75 copied out of the official register in an unidentified hand. These volumes witness attempts to organize or rework Bourdelin's records; Alice Stroup suggests that these are the result of Denis Dodart's work on collating and summarizing the results of the plants project.[13]

4. PARIS, ADS, 1J18, 3 VOLS.

These three heavy folio volumes, bound in leather with ornate gilding, represent a mid-eighteenth-century reworking of Bourdelin's records. The first

---

12. The inventory is transcribed and published in Pierre Chabbert, "Jacques Borelly (16. .–1689): Membre de l'Académie des Sciences," *Revue d'histoire des sciences* 23 (1970): 203–27.

13. Alice Stroup, *A Company of Scientists: Botany, Patronage, and Community at the Seventeenth-Century Parisian Royal Academy of Sciences* (Berkeley: University of California Press, 1990), 339.

volume bears the title "Analyses chymiques de l'Academie mises en order par Mr. Duhamel [du Monceau], 1749"; the flyleaf explains that "Ce recueil manuscrit, composé de 10 parties reliées de memes a eté refondu en 3 volumes in-folio manuscrits, reliés en veau." Here Bourdelin's plant analyses are listed in alphabetical order, beginning with "Abricots," and so these volumes seem to derive from the first four volumes of Carton 3 of the Fonds Bourdelin.

5. ITHACA, NY, CORNELL UNIVERSITY LIBRARY, MSS BD. LAVOISIER QD B76++, 6 VOLS.

Yet another version of Bourdelin's notebooks is preserved at Cornell University, having arrived there along with a collection of materials relating to Lavoisier.[14] This version exists in six folio volumes, entitled "Analises chymiques de Monsieur Bourdelin commencées le trante aoust 1672 et finies le deuxieme septembre 1699." The first five volumes contain the plant analyses in alphabetical order (beginning with "Abricots"), while the sixth contains the "leftovers" (work on animal substances, waters, and other materials)—a division similar to the first four volumes of item 3, above. A prefatory note to the first volume, written in a different hand than the rest of the manuscript, specifies that "this collection of analyses which has been given the form of a dictionary has been drawn and transcribed from the ten original manuscript volumes in which the analyses are found arranged according to the time when they were made and verified in the Académie Royale des Sciences."[15] What remains unclear at this point is how the three folios at the Académie (1J18) relate to the six folios at Cornell. The two sets contain the same material organized in the same order, but it is not clear which is earlier. The inside front cover of the first volume at Cornell bears an autograph note from Jean-Étienne Guettard (1715–86) stating that he purchased these six volumes on 19 January 1770 at the "sale of the manuscripts of the library of M. S. M." held at the "salle des Augustins."[16]

14. On this collection, see David W. Corson and L. Pearce Williams, "La collection Lavoisier à l'Université Cornell," in *Il y a 200 ans Lavoisier*, ed. Christiane Demeulenaere-Douyère (Paris: Tec & Doc, 1995), 229–34.

15. "Ce receuïl d'analises auquel on a donné une forme de dictionnaire a esté tiré et transcrit de dix volumes manuscrits et originaux dans lesquelles ces analises se trouvent placeés selon les tems où elles ont esté faites et verifiées dans l'academie Royale des sciences."

16. Guettard's note must certainly refer to the sale described in *Catalogue des livres de M. *** dont la vente se fera au plus offrant & dernier enchérisseur mardi 17 janvier & jours suivans, trois heures de relevée, dans une des salles des RR. PP. Augustins du grand couvent* (Paris, [1769]). The

Jean Hellot's Notebooks

A persistent problem with studying many of the early academicians is the scanty survival of their personal papers, laboratory notebooks, and correspondence. Part of the problem is that only rarely were such documents given directly to any institutional archive upon the death of their owners; instead, they were considered the property of the heirs. Sometimes the documents were sold relatively quickly to interested parties. In other cases the heirs (or the purchasers) charged fees for the privilege of consulting the papers. Eventually these materials were either auctioned off (and thereafter only rarely found their way into an archive, as occurred with the Bourdelin materials now in the BNF) or simply discarded when they were considered to have no further monetary value. Before resorting too quickly to pessimism or resignation to the vagaries of historical preservation, I should mention that in the course of writing this book, I found that archival mining and "fishing expeditions" did succeed in discovering a substantial number of important manuscripts either thought long lost or never suspected to have existed, and often in highly unexpected places.[17] So there is room to hope that more such documents will be recovered by further diligent searches in the future.

One crucial way in which at least a sampling of otherwise lost early materials has been preserved is through copies made by later academicians of their predecessors' papers. For example, Samuel Cottereau Duclos's vitriolic and illuminating 1676 critique of Denis Dodart's editorship of the *Histoire des plantes* survives thanks only to a copy made by Antoine de Jussieu (1686–1758) in the first half of the eighteenth century; this copy found its way with Jussieu's papers into the Bibliothèque du Muséum National d'Histoire Naturelle.[18] The academician who seems to have collected and copied his predecessors' papers more than any other is Jean Hellot (1685–1766). Hellot apparently sought out these papers wherever they resided, sometimes noting where he found them—often enough in the hands of heirs. Virtually all of the originals have now vanished, but many of Hellot's voluminous notebooks composed from them do survive, particularly in two collections, both of which have contributed a wealth of new and highly significant information to this study.

---

identity of M.S.M. remains unclear, as the sale seems to be a composite sale organized by the bookseller (Prault *fils*) rather than an auction of a single collector.

17. See, for example, chapter 4 of the present book; and Lawrence M. Principe, "Sir Kenelm Digby and His Alchemical Circle in 1650s Paris: Newly Discovered Manuscripts," *Ambix* 60 (2013): 3–24.

18. MS 1278. This document was first identified by Stroup, *Company of Scientists*, 341; see her highly useful list of Académie-relevant manuscripts, 339–42.

1. HELLOT CAEN NOTEBOOKS

These manuscripts are preserved in Caen at the Bibliothèque Municipale under the shelfmark MS in-4to 171. Their existence and utility was first signaled
and their contents briefly described in 1966, and their value for historians
noted again in the 1970s.[19] Nevertheless, they have been used surprisingly little since that time, despite the light they shed on the Académie and the work
of its chymists, including Hellot, as well as on other matters. The manuscripts
are bound into nine volumes in pasteboard covers, and total well over *four
thousand* pages in Hellot's tiny but clear handwriting. A contemporaneous
endorsement records that the academician Philibert Trudaine de Montigny
(1733–77) bought them from Hellot's widow for the very substantial price of
6,000 livres; how they subsequently found their way to Caen remains unknown.[20] Hellot compiled these documents from the late 1740s until about
1764 (the latest date cited), judging by the years most commonly mentioned
in them.

Volumes 1–8 contain bundles of papers—sometimes coherent treatises,
sometimes miscellanies—originally labeled by Hellot with a letter of the alphabet, or a letter and a number. Several volumes include short printed pamphlets, drawings, plans, or manuscripts in other hands. Volume 9 is Hellot's
detailed index to the other volumes. From the index it is clear that some of
Hellot's original material was not bound up into the current volumes, and
is therefore missing from the collection. For example, the index references
bundles or notebooks labeled "L," "S," "T," "X," and "Y," which are described as being "in 12°" (bundles with these letters do not exist in the quarto
volumes 1–8), as well as four others, labeled "E," "F," "G," and "R," also "in

---

19. Guy Thuillier and Arthur Birembaut, "Une source inédite: les cahiers du chimiste Jean
Hellot (1685–1766)," *Annales: Histoire, Sciences Sociales* 21 (1966): 357–64. Doru Todériciu used
these manuscripts in writing a 1975 dissertation from which he published two articles: "Le Traité
de chimie inachevé de Jean Hellot," *Physis* 19 (1977): 355–75; and "La bibliothèque d'un savant
chimiste et technologue parisien du XVIIIe siècle: livres et manuscrits de Jean Hellot (1685–
1766)," *Physis* 18 (1976):198–216. These latter two publications should be approached with some
caution since they contain frequent errors of transcription and some unlikely hypotheses. For
example, Todériciu asserts confidently that Hellot could not *possibly* have had any serious interest in transmutational alchemy ("Bibliothèque," 202–3 and note 16) but provides no evidence
for this claim, which leaves unexplained Hellot's almost obsessive copying and organizing of
voluminous materials on transmutation and his ownership of dozens of chrysopoetic books and
manuscripts. For a less programmatic view, see note 23, below.

20. Details about the manuscripts and their provenance appear in Gaston Lavalley, *Catalogue des manuscrits de la Bibliothèque Municipale de Caen* (Caen, 1880), 205–8, where they are
listed as MS 425.

12°" with contents different from the extant quarto bundles bearing those letters.[21] Thus, Hellot must have maintained an additional collection of papers
or notebooks in duodecimo format that apparently did not arrive in Caen
with the quartos. Furthermore, the index frequently references a manuscript
of roughly 300 pages simply called "Chym.," which is also not to be found
among the nine volumes of Caen manuscripts. Judging from its substantial
contents listed in the index, it would be of particular interest if it still exists
and if it ever reappears.[22]

2. HELLOT ARSENAL NOTEBOOKS

These notebooks are preserved at the Bibliothèque de l'Arsenal in Paris as
MSS 3006, 3007, and 3008. I identified them for the first time as Hellot's (and
in his handwriting) in 2010.[23] Although they are in duodecimo format, they
do not correspond to the missing duodecimo bundles mentioned in the Caen
index. A considerable number of other Hellot manuscripts also exist at the
Arsenal. Like the Caen MSS, these three volumes preserve Hellot's excerpts
from the manuscripts and oral reports of a range of both earlier and contemporaneous Académie chymists (Homberg, the Geoffroys, Danty d'Isnard,
Dufay, and so on), as well as from other now-lost or unidentifiable manuscripts (such as the "MSc Provençal"), some of which seem to have been in
the possession of his colleagues at the Académie. These volumes are of an earlier date than the Caen notebooks; the latest dates cited in these manuscripts
are from the 1730s. Again unlike the Caen MSS, these three documents are
more rigorously organized in their entirety (see chapter 7). The great majority of the entries deal with transmutational chymistry.

21. If Hellot created quarto bundles labeled "S" and "T," they should have been bound
into volume 8 between bundles "R" and "V." But I have found no reference in the index to
bundles bearing those two letters.

22. Todériciu, "Traité de chimie inachevé," 370–75 usefully collects all of the references
to "Chym." in the index and compiles them into a "table of contents" for the lost manuscript.
I am skeptical however of his conclusion that "Chym." represents an unfinished "textbook of
chymistry." Although its contents were clearly organized, the topics are too idiosyncratic for a
typical mid-eighteenth-century textbook, and furthermore Hellot's Arsenal MSS also display
considerable organization showing that Hellot put some of his collections into stricter order
without any view towards publication.

23. I used these Hellot Arsenal MSS (as well as the Hellot Caen MSS) extensively in Lawrence M. Principe, "The End of Alchemy? The Repudiation and Persistence of Chrysopoeia at
the Académie Royale des Sciences in the Eighteenth Century," *Osiris* 29 (2014): 96–116. Part of
this paper is reprised (with revisions) toward the end of chapter 7.

3. HELLOT MAZARINE MSS

Finally, it should be mentioned in passing that another cache of Hellot manuscripts exists at the Bibliothèque Mazarine. They are currently scattered through the collection, most of them not identified as having belonged to Hellot. Nevertheless, the former catalogue numbers fall into a sequence from 2571 to 2578, indicating that they arrived at the library as a coherent set of materials. These documents contain both alchemical treatises and botanical materials. These manuscripts have not been used materially in this study; they require further examination.[24]

24. The manuscripts are now listed as MSS 2576, 2755, 2756, 3586, 3587, 3682, 3683, and 3690. MS 3625 (formerly 2579) is also botanical and might have been part of the acquisition. It is possible that not all of this collection was actually owned by Hellot, but that some of the items were joined to the three unambiguously Hellot MSS by a subsequent owner before they came to the library.

Sources Cited

Archival Sources

FRANCE

Bordeaux, Université de Bordeaux 3—Lettres
Caen, Bibliothèque Municipale de Caen, MS in-4to 171, 9 vols.
Orléans, Médiathèque, MS 1037
Paris, Archives de l'Académie des Sciences
    Fonds Bourdelin, Cartons 1–2, 11 vols., and Carton 3, 10 vols.
    Dossiers biographiques
    Pochettes de séances
    Procès-verbaux
    MS 1J18, 3 vols.
Paris, Archives Nationales
    G/7/468, G/7/470, G/7/472, G/7/473, G/7/474, G/7/475, G/7/477, G/7/1435, G/7/1438, G/7/1463,
    O/1/27, O/1/30, O/1/31, O/1/43, O/1/50, O/1/52, O/1/56, O/1/62, O/1/72, O/1/630, O/1/2124,
    PP//151, X/1A/8682, Y//281, Y//11647; Minutier Central, LX-205, LX-206, LXXIII-639,
    LXXIII-643
Paris, Bibliothèque du Muséum National d'Histoire Naturelle
    MSS 1278 and 1279
Paris, Bibliothèque Interuniversitaire de Santé
    Pharmacie MS 19 and Médecine MS 5022
Paris, Bibliothèque Nationale de France
    MSS français 1333, 9132, 12309, 17051, and 23225; MSS NAF 31, 4302, and 5134–49; MSS Clai-
    rambault 452 and 566
Paris, Bibliothèque de l'Arsenal
    MSS 2064, 2517, 3006, 3007, 3008, 10548, 10555, 10582, 10538, 10588, 10590, 10599

GERMANY

Dresden, Sächsische Staatsarchiv, Geheimes Kabinett
Hannover, Niedersächische Landesbibliothek, Leibniz Briefe
Leipzig, Universitätsbibliothek, Carpzov Correspondence
Magdeburg, Landeshauptarchiv Sachsen-Anhalt

ITALY

Padua, Università di Padova, Archives

NETHERLANDS

Amsterdam, Stadsarchief, Archief van de Burgerlijke Stand
Den Haag, Nationaal Archief, Genealogical Section

POLAND

Wrocław, Wrocław University Library, Akc 1948/0562[1]

RUSSIA

St. Petersburg, Voenno-Meditsinskoi Akademii (Military-Medical Academy)
    Boerhaave Archive, MSS 128 and 130

SWEDEN

Uppsala, Uppsala University Library, Waller MSS

SWITZERLAND

Basel, Universitätsbibliothek, Bernoulli Correspondence

UNITED KINGDOM

London, British Library, Sloane MSS
London, Royal Society of London Archives
London, Wellcome Institute Library, Manuscripts and Autograph Collection

---

1. This important collection of letters to Tschirnhaus was the one previously housed in Gör-
litz, whose contents were described in Curt Reinhardt and Richard Jecht, "Über die Verfasser
der Briefe an Ehrenfried Walther von Tschirnhaus aus der Sammelung der Oberlausitzischen
Gesellschaft der Wissenschaften," *Neues Lausitzisches Magazin* 116 (1940): 100–108.

UNITED STATES

Ithaca, NY, Cornell University Library, Special Collections, Lavoisier Bound Manuscripts QD B76++

Printed Sources

Abot de Bazinghen, François. *Traité des monnoies et de la jurisdiction de la cour des monnoies.* 2 vols. Paris, 1764.

Abou-Nemeh, Catherine. "The Natural Philosopher and the Microscope: Nicolas Hartsoeker Unravels Nature's 'Admirable Oeconomy.'" *History of Science* 51 (2013): 1–28.

———. "Réaumur's Crayfish Experiments in Hartsoeker's *Système.*" In *The Life Sciences in Early Modern Philosophy*, ed. Ohad Nachtomy and Justin E. H. Smith, 157–80. Oxford: Oxford University Press, 2014.

Anon. "A Factitious Stony Matter or Paste, Shining in the Dark like a Glowing Coal." *Philosophical Transactions* 11 (1676–77): 788–89.

Anstey, Peter. "The Coherence of Cohesion in the Later Leibniz." *British Journal for the History of Philosophy* 24 (2016): 594–613.

Arber, Edward, ed. *Term Catalogues, 1668–1709.* 3 vols. London, 1903–6.

Archives of the Bastille. Ed. François Ravaisson. 19 vols. Paris, 1866–1904.

Augarde, Jean-Dominique. "The Scientific Cabinet of Comte d'Ons-en-Bray and a Clock by Domenico Cucci." *Cleveland Studies in the History of Art* 8 (2003): 80–95.

Azzolini, Monica. "The Political Uses of Astrology: Predicting the Illness and Death of Princes, Kings and Popes in the Italian Renaissance." *Studies in History and Philosophy of Biological and Biomedical Sciences* 41 (2010): 136–45.

Balduin, Christian Adolf. *Aurum superius et inferius aurae superioris et inferioris hermeticum.* Amsterdam, 1675.

Barker, Emma. "Mme Geoffrin, Painting and *Galanterie*: Carle Van Loo's *Conversation espagnole* and *Lecture espagnole.*" *Eighteenth-Century Studies* 40 (2007): 587–614.

Becher, Johann Joachim. *Physica subterranea.* Leipzig, 1703.

Beck, Wolfgang. "Michael Maiers Examen Fucorum Pseudo-chymicorum: eine Schrift wider die falschen Alchemisten." PhD diss., Technische Universität München, 1992.

Bedini, Silvio. "Seventeenth Century Italian Compound Microscopes." *Physis* 5 (1963): 383–422.

Beguin, Jean. *Les elemens de chymie.* Paris, 1620.

Bekker, Hans. "Bemerkungen zu den Dr.-Dissertationen des Wilhelm Homberg." *Wissenschaftliche Zeitschrift der Technischen Hochschule "Otto von Guericke" Magdeburg* 31 (1987): 63–64.

Bensaude-Vincent, Bernadette, and Christine Lehman. "Public Lectures of Chemistry in Mid-Eighteenth-Century France." In *New Narratives in Eighteenth-Century Chemistry*, ed. Lawrence M. Principe, 77–96. Dordrecht: Springer, 2007.

Beretta, Marco. "Rinman, Diderot, and Lavoisier: New Evidence Regarding Guillaume-François Rouelle's Private Laboratory and Chemistry Course." *Nuncius* 26 (2011): 355–79.

———. "Transmutations and Frauds in Enlightened Paris: Lavoisier and Alchemy." In *Fakes!? Hoaxes, Counterfeits and Deception in Early Modern Science*, ed. Marco Beretta and Maria Conforti, 69–107. Sagamore Beach, MA: Science History Publications, 2014.

Bernard, August. *Histoire de l'Imprimerie royale au Louvre.* Paris, 1867.

Bernoulli, Johann. "Nouvelle manière de rendre les Baromètres lumineux." *MARS* (1700): 178–90.

Bertucci, Paola. *Artisanal Enlightenment: Science and the Mechanical Arts in Old Regime France.* New Haven, CT: Yale University Press, 2017.

Bibliotheca Carpzoviana, sive Catalogus librorum. Leipzig, 1700.

Birembaut, Arthur, Pierre Costabel, and Suzanne Delorme. "La correspondance Leibniz-Fontenelle et les relations de Leibniz avec l'Académie Royale des Sciences en 1700–1701." *Revue d'histoire des sciences et de leurs applications* 19 (1966): 115–32.

Bléchet, Françoise. "L'abbé Bignon, président de l'Académie royale des sciences: un demi-siècle de direction scientifique." In *Règlement, usages et science dans la France de l'absolutisme.*, ed. Christiane Demeulenaere-Douyère and Éric Brian, 51–69. Paris: Lavoisier Tec & Doc, 2002.

Boantza, Victor D. "Alkahest and Fire: Debating Matter, Chymistry, and Natural History at the Early Parisian Academy of Sciences." In *The Body as Object and Instrument of Knowledge: Embodied Empiricism in Early Modern Science*, ed. Charles T. Wolfe and Ofer Gal, 75–92. Dordrecht: Springer, 2010.

———. *Matter and Method in the Long Chemical Revolution.* Burlington, VT: Ashgate, 2013.

Boas Hall, Marie. "Acid and Alkali in Seventeenth-Century Chemistry." *Archives internationales d'histoire des sciences* 9 (1956): 13–28.

———. "Frederick Slare, F.R.S." *Notes and Records of the Royal Society of London* 46 (1992): 23–41.

———. "Homberg, Guillaume." *Dictionary of Scientific Biography*, ed. Charles Coulston Gillispie. 18 vols. New York: Scribners, 1970–90.

———. *Robert Boyle and Seventeenth-Century Chemistry.* Cambridge: Cambridge University Press, 1958.

Bodemann, Eduard. *Aus der Briefe der Herzogin Elisabeth Charlotte von Orléans an die Kurfürstin Sophie von Hannover.* 2 vols. Hannover, 1891.

———. "Briefwechsel zwischen Leibniz und der Herzogin Elisabeth Charlotte von Orleans 1715/16." *Zeitschrift des historischen Vereins für Niedersachsen* 49 (1884): 1–66.

Boerhaave, Herman. *Elementa chimiae.* 2 vols. Paris, 1733.

———. "Sur le mercure." *MARS* 36 (1734): 539–52.

Bogel, Else, and Elger Blühm. *Die deutsche Zeitungen des 17. Jahrhunderts.* 3 vols. Bremen: Schünemann Universitätsverlag, 1971.

Boislisle, A. M., ed. *Correspondance des contrôleurs généraux des finances.* 3 vols. Paris, 1874–97.

Bond, Henry. *The Longitude Found.* London, 1676.

Borrichius, Olaus. *De ortu et progressu chemiae.* Copenhagen, 1668.

———. "Efficere, ut duo spiritus tactu frigidi, invicem confusi flammam edant." *Acta medica et philosophica Hafniensia* 1 (1671–72): 133–35.

———. *Itinerarium 1660–1665.* Ed. H. D. Schepelern. 4 vols. Copenhagen: Danish Society of Language and Literature, 1983.

Böttcher, Hans-Joachim. *Ehrenfried Walther von Tschirnhaus: Das bewunderte, bekämpfte, und totgeschwiegene Genie.* Dresden: Dresdener Buchverlag, 2014.

Bougard, Michel. *La chimie de Nicolas Lemery.* Turnhout: Brepols, 1999.

Boulduc, Simon. "Analyse d'Ypecacuanha." *MARS* (1700): 1–5.

———. "De la calcination de l'antimoine." PV 17, 112r–113v (26 February 1698).

Bouvet, Maurice. "Les laboratoires autorisés au XVIIe siècle." *Bulletin de la Société d'Histoire de la Pharmacie* 13 (1925): 10–16, 55–60.

Boyle, Robert. *Correspondence of Robert Boyle.* Ed. Michael Hunter, Antonio Clericuzio, and Lawrence M. Principe. 6 vols. London: Pickering & Chatto, 2001.

———. "New Experiments Concerning the Relation Between Light and Air." *Philosophical Transactions* 2 (1668): 581–600, 605–12.

———. "Of the Incalescence of Quicksilver with Gold." *Philosophical Transactions* (1676): 515–33.

———. *Works of Robert Boyle.* Ed. Michael Hunter and Edward B. Davis. 14 vols. London: Pickering & Chatto, 1999–2000.

Breger, Herbert. "Notiz zur Biographie des Phosphor-Entdeckers Henning Brand." *Studia Leibnitiana* 19 (1987): 68–73.

Bret, Patrice. "Les chimistes à l'Académie Royale des Sciences à l'époque des Lémery (1699–1743)." *Revue d'histoire de la pharmacie* 64 (2016): 385–404.

Brian, Éric, and Christiane Demeulenaere-Douyère, eds. *Histoire et Mémoire de l'Académie des Sciences: Guide des recherches.* Paris: Tec & Doc, 1996.

Briggs, Robin. "The Académie Royale des Sciences and the Pursuit of Utility." *Past and Present* 131 (1991): 38–88.

Brodsley, Laurel, Charles Frank, and John W. Steeds. "Prince Rupert's Drops." *Notes and Records of the Royal Society of London* 41 (1986): 1–26.

Brown, Harcourt. *Scientific Organizations of Seventeenth-Century France.* Baltimore: Williams & Wilkins Co., 1934.

Bryden, D. J. "Magnetic Inclinatory Needles Approved by the Royal Society?" *Notes and Records of the Royal Society of London* 47 (1993): 17–31.

Butterfield, Herbert. *The Origins of Modern Science.* New York: Macmillan, 1959.

Cap, Paul-Antoine. *Études biographiques pour servir à l'histoire des sciences.* 2 vols. Paris, 1857–64.

———. "Guillaume Homberg, naturaliste, 1652–1715." *Journal de pharmacie et de chimie* 3d ser., 44 (1863): 406–18.

Catalogue des livres de feu M. Hellot. Paris, 1766.

Catalogus librorum Stephani-Francisci Geoffroy. Paris, 1731.

Cellio, Marco Antonio. *Il fosforo, o'vero la pietra bolognese.* Rome, 1680.

Chabbert, Pierre. "Jacques Borelly (16 . .–1689): Membre de l'Académie des Sciences." *Revue d'histoire des sciences* 23 (1970): 203–27.

Champault, Philippe. "Les Gendron 'médecins des rois et des pauvres.'" *Transactions of the Royal Society of Canada* 2d ser., 6 (1912): 35–120.

Chang, Kevin (Ku-Ming). "Communications of Chemical Knowledge: Georg Ernst Stahl and the Chemists at the French Academy of Sciences in the First Half of the Eighteenth Century." *Osiris* 29 (2014): 135–57.

———. "Fermentation, Phlogiston and Matter Theory: Chemistry and Natural Philosophy in Georg Ernst Stahl's *Zymotechnia fundamentalis.*" *Early Science and Medicine* 7 (2002): 53–57.

———. "From Oral Disputation to Written Text: The Transformation of the Dissertation in Early Modern Europe." *History of Universities* 19 (2004): 129–87.

———. "Phlogiston and Chemical Principles: The Development and Formulation of Georg Ernst Stahl's Principle of Inflammability." In *Bridging Traditions: Alchemy, Chemistry and Paracelsian Practices in the Early Modern Era: Essays in Honor of Allen G. Debus*, ed. Karen Hunger Parshall, Michael T. Walton, and Bruce T. Moran, 101–30. Kirksville, MO: Truman State University Press, 2015.

Charas, Moyse. *Nouvelles experiences sur la vipère.* Paris, 1669.

―――. "Le sel volatile de tartre." AdS, pochette de séances 1692.

Chaussinand-Nogaret, Guy. *Le cardinal Dubois, 1656–1723: une certaine idée de l'Europe.* Paris: Perrin, 2000.

Chevreul, Michel-François. Review of L. P. François Cambriel's *Cours de philosophie hermétique.* *Journal des savants* (May 1851): 284–98; (June 1851): 337–52; (August 1851): 492–506; (December 1851): 752–68.

Ciampini, Giovanni Giustino. *Nuove inventioni di tubi ottici dimostrate nell'Accademia Fisico-matematica Romana l'anno 1686.* Rome, 1687.

Clément, Pierre, ed. *Lettres, instructions, et mémoires de Colbert.* 8 vols. Paris, 1861–70.

Clericuzio, Antonio. *Elements, Principles, and Corpuscles: A Study of Atomism and Chemistry in the Seventeenth Century.* Dordrecht: Springer, 2000.

Cohen, Ernst, and W. A. T. Cohen-De Meester. "Katalog der wiedergefundenen Manuskripte und Briefwechsel von Herman Boerhaave." *Verhandelingen der Nederlandsche Akademie van Wetenschappen, Afdeelig Naturkunde* 40 (1941): 1–45.

Collection Feuardent: Jetons et méreaux, vente aux enchères publiques. Paris, 1928.

Colonna, Francesco Maria Pompeo. *Abrégé de la doctrine de Paracelse.* Paris, 1724.

―――. *Les secrets les plus cachés de la philosophie.* Paris, 1722.

Comiers, Claude. "L'Homme artificiel anemoscope, ou Prophete physique des changements du temps." *Mercure galant* (March 1683): 164–214.

Condorcet, Marie-Jean-Antoine-Nicolas Caritat, Marquis de. "Éloge de M. Malouin." *MARS* (1778): 57–65.

―――. *Éloges des Académiciens de l'Académie Royale des Sciences, morts depuis 1666 jusqu'en 1699.* Paris, 1773.

Contant, Jean-Paul. *L'enseignement de la chimie au Jardin Royal des Plantes.* Cahors: Coueslan, 1952.

Corson, David W., and L. Pearce Williams. "La collection Lavoisier à l'Université Cornell." In *Il y a 200 ans Lavoisier,* ed. Christiane Demeulenaere-Douyère, 229–34. Paris: Tec & Doc, 1995.

Costa, Gustavo. "Un Collaboratore italiano del Conte di Boulainviller: Francesco Maria Pompeo Colonna (1644–1726)." *Atti e memorie dell'Accademia Toscano di Scienze e Lettere* 29 (1964): 207–95.

D'Alencé, Joachim. *Traité des baromètres, thermomètres et notiomètres ou hygromètres.* Amsterdam, 1688.

Debus, Allen G. "Fire Analysis and the Elements in the Sixteenth and Seventeenth Centuries." *Annals of Science* 23 (1967): 128–47.

―――. "French Alchemy in the Early Enlightenment." In *Ésotérisme, gnoses, et imaginaire symbolique,* ed. Richard Caron, Joscelyn Godwin, Wouter J. Hanegraaff, and Jean-Louis Veillard-Baron, 47–59. Leuven: Peeters, 2001.

―――. "Solution Analyses prior to Robert Boyle." *Chymia* 8 (1962): 41–61.

De Clave, Étienne. *Nouvelle lumiere chymique* (1641). Ed. Bernard Joly. Fayard: Paris, 2000.

De la Hire, Philippe. "De la pesanteur de l'air." PV 14, fols. 243r–250r (26 January 1696).

Delorme, Susan. "Des éloges de Fontenelle et de la psychologie des savants." In *Mélanges G. Jamati,* 91–100. Paris: CNRS, 1956.

Demachy, Jacques-François. *Receuil des dissertations.* Amsterdam, 1774.

Demeulenaere-Douyère, Christiane, and Éric Brian, eds. *Règlement, usages et science dans la France de l'absolutisme.* Paris: Lavoisier Tec & Doc, 2002.

Demeulenaere-Douyère, Christiane, and David J. Sturdy. *L'Enquête du Régent 1716–1718: sciences, techniques et politique dans la France pré-industrielle.* Turnhout: Brepols, 2008.

———. "Image versus Reality: The Archives of the French Académie des Sciences." In *Archives of the Scientific Revolution: The Formation and Exchange of Ideas in Seventeenth-Century Europe*, ed. Michael Hunter, 185–208. Woodbridge, UK: Boydell, 1998.

Denna werldennes största tänckwärdigheeter eller dhe så kallade Relationes curiosae. Stockholm, 1682.

De Rochas, Albert. "L'or alchimique." *La Nature* (1886): 339–43.

Des Chene, Dennis. "Life after Descartes: Régis on Generation." *Perspectives on Science* 11 (2003): 410–20.

De Seilhac, Victor. *L'abbe Dubois, premier ministre de Louis XV.* 2 vols. Paris, 1862.

Dictionnaire de l'Académie Françoise. Paris, 1694.

Donato, Maria Pia. *Accademie romane: una storia sociale, 1671–1824.* Naples: Edizioni Scientifiche Italiane, 2000.

Dooley, Brendan. *Morandi's Last Prophecy and the End of Renaissance Politics.* Princeton, NJ: Princeton University Press, 2002.

Dorveaux, Paul. "Apothicaires membres de l'Académie Royale des Sciences: II. Moyse Charas." *Bulletin de la Société d'Histoire de la Pharmacie* 17 (1929): 329–40.

———. "Apothicaires membres de l'Académie Royale des Sciences: III. Simon Boulduc." *Revue d'histoire de la pharmacie* 1 (1930): 5–15.

———. "Les Boulduc, apothicaires de la Princesse Palatine." *Revue d'histoire de la pharmacie* 21 (1933): 110–11.

———. "Jean Grosse, médecin allemand, et l'invention de l'éther sulfurique." *Bulletin de la Société d'Histoire de la Pharmacie* 17 (1929): 182–87.

Drévillon, Herve. *Lire et écrire l'avenir: l'astrologie dans la France du Grand Siècle (1610–1715).* Seyssel: Champ Vallon, 1996.

Du Cerf, Claude. *An hominis primordia, vermis?* Paris, 1704.

Duclos, Samuel Cottereau. *Dissertation sur les principes des mixtes naturels.* Amsterdam, 1680.

———. "Experiences de l'augmentation du poids." PV 1, 40–52 (January 1667).

Dufay, Charles François de Cisternay. "Mémoire sur un grand nombre de phosphores nouveaux." *MARS* (1730): 524–35.

Duhamel, Jean-Baptiste. *Regiae scientiarum academiae historia.* Paris, 1698. 2d ed. Paris, 1701.

Duncan, Alistair. *Laws and Order in Eighteenth-Century Chemistry.* Oxford: Oxford University Press, 1996.

Durbec, J. A. "L'Alchimiste de Saint-Auban." *Annales de la Société Scientifique et Littéraire de Cannes et de l'Arrondissment de Grasse* 15 (1957–61): 131–87.

Edit du roy pour la punition de different crimes qui sont devins, magiciens, sorciers, empoisoneurs. Paris, 1682.

Ehrenfried Walter von Tschirnhaus (1651–1708)—Experimente mit dem Sonnenfeuer. Dresden: Staatliche Kunstsammlungen, 2001.

Erler, Georg, ed. *Die Jüngerer Matrikel der Universität Leipzig, 1559–1809.* 3 vols. Leipzig: Giesecke & Devrient, 1909.

L'Etat de la France. 2 vols. Paris, 1712.

Favino, Federica. "Beyond the 'Moderns'? The Accademia Fisico-matematica of Rome (1677–1698) and the Vacuum." *History of Universities* 23 (2008): 120–58.

Felix, Fred W. "Moyse Charas, maître apothicaire et docteur en médecine." *Revue d'histoire de la pharmacie* 90 (2002): 63–80.

Feuardent, Félix. *Jetons et méreaux.* 4 vols. Paris: Rollin & Feuardent, 1904–15.

Figuier, Louis. *L'alchimie et les alchimistes*. Paris, 1856.

Fontenelle, Bernard de. "Éloge de l'abbé Louvois." *HARS* (1718): 101–4.

———. "Éloge de M. Chirac." *HARS* (1732): 120–30.

———. "Éloge de M. Dodart." *HARS* (1707): 182–92.

———. "Éloge de M. Dufay." *HARS* (1739): 73–83.

———. "Éloge de M. Geoffroy." *HARS* (1731): 93–100.

———. "Éloge de M. Hartsoeker." *HARS* (1725): 137–53.

———. "Éloge de M. Homberg." *HARS* (1715): 82–93.

———. "Éloge de M. Lemery." *HARS* (1715): 73–82.

———. "Éloge de M. Poli." *HARS* (1714): 129–34.

———. "Éloge de M. Thuillier." *HARS* (1704): 139.

———. "Éloge de M. Tournefort." *HARS* (1708): 143–54.

———. "Éloge du P. Sébastien Truchet." *HARS* (1729): 93–101.

———. "Éloge de M. Tschirnaus." *HARS* (1709): 114–24.

———. *Oeuvres diverses*. 3 vols. Paris, 1724.

Fors, Hjalmar. *The Limits of Matter: Chemistry, Mining, and Enlightenment*. Chicago: University of Chicago Press, 2015.

Fors, Hjalmar, Lawrence M. Principe, and H. Otto Sibum. "From the Library to the Laboratory and Back Again: Experiment as a Tool for Historians of Science." *Ambix* 63 (2016): 85–97.

France, Peter. "From Eulogy to Biography: The French Academic Éloge." In *Mapping Lives: The Uses of Biography*, ed. Peter France and William St. Clair, 83–102. Oxford: Oxford University Press, 2002.

Franckowiak, Rémi. "Du Clos, un chimiste post-*Sceptical Chymist*." In *La philosophie naturelle de Robert Boyle*, ed. Miriam Dennehy and Charles Ramond, 361–77. Paris: Vrin, 2009.

Franckowiak, Rémi, and Luc Peterschmitt. "La chimie de Homberg: une vérité certaine dans une physique contestable." *Early Science and Medicine* 10 (2005): 65–90.

Frémontier-Murphy, Camille. "La construction monarchique d'un lieu neutre: l'Académie Royale des Sciences au Palais du Louvre." In *Règlement, usages et science dans la France de l'absolutisme*, ed. Christiane Demeulenaere-Douyère and Éric Brian, 169–203. Paris: Lavoisier Tec & Doc, 2002.

Furetière, Antoine. *Dictionaire universel*. 3 vols. The Hague, 1690.

Gardair, Jean-Michel. *Le Giornale de' letterati di Rome, 1668–1681*. Florence: Olschki, 1984.

Geber. *Summa perfectionis*. Ed. William R. Newman. Leiden: Brill, 1990.

Geoffroy, Étienne-François. "Des supercheries concernant la pierre philosophale." *MARS* (1722): 61–70.

———. "Des Teintures des Métaux et particulièrement des teintures d'or." PV 32 (15 March 1713).

———. "Eclaircissemens sur la production artificielle du Fer, & sur la composition des autres Métaux." *MARS* (1707): 176–88.

———. "Eclaircissemens sur la Table inserée dans les *Memoires* de 1718." *MARS* (1720): 20–34.

———. "Experiences sur les metaux faites avec le verre ardent du Palais Royal." *MARS* (1709): 162–76.

———. "Manière de faire l'alun de roche." PV 21, fols. 11r–19v.

———. "Maniere de recomposer le Souffre commun par la réunion de ses principes." *MARS* (1704): 278–86.

———. "Observations sur les dissolutions & sur les fermentations que l'on peut appeler froides." *MARS* (1700): 110–21.

———. "Problème de chimie: Trouver des cendres qui ne contiennent aucunes parcelles de fer." PV 24, fols. 393r–395r. Published with significant alterations in *MARS* (1705): 362–63.

———. "Table des differens rapports observés en chimie entre differentes substances." *MARS* (1718): 202–12.

———. *Tractatus de materia medica.* 3 vols. Paris, 1741.

Gerhardt, Carl Immanuel. *Der Briefwechsel von Gottfried Wilhelm Leibniz mit Mathematikern.* Berlin, 1899.

———, ed. *Die philosophische Schriften von Gottfried Wilhelm Leibniz.* 7 vols. Berlin, 1887.

Gerland, E. "Die Sammlung von astronomischen, geodätischen und physikalischen Apparaten aus dem 16., 17., 18. Jahrhundert des Königlichen Museum in Kassel." *Repertorium für Experimental-Physik* 12 (1876): 362–75.

Gjörwell, Carl Christoffer. *Det Swenska Biblioteket.* 2 vols. Stockholm, 1757–62.

Glanius, W. *A New Voyage to the East Indies.* London, 1682.

Glaser, Christophle. *Traité de la chymie.* Paris, 1663.

Golinski, Jan V. "A Noble Spectacle: Phosphorus and the Public Culture of Science in the Early Royal Socety." *Isis* 80 (1989): 11–39.

Goulard, Roger. "À propos de l'affaire des poisons: le célèbre édit de 1682." *Bulletin de la Société Française de l'Histoire de la Médecine* 13 (1914): 260–66.

———. "Un mystère à la Bastille: Etienne Vinache, médecin empirique et alchimiste." *Bulletin de la Société Française de l'Histoire de la Médecine* 14 (1920): 360–72.

Goupil, Michelle. *Du flou au clair? Histoire de l'affinité chimique.* Paris: CTHS, 1991.

Grandjean de Fouchy, Jean-Paul. "Éloge de M. Hellot." *HARS* 68 (1766): 167–79.

———. "Éloge de M. d'Onsenbray." *HARS* (1754): 143–54.

———. "Éloge de M. Rouelle." *HARS* (1770): 137–49.

Gröste Denckwürdigkeiten der Welt, oder so genannt Relationes curiosae. Hamburg, 1683.

Die grösten Denkwürdigkeiten dieser Welt, oder so genante Relationes Curiosae. Hamburg: Thomas von Wiering, 1681.

Guenon, Anne-Sylvie. "Les publications de l'Académie des Sciences." In *Histoire et mémoire de l'Académie des Sciences: Guide des recherches,* ed. Christine Demeulenaere-Douyère and Éric Brian, 107–12. Paris: Tec & Doc, 1996.

Guerlac, Henry. *Lavoisier: The Crucial Year.* Ithaca, NY: Cornell University Press, 1961.

———. *Newton on the Continent.* Ithaca, NY: Cornell University Press, 1981.

Guerrini, Anita. *The Courtiers' Anatomists: Animals and Humans in Louis XIV's Paris.* Chicago: University of Chicago Press, 2015.

Guglielmini, Domenico. *De principio sulphureo dissertationes.* Venice, 1710.

———. *De salibus dissertatio epistolaris.* Venice, 1705.

Haase, Gisela. "Tschirnhaus und die sächsischen Glashütten in Pretzsch, Dresden und Glücksberg." In *Ehrenfried Walter von Tschirnhaus (1651–1708)—Experimente mit dem Sonnenfeuer,* 55–67. Dresden: Staatliche Kunstsammlungen, 2001.

Hagendijk, Thijs. "Learning a Craft from Books: Historical Re-enactment of Functional Reading in Gold- and Silversmithing." *Nuncius* 33 (2018): 198–235.

Halleux, Robert. "L'alchimiste et l'essayeur." In *Die Alchemie in der europaeischen Kultur- und Wissenschaftsgeschichte,* ed. Christoph Meinel, 277–91.Wiesbaden: Otto Harrassowitz, 1986.

Halleux, Robert, James McClellan, Daniela Berariu, and Geneviève Xhayet. *Les publications de l'Académie Royale des Sciences (1666–1793)*. 2 vols. Turnhout: Brepols, 2001.

Hannaway, Owen. *Chemists and the Word*. Baltimore: Johns Hopkins University Press, 1975.

Harris, John. *Lexicon technicum*. 2 vols. London, 1704 and 1710.

Hartsoeker, Nicolas. *Cours de physique*. The Hague, 1730.

———. *Eclaircissemens sur les Conjectures physiques*. Amsterdam, 1710.

———. *Principes de physique*. Paris, 1696.

Harvey, E. Newton. *A History of Luminescence*. Philadelphia: American Philosophical Society, 1957.

Heilbron, John L. *Electricity in the Seventeenth and Eighteenth Centuries: A Study in Early Modern Physics*. Berkeley: University of California Press, 1979.

Hellman, S., ed. *Aus den Briefen der Herzogin Elisabeth Charlotte von Orléans an Étienne Polier de Bottens*. Bibliothek des litterarischen Vereins in Stuttgart 231 (1903).

Hellot, Jean, "Analise du zinc. Premier mémoire." *MARS* (1735): 12–31.

Helvetius, Johann Friedrich. *Vitulus aureus*. Amsterdam, 1667.

Hensing, Johann Thomas. "Dissertation sur la pierre philosophale." In *Mémoires littéraires*, ed. Marc Antoine Eidous, 121–54. Paris, 1750.

Hillebrand, Werner, ed. *Die Matrikel der Universität Helmstedt*. 3 vols. Hildesheim: Verlag August Lax, 1981.

Hiscock, W. G. *David Gregory, Isaac Newton and Their Circle: Extracts from David Gregory's Memoranda 1677–1708*. Oxford: Printed for the Editor, 1937.

Hjärne, Urban. *Själfbiografi*. Ed. Henrik Schück. Äldre Svenska Biografier 5, Uppsala Universitet Årsskrift. Uppsala: Almqvist & Wiksells, 1916.

Hoffman, Klaus. *Johann Friedrich Böttger: Von Alchemistengold zum weißen Porzellan*. Berlin: Verlag Neues Leben, 1985.

Hoffmann, Friedrich. *Demonstrationes physicae curiosae*. Halle, 1700.

Hoffmann, Friedrich Wilhelm. *Otto von Guericke, Bürgermeister der Stadt Magdeburg*. Magdeburg, 1874.

Holland, Wilhelm Ludwig, ed. *Briefe der Herzogin Elisabeth Charlotte von Orléans aus dem Jahre 1720*. Bibliothek des litterarischen Vereins in Stuttgart 144 (1879).

Holmes, Frederic L. "Analysis by Fire and Solvent Extractions: The Metamorphosis of a Tradition." *Isis* 62 (1971): 129–48.

———. "Chemistry in the Académie Royale des Sciences." *Historical Studies of the Physical and Biological Sciences* 34 (2003): 41–68.

———. "The Communal Context for Etienne-François Geoffroy's 'Table des rapports.'" *Science in Context* 9 (1996): 289–311.

———. *Eighteenth-Century Chemistry as an Investigative Enterprise*. Berkeley, CA: Office of the History of Science and Technology, 1989.

———. "Investigative and Pedagogical Styles in French Chemistry at the End of the Seventeenth Century." *Historical Studies of the Physical and Biological Sciences* 34 (2004): 277–309.

Homberg, Andreas. *Disputatio inauguralis de fracturis cranii*. Wittenberg, 1673.

———. *Exercitatio medica de cephalgia*. Helmstedt, 1672.

Homberg, Wilhelm. "Analise du souffre commun." PV 22, fols. 85r–88r (21 March 1703).

———. "De l'effet du Siphon dans un lieu vuide d'air." PV 33, fol. 155r–158v (12 May 1714). Originally "De l'effet du Siphon dans un lieu vuide d'air." PV 17, fols. 358v–362r (20 August 1698).

———. *Disputatio juridica de diffidationibus, vulgo vom Befehden.* Leipzig, 1676.

———. "Eclaircissemens touchant la vitrification de l'or au verre ardant." *MARS* (1707): 40–48.

———. "Essais d'elemens de la chimie." PV 21, 61r–73v (15 February 1702).

———. "Essay de l'analyse du souffre commun." *MARS* (1703): 31–40.

———. "Essays de chimie." *MARS* (1702): 33–52.

———. "Essay sur l'adoucissement des acides." PV 16, fols. 40r–43r (6 March 1697).

———. "Essays sur l'analyse des sels des plantes." PV 16, fols. 194r–196r (17 July 1697).

———. "Experience de l'evaporation de l'eau dans le vuide, avec des réflexions." *HMARS 1666–99*, 10:319–23.

———. "Experiences sur la glace dans le vuide." *MARS* (28 February 1693): 19–25. Reprinted in *HMARS 1666–99*, 10:255–62.

———. "Experiences sur la piere de bologne." AdS, pochette de séances 1694.

———. "Maniere de copier sur le verre coloré les pierres gravées." *MARS* (1712): 189–97.

———. "Maniere de faire le phosphore brûlant de Kunkel." *MARS* (30 April 1692). Reprinted in *HMARS 1666–99*, 10:84–90.

———. "Maniere d'extraire un sel volatile acide minérale en forme séche." *MARS* (15 December 1692): 171–76. Reprinted in *HMARS 1666–99*, 10:202–8.

———. "Memoire touchant la volatilisation des sels fixes des plantes." *MARS* (1714): 186–95.

———. "Nouvelle maniere de distiller sans aucun chaleur." PV 16, fols. 126v–129v (15 May 1697).

———. "Nouveau phosphore." *MARS* (31 December 1693): 187–91. Reprinted *HMARS 1666–99*, 10:445–48.

———. "Observation sur la quantité exacte des sels volatiles acides contenus dans tous les differens esprits acides." *MARS* (1699): 44–51.

———. "Observation sur un battement de veines semblable au battement des arteres." *MARS* (1704): 159–73.

———. "Observation sur une separation de l'Or d'avec l'Argent par la Fonte." *MARS* (1713): 67–70.

———. "Observation sur une sublimation de mercure." *MARS* (1713): 265–67.

———. "Observations et considerations touchant les causes des vents." PV 27, fols. 113r–121v (18 April 1708).

———. "Observations faites par le miroir ardent du Palais Royal." PV 21, fols. 227r–231r (3 June 1702).

———. "Observations faites par le moyen du Verre ardant." *MARS* (1702): 141–49.

———. "Observations sur cette sorte d'insects qui s'appellent ordinairement Demoiselles." *MARS* (1699): 145–51.

———. "Observations sur des matières qui pénetrent & qui traversent les métaux sans les fondre." *MARS* (1713): 306–13.

———. "Observations sur l'acide du souffre." AdS, pochette de séances 1694.

———. "Observations sur l'acide qui se trouve dans le sang." *MARS* (1712): 8–15.

———. "Observations sur la addoucissement des acides." PV 18, fols. 97r–99v (28 January 1699).

———. "Observations sur la matiere fecale." *MARS* (1711): 39–46.

———. "Observations sur la quantité d'acides absorbées par les alcalis terreux." *MARS* (1700): 64–71.

———. "Observations sur la quantité exacte des sels volatiles acides." *MARS* (1699): 44–51.

———. "Observations sur le different poids d'un mesme volume d'air selon qu'il est plus ou moins dilaté par les differens dégres de chaleur." PV 14, fols. 296v–298v (14 March 1696).

———. "Observations sur le fer au verre ardent." MARS (1706): 158–65.

———. "Observations sur le rafinage de l'argent." MARS (1701): 40–43.

———. "Observations sur le sel urineux des plantes." PV 17, fols. 40r–43r (27 November 1697).

———. "Observations sur les analyses des plantes." MARS (1701): 113–17.

———. "Observations sur les araignées." MARS (1707): 339–52.

———. "Observations sur les dissolvans du mercure." MARS (1700): 190*–96*.

———. "Observations sur les matieres sulphureuses & sur la facilité de les changer d'une espece de souffre en une autre." MARS (1710): 225–34.

———. "Observations sur les sels fixes des Plantes." PV 18, fols. 37v–39v (3 December 1698).

———. "Observations sur quelques effets de fermentations." MARS (1701): 95–99.

———. "Observations sur quelques effets d'une legére brulure." PV (1708), fols. 85r–88v.

———. "Observations sur quelques effets que l'or produit seul dans le grand feu." PV 14, fols. 274r–277v (22 February 1696).

———. "Observations sur une dissolution de l'argent." MARS (1706): 102–7.

———. "Phosphore nouveau, ou suite des observations sur la matiere fecale." MARS (1711): 238–45.

———. "Reflexions sur differentes vegetations metalliques." Mémoires de mathematique et de physique (30 November 1692): 145–52.

———. "Reflexions sur l'expérience des larmes de verre qui se brisent dans le vuide." MARS (31 December 1692): 183–87. Reprinted in HMARS 1666–99, 10:215–20.

———. "Sel qui travers le fer sans le fondre." PV 14, fols. 212r–213r (14 December 1695).

———. "Suite de l'article trois des essays de chimie." MARS (1706): 260–72.

———. "Suite des essays de chimie: Article troisième: Du souphre principe." MARS (1705): 88–96.

———. "Suite des essais de chimie: Art. IV. du Mercure." MARS (1709): 106–17.

———. "Suite des Expériences du Miroir ardent." PV 21, 399r–401v (15 November 1702).

———. "Suite des observations faites au Miroir ardent." PV 21, 341r–343r (12 August 1702).

———. "Suite des observations sur l'acide qui se trouve dans le sang." MARS (1712): 267–75.

———. "Suite des observations sur une infusion d'antimoine." AdS, pochette de séances 1694.

———. "Sur la lumière du Mercure dans le vuide." PV 20, fols. 28r–34r (26 January 1701).

Hooykaas, Reijer. "Domenico Guglielmini et le développement de la cristallographie." Atti della Fondazione Ronchi 8 (1953): 5–20.

Howarth, Richard J. "Fitting Geomagnetic Fields before the Invention of Least Squares: Henry Bond's Predictions (1636, 1668) of the Change in Magnetic Declination in London." Annals of Science 59 (2002): 391–408.

Huard, Pierre. "L'autopsie de Guillaume Homberg (1652–1715)." In 89e Congrès des societés savantes, 155–56. Lyons: Imprimerie Nationale, 1964.

Hufbauer, Karl. The Formation of the German Chemical Community. Berkeley: University of California Press, 1982.

Hunter, Michael. Boyle: Between God and Science. New Haven, CT: Yale University Press, 2009.

———. Boyle Studies: Aspects of the Life and Thought of Robert Boyle. Burlington, VT: Ashgate, 2015.

Huygens, Christiaan. Oeuvres complètes. 22 vols. The Hague: Nijhoff, 1888–1950.

Imbert, Henri. "Histoire d'un assoupissement extraordinaire." *MARS* (1713): 313–17.

Ince, Joseph. "Ambrose Godfrey Hanckwitz." *Pharmaceutical Journal* 18, ser. 1 (1858): 126–30, 157–62, 215–22.

Index biographique de l'Académie des Sciences. Paris: Gauthier-Villars, 1979.

Jacob, Paul L. [Paul Lacroix]. *La chambre des poisons: Histoire du temps de Louis XIV (1712).* 2 vols. Paris: Victor Magen, 1839.

Jacques, Jean. "Le cours de chimie de G.-F. Rouelle recuelli par Diderot." *Revue d'histoire des sciences* 38 (1985): 43–53.

Jammes, André. "La réforme de la typographie royale sous Louis XIV." Paris: Paul Jammes, 1961. English version: "Académisme et Typographie: The Making of the Romain du Roi." *Journal of the Printing Historical Society* 1 (1965): 71–95.

Jauernig, Reinhold, and Marga Steiger, eds. *Die Matrikel der Universität Jena.* 5 vols. Weimar: Hermann Böhlaus Nachfolger, 1977.

Jensen, William B. "The Origin of the Term 'Base.'" *Journal of Chemical Education* 83 (2006): 1130.

———. "Whatever Happened to Homberg's Pyrophorus?" *Bulletin of the History of Chemistry* 3 (1989): 21–24.

Jolibois, Jean-François. *Histoire de la ville et du canton de Trévoux.* Lyons, 1853.

Joly, Bernard. "À propos d'une querelle concernant la production artificielle du fer. Les divergences entre Nicolas Lémery et son fils Louis." *Revue de l'histoire de la pharmacie* 64 (2016): 375–84.

———. "L'alkahest, dissolvant universel, ou quand la théorie rend pensible une pratique impossible." *Revue de l'histoire des sciences* 49 (1996): 308–30.

———. "L'anti-Newtonianisme dans la chimie française au début du XVIIIe siècle." *Archives internationales d'histoire des sciences* 53 (2003): 213–24.

———. "Could a Practicing Chemical Philosopher Be a Cartesian?" In *Cartesian Empiricisms,* ed. M. Dobre and T. Nyden, 125–48. Dordrecht: Springer, 2013.

———. "De l'alchimie à la chimie: le développement des *Cours de chymie* au XVIIe siècle." In *Aspects de la tradition alchimique au XVIIe siècle,* ed. Frank Grenier, 85–94. Textes et travaux de chrysopoeia 4. Paris: S.É.H.A. and Milan: Archè, 1998.

———. *Descartes et la chimie.* Paris: Vrin, 2011.

———. "Etienne-François Geoffroy (1672–1731), a Chemist on the Frontiers." *Osiris* 29 (2014): 117–31.

———. "Le mécanisme et la chimie dans la nouvelle Académie Royale des Sciences: les débats entre Louis Lémery et Étienne-François Geoffroy." *Methodos* 8 (2008). Special issue: *Chimie et mécanisme à l'âge classique.* http://methodos.revues.org/1403.

———. "Quarrels between Etienne-François Geoffroy and Louis Lémery at the Académie Royale des Sciences in the Early Eighteenth Century: Mechanism and Alchemy." In *Chymists and Chymistry: Studies in the History of Alchemy and Early Modern Chemistry,* ed. Lawrence M. Principe, 203–14. Sagamore Beach, MA: Science History Publications, 2007.

———. "La thèorie des cinq éléments d'Étienne de Clave dans la *Nouvelle Lumière Philosophique.*" Dossier Étienne de Clave. *Corpus: Revue de philosophie* 39 (2001): 9–44.

Journal du Marquis de Dangeau. Ed. Eudore Soulié et al. 19 vols. Paris, 1854–60.

———. 35 vols. Clermont-Ferrand: Éditions Paleo, 2002.

Juntke, Fritz, ed. *Album Academiae Vitebergensis, Jüngere Reihe.* 4 vols. Halle: Selbstverlag der Universitäts- und Landesbibliothek, 1952.

Kahn, Didier. "L'alchimie sur la scène française aux XVIe et XVIIe siècles." *Chrysopoeia* 2 (1988): 62–96.

———, ed. *Le Comte de Gabalis, ou Entretiens sur les sciences secrètes*. Paris: Honoré Champion, 2010.

———. *Le fixe et le volatile*. Paris: CNRS Éditions, 2016.

———. "Helisaeus Röslin, Joseph Du Chesne et la doctrine des cinq éléments et principes." In *Nouveau ciel, nouvelle terre: la révolution copernicienne dans l'Allemagne de la Réforme*, ed. Edouard Mehl and Miguel Granada, 339–54. Paris: Belles Lettres, 2009.

———. "The Significance of Transmutation in Alchemy: The Case of Thurneysser's Half-Gold Nail." In *Fakes!? Hoaxes, Counterfeits and Deception in Early Modern Science*, ed. Marco Beretta and Maria Conforti, 35–68. Sagamore Beach, MA: Science History Publications, 2014.

Kapp, Volker. "Les qualités du scientifique et le prestige des sciences dans les éloges académiques de Fontenelle." In *Fontenelle: Actes du colloque tenu a Rouen*, ed. Alain Niderst, 441–53. Paris: Presses Universitaire de France, 1989.

Keller, Vera. "Hermetic Atomism: Christian Adolf Balduin (1632–1682), *Aurum aurae*, and the 1674 Phosphor." *Ambix* 61 (2014): 366–84.

Kent, Andrew, and Owen Hannaway. "Some New Considerations on Beguin and Libavius." *Annals of Science* 16 (1960): 241–50.

Key, Emil. *Försök till svenska tidningspressen historia*. Stockholm, 1883.

Khunrath, Heinrich. *Trewhertzige Warnungs-Vermahnung*. Magdeburg, 1597.

Kim, Mi Gyung. *Affinity, That Elusive Dream*. Cambridge, MA: MIT Press, 2003.

———. "Chemical Analysis and the Domains of Reality: Wilhelm Homberg's *Essais de Chimie*, 1702–1709." *Studies in the History and Philosophy of Science* 31 (2000): 37–69.

———. "The 'Instrumental' Reality of Phlogiston." *Hyle* 14 (2008): 27–51.

Kirchner, Joachim. *Das deutsches Zeitschriftenwesen: Seine Geschichte und seine Probleme*. 2 vols. Wiesbaden: Harrassowitz, 1958–62.

Kirchvogel, Paul Adolf. "Der Atlas von Falun." *Kunst in Hessen und am Mittelrhein* 22 (1982): 71–75.

Klein, Joel A. "Corporeal Elements and Principles in the Learned German Chymical Tradition." *Ambix* 61 (2014): 345–65.

Klein, Ursula. "The Chemical Workshop Tradition and the Experimental Practice: Discontinuities within Continuities." *Science in Context* 9 (1996): 251–87.

———. "E. F. Geoffroy's Table of Different 'Rapports' Observed between Different Chemical Substances—A Reinterpretation." *Ambix* 42 (1995): 79–100.

Krafft, Fritz. *Otto von Guericke*. Darmstadt: Wissenschaftliche Buchgesellschaft, 1978.

Kraft, Alexander. "On the Discovery and History of Prussian Blue." *Bulletin for the History of Chemistry* 33 (2008): 61–67.

Krampl, Ulrike. "Diplomaten, Kaufleute und ein Mann 'obskurer' Herkunft." In *Nützliche Netzwerke und korrupte Seilschaften*, ed. Arne Karsten and Hillard von Thiessen, 137–62. Göttingen: Vandenhoek & Ruprecht, 2006.

———. *Les secrets de faux sorciers: police, magie, et escroquerie à Paris au XVIIIe siècle*. Paris: Éditions EHESS, 2011.

Krekel, Christoph. *Chemische und kulturhistorische Untersuchungen des Buchmalereifarbstoffs folium und weiterer Inhaltsstoffe aus Chrozophora tinctoria und Mercurialis perennis*. PhD diss., Munich, 1996.

Kristiansson, Sture. *Strömningar till och från Store Kopparberget*. Filipstad: Bronells, 1997.

Krul, R. "Jean-Frédéric Helvétius et sa famille." *Janus* 1 (1896): 564–71.

Kunckel, Johann. *Collegium physico-chymicum experimentale.* Hamburg and Leipzig, 1716.

——. *Öffentliche Zuschrifft von dem Phosphoro mirabili.* Leipzig, 1678.

Kunze, Alfred. "Lebensbeschreibung des Ehrenfried Walther von Tschirnhaus auf Kieslingswalde und Würdigung seiner Verdienste." *Neues Lausitzisches Magazin* 43 (1866): 1–40.

Lafond, Louis. *La dynastie des Helvétius: Les remèdes du roi.* Paris, 1926.

Lafont, Olivier. "Échevins & apothicaires sous Louis XIV: la vie de Matthieu-François Geoffroy, bourgeois de Paris." Paris: Parmathèmes, 2008.

Laissus, J. "Les eaux merveilleuses de Balaruc." *Revue d'histoire de la pharmacie* 53 (1965): 367–77.

Lavalley, Gaston. *Catalogue des manuscrits de la Bibliothèque Municipale de Caen.* Caen, 1880.

Lavoisier, Antoine-Laurent. "Expériences sur la combinaison de l'alun avec les matières charbonneuses." *MARS* (1777): 363–72.

——. *Oeuvres.* 6 vols. Paris: Imprimerie Impériale, 1862–93.

Lebigre, Arlette. *L'affaire des poisons, 1679–1682.* Brussels: Éditions Complexe, 1989.

Le Cat, Claude Nicolas. "A Memoir on the *Lacrymae Batavicae.*" *Philosophical Transactions* 46 (1749–50): 175–88.

Lehman, Christine. "Alchemy Revisited by the Mid-Eighteenth Century Chemists in France: An Unpublished Manuscript by Pierre-Joseph Macquer." *Nuncius* 28 (2013): 165–216.

——. "Innovation in Chemistry Courses in France in the Mid-Eighteenth Century." *Ambix* 57 (2010): 3–26.

——. "Mid-Eighteenth-Century Chemistry in France as Seen through Student Notes from the Courses of Gabriel-François Venel and Guillaume-François Rouelle." *Ambix* 56 (2009): 163–89.

——. "Pierre-Joseph Macquer: Chemistry in the French Enlightenment." *Osiris* 29 (2014): 245–61.

——. "What Is the 'True' Nature of Diamond?" *Nuncius* 31 (2016): 361–407.

Lehman, Christine, and François Pépin, eds. *La chimie et l'*Encyclopédie. *Corpus, revue de philosophie* 56 (2009).

Leibniz, Wilhelm Gottfried. "Historia inventionis phosphori." *Miscellanea Berolinensia* 1 (1710): 91–98.

——. *Sämtliche Schriften und Briefe.* Berlin: Akademie Verlag, 1923–2015.

Lémery, Jean. "Experiences sur la diversité des matieres qui sont propres à faire un phosphore avec l'alun." *MARS* (1714): 402–8.

——. "Reflexions physiques sur un nouveau phosphore." *MARS* (1715): 23–41.

Lémery, Louis. "Conjectures & reflections sur la matiere du feu ou de la lumiere." *MARS* (1709): 400–418.

——. "Explication mechanique . . . de la dissolution des differents sels." *MARS* (1716): 154–72.

——. "Premiere memoire sur le nitre." *MARS* (1717): 31–51.

——. "Quatrième memoire sur les analises ordinaires des plantes et des animaux." *MARS* (1721): 22–44.

——. "Que les plantes contiennent reellement du fer." *MARS* (1706): 411–18.

——. "Reflexions physiques sur le défaut & le peu d'utilité des analises ordinaires des plantes & des animaux." *MARS* (1719): 173–88.

——. "Seconde memoire sur le nitre." *MARS* (1717): 122–46.

——. "Second memoire sur les analises ordinaires de chimie." *MARS* (1720): 98–107.

——. "Sur la volatilisation vraye ou apparente des sels fixes." *MARS* (1717): 246–56.

————. "Troisième memoire sur les analises de chimie." *MARS* (1720): 166–78.

Lémery, Nicolas. *Cours de chymie*. Paris, 1675. 2d ed. Paris, 1677. 5th ed. Paris, 1683. 7th ed. Paris, 1690.

————. "Explication physique et chimique des feux sousterrains, des Tremblement de terre, des ouragans, des eclairs et du tonnerre." *MARS* (1700): 101–10.

Lenglet de Fresnoy, Nicolas. *Histoire de la philosophie hermétique*. 3 vols. Paris, 1742.

Lettres des religieuses de P. R. des Champs . . . touchant les Bulles de nôtre S. P. le Pape Clement XI du 27 mars 1708. Paris, 1709.

Lindroth, Sten. *Christopher Polhem och Stora Kopparberget*. Uppsala: Almqvist & Wiksells, 1951.

————. "Urban Hjärne och Laboratorium Chymicum." *Lychnos* 57–58 (1946–47): 51–116.

Lister, Martin. *Journey to Paris in the Year 1698*. 3d ed. London, 1699. Reprint. Ed. Raymond Phineas Stearnes. Urbana: University of Illinois Press, 1967.

Loibl, Werner. "Ehrenfried Walther von Tschirnhaus und der frühneuzeitliche Glasguss in Sachsen." *Neues Lausitzisches Magazin* 135 (2013): 65–96.

López Pérez, Miguel. "El alquimista Dubois y el Cardenal Richelieu." *Azogue* 7 (2010–13): 327–38.

————. "La trágica historia de Jean Delisle." *Azogue* 7 (2010–13): 402–48.

Lubienietz, Stanislao de. *Theatri cometici*. 3 vols. Amsterdam, 1667–68.

Lundstedt, Bernhard. *Sveriges periodiska litteratur*. 2 vols. Stockholm: Aktiebolag, 1895. Reprint. Stockholm: Rediviva, 1969.

Lunel, Alexandre. *La maison médicale du roi*. Seysell: Champ Vallon, 2008.

Lux, David S. "Colbert's Plan for the *Grande Académie*: Royal Policy towards Science, 1663–67." *Seventeenth-Century French Studies* 12 (1990): 177–88.

Macquer, Pierre-Joseph. *Dictionnaire de chimie*. 2 vols. Paris, 1766.

————. *Dictionnaire de chimie*. 4 vols. Paris, 1778.

————. *Élémens de chymie-théorique*. Paris, 1756.

————. "Mémoire . . . sur des expériences faites en commun, au foyer des grands lentilles de Tchirnhausen." In *Introduction aux observations sur la physique*, 2:612–16. 2 vols. Paris, 1777.

Maddison, R. E. W. "Studies in the Life of Robert Boyle, F.R.S.: Part I, Robert Boyle and His Foreign Visitors." *Notes and Records of the Royal Society* 9 (1951): 1–35.

————. "Studies in the Life of Robert Boyle, F.R.S.: Part V, Boyle's Operator Ambrose Godfrey Hanckwitz." *Notes and Records of the Royal Society of London* 11 (1955): 159–88.

Maier, Michael. *Examen fucorum pseudo-chymicorum detectorum et in gratiam veritatis amantium succincte refutatorum*. Frankfurt, 1617.

Mairan, Jean-Jacques Dortous de. *Dissertation sur la cause de la lumiere des phosphores*. Bordeaux, 1717.

————. "Éloge de M. Boulduc." *HARS* (1742): 167–71.

————. "Éloge de M. Hunauld." *HARS* (1742): 206–12.

Malouin, Paul-Jacques. *Traité de chimie*. Paris, 1734.

Mareschal de Bièvre, Gabriel. *Georges Mareschal, Seigneur de Bièvre*. Paris: Librairie Plon, 1906.

Mariotte, Edmé. *Oeuvres*. 2 vols. The Hague, 1740.

Marsak, Leonard M. *Bernard de Fontenelle: The Idea of Science in the French Enlightenment*. Transactions of the American Philosophical Society 49 (1959).

Matton, Sylvain. "Cartésianisme et alchimie: à propos d'un témoignage ignoré sur les travaux alchimiques de Descartes." In *Aspects de la tradition alchimique au XVIIe siècle*, ed. Frank Grenier, 111–84. Textes et travaux de chrysopoeia 4. Paris: S.É.H.A. and Milan: Archè, 1998.

Maupertuis, Pierre-Louis Moreau de. "Sur la pierre philosophale." In *Lettres de M. Maupertuis*, 81–85. Dresden, 1752.

Mauskopf, Seymour. "Reflections: 'A Likely Story.'" In *New Narratives in Eighteenth-Century Chemistry*, ed. Lawrence M. Principe, 177–93. Dordrecht: Springer, 2007.

Mayer, Jean. "Portrait d'un chimiste: Guillaume-François Rouelle." *Revue d'histoire des sciences et leurs applications* 23 (1970): 305–32.

Mazauric, Simone. *Fontenelle et l'invention de l'histoire de sciences à l'aube des Lumières*. Paris: Fayard, 2007.

Mémoires historiques et authentiques sur la Bastille. 3 vols. London, 1789.

Mémoires pour servir à l'histoire des plantes. Ed. Denis Dodart. Paris, 1676. Reprinted in *HMARS 1666–99*, 4:424–572.

Mesnard, Jean. *Pascal et les Roannez*. 2 vols. Paris: Desclée, 1965.

Metzger, Hélène. *Les doctrines chimiques en France du début du XVIIe siècle à la fin du XVIIIe siècle*. Paris: Les Presses Universitaires, 1923.

———. *Newton, Stahl, Boerhaave et la doctrine chimique*. Paris: Librairie Scientifique Albert Blanchard, 1930.

Middleton, W. E. Knowles. *The History of the Barometer*. Baltimore: Johns Hopkins University Press, 1964.

———. "Science in Rome, 1675–1700, and the Accademia Fisicomatematica of Giovanni Giustino Ciampini." *British Journal for the History of Science* 8 (1975): 138–54.

Miege, Guy. *A New Dictionary, French and English*. London, 1677.

Moewes, Erich. "Irrtümer und Unklarheiten: Bemerckungen zu Beschreibung der Guerickeschen Experimente." *Monumenta Guerickiana* 6 (1999): 41–50.

Monconys, Balthasar de. *Journal des voyages*. 3 vols. Lyons, 1665–66.

Mongin, Jean. *Le chimiste physicien*. Paris, 1704.

Montalbani, Ovidio. *De illuminabili lapide Bononiensi epistolae*. Bologna, 1634.

Montesquieu. *Oeuvres complètes*. 19 vols. Oxford: Voltaire Foundation, 1998.

Moran, Bruce T. *The Alchemical World of the German Court*. Sudhoffs Archiv 29 (1991).

———. *Chemical Pharmacy Enters the University*. Madison, WI: American Institute of the History of Pharmacy, 1991.

Morhof, Daniel Georg. *Polyhistor*. Lübeck, 1688.

Morin de Toulon. "Observations sur l'azur des Cendres bleuës." AdS, pochette de séances 1694.

Mothu, Alain. "L'alchimie en examen: *L'examen des principes des alchymistes sur la pierre philosophale* (1711) d'après les journaux de l'époque." *Chrysopoeia* 5 (1992–96): 739–50.

Mühlpfordt, Günter. *Ehrenfried Walther von Tschirnhaus (1651–1708)*. Leipzig: Leipziger Universitätsverlag, 2008.

Murphy, Antoin E. *John Law: Economic Theorist and Policy-Maker*. Oxford: Clarendon Press, 1997.

Muslow, Martin. "Philalethes in Deutschland: Alchemische Experimente am Gothaer Hof 1679–1683." In *Goldenes Wissen: Die Alchemie, Substanzen, Synthesen, Symbolik*, ed. Petra Feuerstein-Herz and Stefan Laube, 139–54. Wiesbaden: Harrassowitz, 2014.

Narbonne, Pierre. *Journal des règnes de Louis XIV et Louis XV*. Ed. J. A. Le Roi. Paris, 1866.

Neumann, Caspar. *Chymiae medicae dogmatico-experimentalis*. 6 vols. Züllichau, 1749–55.

Newman, William R. "Alchemical and Chymical Principles." In *The Idea of Principles in Early Modern Thought: Interdisciplinary Perspectives*, 77–97. Ed. Peter Anstey. New York: Routledge, 2017.

———. "Alchemical Symbolism and Concealment: The Chemical House of Libavius." In *The Architecture of Science*, ed. Peter Galison and Emily Thompson, 59–77. Cambridge, MA: MIT Press, 1999.

———. "Elective Affinity before Geoffroy: Daniel Sennert's Atomistic Explanation of Vinous and Acetous Fermentation." In *Matter and Form in Early Modern Science and Philosophy*, ed. Gideon Manning, 99–124. Leiden: Brill, 2012.

———. *Gehennical Fire: The Lives of George Starkey, An American Alchemist in the Scientific Revolution*. Cambridge, MA: Harvard University Press, 1994.

———. "Mercury and Sulphur among the High Medieval Alchemists: From Rāzī and Avicenna to Albertus Magnus and Pseudo-Roger Bacon." *Ambix* 61 (2014): 327–44.

———. *Newton the Alchemist: Science, Enigma, and the Quest for Nature's "Secret Fire."* Princeton, NJ: Princeton University Press, 2018.

———. "Newton's *Clavis* as Starkey's *Key.*" *Isis* 78 (1987): 564–74.

———. *Promethean Ambitions: Alchemy and the Quest to Perfect Nature*. Chicago: University of Chicago Press, 2004.

———, ed. and trans. *The Summa Perfectionis of the Pseudo-Geber*. Leiden: Brill, 1991.

Newman, William R., and Lawrence M. Principe. "Alchemy and the Changing Significance of Analysis." In *Wrong for the Right Reasons*, ed. Jed. Z. Buchwald and Allan Franklin, 73–89. Dordrecht: Springer, 2005.

———. *Alchemy Tried in the Fire: Starkey, Boyle, and the Fate of Helmontian Chymistry*. Chicago: University of Chicago Press, 2004.

———. "Alchemy vs. Chemistry: The Etymological Origins of a Historiographic Mistake." *Early Science and Medicine* 3 (1998): 32–65.

Newton, Isaac. *The Correspondence of Isaac Newton*. 7 vols. Cambridge: Cambridge University Press, 1959–77.

———. *Optice*. London, 1706.

Niceron, Jean-Pierre. *Mémoires des hommes illustres dans la république des lettres*. 42 vols. Paris, 1727.

Nicholson, Marjorie. "The Scientific Background to Swift's *Voyage to Laputa.*" *Annals of Science* 2 (1937): 299–334.

Niderst, Alain. *Fontenelle à la recherche de lui-même (1657–1702)*. Paris: Nizet, 1972.

———, ed. *Fontenelle: Actes du colloque tenu à Rouen*. Paris: Presses Universitaires de France, 1989.

Nierenstein, M. "Helvetius, Spinoza, and Transmutation." *Isis* 17 (1932): 408–11.

Nummedal, Tara. *Alchemy and Authority in the Holy Roman Empire*. Chicago: University of Chicago Press, 2007.

———. "The Problem of Fraud in Early Modern Alchemy." In *Shell Games: Studies in Scams, Frauds, and Deceits (1300–1650)*, ed. Mark Crane, Richard Raiswell, and Margaret Reeves, 37–58. Toronto: Centre for Reformation and Renaissance Studies, 2004.

Oldenburg, Henry. *Correspondence*. Ed. A. Rupert Hall and Marie Boas Hall. 13 vols. Vols 1–9. Madison: University of Wisconsin Press, 1965–73. Vols. 10–11. London: Mansell Press, 1975–77. Vols. 12–13. London: Taylor & Francis, 1986.

Olivier, Eugène. "Bernard Penot (Du Port), médicin et alchimiste (1519–1617)." *Chrysopoeia* 5 (1992–96): 571–673.

Olsson, Daniels Sven. *Falun Mine*. Falun: Stiftelsen Stora Kopparberget, 2010.

O Nauke i Uchenykh: Arkhiv Burgave v Voenno-Meditsinskoi Akademii. St. Petersburg: Voenno-Meditsinskoi Akademii, 2003.

Orain, Arnaud. *La politique du merveilleux: une autre histoire du système de Law (1695–1795)*. Paris: Fayard, 2018.

Pain, Stephanie. "The Winter of Incomparable Cold." *New Scientist* (7 February 2009): 46–47.

Papon, Jean-Pierre. *Histoire générale de Provence*. 4 vols. Paris, 1777–86.

Partington, James R. *A History of Chemistry*. 4 vols. London: Macmillan, 1961.

Paul, Charles B. *Science and Immortality: The Éloges of the Paris Academy of Sciences (1699–1791)*. Berkeley: University of California Press, 1980.

Pépin, François. "Fontenelle, l'Académie et le devenir scientifique de la chimie." *Methodos* 12 (2012). https://journals.openedition.org/methodos/2898.

Perrault, Charles. *Mémoires de ma vie*. Ed. Paul Bonnefon. Paris: Renouard, 1909.

Pertz, Georg Heinrich, ed. *Leibnizens gesammelte Werke*. 3d ser. 7 vols. Halle, 1859.

Peterschmitt, Luc. "The Cartesians and Chemistry: Cordemoy, Rohault, Régis." In *Chymists and Chymistry: Studies in the History of Alchemy and Early Modern Chemistry*, ed. Lawrence M. Principe, 193-202. Sagamore Beach, MA: Science History Publications, 2007.

———. "Fontenelle et la chimie: la recherche d'une 'loi fondamentale' pour la chimie." *Methodos* 12 (2012). https://journals.openedition.org/methodos/2873.

———. "Fontenelle, the Idea of Science and the Spirit of Chemistry." In *Chymia: Science and Nature in Early Modern Europe*, ed. Miguel Lopez-Perez, Didier Kahn, and Mar Rey Bueno, 367–85. Cambridge: Cambridge Scholars Publishing, 2010.

Petitfils, Jean-Christian. *L'affaire des poisons: alchimistes et sorciers sous Louis XIV*. Paris: Michel, 1977. Rev. ed. Paris: Perrin, 2010.

Peumery, Jean-Jacques. "Les Dodart, père et fils, médecins du roi." *Histoire des sciences médicales* 34 (2000): 39–46.

Philalethes, Eirenaeus [George Starkey]. *Introitus apertus ad occlusum regis palatium*. In *Museum hermeticum*. Frankfurt, 1678; reprint, Graz: Akademische Druck, 1970, 647–99.

———. *The Marrow of Alchemy*. London, 1654–55.

———. *Secrets Reveal'd, or An Open Entrance to the Shut Palace of the King*. London, 1669.

Pietsch, Ulrich. "Tschirnhaus und das europäische Porzellan." In *Ehrenfried Walter von Tschirnhaus (1651–1708)—Experimente mit dem Sonnenfeuer*, 68–74. Dresden: Staatliche Kunstsammlungen, 2001.

Piganiol de la Force, Jean-Aymar. *Description historique de la ville de Paris*. 10 vols. Paris, 1765.

Pluche, Noël Antoine. *Histoire du ciel*. 2 vols. Paris, 1739.

Poli, Martino. *Il trionfo degli acidi vendicati dalle calunnie di molti moderni*. Rome, 1706.

Porto, Paulo A. "*Summus atque felicissimus salium*: The Medical Relevance of the *Liquor Alkahest*." *Bulletin of the History of Medicine* 76 (2002): 1–29.

Pousse, François. *L'examen des principes des alchymistes sur la pierre philosophale*. Paris, 1711.

Powers, John C. "Fire Analysis in the Eighteenth-Century: Herman Boerhaave and Scepticism about the Elements." *Ambix* 61 (2014): 385–406.

———. *Inventing Chemistry: Herman Boerhaave and the Reform of the Chemical Arts*. Chicago: University of Chicago Press, 2012.

———. "Scrutinizing the Alchemists: Herman Boerhaave and the Testing of Chymistry." In *Chymists and Chymistry: Studies in the History of Alchemy and Early Modern Chemistry*, ed. Lawrence M. Principe, 227–38. Canton, MA.: Science History Publications/Chemical Heritage Foundation, 2007.

Principe, Lawrence M. *Alchemy and Chemistry: Breaking Up and Making Up (Again and Again).* Washington, DC: Dibner Library Publications, Smithsonian Institution, 2017.

———. "Apparatus and Reproducibility in Alchemy." In *Instruments and Experimentation in the History of Chemistry,* ed. Frederic L. Holmes and Trevor Levere, 55–74. Cambridge, MA: MIT Press, 2000.

———. *The Aspiring Adept: Robert Boyle and His Alchemical Quest.* Princeton, NJ: Princeton University Press, 1998.

———. "Chymical Exotica in the Seventeenth Century, or, How to Make the Bologna Stone." *Ambix* 63 (2016): 118–44.

———. "The Chymist and the Physician: Rivalry and Conflict at the Académie Royale des Sciences." In *Alchemy and Medicine from Antiquity to the Enlightenment,* ed. Jennifer M. Rampling and Peter M. Jones. London: Routledge, forthcoming 2020.

———. "The End of Alchemy? The Repudiation and Persistence of Chrysopoeia at the Académie Royale des Sciences in the Eighteenth Century." *Osiris* 29 (2014): 96–116.

———. "The Gold Process: Directions in the Study of Robert Boyle's Alchemy." In *Alchemy Revisited,* ed. Z. R. W. M. van Martels, 200–205. Leiden: Brill, 1990.

———. "Goldsmiths and Chymists: The Activity of Artisans in Alchemical Circles." In *Laboratories of Art: Alchemy and Art Technology from Antiquity to the Eighteenth Century,* ed. Sven Dupré, 157–79. Dordrecht: Springer, 2014.

———. "Newly-Discovered Boyle Documents in the Royal Society Archive: Alchemical Tracts and His Student Notebook." *Notes and Records of the Royal Society of London* 49 (1995): 57–70.

———. "Revealing Analogies: The Descriptive and Deceptive Roles of Sexuality and Gender in Latin Alchemy." In *Hidden Intercourse: Eros and Sexuality in Western Esotericism,* ed. Wouter Hanegraaff and Jeffrey J. Kripal, 209–29. Leiden: Brill, 2008.

———. "A Revolution Nobody Noticed? Changes in Early Eighteenth-Century Chymistry." In *New Narratives in Eighteenth-Century Chemistry,* ed. Lawrence M. Principe, 1–22. Dordrecht: Kluwer, 2007.

———. "Robert Boyle's Alchemical Secrecy: Codes, Ciphers, and Concealments." *Ambix* 39 (1992): 63–74.

———. *The Secrets of Alchemy.* Chicago: University of Chicago Press, 2013.

———. "Sir Kenelm Digby and His Alchemical Circle in 1650s Paris: Newly Discovered Manuscripts." *Ambix* 60 (2013): 3–24.

———. "Sir Kenelm Digby et son cercle alchimique parisien des années 1650." *Textes et travaux de chrysopoeia* 16 (2015):155–82.

———. "Wilhelm Homberg: Chymical Corpuscularianism and Chrysopoeia in the Early Eighteenth Century." In *Late Medieval and Early Modern Corpuscular Matter Theories,* ed. C. Lüthy, J. E. Murdoch, and W. R. Newman, 535–56. Leiden: Brill, 2001.

———. "Wilhelm Homberg et la chimie de la lumière." *Methodos* 8 (2008). https://journals .openedition.org/methodos/1223.

Quedlimburgica officina pharmaceutica. Quedlinburg, 1665.

Rampling, Jennifer M. *The Making of English Alchemy.* Chicago: University of Chicago Press, forthcoming 2020.

Rampling, Jennifer M., and Peter M. Jones, eds. *Alchemy and Medicine from Antiquity to the Enlightenment.* London: Routledge, forthcoming 2020.

Rappaport, Rhoda. "G.-F. Rouelle: An Eighteenth Century Teacher and Chemist." *Chymia* 6 (1960): 68–101.

Ray, John. *Philosophical Letters*. Ed. William Derham. London, 1718.

Réaumur, René-Antoine Ferchault de. "Idée generale des differentes manières dont on peut faire la Porcelaine." *MARS* (1727): 185–203.

Régis, Pierre-Sylvain. *Cours entier de la philosophie, ou Système générale selon les principes de M. Descartes*. 3 vols. Paris, 1690.

Reinhardt, Curt. "Tschirnhaus oder Böttger? Eine urkundliche Geschichte der Erfindung des Meissener Porzellans." *Neues Lausitzisches Magazin* 88 (1912).

Reinhardt, Curt, and Richard Jecht. "Über die Verfasser der Briefe an Ehrenfried Walther von Tschirnhaus aus der Sammelung der Oberlausitzischen Gesellschaft der Wissenschaften." *Neues Lausitzisches Magazin* 116 (1940): 100–108.

Respour, P. M. *Besondere Versuche vom Mineralgeist*. Trans. Johann Friedrich Henckel. Dresden and Leipzig, 1742.

———. *Rares experiences sur l'esprit minerale pour la preparation et transmutation des corps metaliques*. Paris, 1668.

Rey, Anne-Lise. "Alchemy and Chemistry." In *The Oxford Handbook of Leibniz*, ed. Maria Rosa Antognazza. Oxford: Oxford University Press, 2013 (online), 2018 (print). http://www.oxfordhandbooks.com/view/10.1093/oxfordhb/9780199744725.001.0001/oxfordhb-9780199744725-e-32.

Rigaud, Hyacinthe. *Le livre de raison du peintre Hyacinthe Rigaud*. Edited by J. Roman. Paris: Henri Laurens, 1919.

Rivest, Justin. "Secret Remedies and the Medical Needs of the French State: The Career of Adrien Helvétius, 1662–1727." *Canadian Journal of History* 51 (2016): 473–99.

Roberts, Lissa. "The Death of the Sensous Chemist: The 'New' Chemistry and the Transformation of Sensuous Technology." *Studies in the History and Philosophy of Science* 26 (1995): 503–29.

Robinet, André. *Malebranche et Leibniz: Relations personelles*. Paris: Vrin, 1955.

Roger, Jacques. *The Life Sciences in Eighteenth-Century French Thought*. Ed. Keith Benson; trans. Robert Ellrich. Stanford, CA: Stanford University Press, 1997.

Rossetti, Lucia, ed. *Matricula nationis Germanicae artistarum in gymnasio Patavino (1553–1721)*. Padua: Editrice Antenore, 1986.

Rossetti, Lucia, and Antonio Gauba, eds. *Acta nationis germanicae artistarum (1663–1694)*. Padua: Editrice Antenore, 1999.

Saint-Simon, Louis de Rouvroy, duc de. *Mémoires*. Ed. Yves Coirault. 8 vols. Paris: Gallimard, 1983–88.

———. *Mémoires*. Ed. A. de Boislisle. 41 vols. Paris: Hachette, 1879–1930.

Sarmant, Thierry. "L'abbé de Louvois: bibliothécaire du roi, 1684–1718." *Revue de la BNF* 41 (2012): 76–83.

Saunders, Elmo Stewart. *The Decline and Reform of the Académie des Sciences à Paris, 1676–1699*. PhD diss., Ohio State University, 1980.

Scagliola, Robert. *Les apothicaires de Paris et les distillateurs*. Thèse de pharmacie, Université de Clermont-Ferrand, 1943.

Schelstrate, Emmanuel. *La correspondance*. Ed. Lucien Ceyssens. Rome: Academia Belgica, 1949.

Schillinger, Klaus. "Die Herstellung von Brennspiegeln und Brenngläsern durch Ehrenfried Walther von Tschirnhaus und ihre Widerspiegelung in ausgewählten Briefen." In *Europa in frühen Neuzeit*, ed. Erich Donnert, 4:97–114. 6 vols. Weimar: Böhlau, 1997.

————. "Herstellung und Anwendung von Brennspiegeln und Brenngläsern durch Ehrenfried Walther von Tschirnhaus." In *Ehrenfried Walter von Tschirnhaus (1651–1708)—Experimente mit dem Sonnenfeuer*, 43–54. Dresden: Staatliche Kunstsammlungen, 2001.

————. *Solare Brenngeräte*. Dresden: Staatliche Mathematisch-Physikalischer Salon, 1992.

Schimank, Hans, ed. With Hans Gossen, Gregor Maurach, and Fritz Krafft. *Otto von Guerickes Neue (sogenanntes Magdeburger Versuche über den leeren Raum)*. Düsseldorf: VDI Verlag, 1968.

Schneider, Ditmar. *Otto von Guericke: Ein Leben für die Alte Stadt Magdeburg*. Stuttgart: Teubner, 1995.

Schönfeld, Martin. "Was There a Western Inventor of Porcelain?" *Technology and Culture* 39 (1998): 716–27.

Schröcker, Alfred. "Gabriel d'Artis, Leibniz, und das *Journal de Hambourg*." *Niedersächsisches Jahrbuch für Landesgeschichte* 49 (1977): 109–29.

Schulte, B. P. M. *Hermanni Boerhaave Praelectiones de morbis nervorum*. Analecta Boerhaaviana 2. Leiden: Brill, 1959.

Schuwirth, Theo. *Eberhard Werner Happel (1647–1690): Ein Beitrag zur deutschen Literaturgeschichte des siebzehnten Jahrhunderts*. Marburg, 1908.

Siegfried, Robert. *From Elements to Atoms: A History of Chemical Composition*. Philadelphia: American Philosophical Society, 2002.

Simms, D. L., and P. L. Hinckley. "Brighter Than How Many Suns? Sir Isaac Newton's Burning Mirror." *Notes and Records of the Royal Society of London* 43 (1989): 31–51.

————. "David Gregory on Newton's Burning Mirror." *Notes and Records of the Royal Society of London* 55 (2001): 185–90.

Slare, Frederick. "An Account of Some Experiments Relating to the Production of Fire and Flame." *Philosophical Transactions* 18 (1694): 201–18.

Smeaton, William A. "E. F. Geoffroy Was Not a Newtonian Chemist." *Ambix* 18 (1971): 212–14.

————. "Macquer on the Composition of Metals and the Artificial Production of Gold and Silver." *Chymia* 11 (1966): 81–88.

Smith, Pamela. *The Business of Alchemy: Science and Culture in the Holy Roman Empire*. Princeton, NJ: Princeton University Press, 1994.

Soll, Jacob. *The Information Master: Jean-Baptiste Colbert's Secret State Intelligence System*. Ann Arbor: University of Michigan Press, 2009.

Stahl, Georg Ernst. *Alphabetum minerale*. In *Tripus hermeticus fatidicus*. Frankfurt, 1689.

————. *Specimina Beccherianum*. Leipzig, 1703.

————. *Zufällige Gedanken und nützliche Bedencken über den Streit von dem so genannten Sulphure*. Halle, 1718.

Starkey, George (*see also* Philalethes, Eirenaeus). *The Alchemical Laboratory Notebooks and Correspondence*. Ed. and trans. William R. Newman and Lawrence M. Principe. Chicago: University of Chicago Press, 2004.

————. *Pyrotechny Asserted*. London, 1658.

Steinhauser, Georg, Jürgen Evers, Stefanie Jakob, Thomas M. Klapötke, and Gilbert Oehlinger. "A Review on Fulminating Gold (Knallgold)." *Gold Bulletin* 41 (2008): 305–17.

Stroup, Alice. "Censure ou querelles savantes: l'Affaire Duclos (1666–1685)." In *Règlement, usages et science dans la France de l'absolutisme*, ed. Christiane Demeulenaere-Douyère and Éric Brian, 435–52. Paris: Lavoisier Tec & Doc, 2002.

————. *A Company of Scientists: Botany, Patronage, and Community at the Seventeenth-Century Parisian Royal Academy of Sciences*. Berkeley: University of California Press, 1990.

————. "Nicolas Hartsoeker, savant hollandais associé de l'Académie et espion de Louis XIV." In *De la diffusion des sciences à l'espionnage industriel, XVe–XXe siècle*, ed. André Guillerme, 205–28. Lyon: ENS Editions, 1999.

————. *Royal Funding of the Parisian Académie Royale des Sciences during the 1690s. Transactions of the American Philosophical Society* 77 (1987).

————. "Science, politique et conscience aux débuts de l'Académie Royale des Sciences." *Revue de synthèse* 114 (1993): 423–53.

————. "Wilhelm Homberg and the Search for the Constituents of Plants at the 17th-Century Académie Royale des Sciences." *Ambix* 26 (1979): 184–202.

Sturdy, David. *Science and Social Status: The Members of the Académie des Sciences, 1666–1750*. Woodbridge, UK: Boydell Press, 1995.

Swart, Jan. *Bibliotheque de feu Mr. N. Hartsoeker*. The Hague, 1727.

Take, Heinz-Herbert. *Otto Tachenius, 1610–1680: Ein Wegbereiter der Chemie zwischen Herford und Venedig*. Bielefeld: Verlag für Regionalgeschichte, 2002.

Taton, René. *Les origines de l'Académie Royale des Sciences*. Paris: Palais de la Découverte, 1965.

Thuillier, Guy, and Arthur Birembaut. "Une source inédite: les cahiers du chimiste Jean Hellot (1685–1766)." *Annales: Histoire, Sciences Sociales* 21 (1966): 357–64.

Thuillier, Jacques. *Sébastien Bourdon 1616–1671: Catalogue critique et chronologique*. Paris: Réunion des Musées Nationaux, 2000.

Tits-Dieuaide, Jeanne-Marie. "L''affection' de Louis XIV pour l'Académie Royale des Sciences: sur les raisons d'être du règlement de 1699." In *Règlement, usages et science dans la France de l'absolutisme*, ed. Christiane Demeulenaere-Douyère and Éric Brian, 37–50. Paris: Lavoisier Tec & Doc, 2002.

————. "Une institution sans statuts: l'Académie Royale des Sciences de 1666 à 1699." In *Histoire et mémoire de l'Académie des Sciences: Guide des recherches*, ed. Éric Brian and Christiane Demeulenaere-Douyère, 3–13. Paris: Tec & Doc, 1996.

Todériciu, Doru. "La bibliothèque d'un savant chimistre et technologue parisien du XVIIIe siècle: livres et manuscrits de Jean Hellot (1685–1766)." *Physis* 18 (1976): 198–216.

————. "Sur la vraie biographie de Samuel Duclos (Du Clos) Cotreau." *Revue d'histoire des sciences* 27 (1974): 64–67.

————. "Le Traité de chimie inachevé de Jean Hellot." *Physis* 19 (1977): 355–75.

Tschirnhaus, Ehrenfried Walther von. "Additio ad D. T. intimationem de emendatione artis vitriariae." *Acta eruditorum* (December 1696): 554.

————. *Gesamtausgabe*. Ed. Eberhard Knobloch. Leipzig: Sächsischen Academie der Wissenschaften, 2000–2006.

————. "Intimatio singularis novaeque emendationis artis vitriariae." *Acta eruditorum* (August 1696): 345–67.

Urban VIII. *Constitutio contra astrologos judiciarios*. Rome, 1631.

Van Helmont, Joan Baptista. *Opuscula medica inaudita*. Amsterdam, 1648. Reprint. Brussels: Culture & Civilisation, 1966.

————. *Ortus medicinae*. Amsterdam, 1648. Reprint. Brussels: Culture & Civilisation, 1966.

Vanzo, Alberto. "Corpuscularism and Experimental Philosophy in Domenico Guglielmini's 'Reflections' on Salts." In *The Idea of Principles in Early Modern Thought*, ed. Peter R. Anstey, 147–71. New York: Routledge, 2017.

La vie et les lettres de Messire Jean Soanen. 2 vols. Cologne, 1750.

Voltaire. *Siècle de Louis XIV.* Ed. Émile Bourgeois. Paris: Hachette, 1910.

von Guericke, Otto. *Experimenta nova (ut vocantur) Magdeburgica de vacuo spatio.* Amsterdam, 1672.

von Mackensen, Ludolf. *Die naturwissenschaftlich-technische Sammelung in Kassel.* Kassel: Georg Wenderoth Verlag, 1991.

von Uffenbach, Zacharias Conrad. *Merkwürdige Reisen durch Niedersachsen, Holland, und Engelland.* 3 vols. Frankfurt and Leipzig, 1753–54.

Waddell, Mark. *Jesuit Science and the End of Nature's Secrets.* Burlington, VT: Ashgate, 2015.

Warolin, Christian. "La dynastie des Boulduc, apothicaires à Paris aux XVIIe et XVIIIe siècles." *Revue d'histoire de la pharmacie* 49 (2001): 333–54; 50 (2002): 439–50; 51 (2003): 103–10.

Welling, Georg von. *Opus mago-cabbalisticum et theosophicum.* Homburg, 1735.

Die Welt im leeren Raum: Otto von Guericke, 1602–1686. Munich: Deutscher Kunstverlag, 2002.

Westfall, Richard S. *The Construction of Modern Science.* Cambridge: Cambridge University Press, 1971.

Wieseman, Marjorie E. *Caspar Netscher and Late Seventeenth Century Dutch Painting.* Doornspijk: Devaco, 2002.

Willemse, David. *António Nunes Ribeiro Sanches, élève de Boerhaave, et son importance pour la Russie.* Leiden: Brill, 1966.

Willis, Thomas. *Diatribe duae medico-philosophicae.* Amsterdam, 1663.

Winter, Eduard, ed. *E. W. von Tschirnhaus und die Fruhaufklärung in Mittel- und Osteuropa.* Berlin: Akademie Verlag, 1960.

Index